Drawing and Detailing with SolidWorks 2003

By David C. Planchard and Marie P. Planchard

Referencing the ASME Y14 Engineering Drawing and Related Documentation Practices

ISBN: 1-58503-129-1

www.schroff.com
www.schroff-europe.com

Trademarks and Disclaimer

SolidWorks and its family of products are registered trademarks of the Dassault Systemes. Microsoft Windows and its family of products are trademarks of the Microsoft Corporation. Other software applications and parts described in this book are trademarks or registered trademarks of their respective owners.

When parts from companies are used, dimensions may be modified for illustration purposes. Every effort has been made to provide an accurate text. The authors and the manufacturers shall not be held liable for any parts developed or designed with this book or any responsibility for inaccuracies that appear in the book.

About the Cover

The drawing and assembly on the cover are based on components designed and manufactured by Emerson Power Transmission Corporation, a division of Emerson, Ithaca, NY, USA.

Emerson Power Transmission Corporation,

www.emerson-ept.com is a global leader in transmission systems and equipment.

The assembly and drawings on the cover are modified for instructional purposes.

Models and Images
Courtesy of
Emerson Power Transmission Corporation,
a division of Emerson
Ithaca, NY, USA

About the Projects

SMC Corporation of American provided the components utilized in Project 2 through Project 4.

SMC is the world largest manufacturer of pneumatic automation products.

Drawings created in this book are based on the CQ2 series compact air cylinder. Dimensions and features are modified for instructional purposes.

Models and Images
Courtesy of
SMC Corporation of America
Indianapolis, IN, USA

About the Authors

Marie Planchard is CAD department manager at Mass Bay College in Wellesley Hills, MA. Before developing the CAD engineering design program, she spent 13 years in industry and held a variety of High Technology management positions including Beta Test Manager for CAD software at Computervision Corporation. She has written and presented numerous technical papers on 3D modeling. She was Vice-President of the New England Pro/Users Group for 6 years, an active member of the SolidWorks Educational Advisory Board, a SolidWorks Research Partner, CSWP, SolidWorks World 2003 Presenter and coordinator for the New England SolidWorks Users Group.

David Planchard is the President of D & M Education, LLC. Before starting D & M Education LLC, he spent over 23 years in industry and academia holding various Engineering and Marketing positions. He has five U.S. and one International patents. He has published and authored numerous papers on equipment design. He is a member of the New England Pro/Users Group, New England SolidWorks Users Group and the Cisco CCNA Regional Academy Users Group.

David and Marie Planchard are Co-founders of D & M Education LLC and are active industry and education consultants. They are Co-authors of the following SDC Publication books:

- **Assembly Modeling with SolidWorks 2001Plus/2003.**
- **Engineering Design with SolidWorks 1999, 2000, 2001, 2001Plus, 2003.**
- **Drawing and Detailing with SolidWorks 2001/2001Plus.**
- **SolidWorks Tutorial 2001/2001Plus.**
- **Applications in Sheet Metal Using Pro/SHEETMETAL & Pro/ENGINEER.**
- **An Introduction to Pro/SHEETMETAL.**

INTRODUCTION

Drawing and Detailing with SolidWorks 2003 is written to educate and assist students, designers, engineers and professionals in the Drawing and Detailing Options of SolidWorks.

The book is designed to compliment the On-line tutorials contained within SolidWorks, Drawing Standards, Engineering Drawing/Design and Graphics Communications reference books.

Commands are presented in a step-by-step progressive approach. The learning process is explored through a series of design situations, design intents, industry scenarios, projects and objectives.

Drawing and Detailing with SolidWorks 2003 is *not* a reference book for all drafting and drawing techniques. The book provides examples to:

- Apply Standards to a drawing.
- Drawing interaction with other SolidWorks documents.
- Modify drawing and detailing information.

The authors recognize that companies may have additional drawing standards. The authors developed the industry scenarios by combining industry experience with their knowledge of engineers, sales, vendors and manufacturers.

These professionals are directly involved with SolidWorks everyday. Their work goes far beyond a simple drawing with a few dimensions. They create detailed drawings, assembly drawings, marketing drawings, customer drawings and many other documents to complete a project on time.

Trademarks, Disclaimer and Copyrighted Material

SolidWorks and its family of products are registered trademarks of the SolidWorks Corporation. Microsoft Windows, Microsoft Office and its family of products are registered trademarks of the Microsoft Corporation. Pro/ENGINEER is a registered trademark of PTC. AutoCAD is a registered trademark of AutoDesk.

Other software applications and parts described in this book are trademarks or registered trademarks of their respective owners.

Dimensions of parts are modified for illustration purposes. Every effort is made to provide an accurate text.

The authors and the manufacturers shall not be held liable for any parts or drawings developed or designed with this book or any responsibility for inaccuracies that appear in the book.

Information in this text is provided from the ASME Engineering Drawing and Related Documentation Publications:

ASME Y14.1 1995

ASME Y14.2M-1992 (R1998)

ASME Y14.3M-1994 (R1999)

ASME Y14.5-1982

ASME Y14.5M-1994

ASME B4.2

The illustrations and part documents were recreated in SolidWorks.

Additional information is provided from the American Welding Society, AWS 2.4:1997 Standard Symbols for Welding, Braising and Non-Destructive Examinations, Miami, Florida, USA.

Acknowledgements

The authors would like to acknowledge the following professionals for their contribution to the design and content of this book. Their assistance has been invaluable.

- Dave Kempski, Steve Hoffer and the Etech Team, SMC Corp. of America.
- Richard Barber and the EPT Team, Emerson-EPT Corporation.
- Rosanne Kramer, Pierre Devaux and the SolidWorks Educational Team.
- Computer Aided Products Application Engineers: Jason Pancoast, Keith Pederson, Adam Snow, Joe St Cyr.
- Dave Pancoast and the SolidWorks Training Team.
- CSWPs: Mike J. Wilson, Devon Sowell, Scott Baugh, Paul Salvador and Matt Lombard for their support of engineering education.
- Ivette Rodriguez, ASME International.
- Leonard Connor, American Welding Society.
- SDC Publications: Stephen Schroff, Mary Schmidt and the SDC team.

For this 2nd edition of Drawing and Detailing with SolidWorks we realize that keeping software application books up to date is very important to our customers.

We value the hundreds of professors, students, designers and engineers that have provided us input to enhance our books. We value your suggestions and comments.

Please contact us with any comments, questions or suggestions on this book or any of our other SolidWorks SDC Publications.

Marie P. Planchard
Mass Bay College
mplanchard@massbay.edu

David C. Planchard
D & M Education, LLC
dplanchard@msn.com

References

- SolidWorks Users Guide, SolidWorks Corporation, 2003.
- ASME Y14 Engineering Drawing and Related Documentation Practices, ASME, NY[1].
- ASME B4.2 Dimensions Preferred Metric Limits and Fits, ASME, NY[1].
- AWS 2.4: 1997 Standard Symbols for Welding, Braising and Non-Destructive Examinations, American Weld Society, Miami, Florida[4].
- Mark's Standard Handbook for Mechanical Engineers, Beumeister et al, 8th 3d. McGraw Hill, 1978.
- Betoline, Wiebe, Miller, Fundamentals of Graphics Communication, Irwin, 1995.
- Earle, James, Engineering Design Graphics, Addison Wesley, 1999.
- French, et al, Engineering Drawing & Graphic Technology, McGraw Hill 1993.
- Giesecke et al. Modern Graphics Communication, Prentice Hall, 1998.
- Hoelscher, Springer, Dobrovolny, Graphics for Engineers, John Wiley, 1968.
- Jensel & Helsel, Engineering Drawing and Design, Glencoe, 1990.
- Jensen, Cecil, Interpreting Engineering Drawings, Delmar-Thomson Learning, 2002.
- Lockhart & Johnson, Engineering Design Communications, Addison Wesley, 1999.
- Madsen, David et al. Engineering Drawing and Design, Delmar Thomson Learning, 2002.
- Planchard & Planchard, Engineering Design with SolidWorks 2003, SDC Publications, Mission, KS 2001.
- Planchard & Planchard, Applications in SheetMetal Using Pro/ENGINEER, SDC Publications, Mission, KS 2001.
- SMC Corporation, Compact Guide Cylinder Product Manual, SMC Corporation.[2]

- Emerson-EPT Corporation, Shaft Mount Reducer Product Manual, Emerson-EPT Corporation, a division of Emerson[3].
- Walker, James, Machining Fundamentals, Goodheart Wilcox, 1999.

[1] An on-line catalog of ASME Codes and Standards is available on their web site www.asme.org.

[2] An on-line catalog of SMC parts and documents is available on their web site www.smcusa.com. Instructions to down load additional SMC components are available in the Appendix.

[3] An on-line catalog of Emerson-EPT parts and documents is available on their web site www.emerson-ept.com.

[4] An on-line catalog of AWS Standards is available on their web site www.aws.org.

Table of Contents

Project 5 – Geometric Tolerancing and Various Symbols(Continued)

Appendix

Index

What is SolidWorks?

SolidWorks is a design automation software package used to produce parts, assemblies and drawings. SolidWorks is a Windows native 3D solid modeling CAD program. SolidWorks provides easy to use, highest quality design software for engineers and designers who create 3D models and 2D drawings ranging from individual parts to assemblies with thousands of parts.

SolidWorks Corporation, headquartered in Concord, Massachusetts, develops and markets innovative design solutions for the Microsoft Windows platform. Additional information on SolidWorks and its family of products can be found at their URL, www.SolidWorks.com.

In SolidWorks, you create 3D parts, assemblies and 2D drawings. The part, assembly and drawing documents are all related. Make a change in the part. The drawing and assembly are linked documents. The changes are reflected in the other documents.

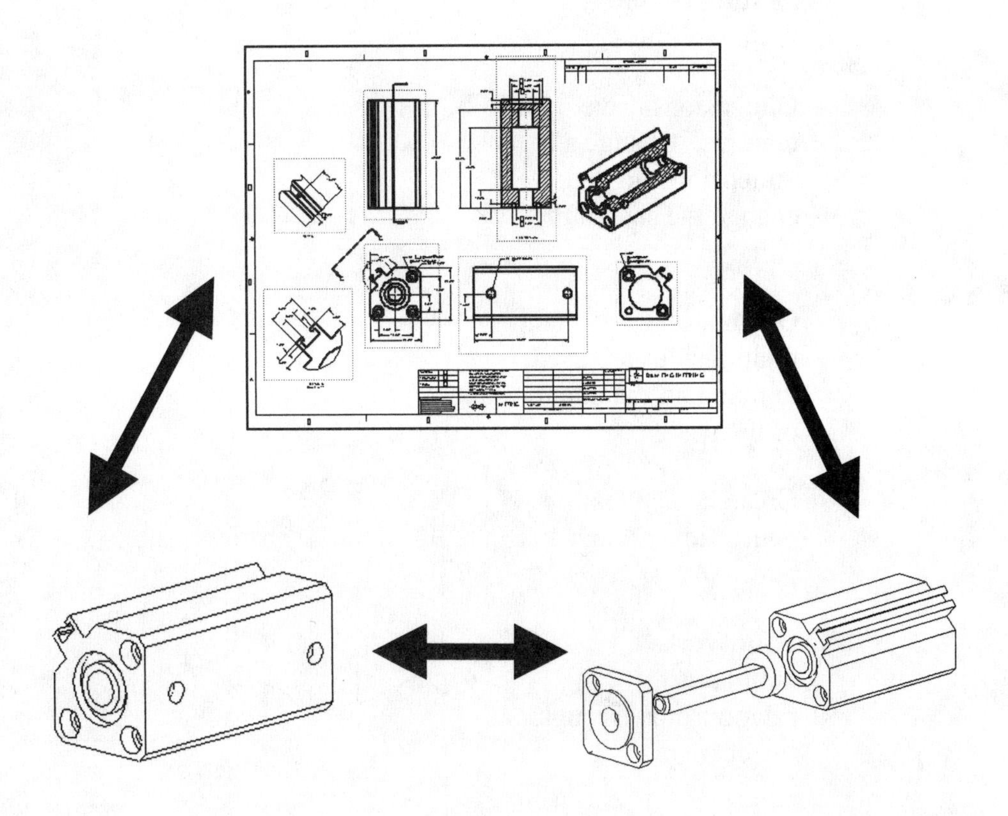

Drawing refers to the SolidWorks module used to insert, add and modify views in an engineering drawing.

Detailing refers to the dimensions, notes, symbols and Bill of Materials used to document the drawing.

The building blocks of parts are called features. Features such as the Extruded-Boss, Cut, Hole, Fillet, Chamfer and others are used to create the parts.

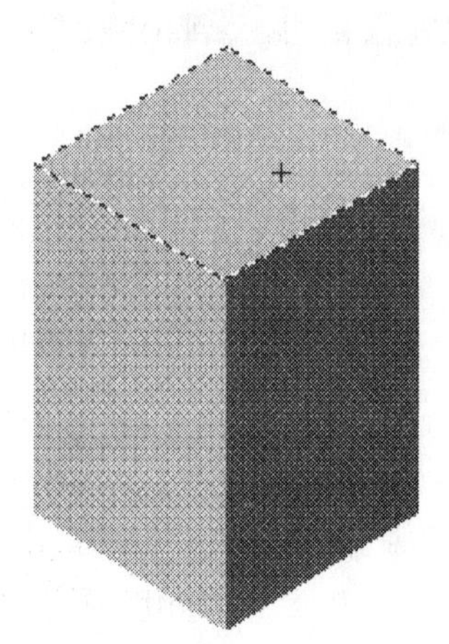

Some features are sketched, such as an Extruded-Boss.

Other features are created by selecting edges or faces of existing features, such as a Fillet.

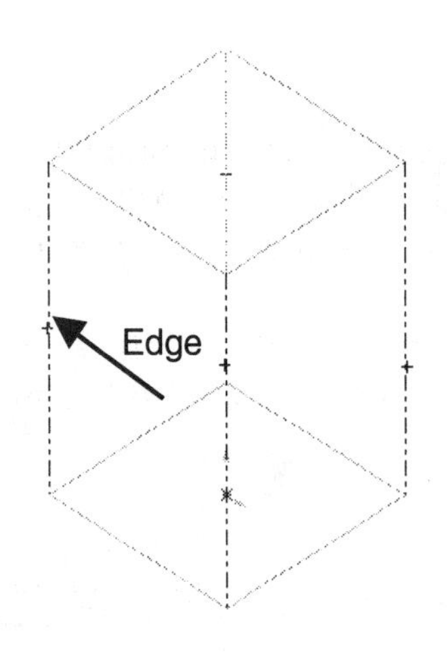

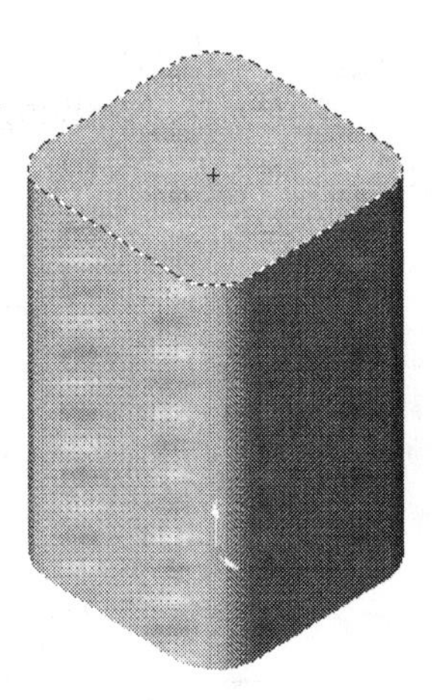

Dimensions drive features. Change a dimension and you change the size of the part.

Geometric relationships are used to maintain the intent of the design.

Create a hole that penetrates through a part.

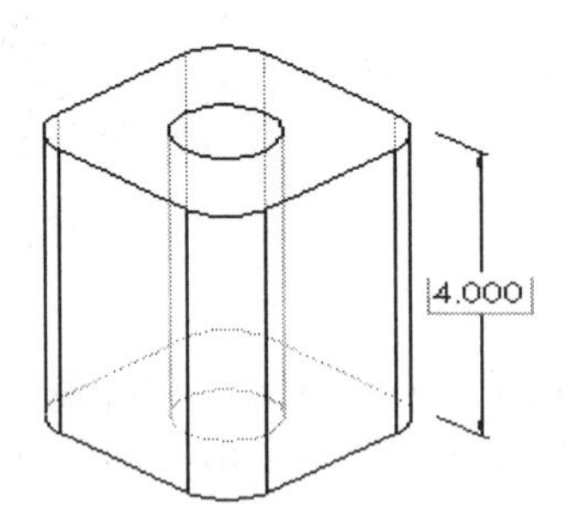

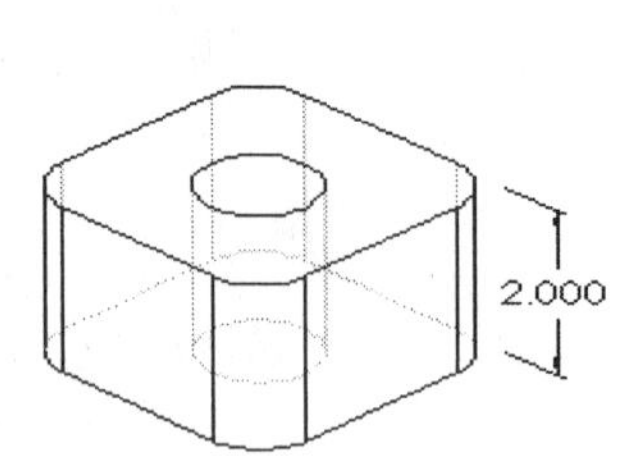

SolidWorks maintains relationships through the change.

The drawing reflects the changes of the part.

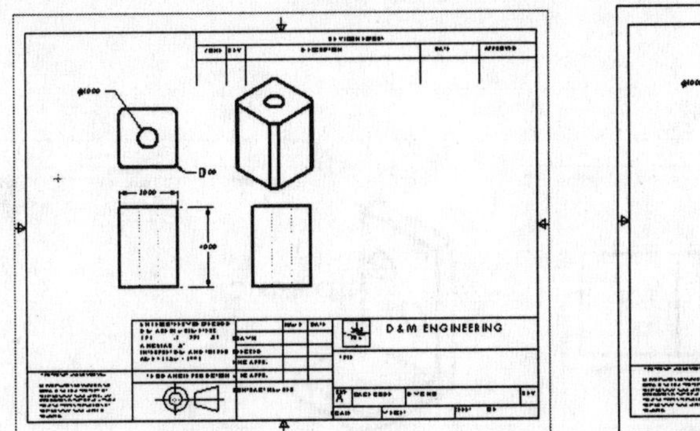

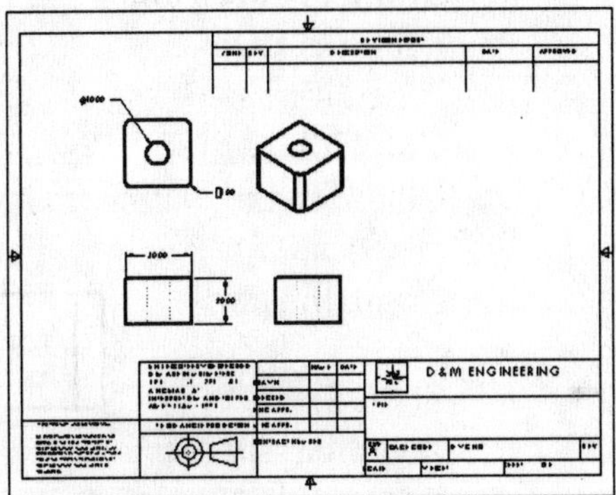

A Drawing Template is the foundation for drawing information. Specify drawing standards and size, company information, manufacturing and or assembly requirements and more are included in a Drawing Template.

Drawing Templates contain Drawing Properties settings such as millimeter or inch units and ANSI or ISO Drawing Standards.

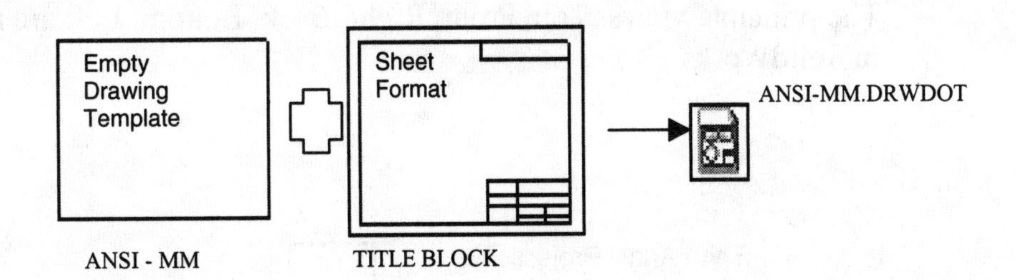

Drawing Templates also contain information included in the Sheet Format such as a title block, company name, company logo and custom properties.

A drawing is a 2D representation of a 3D part or assembly. SolidWorks named views such as Top, Front, Right and Isometric to display the 3D model on the 2D drawing.

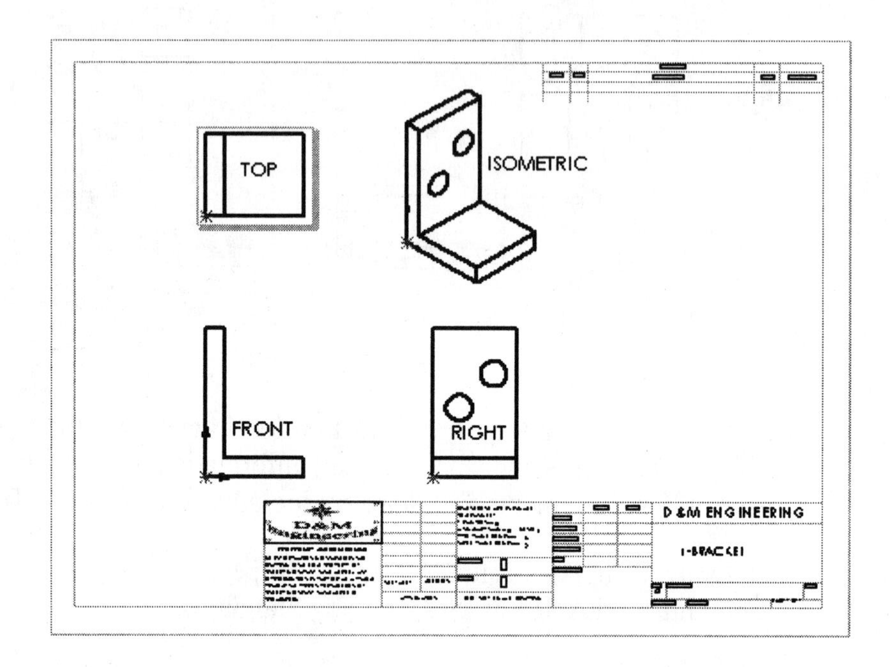

The principle views: Top, Front, Right, Back, Bottom, Left are named views in SolidWorks.

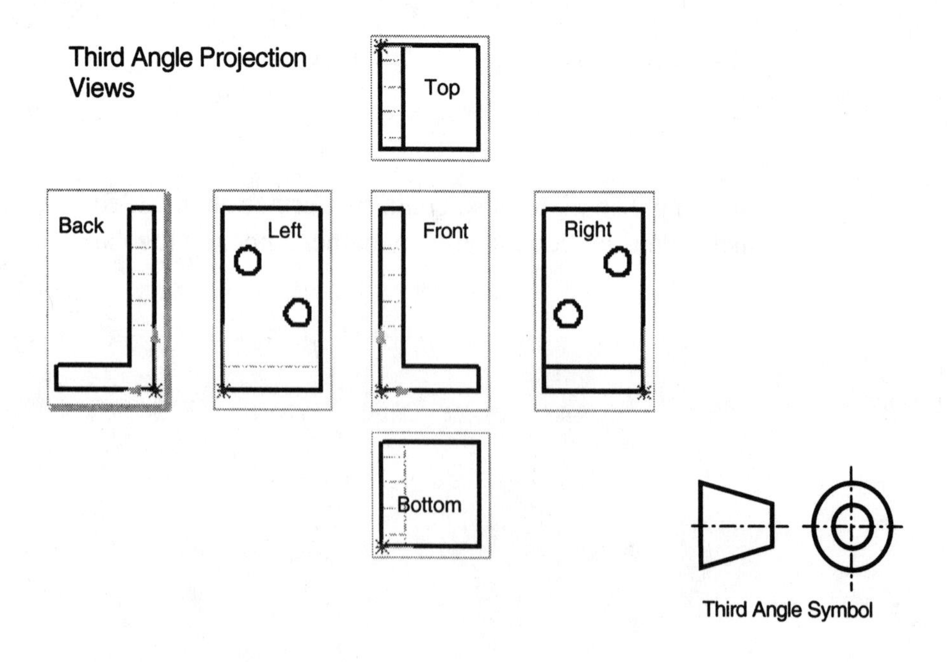

Third Angle Symbol

In a drawing, there are many types of views to represent a 3D model or assembly. Some views are created from the Drawing tools in SolidWorks such as a Section view, Auxiliary view or Detail view. Create additional views by combining Drawing tools with different part configurations.

Example: Half Section Isometric view.

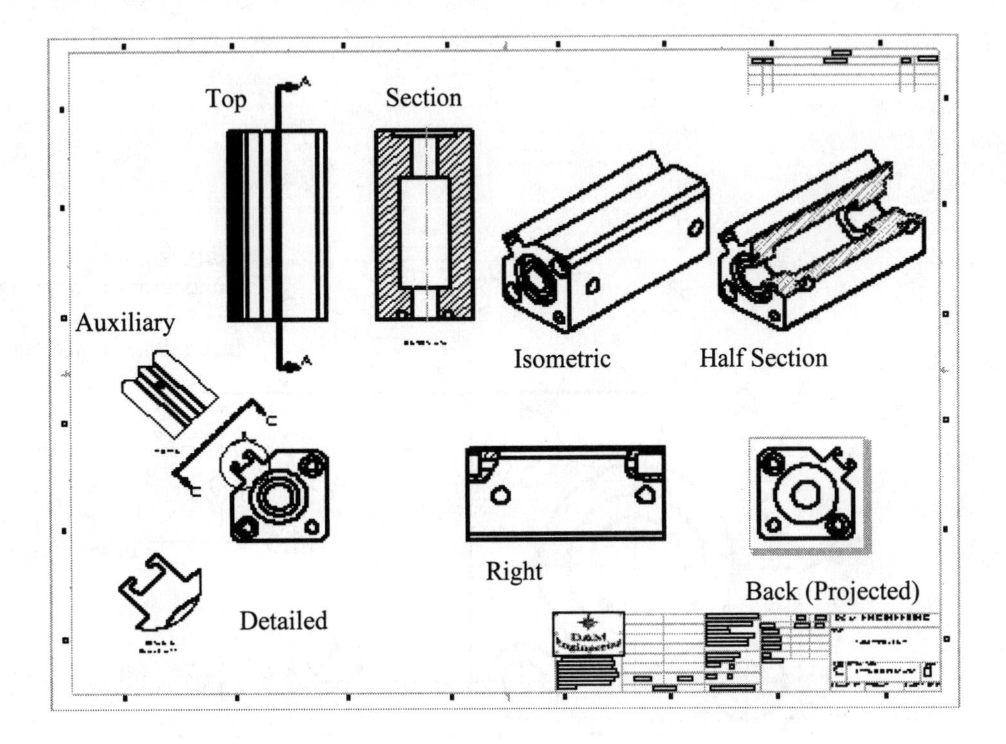

Create views by combining Drawing tools outside the Sheet area.

Example: Broken Isometric view.

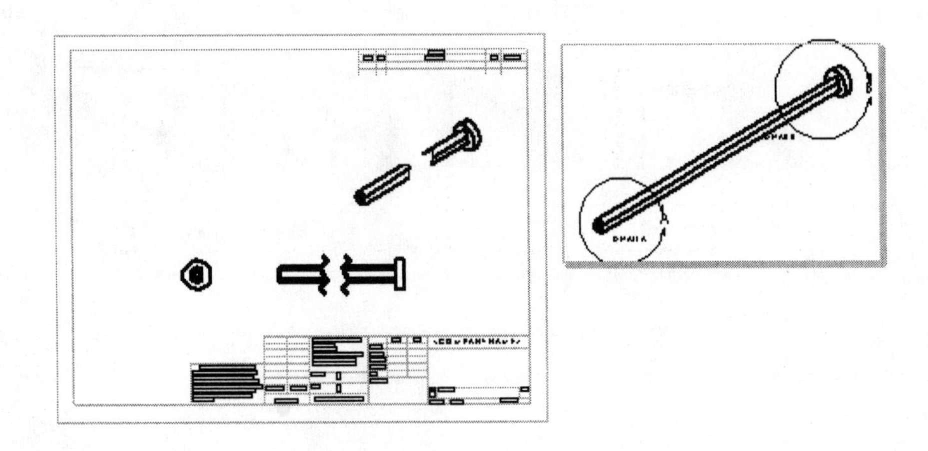

Annotations represent a text note or symbol that documents a part, assembly or drawing.

Dimensions and annotations created in a part or assembly are inserted into the drawing.

Additional reference dimensions and annotations are created in the drawing.

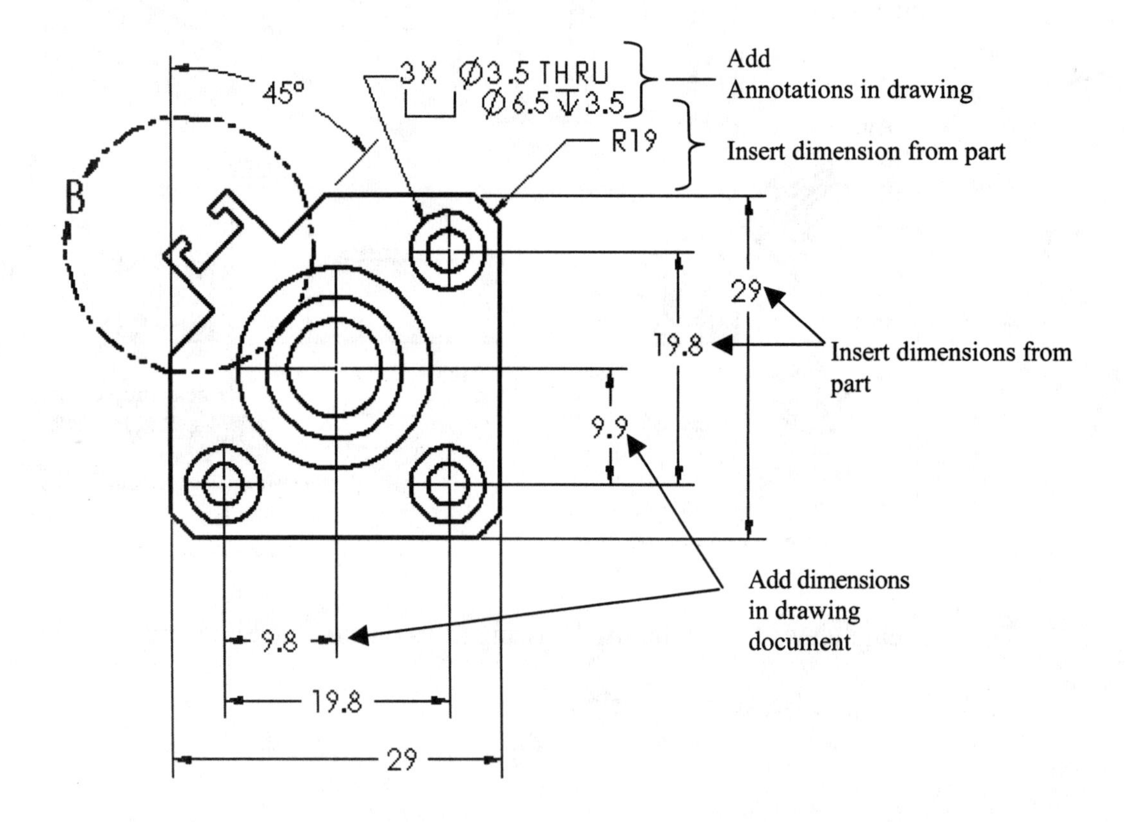

Dimensions and geometric relations are created in a part to represent design intent. Dimensions and relations may be redefined as reference dimensions in the drawing for manufacturing requirements.

Example: The hole in the part is located with a diagonal centerline and a midpoint relation.

The hole in the Design Tables creates various configurations of parts and assemblies.

Apply different configurations in a drawing. Assign properties such as material, mass and cost to individual parts. Incorporate properties into the Bill of Materials.

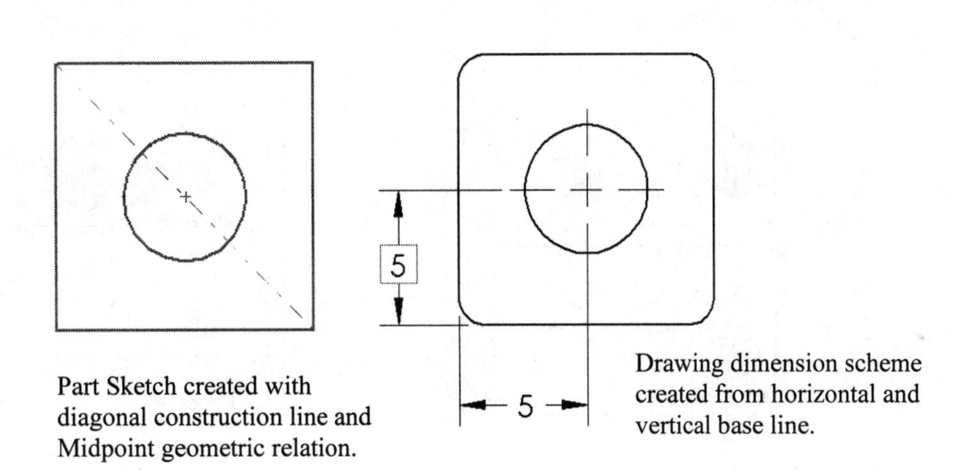

Part Sketch created with diagonal construction line and Midpoint geometric relation.

Drawing dimension scheme created from horizontal and vertical base line.

Design Tables and Bill of Materials are interwoven in SolidWorks.

A	B	C	D	E	F	G
Design Table for: CYLINDER						
	$CONFIGURATION @COVERPLATE<1>	$STATE@COVER PLATE<1>	stroke@ distance1	$CONFIGURATION @TUBE<1>	$STATE@CON CENTRIC2	$USER NOTES
NO COVERPLATE	DEFAULT	S	0	DEFAULT	U	BASIC
COVERPLATE-HOI	With Nose Holes	R	0	DEFAULT	U	ANODIZED COVER
COVERPLATE-NOI	Without Nose Holes	R	0	DEFAULT	U	ANODIZED COVER
STROKE 0	DEFAULT	S	0	DEFAULT	U	NO COVER ROD AT HOME
STROKE 30	DEFAULT	S	30	DEFAULT	U	NO COVER ROD FULLY EX
CUT AWAY	DEFAULT	R	0	Section Cut	S	SEE INTERNAL FEATURES

CYLINDER Design Table
Create in CYLINDER assembly

ITEM NO.	QTY.	PART NO.	MATERIAL	DESCRIPTION	MASS
1	1	10-0408	ALUMINUM	TUBE 16 MM	94.36
2	1	10-0409	STEEL	ROD-16MM	23.34
3	1	10-0410	ALUMINUM	COVERPLATE 16MM	10.18
				Total Mass (g)	127.88

Custom Bill of Materials
Create in CYLINDER drawing.
Use Custom Properties.

Default Bill of Materials
Create in CYLINDER drawing.

ITEM NO.	QTY.	PART NO.	DESCRIPTION
1	1	TUBE	
2	1	ROD	
3	1	COVERPLATE	

CYLINDER assembly
Collapsed and Exploded

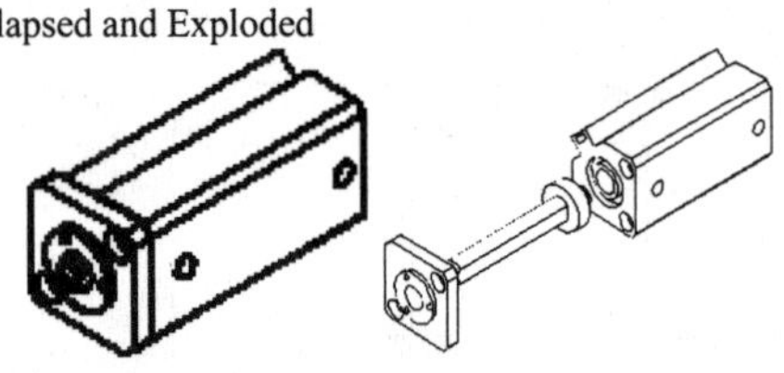

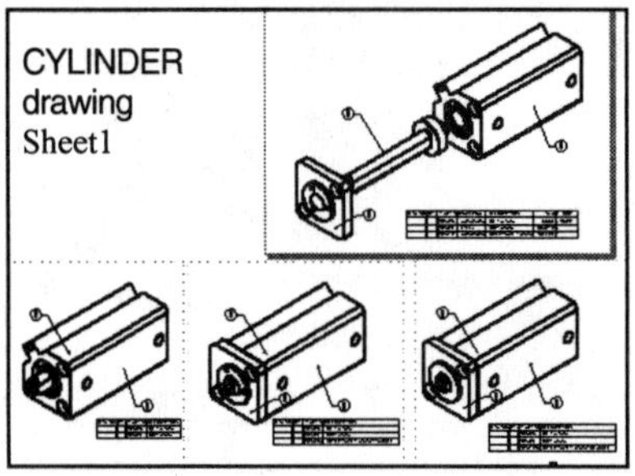

CYLINDER drawing
Sheet1

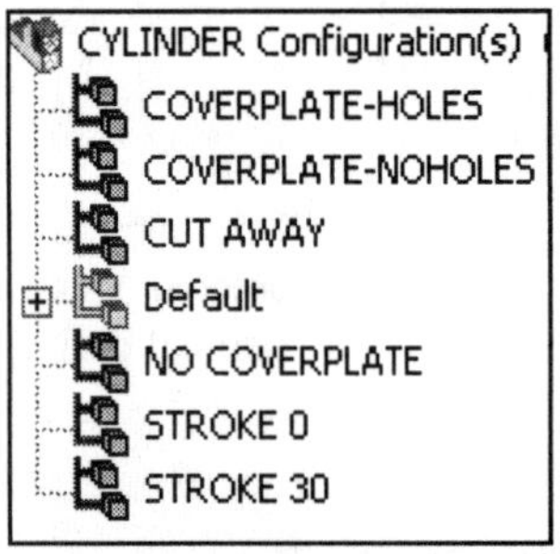

CYLINDER Configuration(s)
- COVERPLATE-HOLES
- COVERPLATE-NOHOLES
- CUT AWAY
- Default
- NO COVERPLATE
- STROKE 0
- STROKE 30

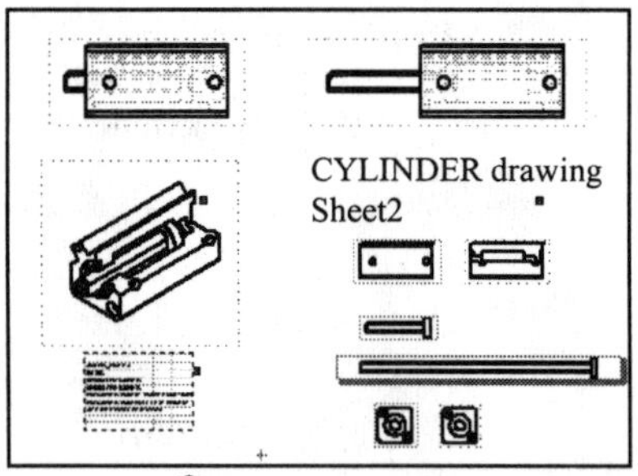

CYLINDER assembly configurations created with Design Table and specified in CYLINDER drawing.

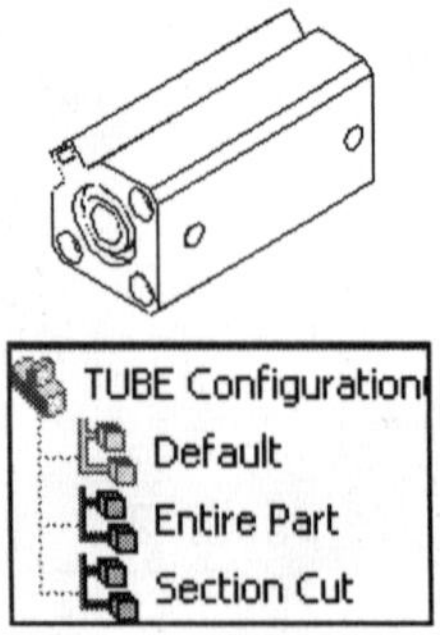

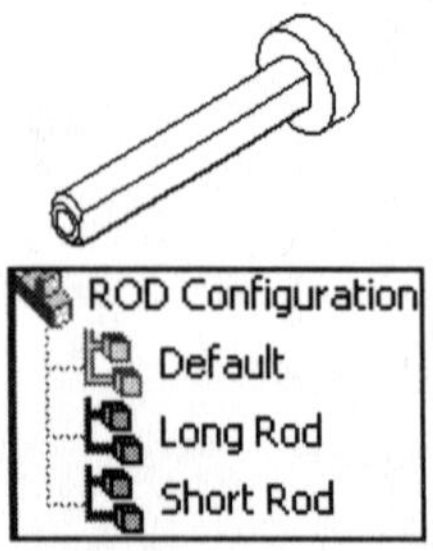

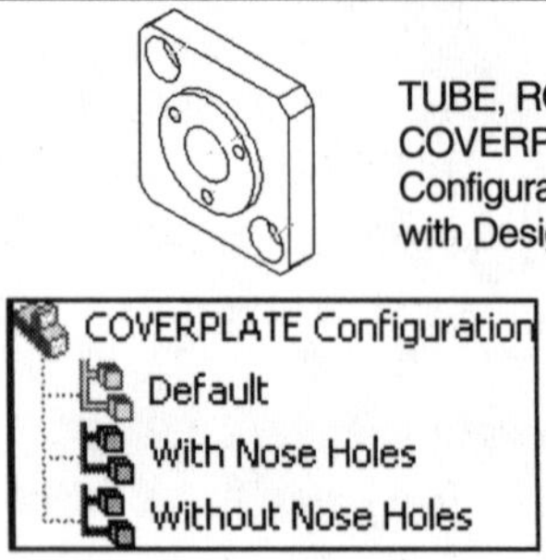

CYLINDER drawing
Sheet2

TUBE, ROD and COVERPLATE Parts Configurations created with Design Tables.

TUBE Configuration
- Default
- Entire Part
- Section Cut

ROD Configuration
- Default
- Long Rod
- Short Rod

COVERPLATE Configuration
- Default
- With Nose Holes
- Without Nose Holes

The step-by-step approach used in this text allows you to apply different parts and assemblies to create engineering drawings. Change is an integral part of design.

Overview of Projects

Project 1: Drawing Template and Sheet Format

Explore the SolidWorks Drawing Template. Apply Document Properties to reflect the ASME Y14 Engineering Drawing Standards.

Investigate the difference between a Sheet Format and a Drawing Template. Import an AutoCAD drawing to create a new Sheet Format. Combine the Sheet Format with and empty Drawing Template to create a Custom Drawing Template.

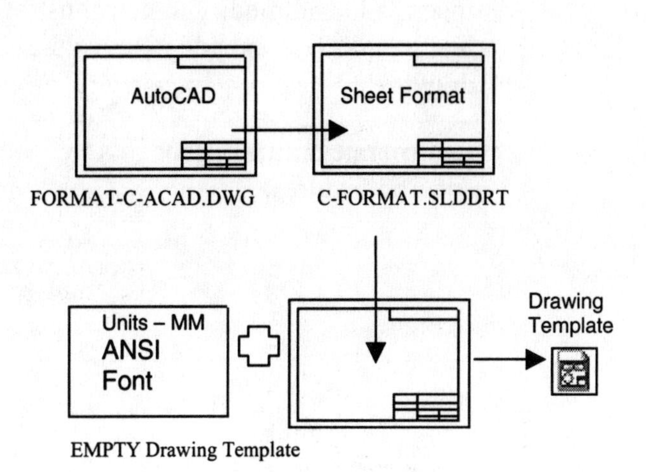

Project 2: Drawing Views

Insert multiple views to create the TUBE, ROD and COVERPLATE drawings.

Explore Standard views, Isometric, Auxiliary, Section, Broken Section, Detail and Half Section (Cut-away) views.

Create multi-sheet drawings from various part configurations.

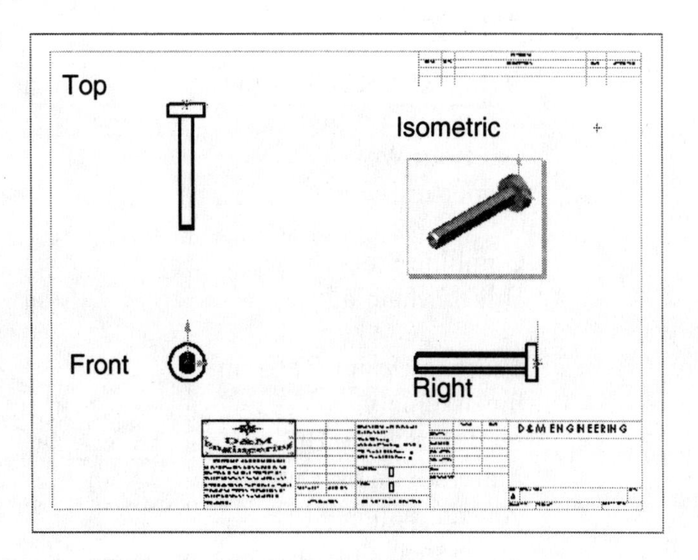

Project 3: Fundamentals of Detailing

Insert dimensions and annotations required to detail the TUBE, ROD and COVERPLATE drawings.

Insert, add and modify dimensions for part features. Insert and add notes to the drawing.

Incorporate drawing standards to document specific features.

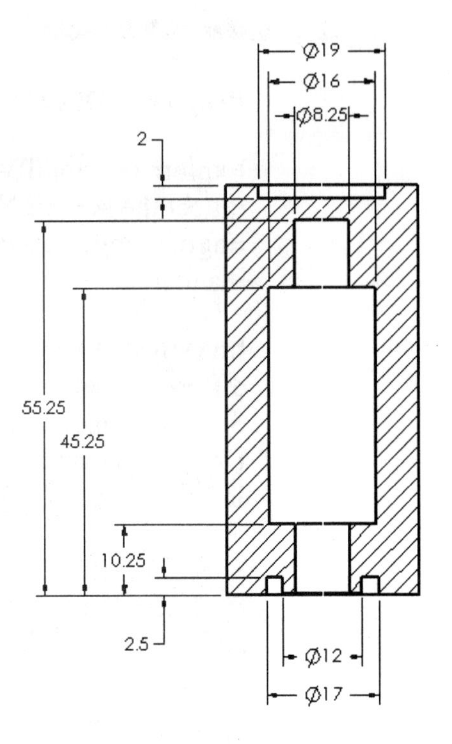

Project 4: Assembly Drawing

Develop the CYLINDER Assembly.

Combine configurations of the TUBE, ROD, and COVERPLATE components.

Obtain an understanding of Custom Properties and SolidWorks Properties.

Combine Properties in a Bill of Materials.

Create a Design Table in the assembly. Incorporate the Bill of Materials and different configurations into a multi-sheet drawing.

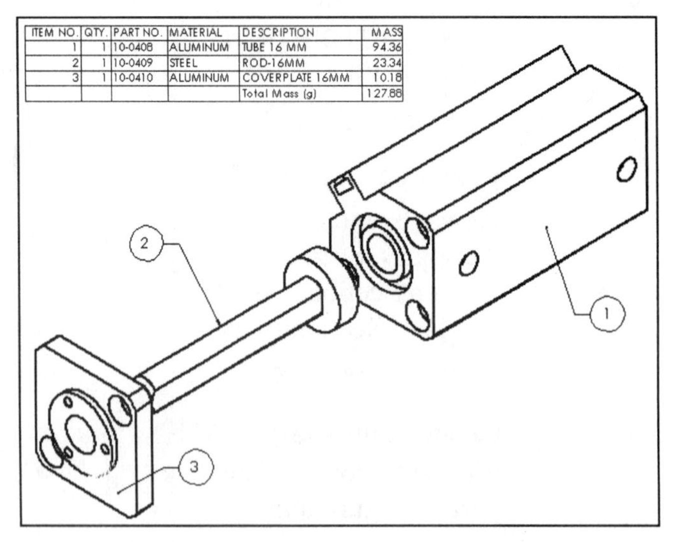

ITEM NO.	QTY.	PART NO.	MATERIAL	DESCRIPTION	MASS
1	1	10-0408	ALUMINUM	TUBE 16 MM	94.36
2	1	10-0409	STEEL	ROD-16MM	23.34
3	1	10-0410	ALUMINUM	COVERPLATE 16MM	10.18
				Total Mass (g)	127.88

Project 5: Applied Geometric Tolerancing and Various Drawing Symbols

Develop three drawings:

1. VALVEPLATE.
2. PLATE-TUBE.
3. PLATE-CATALOG.

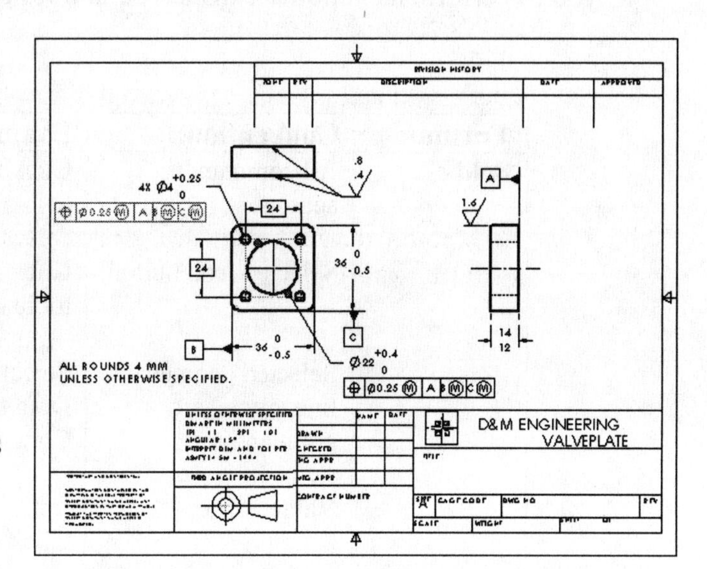

Apply Datum Feature Symbols, Geometric Tolerance Symbols, Surface Finish Symbols and Weld Symbols.

Modify a Design Table in EXCEL.

Command Syntax

The following command syntax is used throughout the text. Commands that require you to perform an action are displayed in **bold** text.

Format:	Convention:	Example:
Bold	All commands actions.	Click **Save**. Click **Tools, Options** from the Main menu.
	Selected icon button.	Click **Rectangle** from the Sketch Tools toolbar.
	Selected geometry: line, circle, arc, point and text.	Select the **center point**. Drag the **circle** downward. Click the **arc**.
	Value entries.	Enter **3mm** for Radius. Click **60mm** from the Depth spin box.
Capitalized	Filenames, part names, assembly, component and drawing names.	The TUBE is contained inside the CYLINDER assembly.
	First letter in a feature name.	Click the **Fillet** feature. Click the **Extrude Base** feature.

Note: Dimension text, notes and symbols are enlarged for easier viewing.

Windows Terminology

The mouse pointer provides an integral role in executing SolidWorks commands. The mouse pointer executes commands, selects geometry, displays Pop-Up menus and provides information feedback. A summary of mouse pointer terminology is displayed below:

Item:	Description:
Click	Press and release the left mouse button.
Double-click	Double press and release the left mouse button.
Click inside	Press the left mouse button. Wait a second and then press the left mouse button inside the text box. This technique is used to modify Feature names in the FeatureManager design tree.
Drag	Point to an object, press and hold down the left mouse button. Move the mouse pointer to a new location, release the left mouse button.
Right-click	Press and release the right mouse button. A Pop-up menu is displayed. Use the left mouse button to select a menu command.
ToolTip	Position the mouse pointer over an Icon (button). The command is displayed below the mouse pointer.
Mouse pointer feedback	Position the mouse pointer over various areas of the sketch: part, assembly or drawing. The cursor provides feedback depending upon the geometry.

Review various Windows terminology that describes: menus, toolbars and commands that constitute the graphical user interface in SolidWorks.

The following Windows terminology is used throughout the text:

Item:	Description:
Dialog box name	Name of a window to enter information in order to carry out a command.
Box name	Name of a sub-window area inside the dialog box.
Check box	Square box, click to turn on/off an option.
Spin box	Box containing up/down arrow to scroll or type by numerical increments.
Dimmed command	Menu command not currently available (light gray).
Tab	Dialog box sub-headings to simplify complex menus.
Option button	Small circle to activate/deactivate a single dialog box option.
List box	Box containing a list of items. Click the list drop down arrow. Click the desired option.
Text box	Box to type text.
Drop down arrow	Opens a cascading list containing additional options.
OK	Executes the command and closes the dialog box.
CANCEL	Closes the dialog box and leaves the original dialog box settings.
APPLY	Executes the command. The dialog box remains open.

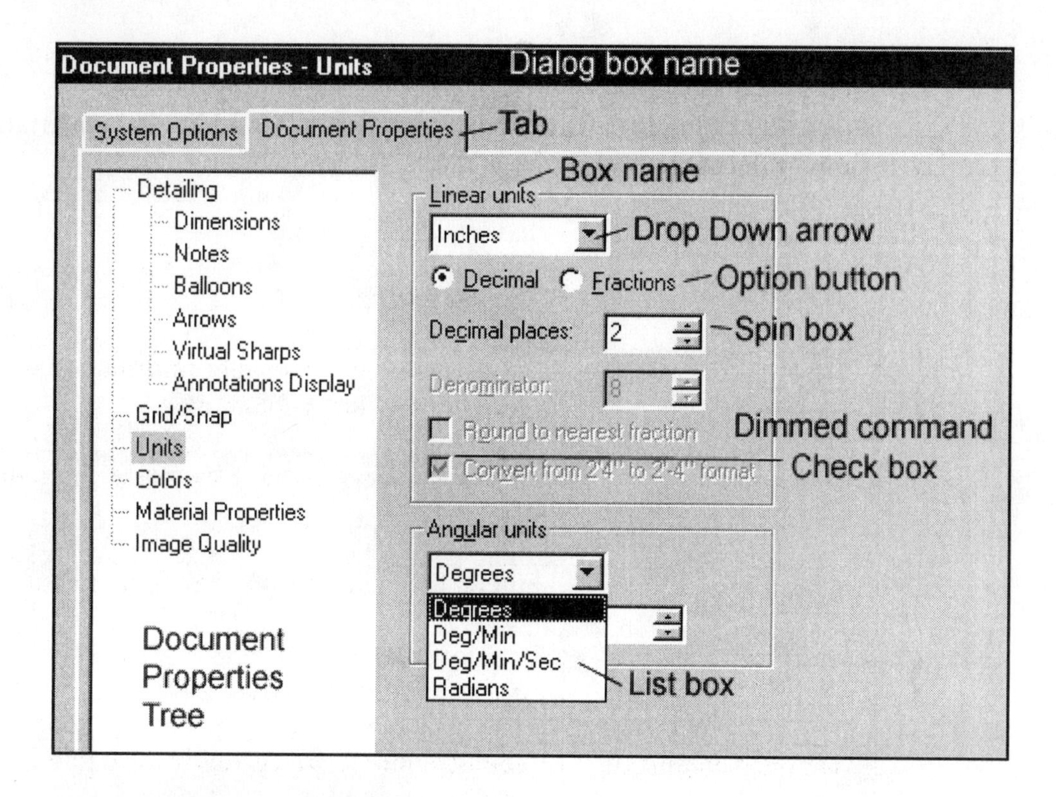
Document Properties - Units
Dialog box name
System Options
Document Properties
Tab
Box name
Detailing
Dimensions
Notes
Balloons
Arrows
Virtual Sharps
Annotations Display
Grid/Snap
Units
Colors
Material Properties
Image Quality
Linear units
Inches
Drop Down arrow
Decimal
Fractions
Option button
Decimal places:
2
Spin box
Denominator:
8
Round to nearest fraction
Dimmed command
Convert from 2'4" to 2'-4" format
Check box
Angular units
Degrees
Degrees
Deg/Min
Deg/Min/Sec
Radians
List box
Document Properties Tree

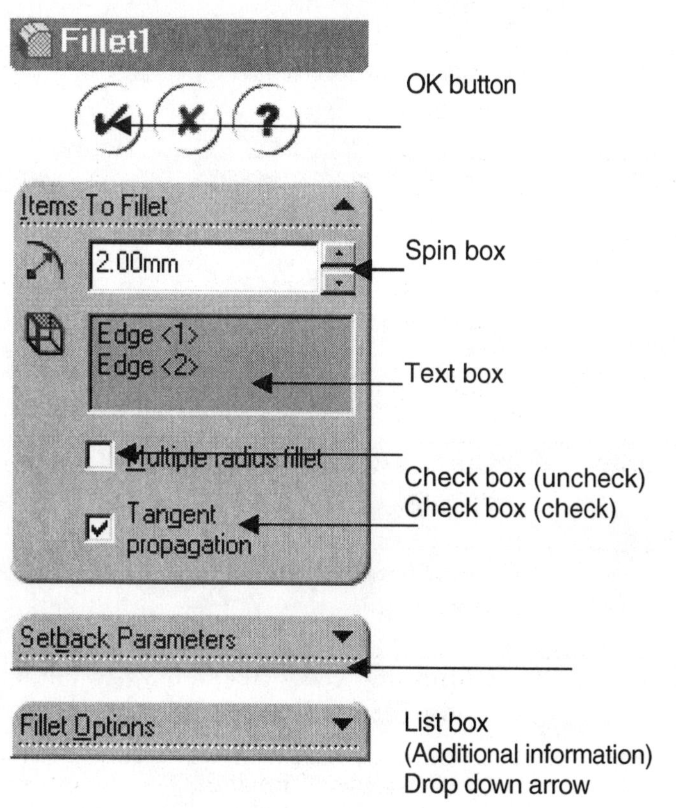
Fillet1
OK button
Items To Fillet
2.00mm
Spin box
Edge <1>
Edge <2>
Text box
Multiple radius fillet
Check box (uncheck)
Check box (check)
Tangent propagation
Setback Parameters
Fillet Options
List box
(Additional information)
Drop down arrow

Note to Instructors

Please contact the publisher www.schroff.com for additional materials that support the usage of this text in the classroom.

Project 1

Drawing Template and Sheet Format

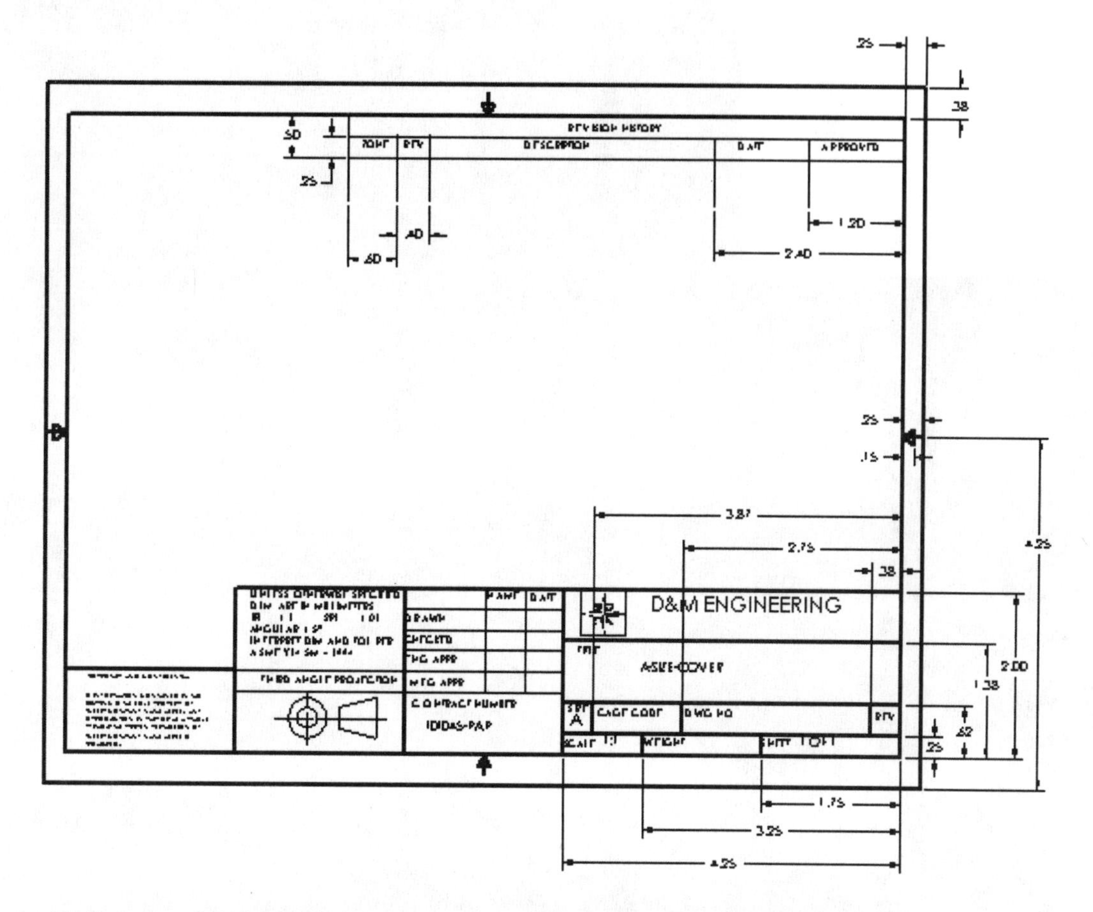

Below are the desired outcomes and usage competencies based on the completion of this Project. Note: The foundation of a SolidWorks drawing is the Drawing Template.

Project Desired Outcomes:	Usage Competencies:
Empty Drawing Templates. Custom Sheet Format. Custom Drawing Template.	Apply Drawing Properties to reflect the ASME Y14 Engineering Drawing and Related Drawing Practices.
	Knowledge and understanding of Drawing Templates and Sheet Formats.
	Wisdom of importing an AutoCAD drawing to create and modify a custom Sheet Format.

Project 1 – Drawing Template and Sheet Format

Project Objective

Obtain and apply drawing properties that reflect the ASME Y14 Engineering Drawing and Related Drawing Practices.

Knowledge and understanding of the Drawing Templates and Sheet Formats.

Create a custom C-size and A-size Drawing Template and Sheet Format. The Drawing Template and Sheet Format contain global drawing and detailing standards.

Provide a comprehensive understanding of importing an AutoCAD drawing to create and modify a custom Sheet Format.

On the completion of this project, you will be able to:

- Create an empty C-size Drawing Template.
- Import an AutoCAD drawing and save the drawing as a C-size Sheet Format.
- Combine the empty Drawing Template and the Sheet Format to create a C-ANSI-MM Drawing Template.
- Create an empty A-size Drawing Template.
- Modify an existing SolidWorks A-size Sheet Format.
- Combine the empty Drawing Template and the Sheet Format to create an A-ANSI-MM Drawing Template.

Project Situation

As the designer, your responsibilities include developing drawings that adhere to the ASME Y14 American National Standard for Engineering Drawing and Related Documentation Practices.

The foundation for a SolidWorks drawing is the Drawing Template. Drawing size, drawing standards, units and other properties are defined in the Drawing Template.

Sheet Formats contain the following: border, title block, revision block, company name, logo, SolidWorks Properties and Custom Properties.

You are under time constraints to complete the project on schedule. Create a SolidWorks custom Sheet Format.

Import an existing AutoCAD C-size drawing.

Create a custom C-size Drawing Template and an A-size Drawing Template.

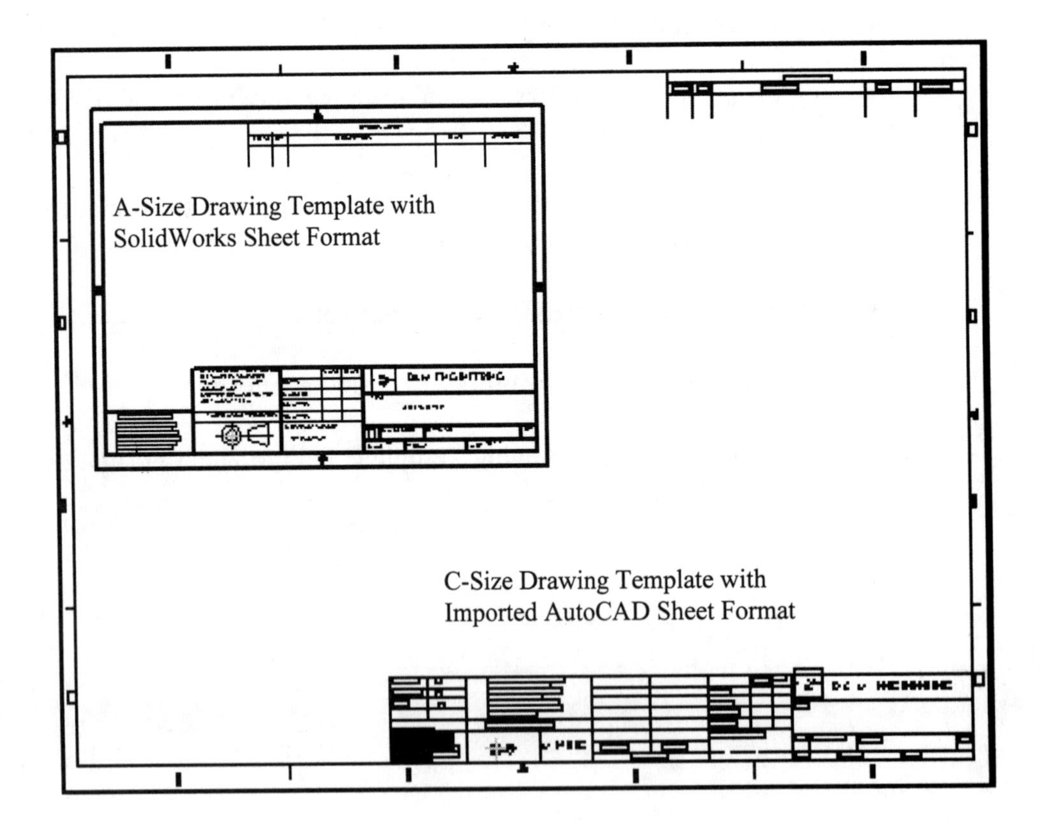

Project Overview

Perform the following tasks in this Project:

- Create an empty C-size Drawing Template.
- Import an AutoCAD drawing and save the drawing as a C-size Sheet Format.
- Combine the empty Drawing Template and the Sheet Format to create a C-ANSI-MM Drawing Template.
- Create an empty A-size Drawing Template.
- Modify an existing SolidWorks A-size Sheet Format.
- Combine the empty Drawing Template and the Sheet Format to create an A-ANSI-MM Drawing Template.

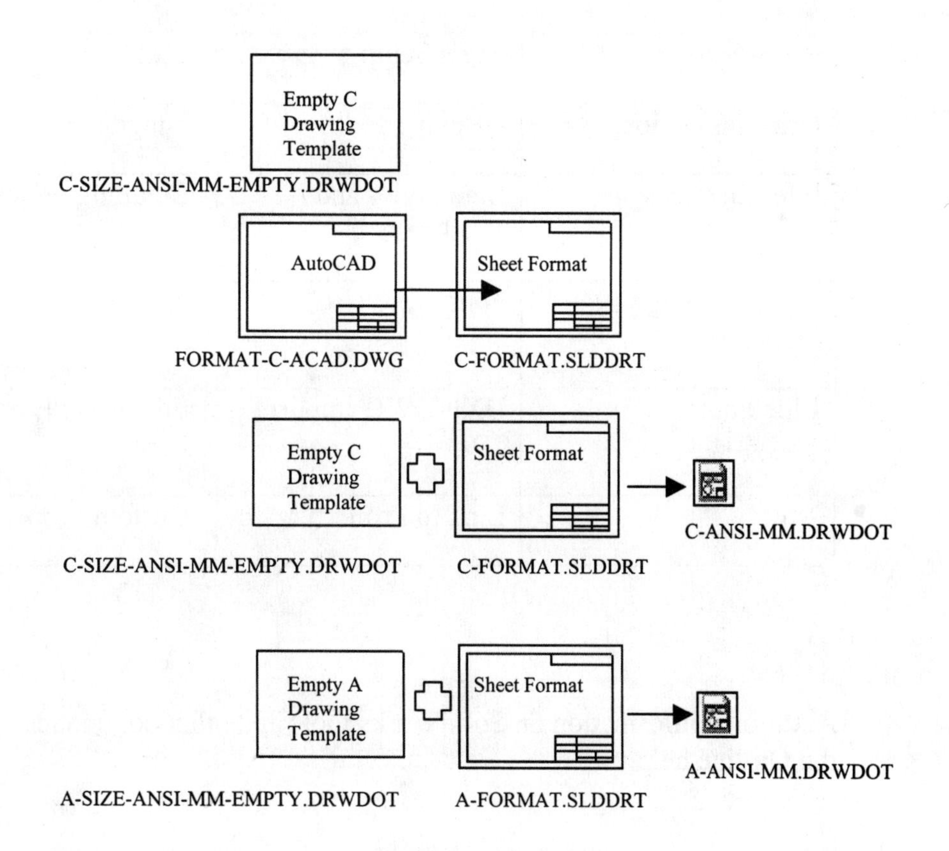

Conserve drawing time. Create a custom Drawing Template and Sheet Format. The Drawing Template and Sheet Format contain global drawing and detailing standards.

Note: Dimensioning techniques are similar for non-ANSI dimension standards.

SolidWorks Tools and Commands

The following SolidWorks tools and commands are utilized in this Project:

SolidWorks Tools and Commands:		
Drawing Template	Tools, Options, System Options	Tools, Options, Document Properties
Standard Sheet Format	Custom Sheet Format	No Sheet Format
Paper Size	Sheet Setup	Scale
Drawing Options	Display Modes	Tangent Edge
File Locations	Line Styles and Thickness	Detailing options
Dimensioning Standard	Font	Arrows
Line Font	DXF/DWG Import	Edit Sheet/Edit Sheet Format
Note	Link to Property	Custom Property

Additional information on SolidWorks tools and other commands are found in the On-line help.

Engineering Drawing and Related Documentation Practices

Drawing Templates in this section are based upon the American Society of Mechanical Engineers ASME Y14 American National Standard for Engineering Drawing and Related Documentation Practices.

These standards represent the drawing practices used by U.S. industry. The ASME Y14 practices supersede the American National Standards Institute ANSI standards.

The ASME Y14 Engineering Drawing and Related Documentation Practices are published by The American Society of Mechanical Engineers, New York, NY. References to the current ASME Y14 standards are used with permission.

ASME Y14 Standard Name	American National Standard Engineering Drawing and Related Documentation	Revision of the Standard
ASME Y14.100M-1998	Engineering Drawing Practices	DOD-STD-100
ASME Y14.1-1995	Decimal Inch Drawing Sheet Size and Format	ANSI Y14.1
ASME Y14.1M-1995	Metric Drawing Sheet Size and Format	ANSI Y14.1M
ASME Y14.24M	Types and Applications of Engineering Drawings	ANSI Y14.24M
ASME Y14.2M (Reaffirmed 1998)	Line Conventions and Lettering	ANSI Y14.2M
ASME Y14.3M-1994	Multiview and Sectional View Drawings	ANSI Y14.3
ASME Y14.5M – 1994 (Reaffirmed 1999)	Dimensioning and Tolerancing	ANSI Y14.5-1982 (R1988)

A portion of the ASME Y14 American National Standard for Engineering Drawing and Related Documentation Practices are presented in this book.

Information presented in Projects 1 - 5 represent sample illustrations of a drawing, view and or dimension type.

The ASME Y14 Standards Committee develops and maintains additional Drawing Standards. Members of these committees are from Industry, Department of Defense and Academia.

Companies create their own drawing standards based upon one or more of the following:

- ASME Y14.
- ISO or other International drawing standards.
- Older ANSI standards.
- Military standards.

Of course there is also the "We've always done it this way" drawing standard or "Go ask the Drafting Supervisor" drawing standard.

Drawing Template

The foundation of a SolidWorks drawing is the Drawing Template.

Drawing size, drawing standards, company information, manufacturing and or assembly requirements, units and other properties are defined in the Drawing Template.

The Sheet Format is incorporated into the Drawing Template. The Sheet Format can contain border, title block and revision block information, company name and or logo information, Custom Properties and or SolidWorks Properties.

Create a custom Drawing Template. SolidWorks starts with a default Drawing Template. Select the No. Sheet Format. Create a custom Sheet Format from the default drawing template.

The default SolidWorks Standard Sheet Format is A-Landscape.

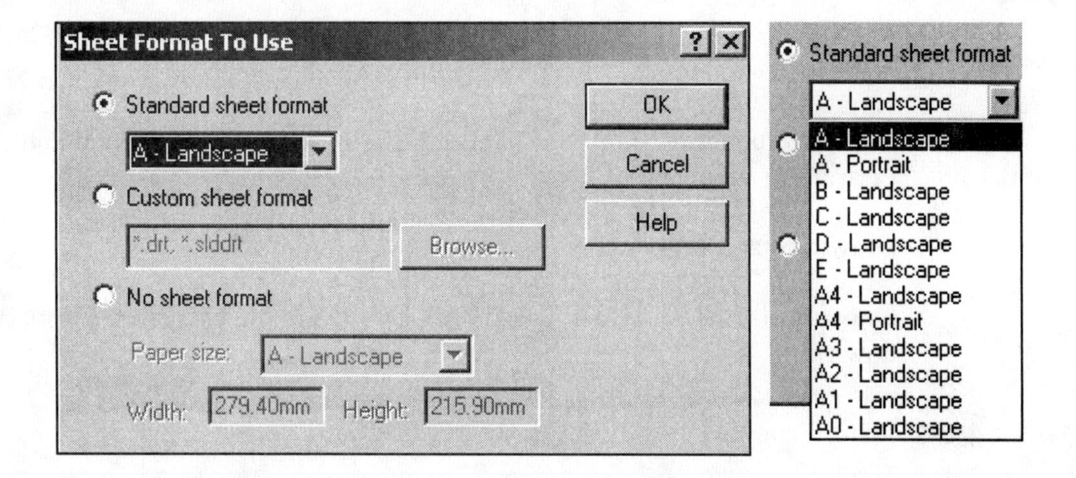

Note: The ASME Y14.1-1995 Decimal Inch Drawing Sheet Size, Format, ASME Y14.1M-1995 Metric Drawing Sheet Size, and format standard define the sheet size specification in inch and metric units respectively.

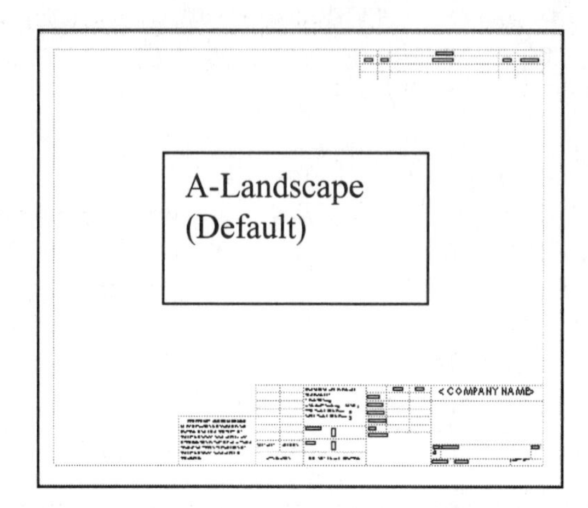

Drawing Size refers to the physical paper size used to create the drawing. The most common paper size in the U.S. is A size: (8.5in. x 11in.).

The most common paper size internationally is A4 size: (210mm x 297mm).

The ASME Y14.1-1995 and ASME Y14.1M-1995 standards contain both a horizontal and vertical format for A and A4 size, respectively.

The corresponding SolidWorks format is Landscape for horizontal and Portrait for vertical.

Drawing sizes A through E are predefined in SolidWorks.

Drawing sizes F, G, H, J & K are User Defined in the No. Sheet Format drop down list.

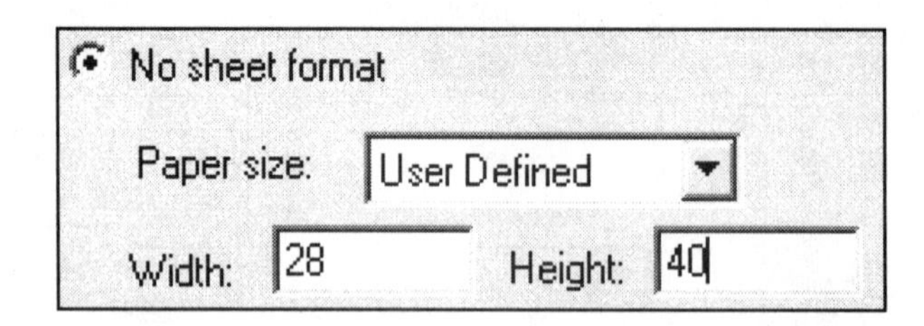

Metric drawing sizes A4 through A0 are predefined in SolidWorks.

Metric roll paper sizes are User Defined in the No Sheet Format drop down list.

The ASME Y14.1-1995 Decimal Inch Drawing Sheet Size standard are as follows:

Drawing Size:	**Size in inches:**	
"Physical Paper"	**Vertical**	**Horizontal**
A horizontal (landscape)	8.5	11.0
A vertical (portrait)	11.0	8.5
B	11.0	17.0
C	17.0	22.0
D	22.0	34.0
E	34.0	44.0
F	28.0	40.0
G, H, J and K apply to roll sizes, User Defined		

The ASME Y14.1M-1995 Metric Drawing Sheet Sizes standard are as follows:

Drawing Size:	**Size in Millimeters:**	
"Physical Paper"	**Vertical**	**Horizontal**
A0	841	1189
A1	594	841
A2	420	594
A3	297	420
A4 horizontal (landscape)	210	297
A4 vertical (portrait)	297	210

Caution should be used when sending electronic drawings between U.S. and International colleagues. Drawing paper sizes vary.

Example: An A-size (11in. x 8.5in.) drawing (280mm x 216mm) does not fit a A4 metric drawing (297mm x 210mm). Use a larger paper size or scale the drawing using the printer setup options.

Note: The Sheet Formats, parts and assemblies required to complete the projects in **Drawing and Detailing with SolidWorks 2003** are ***only available*** On-line at: www.schroff1.com.

Download the 2003drwparts file folder from www.schroff1.com.

1) Enter www.schroff1.com from your web browser.

2) Click the hypertext: **Drawing and Detailing with SolidWorks 2003**. The file folder, 2003drwparts is downloaded.

3) Right-click the 2003drwparts file folder. Click Properties. Uncheck Read Only. Click Apply. Click Apply changes to folders, subfolders and files.

4) Click **OK**.

Start a SolidWorks session.

5) Click **Start** on the Windows Taskbar. Click **Programs**. Click the **SolidWorks** folder.

6) Click the **SolidWorks** application. The SolidWorks program window opens.

Create an Empty C-size Drawing Template.

7) Click **New**. Click **Drawing**.

8) Click **OK**.

9) Select **No Sheet Format** from the Sheet format to Use dialog box.

10) Select **C-Landscape** from the Paper size drop down list.

11) Click **OK**.

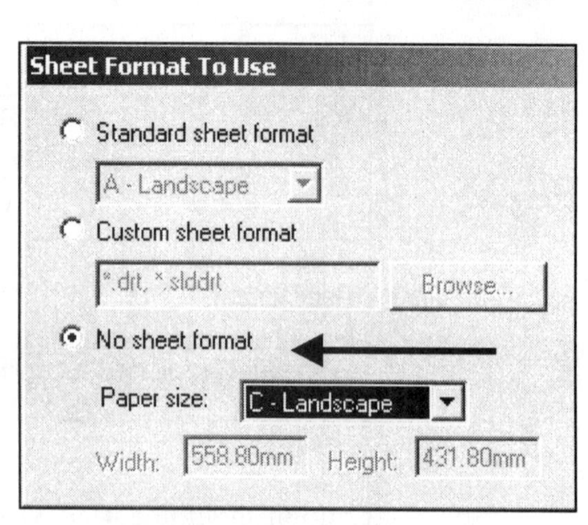

The C-Landscape Drawing Template is displayed in a new Graphics window.

The sheet border defines the C drawing size, (22in. x 17in.). Landscape indicates that the larger dimension is along the horizontal. Portrait indicates that the larger dimension is along the vertical.

Landscape

Portrait

Note: Portrait is only an option for A and A4 paper size.

The Drawing toolbar and Annotations toolbar are displayed left of the Graphics window. The Feature Manager is displayed to the left of the Graphics window.

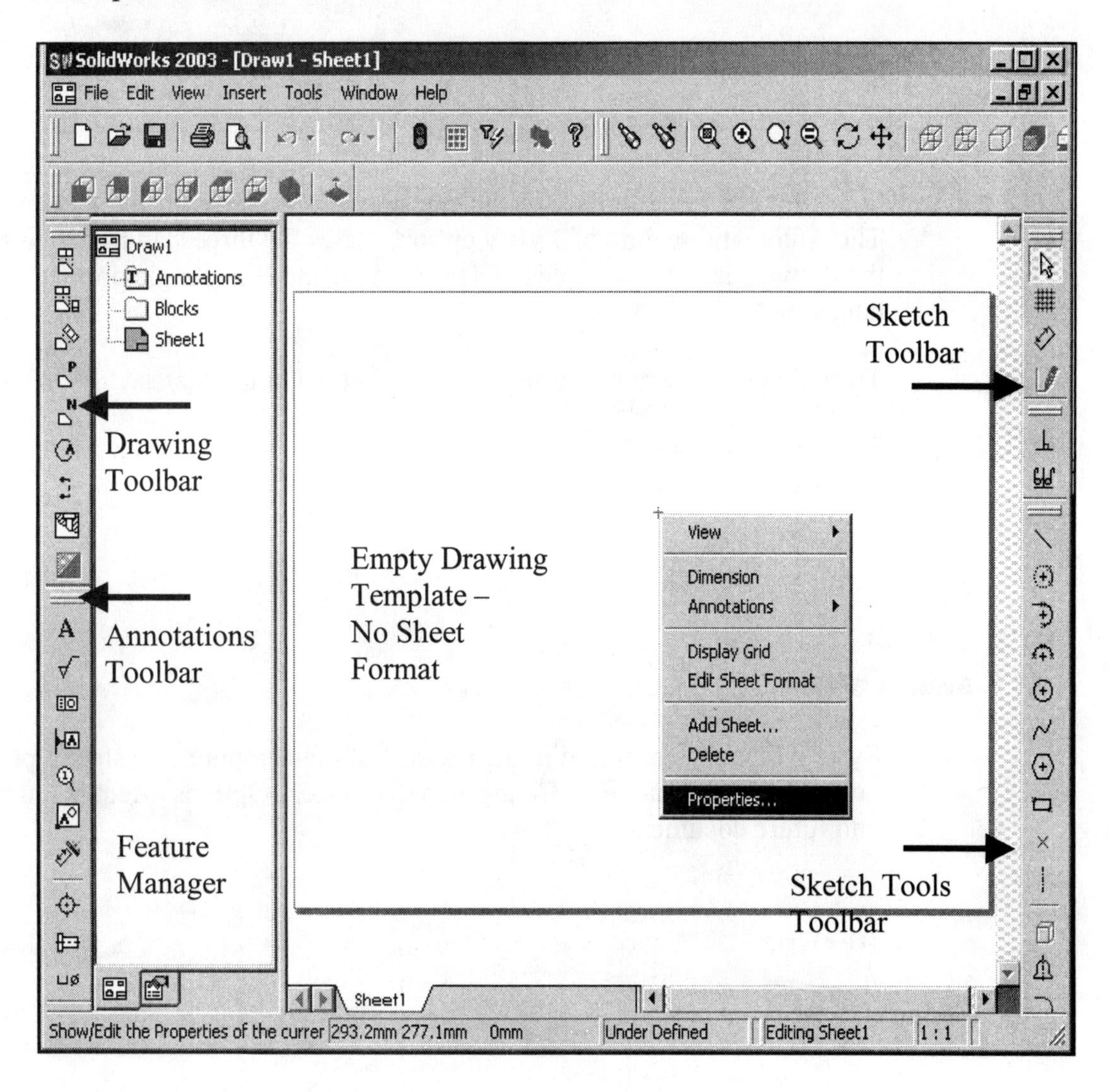

12) The Sketch and Sketch Tools toolbars are displayed to the right of the Graphics window. Right-click in the **Graphics window**.

13) Click **Properties**. The Sheet Setup Properties are displayed.

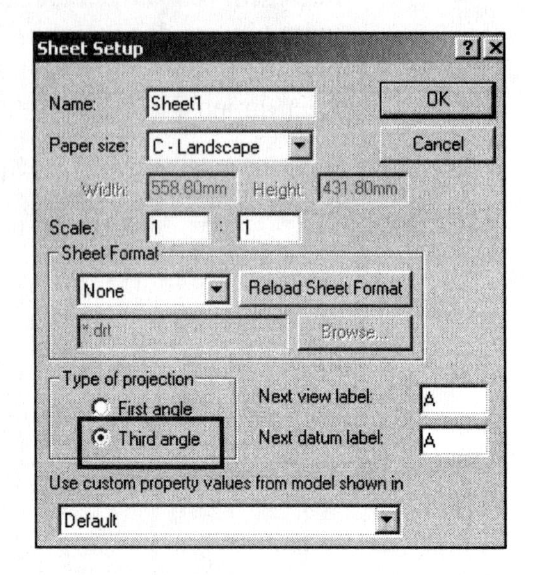

Set the Sheet Properties.

14) The default sheet Name is Sheet1. The Paper size is C-Landscape. A drawing can contain one or more sheets. Sheet scale controls the default scale. The default Sheet Scale is 1:1. Click **Third Angle** for Type of Projection.

15) Click **OK**.

The Automatic scaling of 3 view option, scales the three standard views to fit the drawing sheet. Examples of Third Angle and First Angle projection are developed in Project 2.

Third Angle projection is primarily used in the United States.

For company's supporting a First Angle projection scheme, views in Project 2 are placed in different locations.

System Options and Document Properties

System Options are stored in the registry of the computer. System Options is not part of the document. Changes to the System Options affect all current and future documents.

ANSI or ISO Dimension Standard, Units and other Properties are set in Document Properties.

Document Properties apply only to the current document. When you save the current document as a template, the current parameters are stored with the template.

New documents that utilize the same template contain these set parameters.

Conserve drawing time. Set the System Options and Document Properties before you begin a drawing.

Set System Options.

16) Set the Drawings options used in this text. Click **Tools**, **Options**, **System Options**, **Drawings**. Note: Drawing options can be turned on or off.

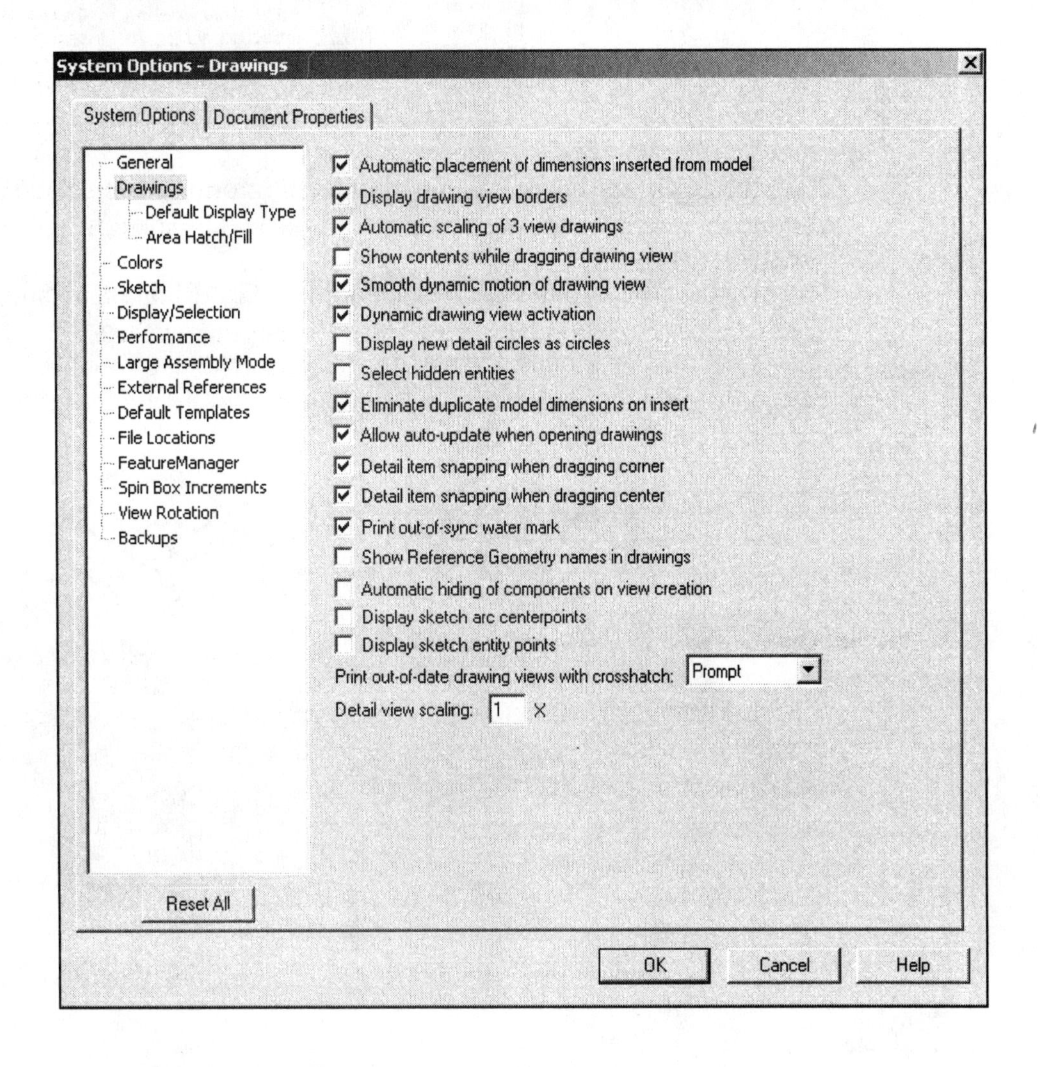

Drawings Options are available from the On-line help.

17) Click the **Help** button in the System Options dialog box. The Drawings Options help is displayed.

18) Review each Drawing option. Drag the **Scroll bar** downward.

19) **Minimize** the Help window.

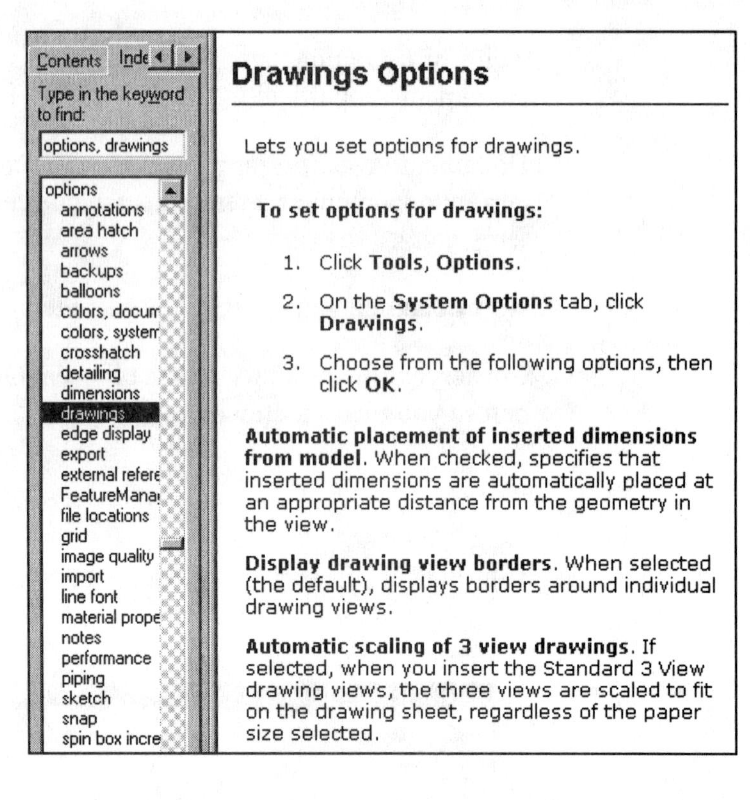

On-line help is a great resource for additional information on SolidWorks functions.

Help is accessible through the Help button, F1 key, Main menu and "?" icon.

Review the display modes settings for a new drawing.

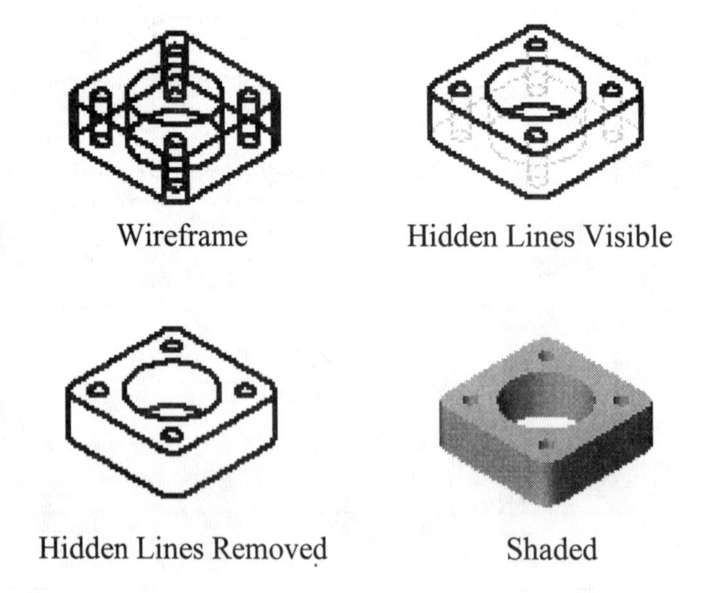

Default Display Modes for New Drawing View

Review the tangent edges setting for a new drawing.

Displayed modes and tangent edge settings can be changed in the individual drawing view.

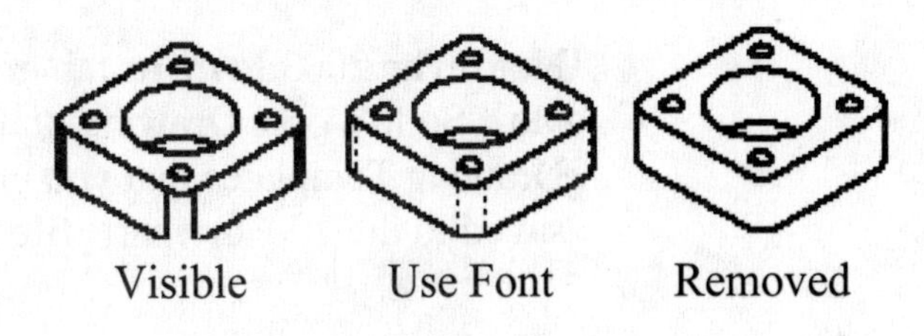

Tangent Edges

20) Set the Default Display Type. Click **Default Display Type** below the Drawings text. Click **Hidden removed** for the Default display mode for new drawing views. Click **Removed** for the Default display of tangent edges in the new drawing views.

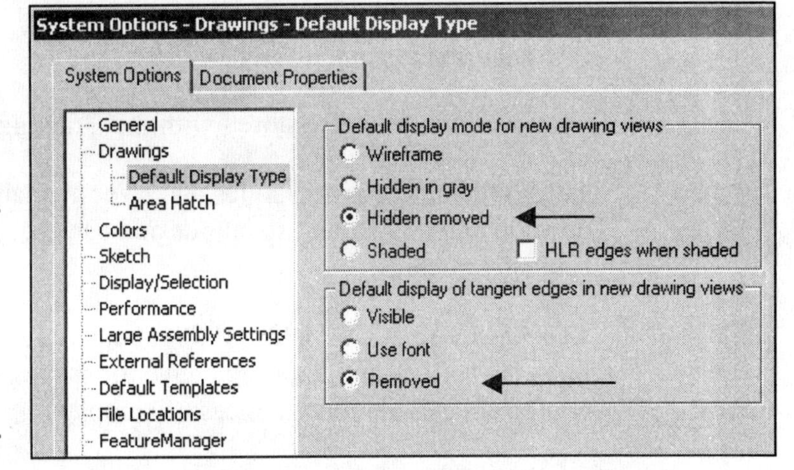

21) Click **OK**.

Set File Locations to the 2003drwparts folder for Drawing Templates.

22) Click **File Locations** from the System Options tab.

23) Select **Document Templates** from the Show Folders for Drop down list.

24) Click the **Add** button.

25) Click **Browse**.

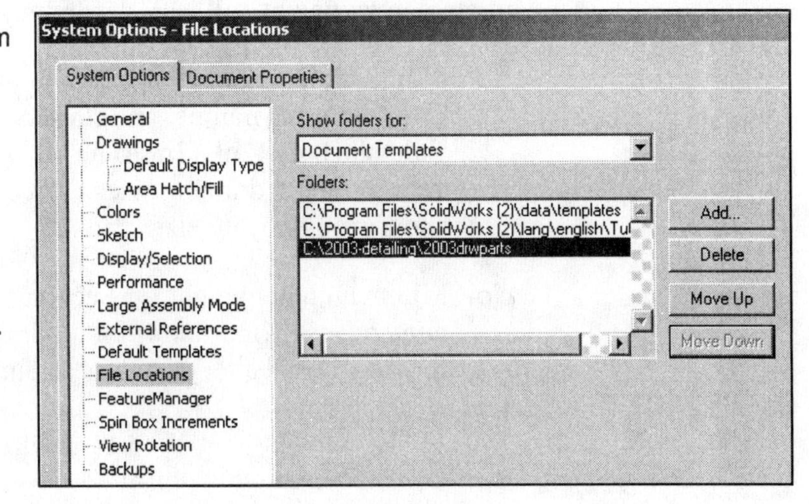

26) Select the **2003drwparts** folder that you downloaded from www.Schroff1.com.

27) Click **OK**.

Note: The 2003drwparts tab appears in the New SolidWorks Drawing dialog box. The Drawing Templates that you create will be saved to the 2003drwparts file folder.

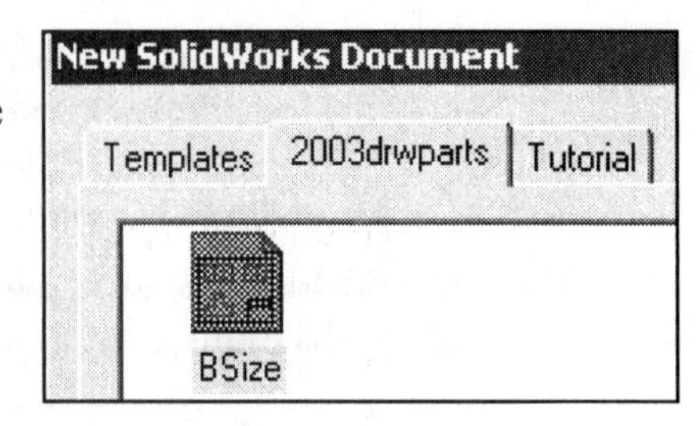

The Drawing Properties Detailing options provide the ability to address: dimensioning standards, text style, center marks, witness lines, arrow styles, tolerance and precision.

Drawing Properties are stored with the Drawing Template.

There are numerous text styles and sizes available in SolidWorks. Companies develop drawing format standards and use specific text height for Metric and English drawings.

The ASME Y14.2M-1992(R1998) standard lists the lettering, arrowhead and line conventions and lettering conventions for engineering drawings and related documentation practices. Examples:

- Font: Utilize a single stroke, gothic lettering in all upper case letters. Use a single font. Century Gothic is the default SolidWorks font. Create a test page to insure that both Windows and your particular Printer/Plotter drivers support the selected font.

- Minimum letter height will vary depending upon usage on a drawing:

    - Minimum letter height used for drawing title, drawing size, CAGE Code, drawing number and revision letter positioned inside the Title block is .12in. (3mm) for A, B and C inch sizes and A2, A3 and A4 metric drawing sizes: Text height is .24in. (6mm) for D and E inch drawing sizes and A0, A1 metric drawing sizes.

    - Minimum letter height for Section views, Zone letters and numerals is .24in. (6mm) for all drawing sizes. Set Text size for Section, Detail and View font to 6mm.

    - Minimum letter height for drawing block headings is .10in. (2.5mm) for all drawing sizes.

    - Minimum letter height for all other characters is .12in. (3mm) for all drawing sizes. Set Text size for Dimension and Note Font to 3mm.

- Arrowheads: Utilize solid filled single style arrowhead, with a 3:1 ratio of arrow length to arrow width. The arrowhead width is proportionate to the line thickness. The Dimension line thickness is 0.3mm. In this project, the arrow length is 3mm. Arrow width is 1mm. SolidWorks defines arrow size with three options: Height, Width and Length. Height corresponds to arrow width. Width corresponds to arrow length. Length corresponds to the distance from the tip of the arrow to the end of the tail.

- The Section line thickness is 0.6mm. The arrow length is 6mm. The arrow width is 2mm.

- Line Widths: The ASME Y14.2M-1992 (R1998) standard recommends two line widths with a 2:1 ratio. The minimum width of a thin line is 0.3mm. The minimum width of a thick, "normal" line is 0.6mm. Note: A single width line is acceptable on CAD drawings. Two line widths are used in this Project; Thin: 0.3mm and Normal: 0.6mm. Apply Line Styles in the Line Font Document Properties. Line Font determines the appearance of a line in the Graphics window. SolidWorks styles utilized in this Project are as follows:

SolidWorks Line Style	Thin (0.3mm)	Normal (0.6mm)
Solid		
Dashed		
Phantom		
Chain		
Center		
Stitch		
Thin/Thick Chain		

Various printers/plotters allow variable Line Weight settings.

Example: Thin (0.3mm), Normal (0.6mm) and Thick (0.6mm).
Refer the printer/plotter owner's manual for Line Weight setting.

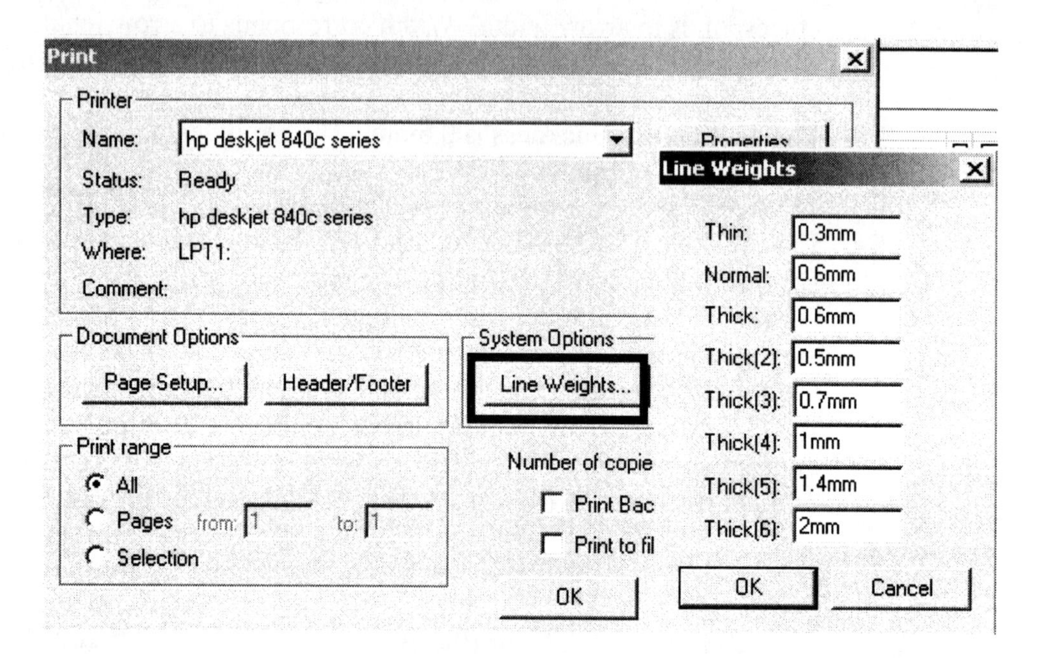

The Page Setup button contains the Scale options:

- Same as window prints the current view of the graphics area. Option only valid for a part or assembly.
- Scale sheet to fit paper prints the drawing sheet to fit the paper size
- Scale prints the document at a scaled value (in percent) that you specify.

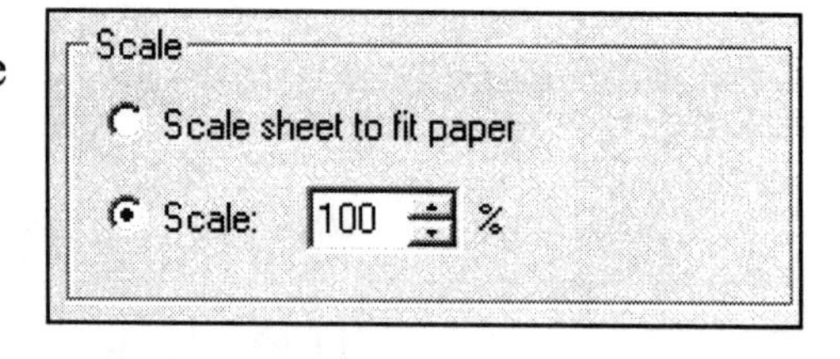

Line Font: The ASME Y14.2M-1992(R1998) standard address the type and style of lines used on engineering drawings. Combine different styles and use drawing Layers to achieve the following types of lines:

ASME Y14.2-1992(R1998) TYPE of LINE and an example	SolidWorks Line Font Type of Edge	Style	Thickness
Visible line displays the visible edges or contours of a part.	Visible Edge	Solid	Thick "Normal"
Hidden line displays the hidden edges or contours of a part.	Hidden Edge	Dashed	Thin
Section lining displays the cut surface of a part/assembly in a section view.	Crosshatch	Solid	Thin Different Hatch patterns relate to different materials
Center line displays the axes of center planes of symmetrical parts/features.	Construction Curves	Center	Thin
Symmetry line displays an axis of symmetry for a partial view.			Sketch Thin Center Line and Thick Visible lines on drawing Layer .
Dimension lines/Extension lines/Leader lines combine to dimension drawings.	Dimensions DIMENSION LINE 100 Extension Line Leader Line	Solid	Thin
Cutting plane line or Viewing plane line display the location of a cutting plane for sectional views and the viewing position for removed views.	Section Line View Arrows D D	Phantom Solid	Thick Thick, "Normal"

ASME Y14.2-1992(R1998) TYPE of LINE and an example	SolidWorks Line Font Type of Edge	Style	Thickness
Break line displays an incomplete view. Short Breaks Long Breaks		Curved Small Zig Zag	Broken view Use Curved for Short Breaks Use Small Zig Zag for Long Breaks
Phantom line displays alternative position of moving parts.	—··—··—··—··—··—		Sketch Thin Phantom Line on drawing Layer
Stitch line displays a sewing or stitching process.	....		Sketch Thin Stitch Line on drawing Layer
Chain line displays a surface that requires more consideration or the location of a projected tolerance zone.	— - —— - —— - —		Sketch Thick Chain Line on drawing Layer

Note: The following lines are not predefined in SolidWorks: Symmetry line, Phantom line, Stitch line and Chain line.

The line style and thickness for the above line types are defined on a separate drawing layer.

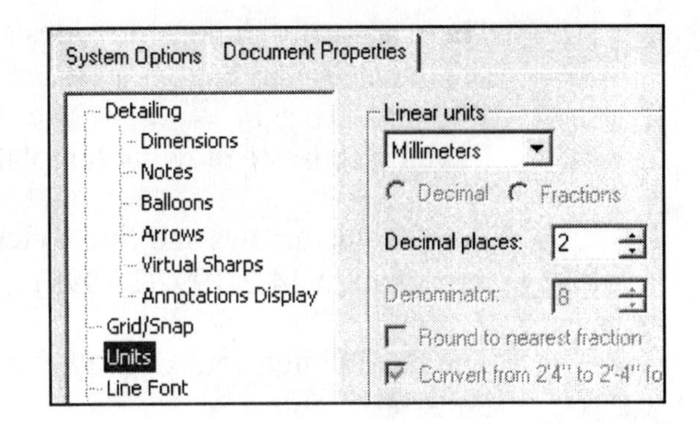

Set Drawing Properties.

28) Click **Tools**, **Options**.

29) Click **Document Properties tab**.

30) Select **Units** from the left text box.

31) Click **Millimeters** from the Linear Units drop down list.

32) Enter **2** for Decimal places.

Note: Set units before entering values for Detailing options.

33) Click **Detailing**. Select **ANSI** from the Dimensioning standard drop down list. Detailing options are available depending upon the selected standard.

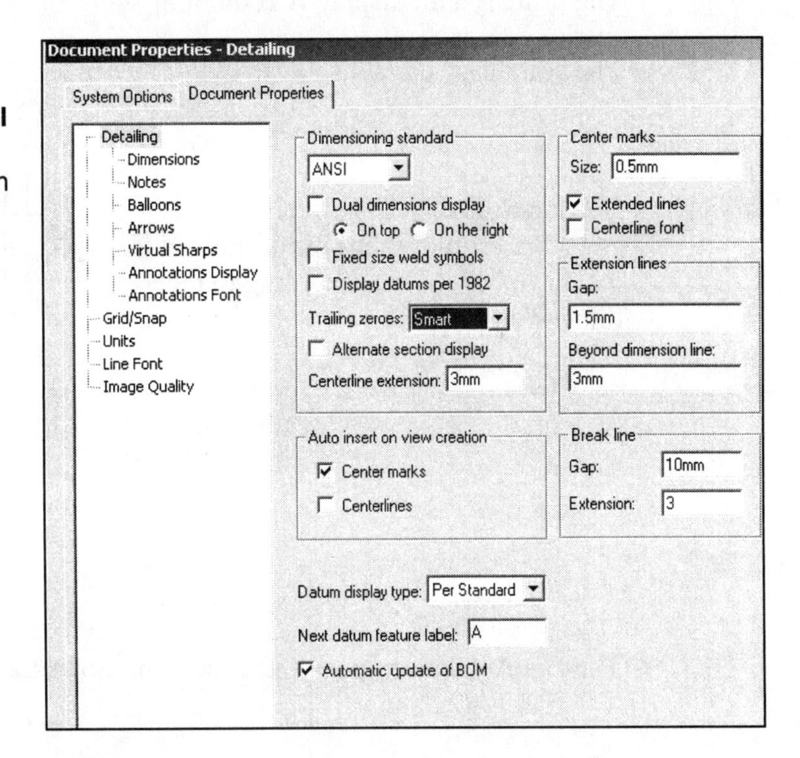

Drawing and option availabilities are affected by various Drawing Properties.

The Dimensioning standard options are: ISO, DIN, JIS, BSI, GOST and GB.

Obtain additional drawing options through the On-line help.

Review the Detailing options function before entering their values.

Millimeter dimensioning and decimal inch dimensioning are the two types of units specified on engineering drawings.

There are other dimension types specified for commercial commodities such as pipe sizes and lumber sizes.

Develop separate drawing templates for decimal inch units.

Text height, arrows and line styles are defined with inch values according to the ASME Y14.2-1992(R1998) Line Conventions and Lettering standard.

The Dual dimensions display check box shows dimensions in two types of units. Example:

Select Dual dimensions display.

Select the On top option.

The primary unit display is 100mm.

The secondary units display is [3.94] inches.

[3.94]
100

The Fixed size weld symbols checkbox displays the size of the weld symbol. Scale according to the dimension font size.

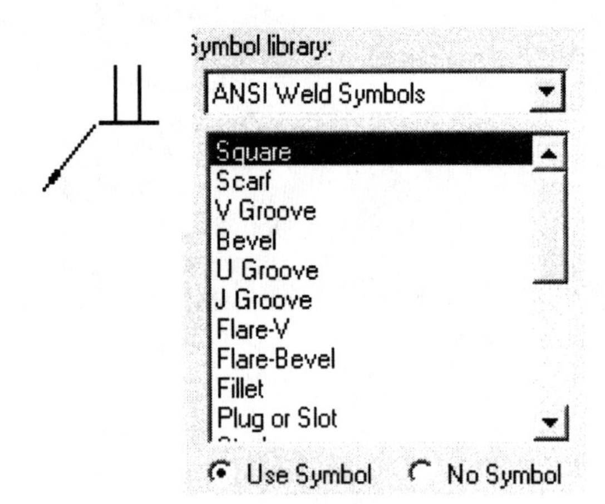

The Display datums per 1982 checkbox shows the ANSI Y14.5M-1982 datums.

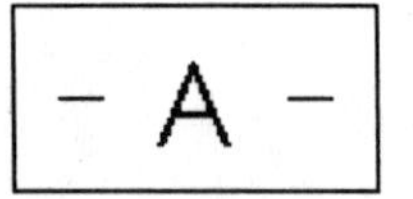

The ASME Y14.5M-1994(R1999) datums are used in this text.

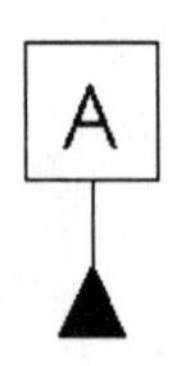

Trailing Zeros list box contains three options:

- Smart.
- Show.
- Removed.

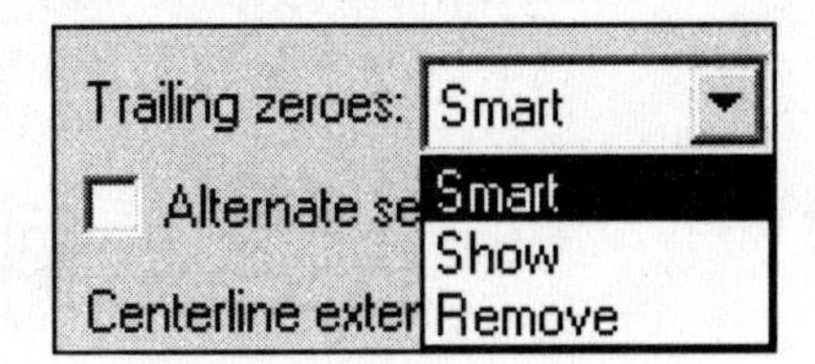

The default Smart option removes trailing zeros based upon the ASME Y14 rules for trailing zeros for dimension values.

The ASME Y14.2M-1992(R1998) standard supports two display styles for the Cutting-plane line or Viewing-plane line. The default section line displays with a continuous Phantom line type (D-D).

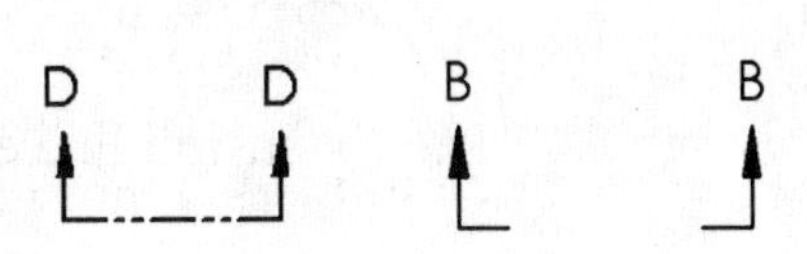

Check the Alternate section display to allow the arrow ends to stop at the ends of the section cut (B-B).

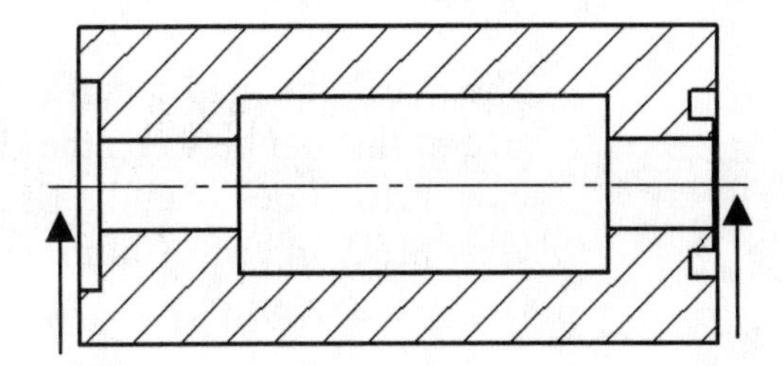

Auto insert on view creation places center marks and centerlines on all appropriate entities when a new view is inserted into a drawing.

The Centerline extension value controls the extension length beyond the section geometry.

Set the extension length to 3mm. Center marks specifies the default center mark size used with arcs and circles. Center marks are displayed with or without center mark lines.

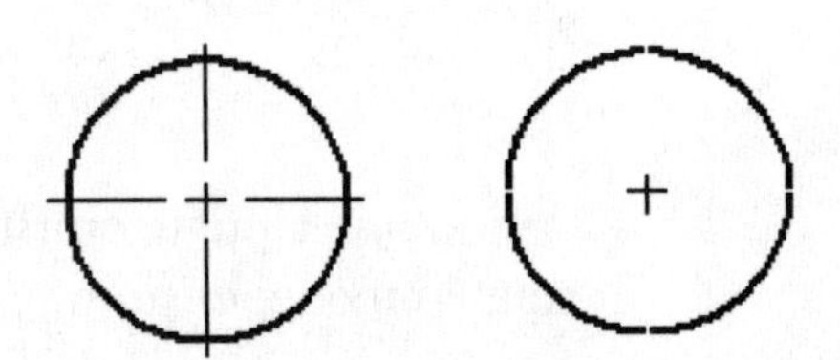

The center mark lines extend just beyond the circumference of the selected circle. Set the default center mark size to 0.5mm. Base the center mark size on the drawing size and scale.

Extension lines are defined in the ASME Y14.2M-1992(R1998) and ASME Y14.5M-1994(R1999) standard.

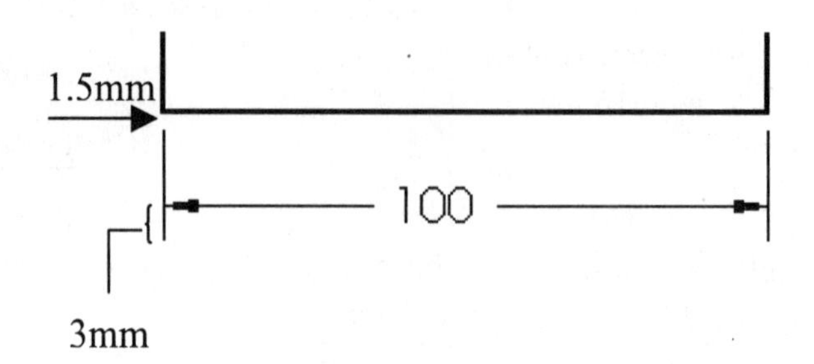

A visible Gap exists between the Extension line and the Visible line.

The Extension line extends 3mm beyond the Dimension line.

Set Gap to 1.5mm. Set the Extension to 3mm. Note: The values 1.5mm and 3mm are a guide. Base the Gap and Extension line on the drawing size and scale.

The Break line gap specifies the size of the gap between the Broken view break lines. Set the Broken view break lines to 10mm. Set the Extension to 3mm.

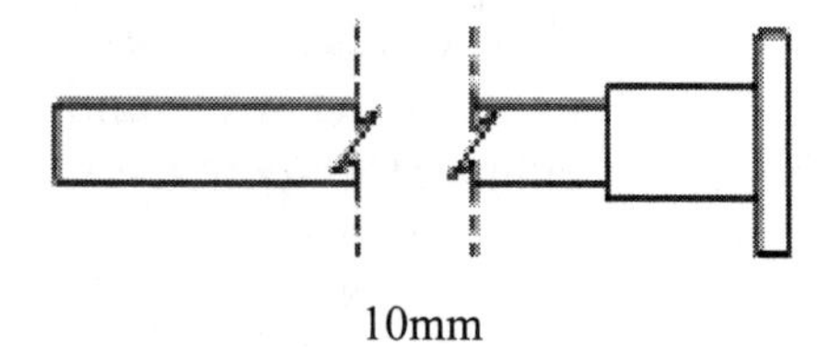

The Next datum feature label specifies the next upper case letter used for the Datum Feature Symbol.

The default value is A. Successive labels are in alphabetical order.

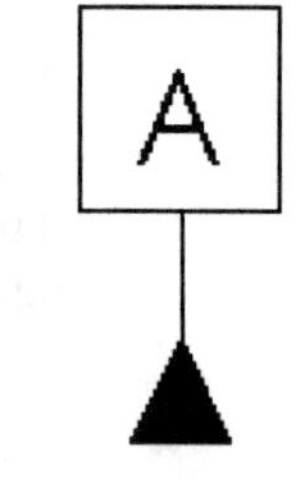

The Datum display type Per Standard shows a filled triangular symbol on the Datum Feature.

Automatic Update on BOM option updates the Bill of Material in a drawing if related model custom properties change.

ITEM NO.	QTY.	PART NO.	MATERIAL
1	1	10-0408	ALUMINUM
2	1	10-0409	STEEL

Set the values in SolidWorks to meet the ASME standard.

Set Detail Options.

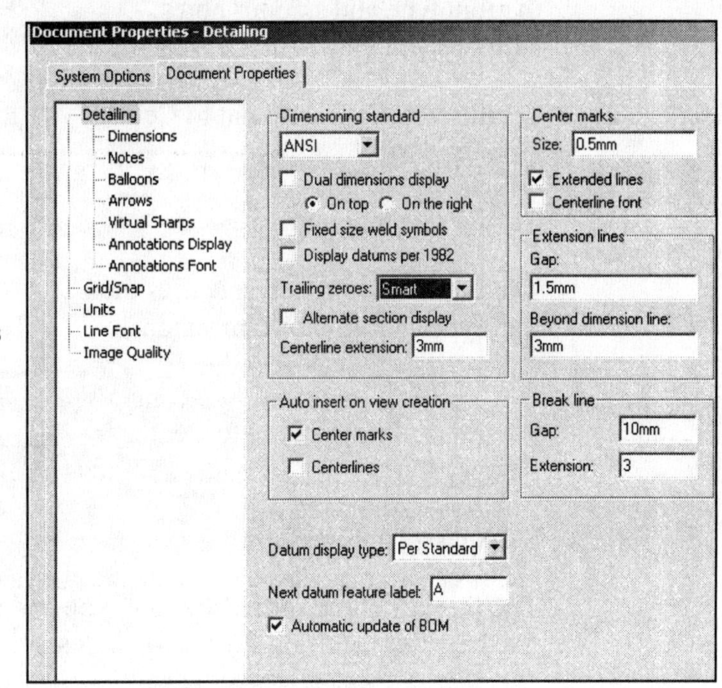

34) Enter **3mm** for the Centerline extension.

35) Enter **0.5mm** for the Center marks.

36) Modify the Witness lines (Extension line) values. Enter **1.5mm** for Gap.

37) Enter **3mm** for Extension.

38) Enter **10mm** for the Break line gap. Enter **3mm** for Extension for the Break line. Note: There is no set value for the Break line gap. Increase the value to accommodate a revolved section.

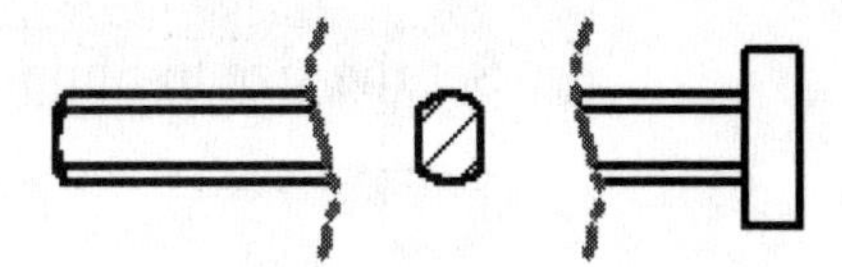

The Annotations Font option controls the text height for:

- Note/Balloon.
- Dimension.
- Detail.
- Section.
- View Arrow.
- Surface Finish.
- Weld Symbol.

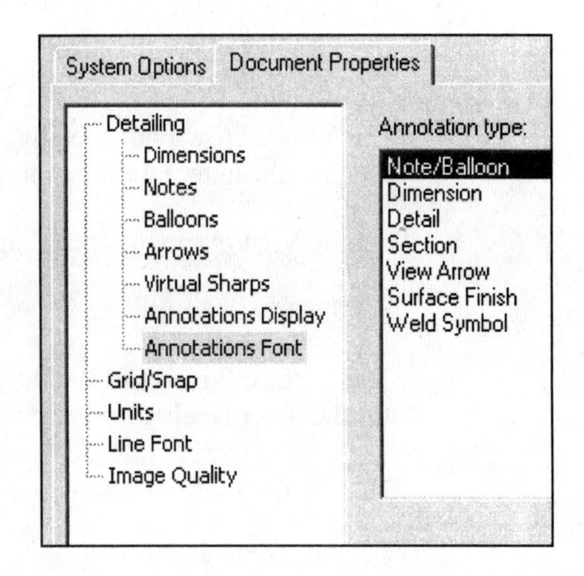

Set each Annotation type font height.

The Note/Balloon option specifies the font type and size for notes, balloons and view labels.

SECTION A-A

1

Set the Note/Balloon Font to Century Gothic.

Set the size to 3 mm.

The Dimension option specifies the font type and size for dimension text.

100

Set the Dimension Font to Century Gothic.

Set the size to 3 mm.

The Detail Font specifies the font type and size used for the letter labels on the detail circles.

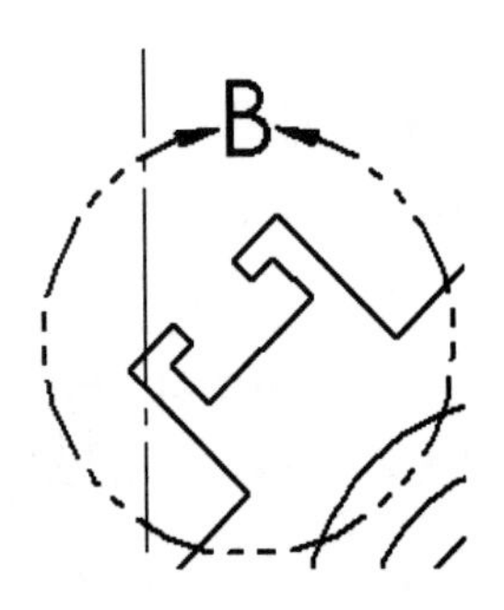

Set the Detail font to Century Gothic.

Set the size to 6mm.

The Section Font specifies the font type and size used for the letter labels on the section lines.

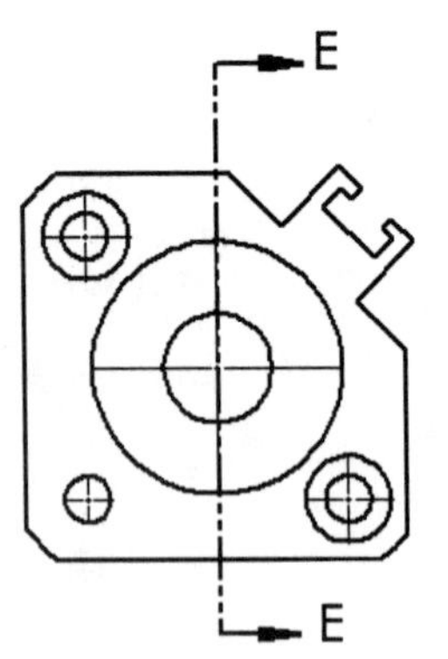

Set the Section font to Century Gothic.

Set the size to 6mm.

The View Arrow Font specifies the font type and size used for the letter labels on the view arrows.

Set the View Arrow font to Century Gothic.

Set the size to 6mm.

The Surface Finish and Weld Symbol fonts specify the font type and size used for the letter labels for Surface Finish and Weld Symbols, respectfully.

Set the View Arrow font to Century Gothic.

Set the size to 3mm.

39) Set the Detail Font. Click the **Note / Balloon** option button.

40) Enter **3mm** for text.

41) Repeat for **Dimension Font**, **Surface Finish**, and **Weld Font**

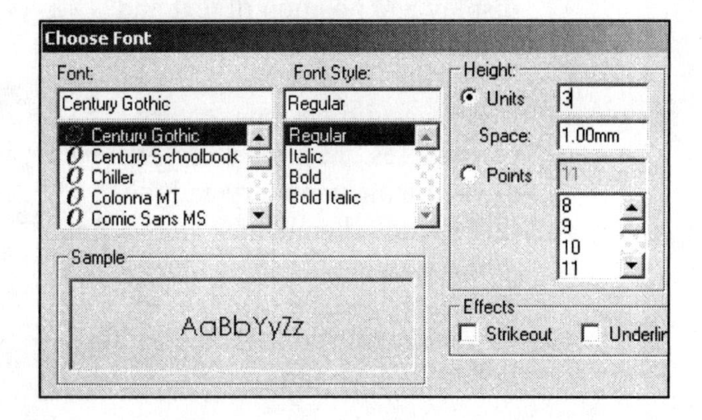

42) Set the Detail Font. Click the **Detail Font** button.

43) Enter **6mm** for text.

44) Repeat for **Section Font** and **View Arrow Font**.

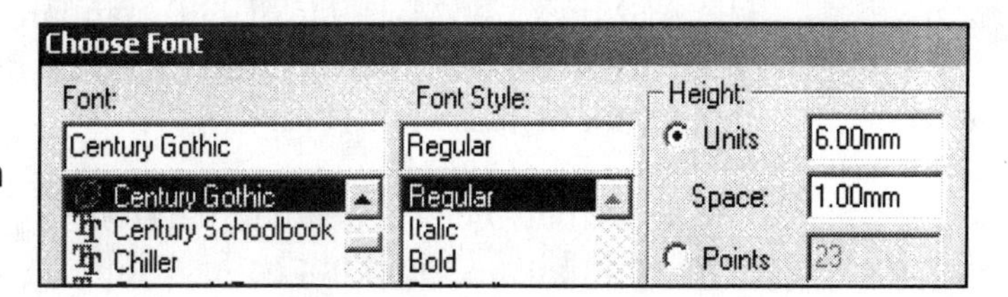

45) Review the Dimension options. Click **Dimensions** from the left side of the Detailing text box.

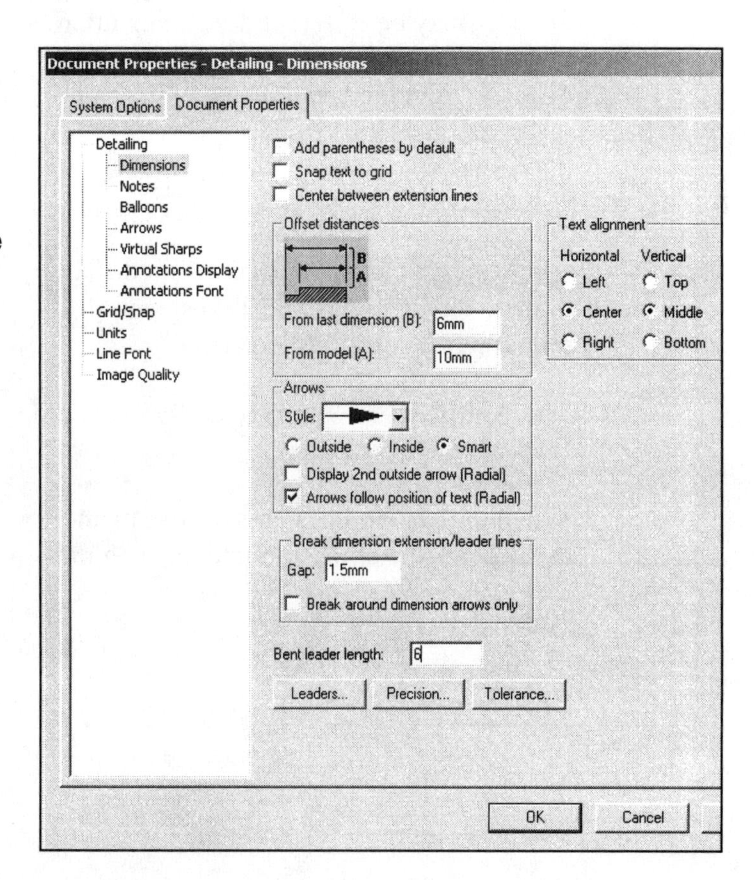

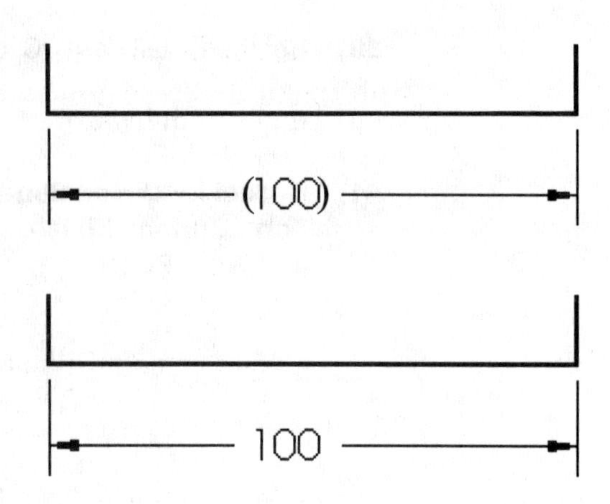

The Dimension options determine the display and position of text and extension lines.

Reference dimensions require parentheses. Many features were created with symmetry and the dimension scheme must be redefined in the drawing.

Uncheck the Add parentheses by default to save time.

Parenthesis can be added to a dimension at anytime through the Property option.

The ASME Y14.5M-1994(R1999) standard set guidelines for dimension spacing. The space between the first dimension line and the part outline should not be less than 10mm.

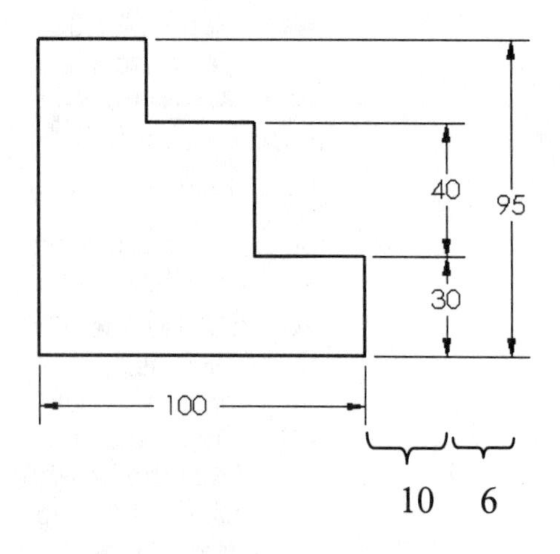

The space between subsequent parallel dimension lines should not be less than 6mm.

Spacing may be different depending upon drawing size and scale. Set the offset distance from the last dimension to 6mm.

Set the offset distance from the model to 10mm.

Arrow heads can be opened or filled. The ASME Y14.2M-1992(R1998) standard recommends a solid filled arrow.

The ASME Y14.5M-1994(R1999) standard states that crossing dimension lines should be avoided.

When dimension lines cross, close to an arrowhead, the extension line must be broken.

Ø8.25

Ø12

Drag the extension line above the arrowhead. Sketch a new line collinear with the extension line below the arrowhead.

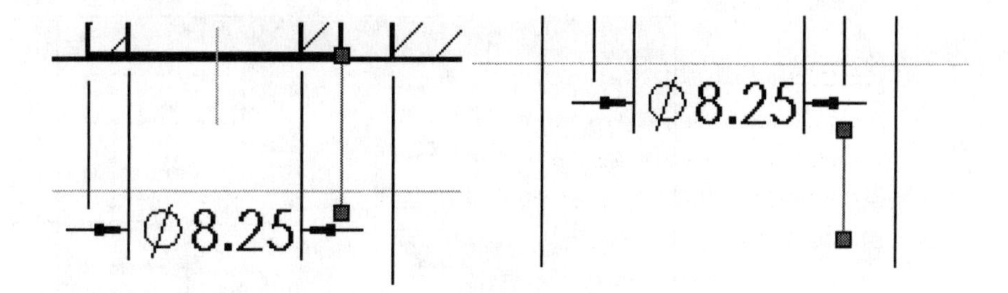

Set the Break Dimension Line Gap to 1.5mm.

Uncheck the Break around the dimension arrows. Control individual breaks on dimensions for this project.

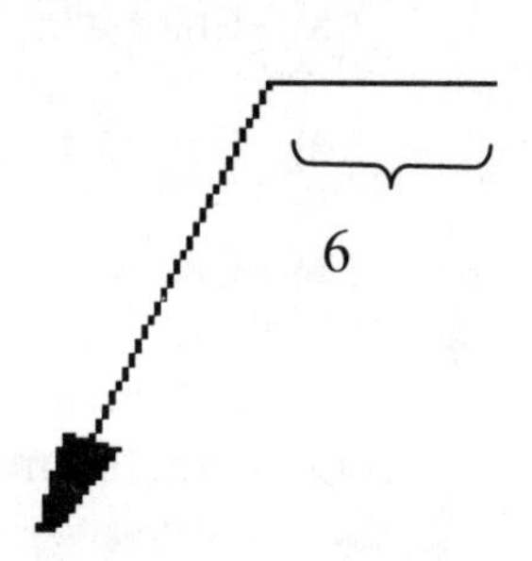

Leader lines are created with a small horizontal segment. This is called the Bent Leader line length. Set the Bent Leader line length to 6mm.

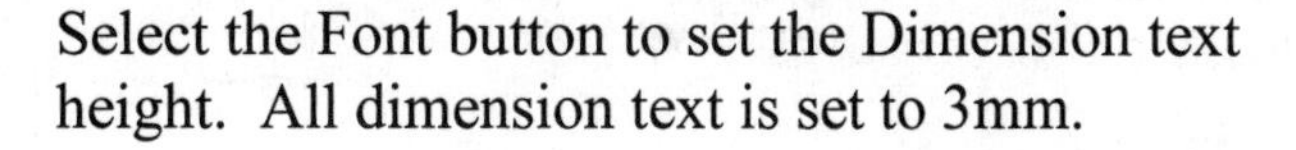

Select the Font button to set the Dimension text height. All dimension text is set to 3mm.

Set Dimensions options.

46) Uncheck the **Add Parentheses by Default** check box.

47) Set the Offset distances to **6mm** and **10mm**.

48) Set the Arrow style to **Solid**.

49) Enter **1.5mm** for the Gap in the Break Dimension Witness/Leader Lines.

50) Uncheck the **Break around dimension arrows only**.

51) Enter **6mm** for the Bent leader length.

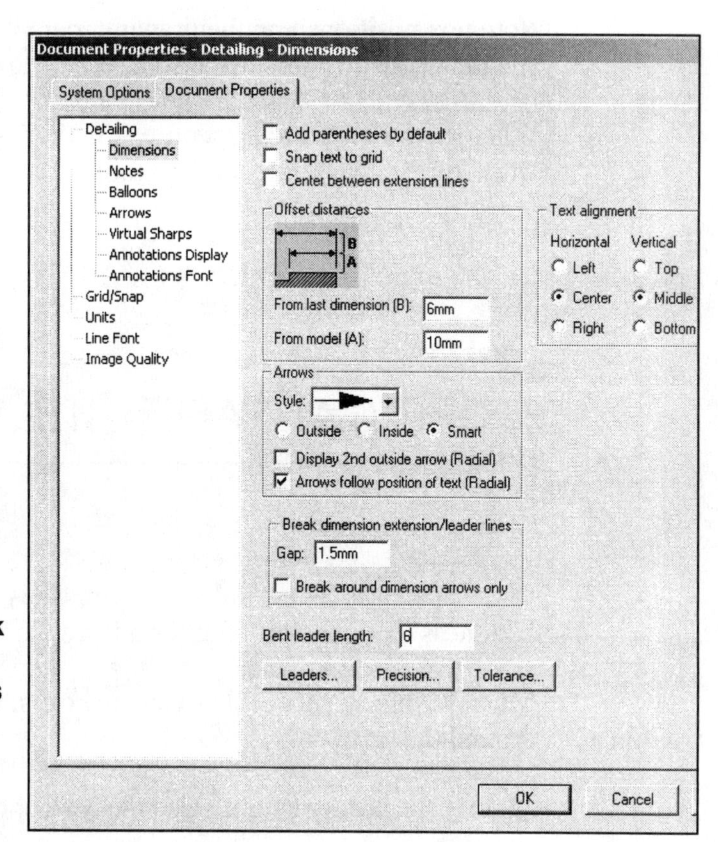

52) Click the **Precision** button. The Primary Units are Millimeters.

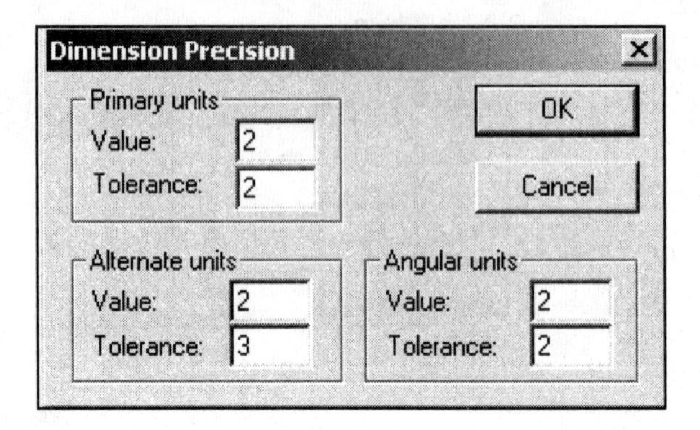

53) Enter **2** for Primary Units Value.

54) Enter **2** for Tolerance.

55) Click **OK.**

The Dimension Precision Value and Tolerance entries depend upon drawing units and the manufacturing requirements.

Note text positioned on the drawing, outside the Title block, are the same font and height as the Dimension font. There are exceptions to the rule.

When a Note refers to a specific ASME Y14.100M-1998 Engineering Drawing Practices extended symbol.

Example:

Use Upper case letters unless lower case is required. Example: HCl – Hardness Critical Item requires a lower case "l".

Modify Note Border Style to create boxes, circles, triangles and other shapes around the text.

Modify the border height. Use the Size option.

Set Notes options.

56) Click **Notes** from the left side of the Detailing text box.

57) Check **Use Bent leaders**.

58) Enter **6mm** for the Leader Length.

Balloon callouts label the parts in an assembly and relate them to the item numbers in the Bill of Materials.

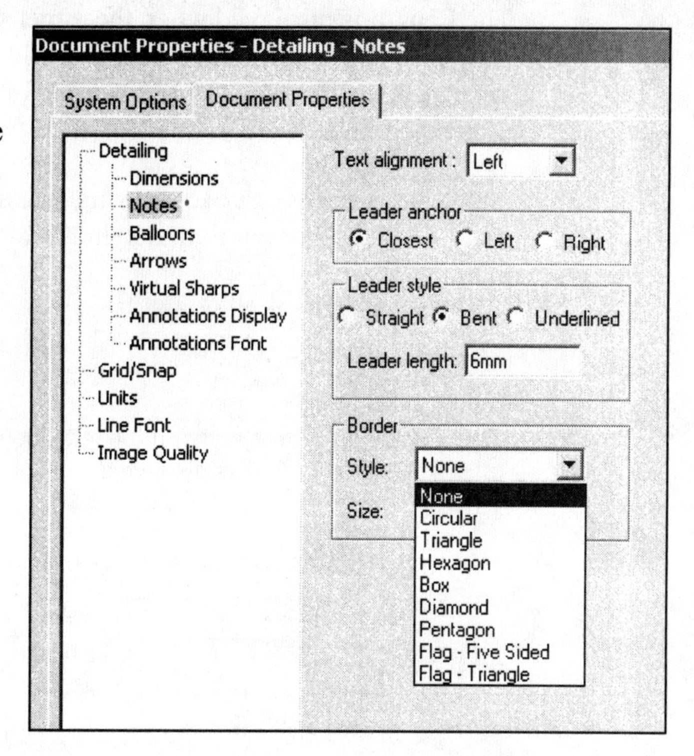

Set the drawing Balloon Properties.

59) Click **Balloons** from the left side of the Detailing text box.

60) Check **Use bent leaders**.

61) Enter **6mm** for the Leader length.

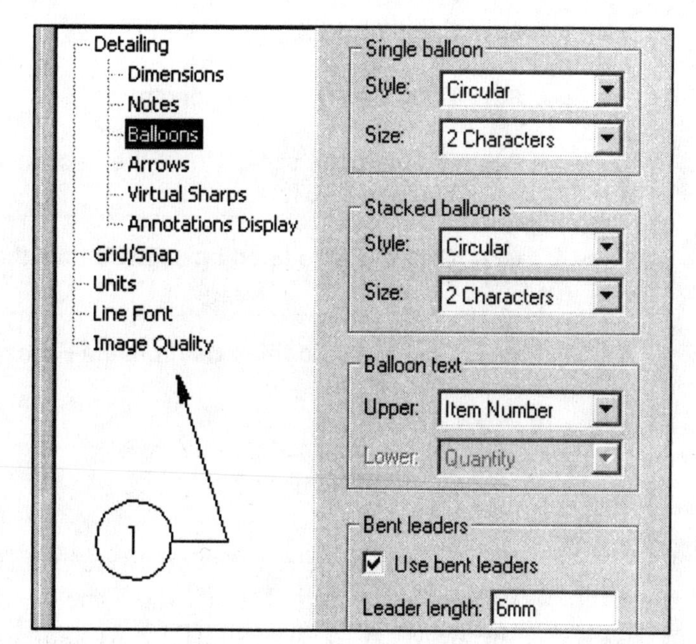

Set Arrows Properties according to the ASME Y14.2M-1992(R1998) standard at a 3:1 ratio for Width:Height. The Length value is the overall length of the arrow from the tip of the head to the end of the tail.

The Length is displayed when the dimension text is flipped to the inside. A Solid filled arrowhead is the preferred arrow type for dimension lines.

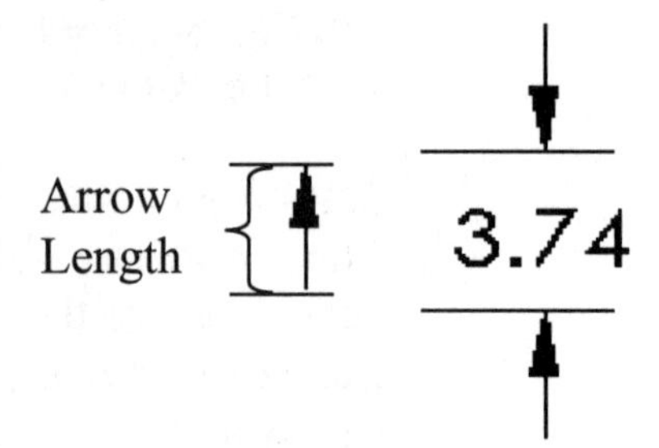

Arrow sizes change due to drawing size and scale. The ratio of width to height remains at 3:1.

Set Arrow Properties.

62) Click the **Arrows** entry on the left side of the Detailing text box. The Detailing - Arrows dialog box is displayed.

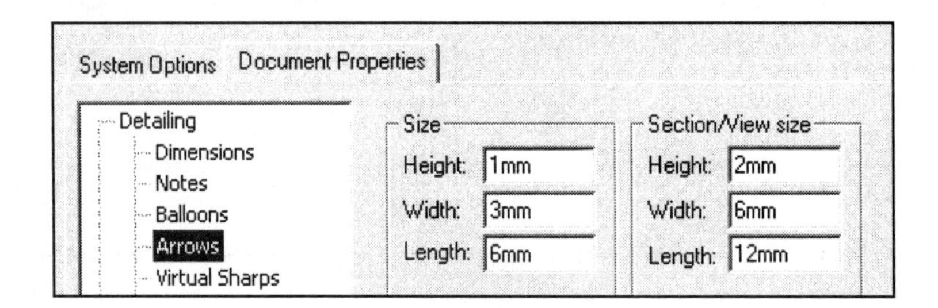

63) Enter **1** for the arrow Height in the Size text box.

64) Enter **3** for the arrow Width.

65) Enter **6** for the arrow Length.

66) Set the arrow style. Under the Section/View size, Enter **2** for Height, **6** for Width and **12** for Length.

67) Click the solid **filled arrowhead** from the Edge/vertex list box.

68) Click the solid **filled dot** from the Face/surface list box.

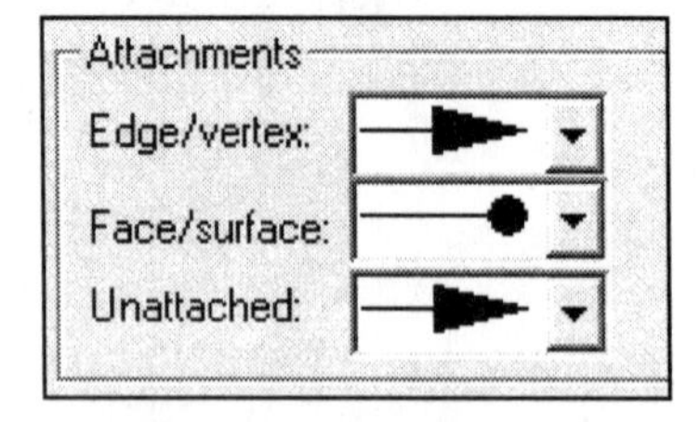

The Line Font determines the Style and Thickness for a particular type of edge in a drawing. Modify the Type of edge, Style and Thickness to reflect the ASME Y14.2M-1992(R1998) standard.

Recall that two line weights are defined in the ASME Y14.2M-1992(R1998) standard; namely 0.3mm and 0.6mm.

Thin Thickness is 0.3mm. Thick (Normal) Thickness is 0.6mm. Review line weights as defined in the File, Page Setup or in File, Print, System Options for your particular printer/plotter.

SolidWorks controls the line weight display in the Graphics window. Use Thin Thickness and Normal Thickness in the Graphics window.

Change all Thick Thickness settings to Normal Thickness. Change Detail Circle Style to Phantom.

Change View Arrows Style to Phantom.

Set the Line Font Properties.

69) Click **Line Font** from the left side of the Detailing text box.

70) Click **Detail Circle** for the Type of edge.

71) Select **Phantom** for Style. Select **Normal** for Thickness.

72) Click **Section line** for the Type of edge. Click **Normal** for Thickness.

73) Click **View Arrows** for the Type of edge. Click **Solid** for Style. Click **Normal** for Thickness.

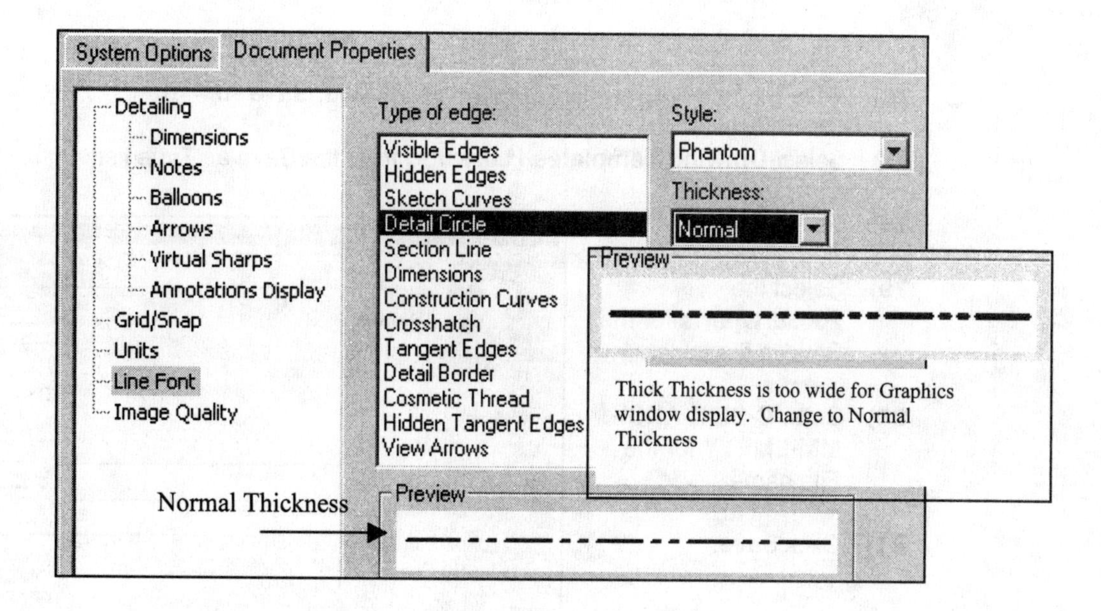

74) Exit Drawing Properties. Click **OK**.

75) Click the **Graphics window**. The drawing border is displayed in green.

The empty Drawing Template contains no geometry. The empty Drawing Template contains the Document Properties and the Sheet Properties: Sheet name, Paper size, No Sheet Format and Third Angle Projection.

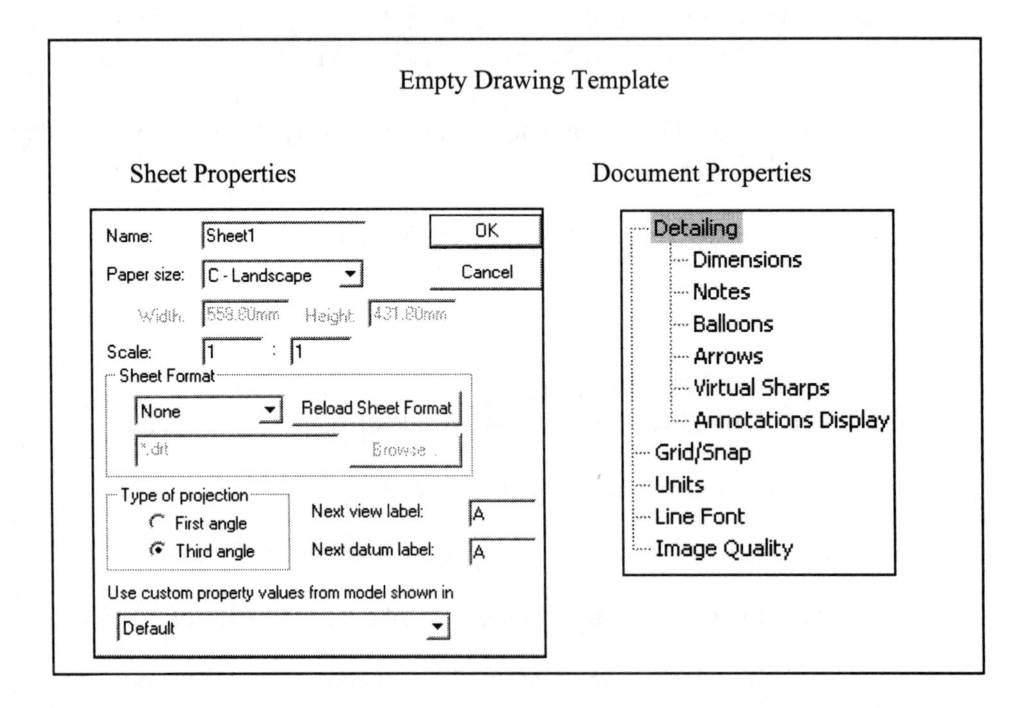

76) Save the empty Drawing Template. Click **File**, **Save As**.

77) Select **Drawing Templates (*.drwdot)** from the Save as Type list.

78) Select **Browse**.

79) Select the **2003drwparts** for the Save in file folder.

80) Enter **C-SIZE-ANSI-MM-EMPTY** for the File name.

81) Click **Save**.

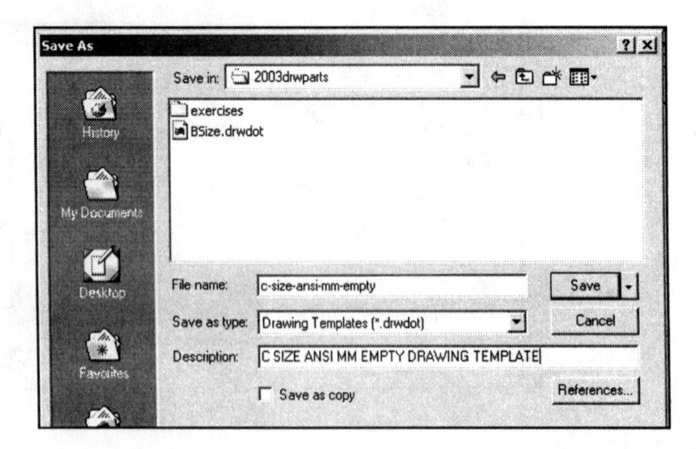

Sheet Format

Customize drawing Sheet Formats to create and match your company's drawing standards.

A customer requests a new product. The engineer designs the product in one location, the company produces the product in a second location and the field engineer supports the customer in a third location.

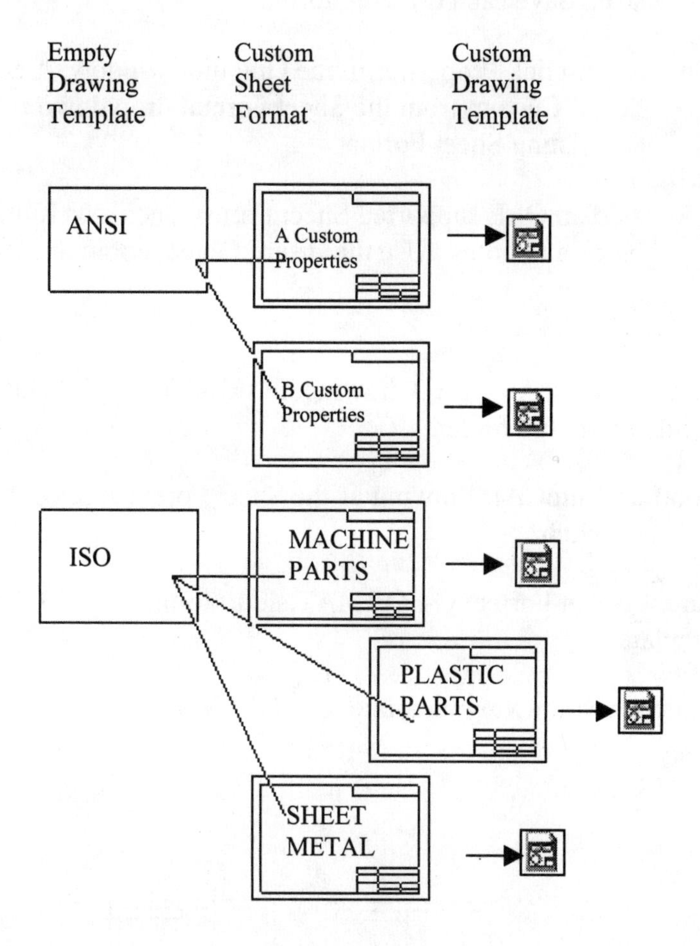

The ASME Y14.24M standard describes various types of drawings.

Example: Engineering produces detailed and assembly drawings. The drawings are used for machined, plastic and sheet metal parts that contain specific tolerances and notes used in fabrication.

Manufacturing adds vendor item drawings with tables and notes. Field Service requires installation drawings that are provided to the customer. Sheet formats are created to support various standards and drawing types.

There are numerous ways to create a custom Sheet Format:

- Open a SolidWorks, AutoCAD, Pro/ENGINEER or other CAD software saved as file type, ".dwg". Save the ".dwg" file as a Sheet Format.
- Right-click in the Graphics window. Select Edit Sheet Format. Create drawing borders, title block, notes and zone locations for each drawing size. Save each drawing format.
- Right-click Properties in the Graphics window. Select Properties. Select Custom from the Sheet Format drop down list. Browse to select an existing Sheet Format.
- Add an OLE supported Sheet Format such as a bitmap file of the title block and notes. Use the Insert, Object command.

Use an existing AutoCAD drawing, FORMAT-C-ACAD.dwg in the 2003drwparts file folder.

Import an AutoCAD drawing as the Sheet Format. Save the Sheet Format, C-FORMAT.slddrt.

Add the Sheet Format C-FORMAT.slddrt to the empty C-size Drawing Template.

Create a new drawing template; C-ANSI-MM.drwdot.

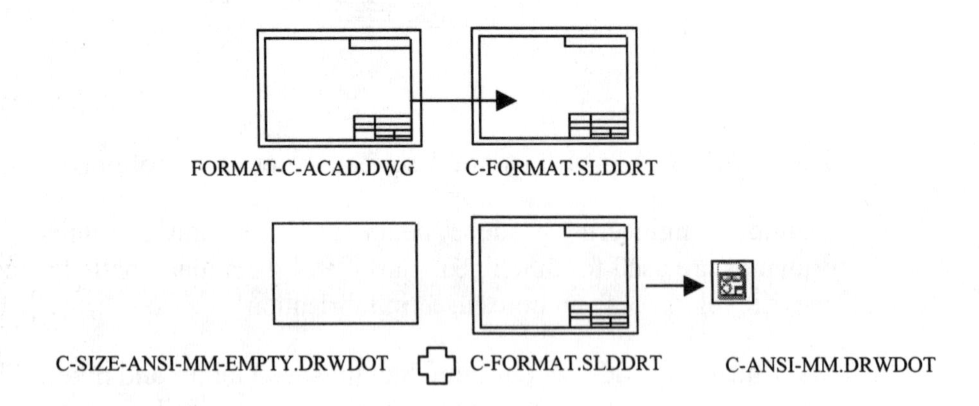

Add an A-size Sheet Format, A-FORMAT.slddrt to an empty A-size Drawing Template. Create an A-ANSI-MM.drwdot Drawing Template.

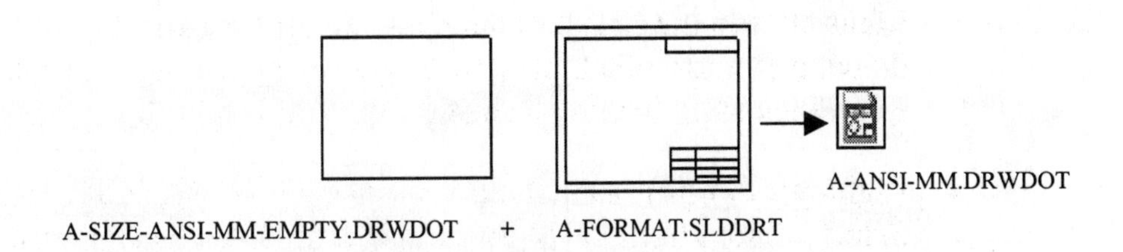

Views from the part or assembly are inserted into the SolidWorks Drawing.

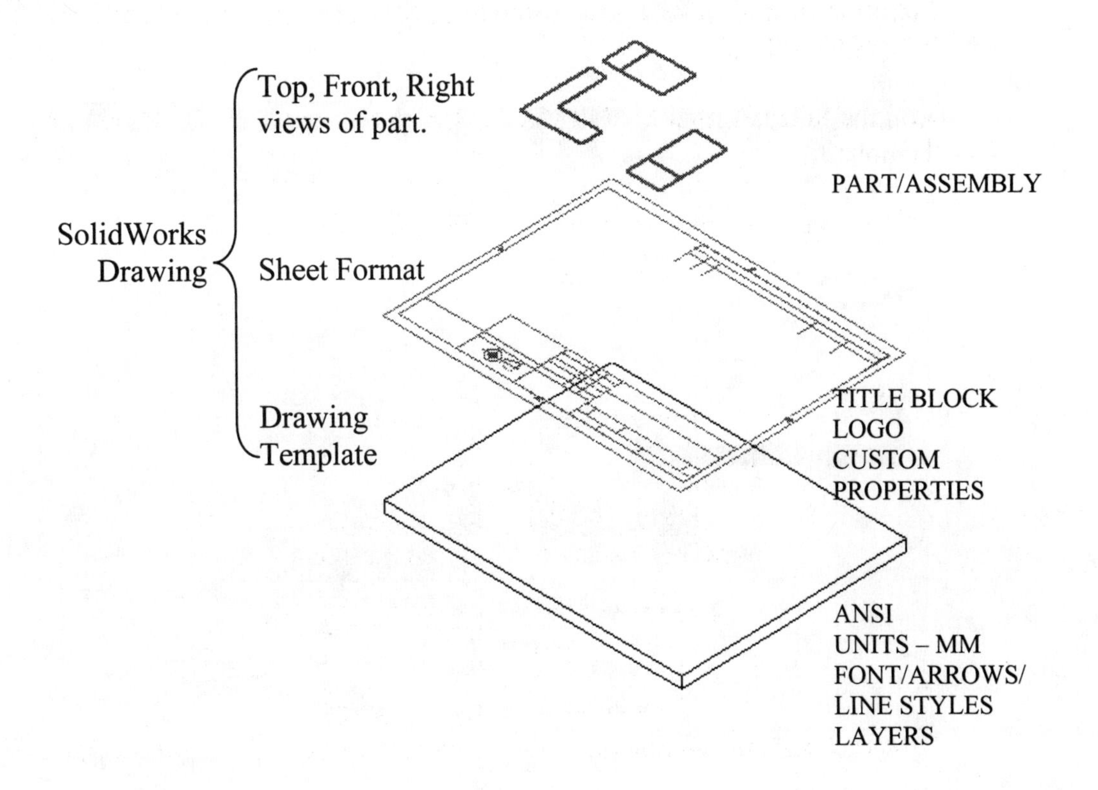

Open the AutoCAD drawing C-FORMAT.dwg.

82) Click **File**, **Open**.

83) Select **DWG (*.dwg)** from the Files of type drop down list.

84) Click **Browse**.

85) Select **FORMAT-C-ACAD** from the 2003drwparts file folder.

86) Click **Open**.

87) Click **Import to a new drawing** from the DXF/DWG Import dialog box.

88) Click **Layers** selected for sheet format.

89) Uncheck **DEFPOINTS**, a non-printable layer in AUTOCAD.

90) Click **Next**.

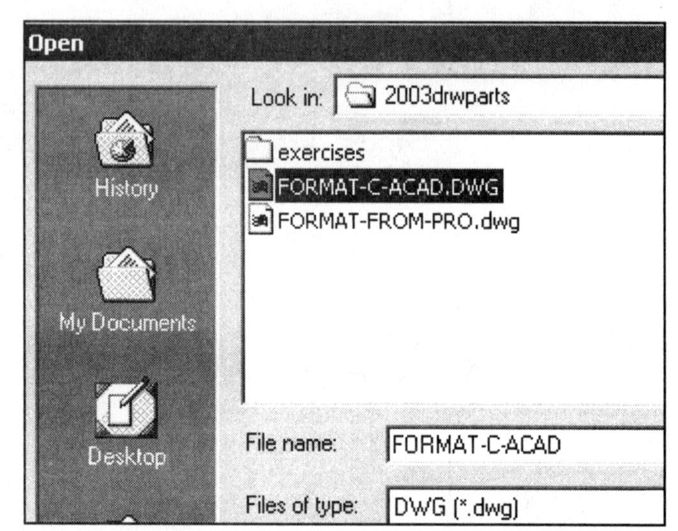

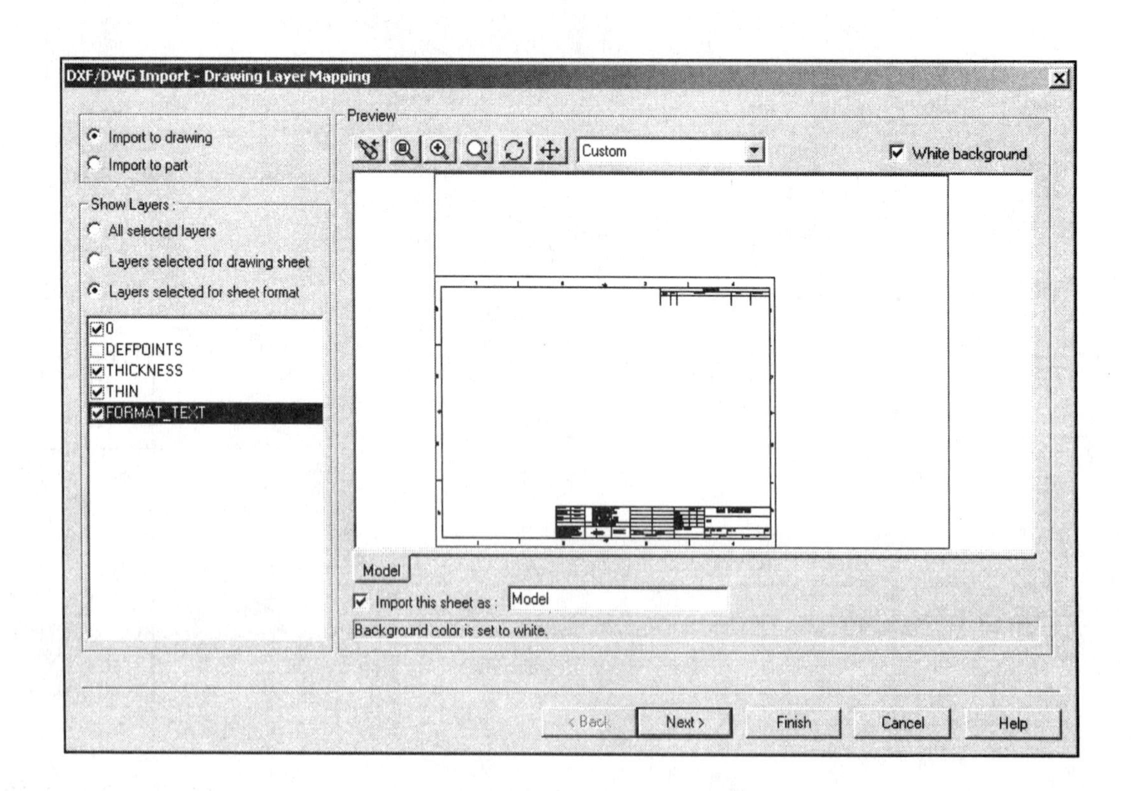

91) Select **Millimeters** for Data units.

92) Select **C-Landscape** for Paper Size. Select **Browse.**

93) Select the **2003drwparts** for the Save in file folder.

94) Select the **C-SIZE-ANSI-MM-EMPTY** for Drawing Template.

95) Click **Open** button.

96) View the Sheet Format. Click **Finish.**

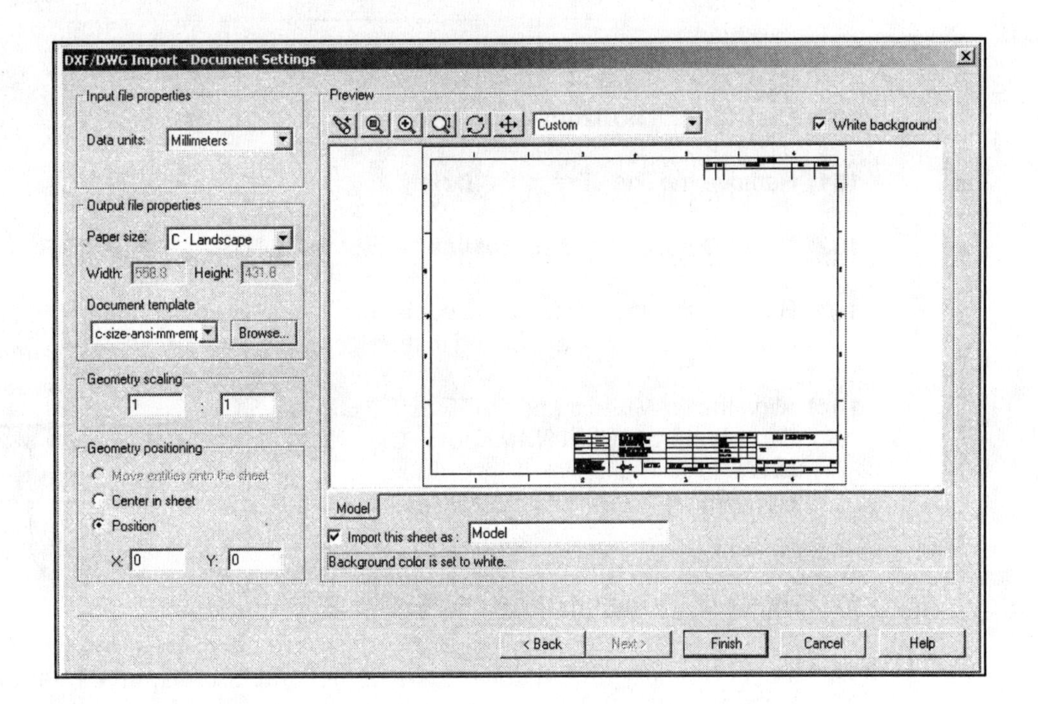

Data imported from other CAD systems may require editing in SolidWorks to produce desired results.

97) Right-click in the **Graphics window**.

98) Click **Edit Sheet Format.**

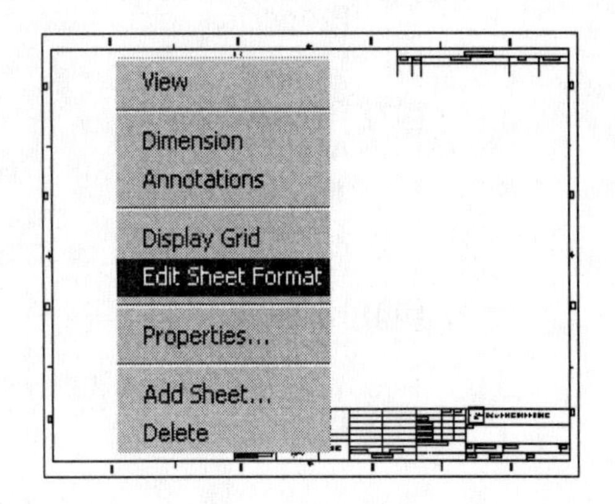

99) Click **Zoom in** on the title block. There are two coincident horizontal lines below the CONTRACT NUMBER text.

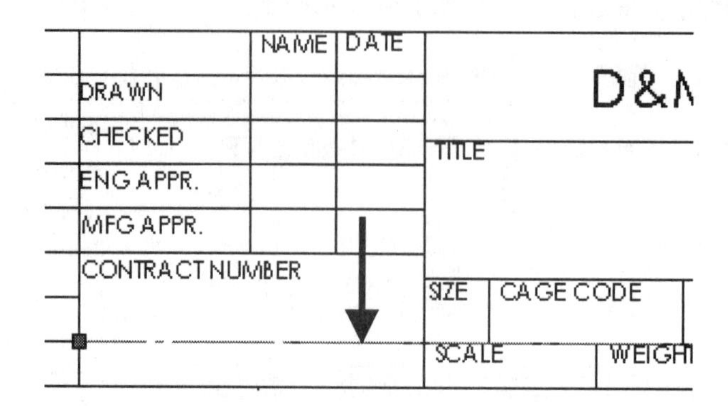

100) Click the first **horizontal line** below the CONTRACT NUMBER.

101) Remove the line. Press the **Delete** key.

102) Click the second **horizontal line** below the CONTRACT NUMBER.

103) Remove the line. Press the **Delete** key. Lines and text created from the AutoCAD title block are edited in the Edit Sheet Format.

104) Align the NAME text and DATE text. Hold the **Ctrl** key down. Click **NAME** text. Click the **DATE** text. Right-click **Align**. Click **Uppermost**. Release the **Ctrl** key.

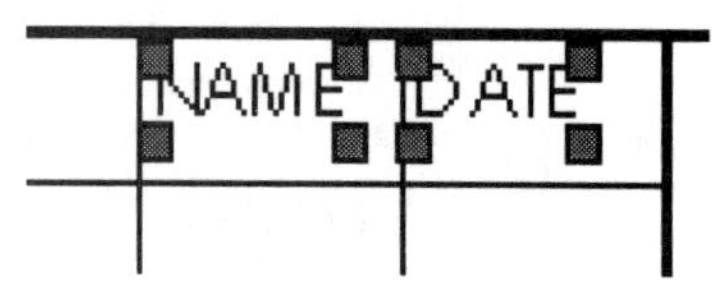

Add drawing notes and title block information in the Edit Sheet Format mode. This saves on rebuild time.

The sheet boundary and major title block heading are displayed with a THICK line style. Modify the drawing layer THICKNESS.

105) Display the Layer toolbar. Right-click a **position** in the gray toolbar main area to the right of the Help menu. Check **Layers**.

106) Display the Layers dialog box. Click the **Layer Properties** folder from the Layer toolbar.

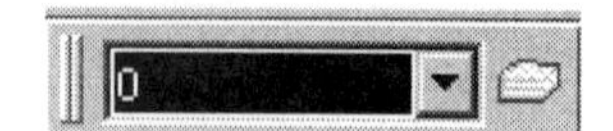

107) Rename the AutoCAD layer **THICKNESS** to **THICK**.

108) Rename description from **THICK** to **THICK BORDER**.

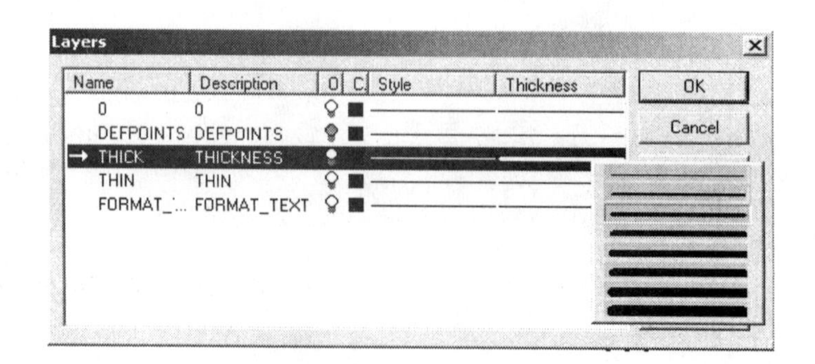

109) Click the **line Thickness** in the **THICK** layer.

110) Select the **second line thickness**.

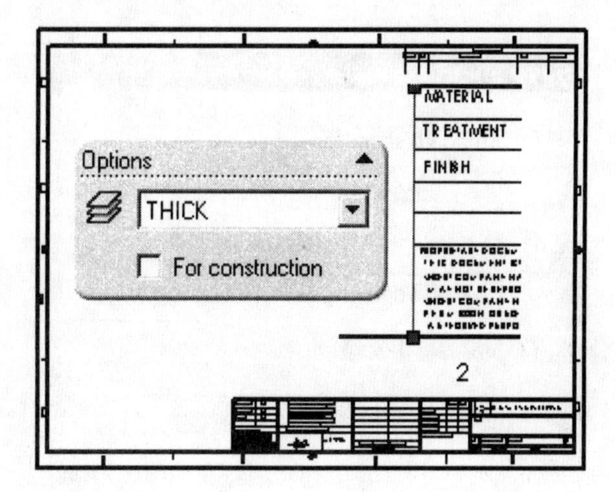

111) Display the Thick line. Click **OK**.

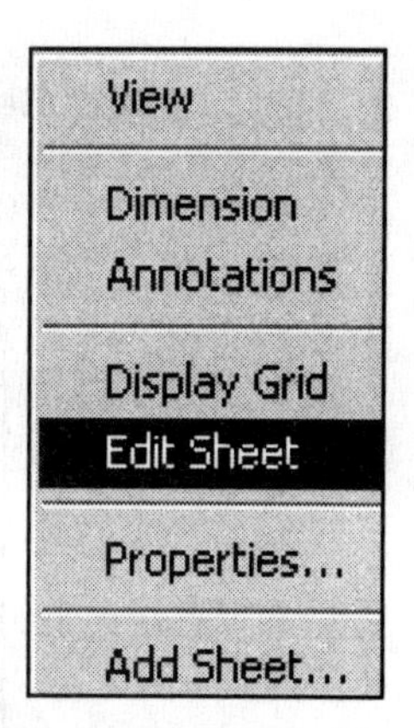

112) The border and title block display the Thick line. The left line in the title block is on the Thin layer. Click on the **left line.**

113) Click **Thick** layer.

Some printers cannot display the outside sheet boundary and or the Zone text.

114) Return to the Edit Sheet. Right-click in the **Graphics window**.

115) Click **Edit Sheet**.

116) Fit the drawing to the Graphics window. Press the **f** key.

117) Click the drop down **arrow** in the Layer text box.

118) Click **None** for Layer.

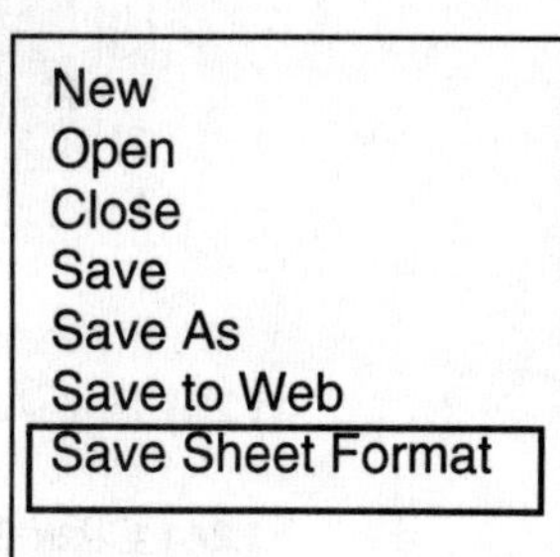

Save Sheet Formats and Drawing Templates in the Edit Sheet mode. Drawing views are not displayed in the Edit Sheet Format mode. The Layer None is saved with the Drawing Template.

119) Save the Sheet Format. Click **File**, **Save Sheet Format**.

120) Click **Custom Sheet Format**.

121) Click **Browse**.

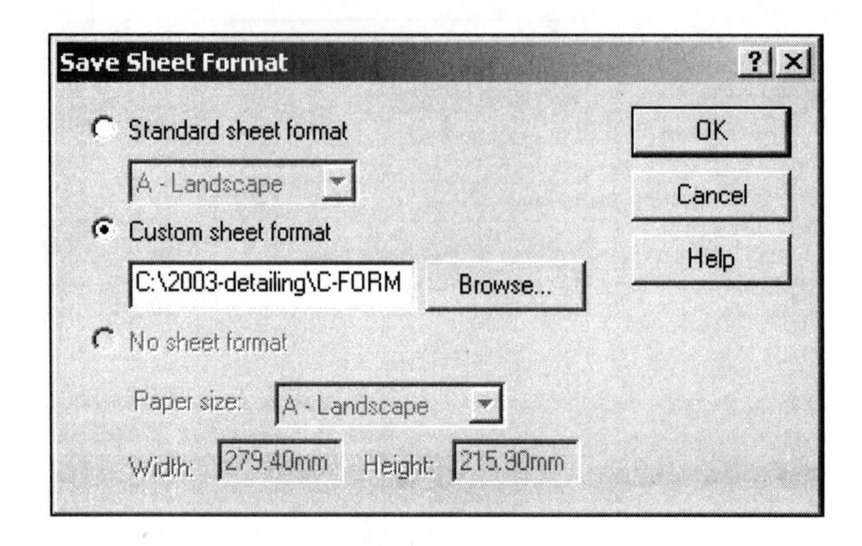

122) Select the **2003drwparts** file folder.

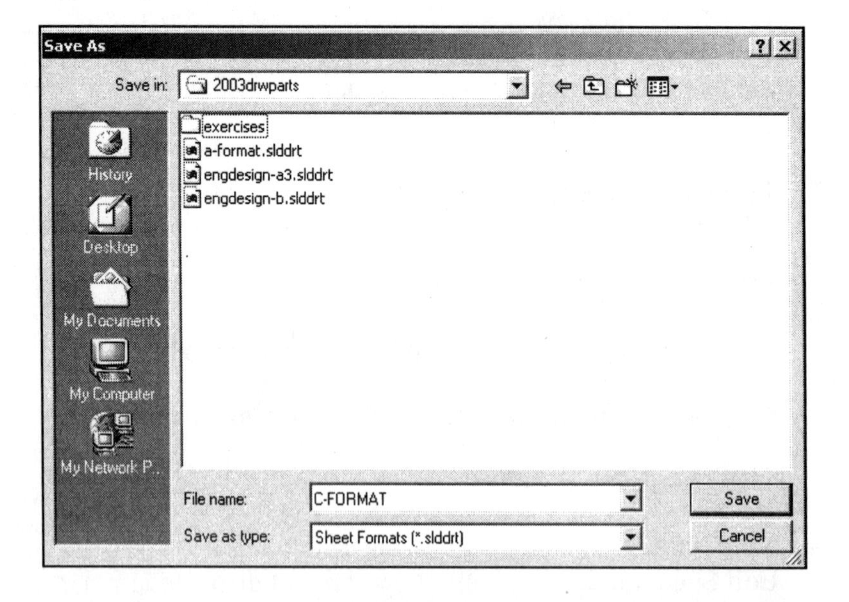

123) Enter **C-FORMAT**. The Sheet Formats file extension is ".slddrt".

124) Click **Save**.

125) Click **OK**.

Title Block Notes and Properties

Title blocks contain vital part and assembly information. Each company creates a unique version of a title block. Most title blocks contain the following type of information:

Company Name/Logo	Part number
Part name	Drawing number
Drawing description	Revision number
Sheet number	Material & finish
Tolerance	Drawing scale
Sheet size	Revision block
CAD file name	Engineering Change Orders
Quantity required	Drawn by
Checked by	Approved by

A title block is normally located in the lower right hand corner of the drawing.

You need to be in the Edit Sheet Format mode to modify the Sheet Format text, lines or title block information.

You need to be in the Edit Sheet mode to insert model views. Edit Sheet and Edit Sheet Format are the two major design modes used to develop a drawing.

The Edit Sheet Format mode provides the ability to:

- Create or change the title block size and text headings.
- Incorporate a logo.
- Add drawing, design or company text, and Custom Properties.

The Edit Sheet mode provides the ability to:

- Add or modify views.
- Add or modify dimensions.
- Add or modify text.

Notes can be created or modified in a title block. Notes can also be linked to SolidWorks Properties and Custom Properties.

Linked notes reflect information in a title block such as file name, sheet name and sheet number.

Edit Sheet Format - Title block.

126) Edit company name. Right-click **Edit Sheet Format** from the Pop-up menu in the Graphics window.

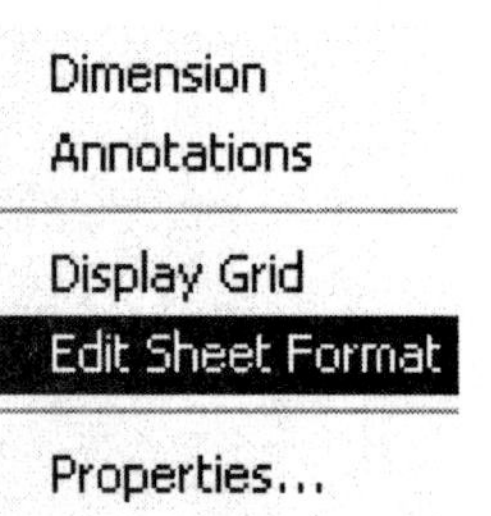

127) View the right side of the title block. Click **Zoom to Area** on the Sheet Format title block.

128) Double-click the **D&M Engineering** text.

129) Enter a **new company name if desired**. Change the font height to fit your company name inside title block if required.

130) Right-click **Properties** on the selected text.

131) Uncheck the **Use document's font** check box from the Note PropertyManager. Change the font size.

132) Click the **Font** button.

133) Click **OK**. The text is displayed in the title block.

134) Click the **Font** button in the Text Format box to access the Property Manager on the left side of the Graphics window.

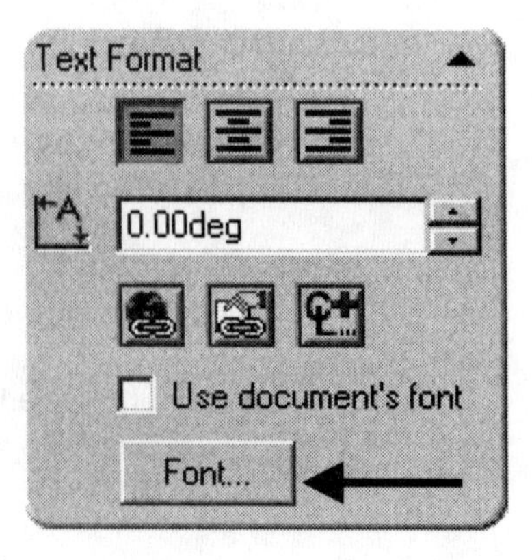

A company logo is normally located in the title block. Create a company logo by copying a picture file from Microsoft ClipArt using Microsoft Word. Copy/Paste the logo into the title block

The following logo example was created in Microsoft Word 2000 using the COMPASS.wmf and WordArt text. Any ClipArt picture, scanned image or bitmap can be used.

Create a logo.

135) Create a New Microsoft Word Document. Click **New** from the Standard toolbar in MS Word.

136) Click **ClipArt** from the Draw toolbar.

137) Drag the **COMPASS.wmf** file in the WORD document. The **COMPASS.wmf** picture file is displayed in the WORD document.

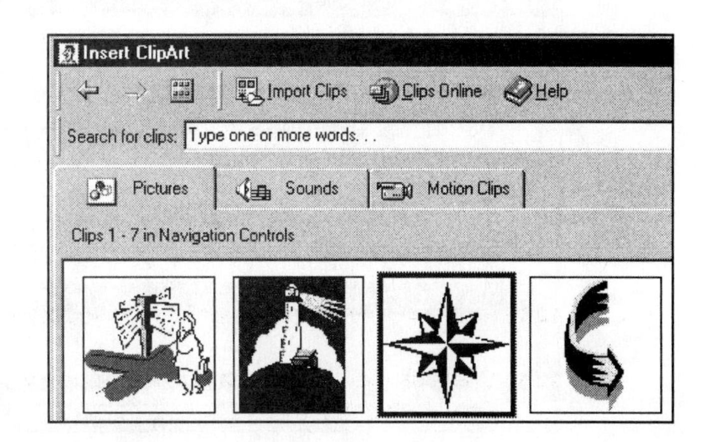

138) Copy the picture. Select the **compass picture**.

139) Click **Copy** .

140) The logo is placed into the Clipboard. The logo is used again to create an A-size Drawing Template. Save the logo. Click **Save**.

141) Enter **Logo** for the WORD filename.

142) Place the logo into the title block. Press **Ctrl-Tab** to display the SolidWorks Graphics window.

143) Click a **position** to the left of the company name in the title block.

144) Click **Edit**, **Paste**.

145) Size the **logo** to the SolidWorks title block by dragging the picture handles.

146) Close Microsoft Word. Click **File**, **Exit**.

Link notes in the title block to the SolidWorks Properties. The drawing TITLE text describes the drawing.

Create a note for the title of the drawing that is linked to the SolidWorks file name. Complete the drawing. Create additional notes.

Create a new Layer for the Title Block notes.

147) Click the **Layer Property Manager**.

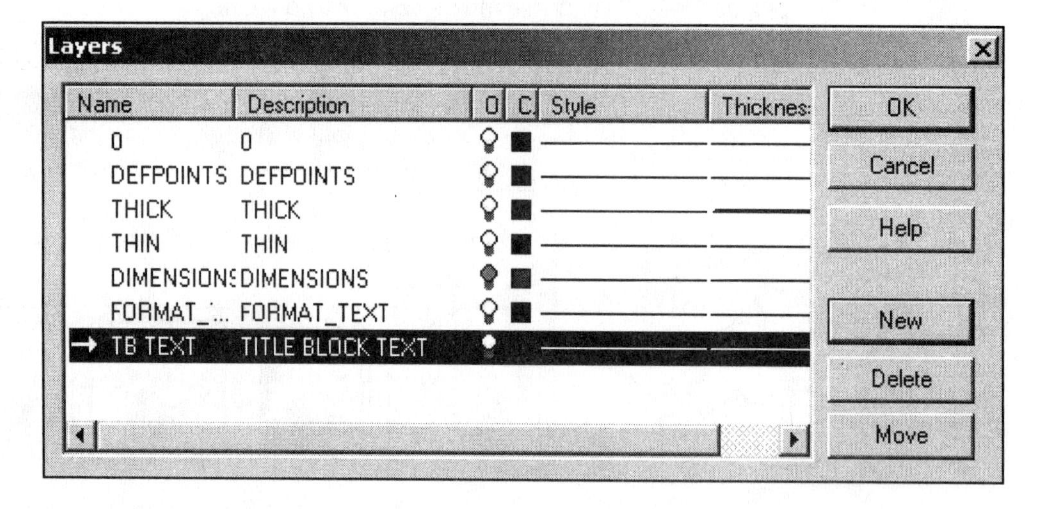

148) Click the **New** button.

149) Enter **TB Text** for Name.

150) Enter **TITLE BLOCK TEXT** for Description.

151) Click **OK**. Note: The larger arrow next to TB TEXT indicates the current layer.

Create a Linked Note.

152) Click **Zoom to Area** on the TITLE section of the title block.

153) Display the Annotations toolbar. Click **View**, **Toolbars**, **Annotations**.

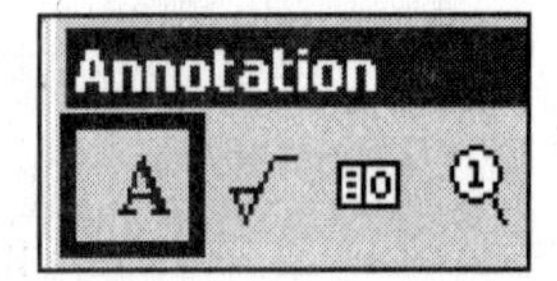

154) Click **Note** from the Annotations toolbar. Click a **start point** to the lower right the TITLE text.

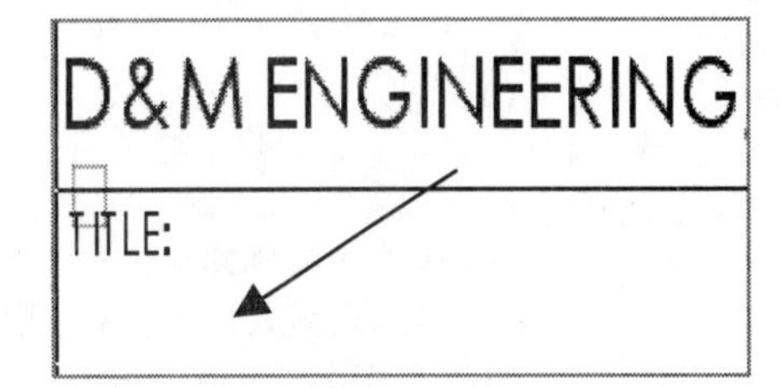

155) The Note Property dialog box is displayed. Click **No leaders** in the Leader text box. Click **Link to Property** from the Text Format box. The Link to Property dialog box is displayed.

156) Click **No leaders** in the Leader text box.

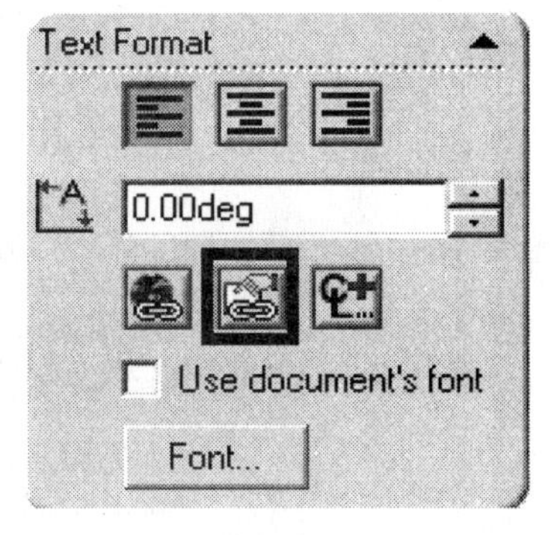

157) Select **SW-File Name** from the drop down list.

158) Click **OK**.

159) The variable $PRP"SW-File Name" is displayed in the Note text box.

160) Uncheck the **Use document's font**.

161) Click the **Font** button. Enter **6mm** for text height.

162) Click **OK**. Draw1 is the current file name.

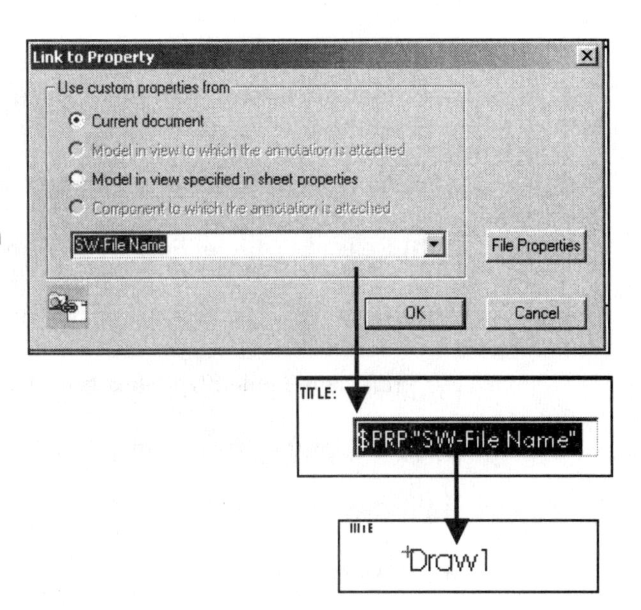

Note: The $PRP"SW-File Name" property will update to contain the part filename.

Example: Insert the part TUBE into a Drawing Template in Project 2.

The text TUBE will replace the SW-FileName.

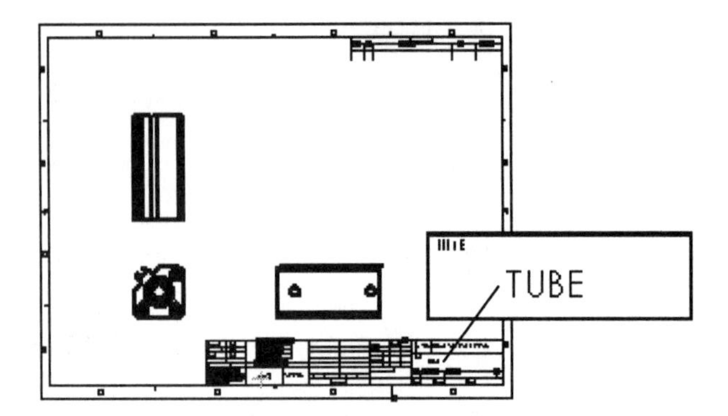

Additional notes are required in the title block. The text box headings: SIZE C, DWG. NO., REV., SCALE, WEIGHT and SHEET 1 OF 1 are entered in the SolidWorks default Sheet Format.

Create SIZE, SHEET and SCALE text with Linked Properties. Change the Sheet Scale. The new value updates in the title block. Add a new sheet. The drawing and the SHEET text values increment.

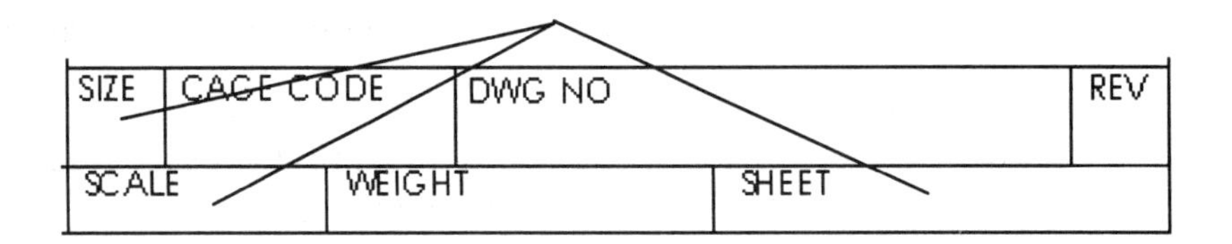

163) Create a Linked Property to the SIZE text. Click **Note** from the Annotations toolbar. Click a **start point** in the upper left hand corner below the SIZE text.

164) Click **Link to Property** from the Text Format box.

165) Select **SW-Sheet Format Size** from the drop down list.

166) Click **OK**. The variable $PRP"SW-Sheet Format Size" is displayed in the Note text box. Click **No leaders** .

167) Display the Sheet Format Size. Click **OK**.

168) Click the **OF** text in the lower right corner of the title block.

169) Press the **Delete** key.

170) Combine Link Properties for the SHEET text. Click **Note** from the Annotations toolbar.

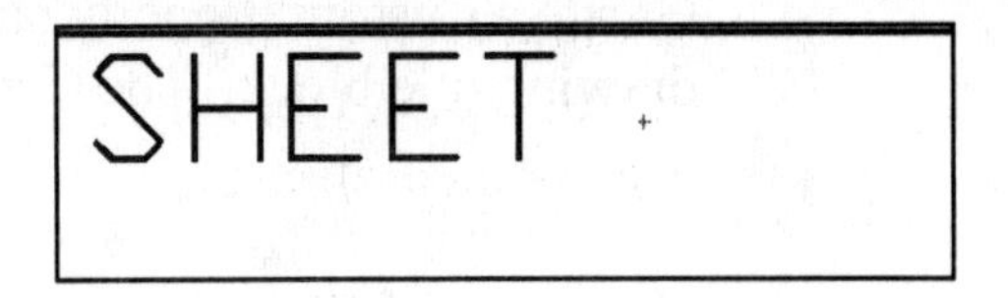

171) Click a **start point** in the upper left hand corner below the SHEET text.

172) Click **Link to Property** from the Text Format box.

173) Select **SW-Current Sheet** from the drop down list.

$PRP:"SW-Current Sheet" OF $PRP:"SW-Total Sheets"

174) Click **OK**.

175) Enter the text **OF**.

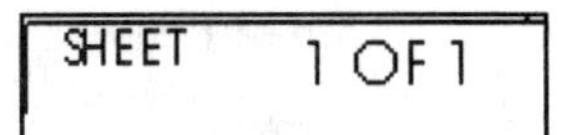

176) Click **Link to Property** from the Text Format box.

177) Select **SW-Total Sheets** from the drop down list. The variable $PRP"SW-Sheet Format Size" is displayed in the Note text box. Display the Sheet Format Size.

178) Click **OK**. The Current Sheet value and Total Sheets value change as additional sheets are added to the drawing.

179) Create a Linked Property to SCALE. Click **Note** from the Annotations toolbar.

180) Click a **start point** in the upper left hand corner below the SCALE text.

$PRP:"SW-Sheet Scale"

181) Click **Link to Property** from the Text Format box.

182) Select **SW-Sheet Scale** from the drop down list.

183) Click **OK**. The variable $PRP "SW-Sheet Scale" is displayed in the Note text box.

184) Click **OK**. The Sheet Scale value changes to reflect the sheet scale properties in the drawing.

Your company has a policy that a contract number must be contained in the title block for all associated drawings in a project.

Create a Custom Property named CONTRACT NUMBER. Add it to the drawing title block. The Custom Property is contained in the Sheet Format.

185) Create a Custom Property for the CONTRACT NUMBER text. Click **Note** from the Annotations toolbar.

186) Click a **start point** in the upper left hand corner below the CONTRACT NUMBER text. Click **No leaders**.

CONTRACT NUMBER

187) Click **Link to Property** from the Text Format box.

188) Select the **File Properties** button.

189) Click the **Custom** tab.

190) Enter the **CONTRACT NUMBER** for Name. Text is the default type.

191) Click **101045-PAP** for Value. Click **Add**. The Custom Property is displayed in the Properties text box.

192) Click **OK**.

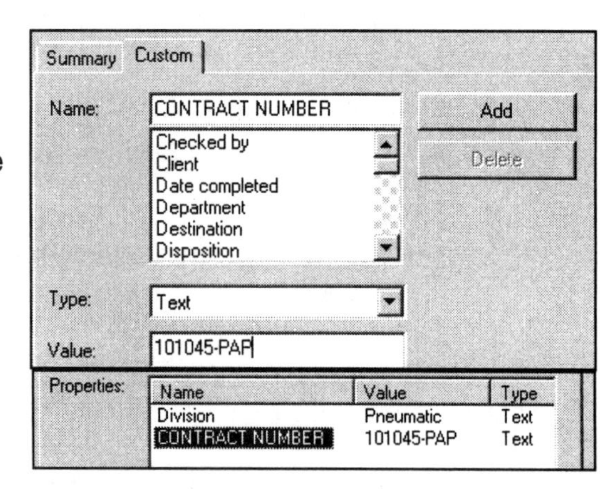

193) Enter the **CONTRACT NUMBER** in the Property Name text box.

194) Click **OK**.

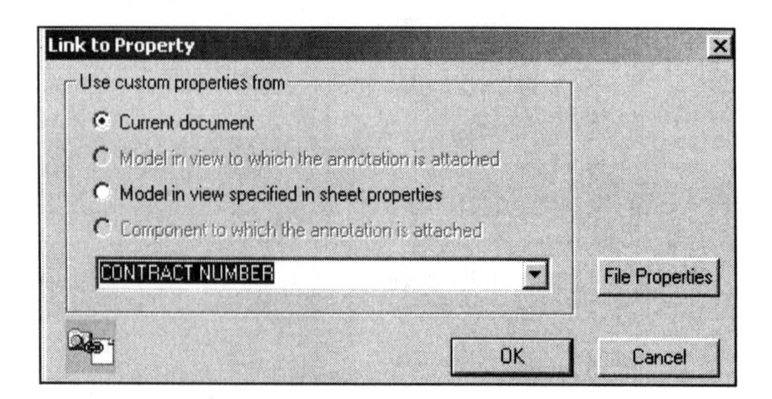

195) The Note text box displays: $PRP: "CONTRACT NUMBER". Display 101045-PAP. Click **OK**.

$PRP:"CONTRACT NUMBER"

196) Fit the drawing to the Graphics window. Press the **f** key.

CONTRACT NUMBER
101045-PAP

Conserve drawing time. Place common general notes in the Sheet Format.

The Engineering department stores general notes in a Notepad file, GENERALNOTES.TXT. General notes are usually located in a corner of a drawing.

197) Create general notes from a text file. Double-click on the Notepad file, **GENERALNOTES.TXT** in the 2003drwparts file folder.

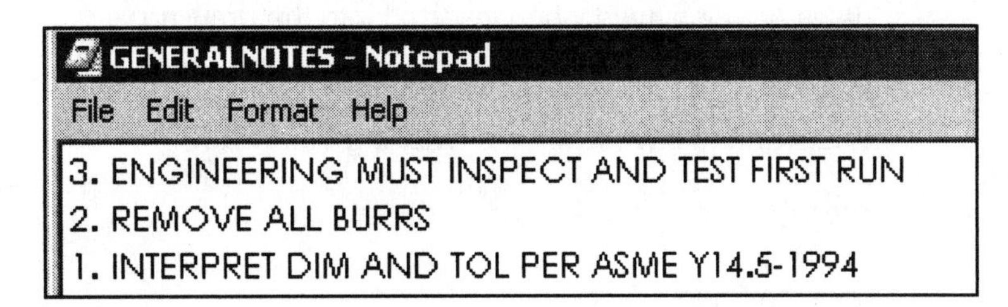

198) Select the text. Click **Edit**, **Select All**.

199) Copy the text into the windows clipboard. Click **Ctrl C**.

200) Display the SolidWorks Graphics window. Click **Ctrl tab**.

201) Click **Note** from the Annotations toolbar.

3. ENGINEERING MUST INSPECT AND TEST FIRST R
2. REMOVE ALL BURRS
1. INTERPRET DIM AND TOL PER ASME Y14.5-1994

202) Click a **start point** in the lower left hand corner of the title block.

203) Click **inside** the Note text box.

204) Paste the three lines of text. Click **Ctrl V**.

205) Display the general notes on the drawing. Click **OK**.

A
3. ENGINEERING MUST INSPECT AND TEST FIRST RUN
2. REMOVE ALL BURRS
1. INTERPRET DIM AND TOL PER ASME Y14.5-1994
1

206) Return to the drawing sheet. Right-click in the **Graphics window**.

207) Click **Edit Sheet**. The drawing boarder is displayed in gray.

208) Fit the drawing to the Graphics window. Press the **f** key.

209) Click **None** from the Layer text box.

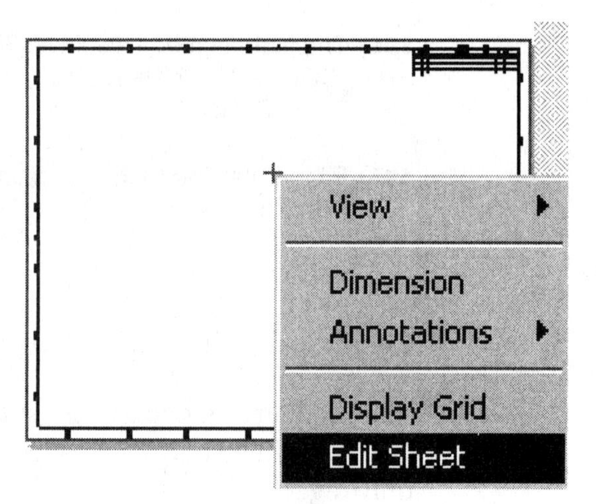

Note: Save your Sheet Format and Drawing Templates in the Edit Sheet mode. Views are displayed when inserted into the drawing.

Views cannot be displayed in the Edit Sheet Format mode. The None option is set for Layer and saved with the Drawing Template.

Save the Sheet Format.
210) Click **File**, **Save Sheet Format**.

211) Select the **Custom Sheet Format** button.

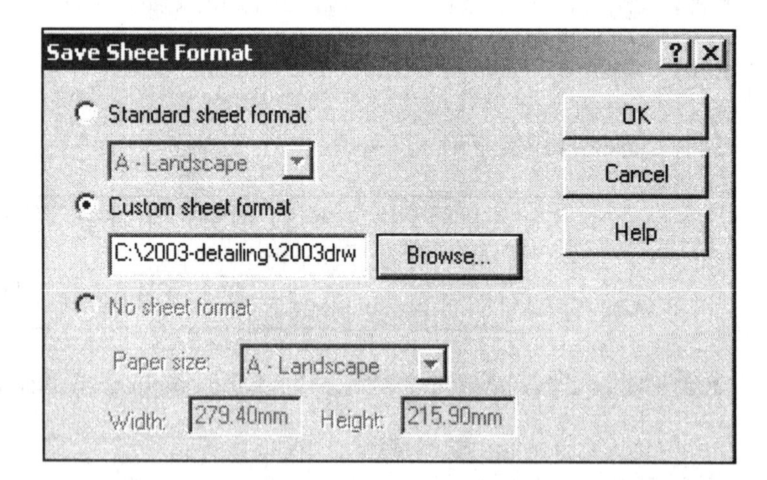

212) Click **Browse**.

213) Select the **C-FORMAT.slddrt** sheet format from the 2003drwparts file folder.

214) Click **Save**.

215) Click **Yes** to overwrite the existing sheet format.

216) Click **OK**.

The Sheet Formats1 icon is displayed in the Feature Manager.

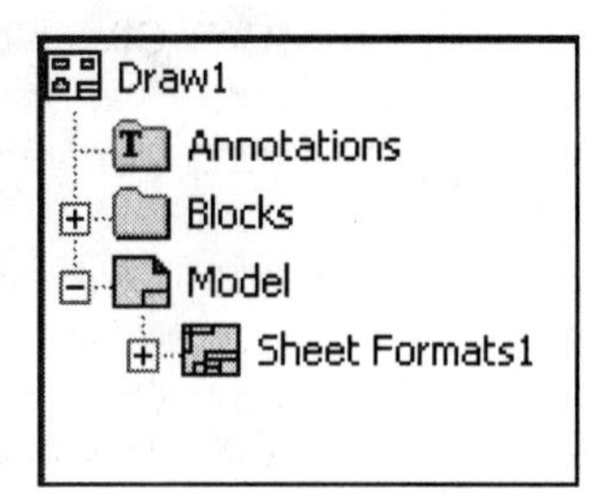

Combine the C-FORMAT Sheet Format with the empty Drawing Template. The C-FORMAT Sheet Format is contained in every Sheet of the drawing in the C-ANSI-MM Drawing Template. Utilize Layers to hide Sheet Format lines and text.

Create a new Drawing Template: C-ANSI-MM.

217) Click **New**.

218) Select the **C-SIZE-ANSI-MM-EMPTY** Drawing Template.

219) Click the **Custom sheet format** option.

220) **Browse** the 2003drwparts file folder.

221) Select **C-FORMAT** for sheet format.

222) Click **Open**.

223) Click **OK**.

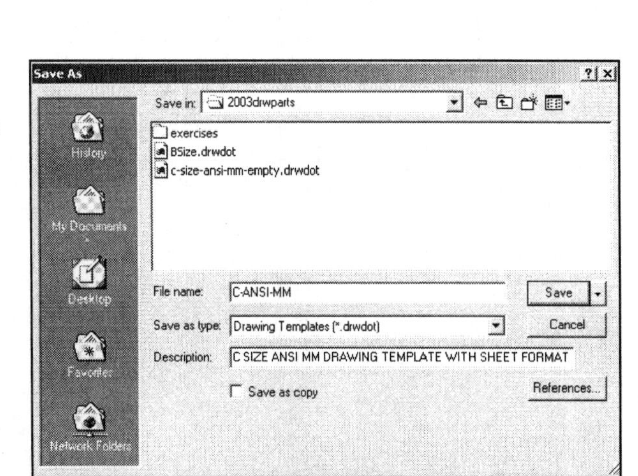

224) Save the Drawing Template. Click **File**, **Save As**.

225) Select **Drawing Template (*drwdot)** for Save as Type.

226) Select **2003drwparts** for Save in folder.

227) Enter **C-ANSI-MM** for File name.

228) Enter **C SIZE ANSI MM DRAWING TEMPLATE WITH SHEET FORMAT** for Description.

229) Click **Save**.

230) Close all documents. Click **Windows**, **Close All**.

231) Click **No** to the questions: "Save DRAW1 and Save DRAW2."

232) Verify the template. Click **New**.

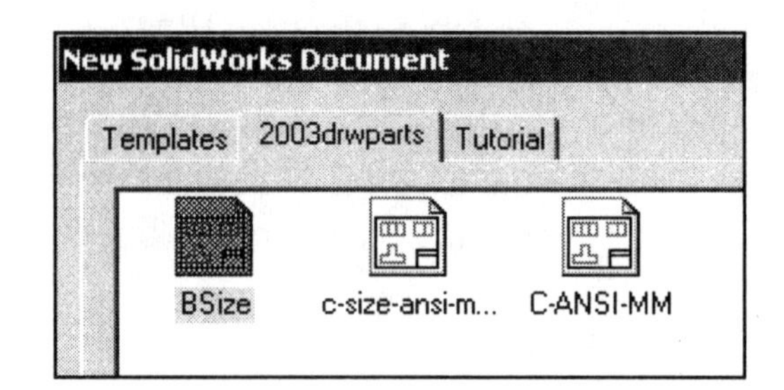

233) Click the **2003drwparts** tab.

234) Click the **C-ANSI-MM** template. The C-ANSI-MM Drawing Template is displayed with the Sheet Format.

235) Click **OK**.

236) Open a new Drawing. Click **New**.

237) Click the **C-ANSI-MM** Drawing Template.

238) Display Sheet2. Right-click the **Sheet1 tab**.

239) Click **Add Sheet**. Sheet2 contains Sheet Formats2.

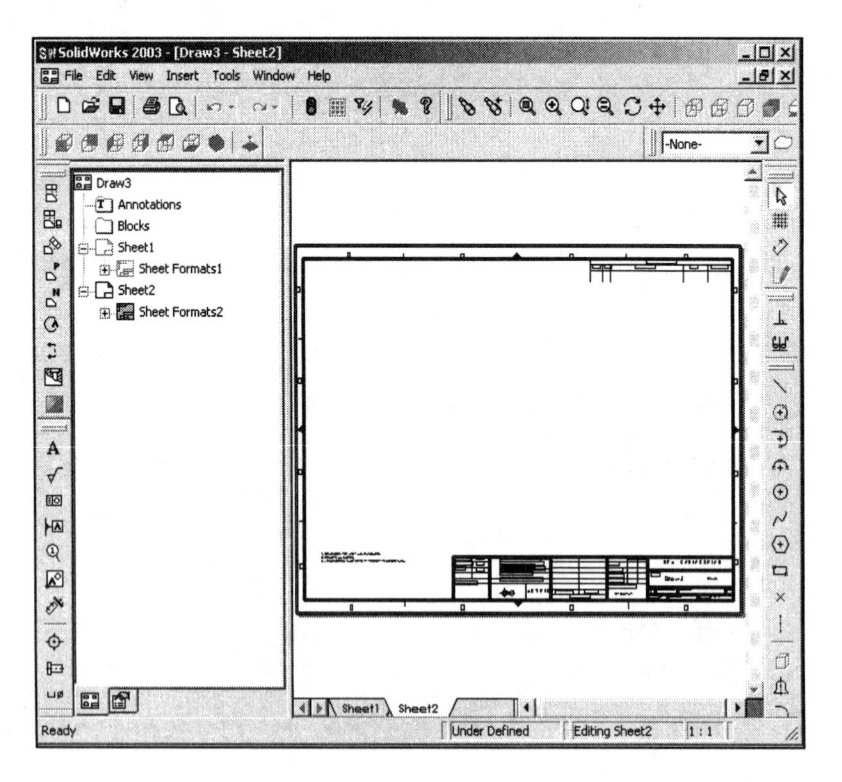

240) Close all files. Click **Windows**.

241) Click **Close All**.

A - Size Drawing Template

Create an A size Drawing Template and an A size Sheet Format. Text size for an A-size drawing is the same as a C-size drawing.

Utilize the empty C-size Drawing Template.

Create an A-ANSI-MM Drawing Template. Add an A-size Sheet Format.

Create a new A-size drawing template.

242) Create a new Drawing Template from an existing Drawing Template. Click **New**.

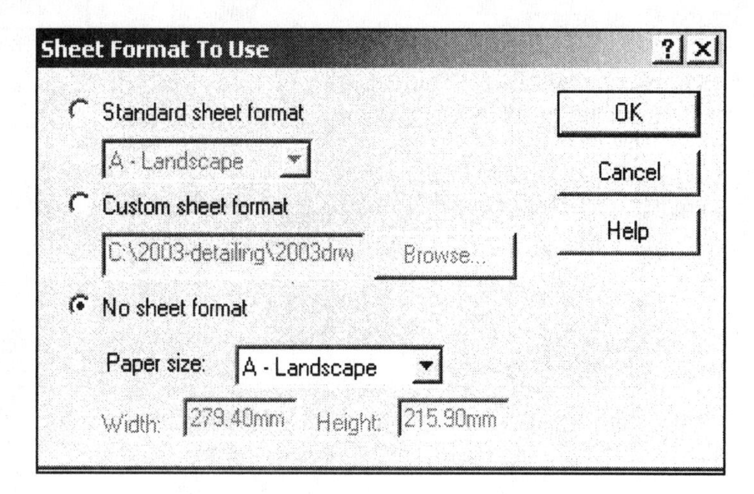

243) Select **C-SIZE-ANSI-MM-EMPTY**.

244) Click **No Sheet Format**.

245) Select **A-Landscape** for Paper size.

246) Click **OK**. Note: The Document Properties set for the C-Size Drawing Template are copied to the A-size Drawing Template.

247) Fit the template to the Graphics window. Press the **f** key.

248) Save the A-size Drawing Template. Click **File**, **Save As**.

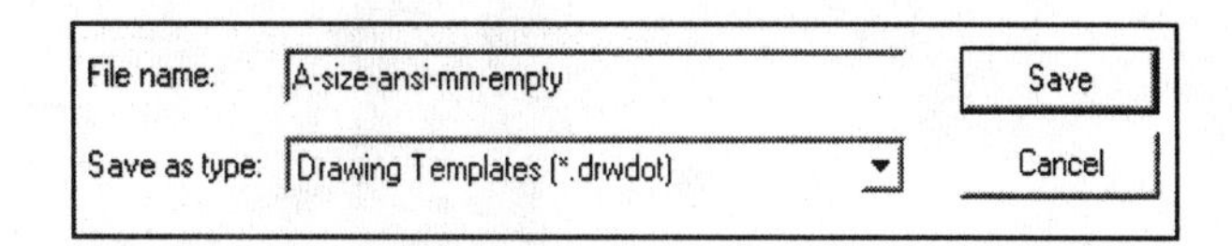

249) Select **Drawing Templates** for Save as type.

250) **Browse** to the 2003drwparts file folder.

251) Enter **A-SIZE-ANSI-MM-EMPTY** for File name.

252) Click the **Save** button.

Load the Custom A-size sheet format.

253) Right-click in the **Graphics window**.

254) Click **Properties**.

255) Click **Custom** for the Sheet Format.

256) Click **Browse**.

257) Select **A-FORMAT.slddrt** from the 2003drwparts file folder.

258) Click **OK**.

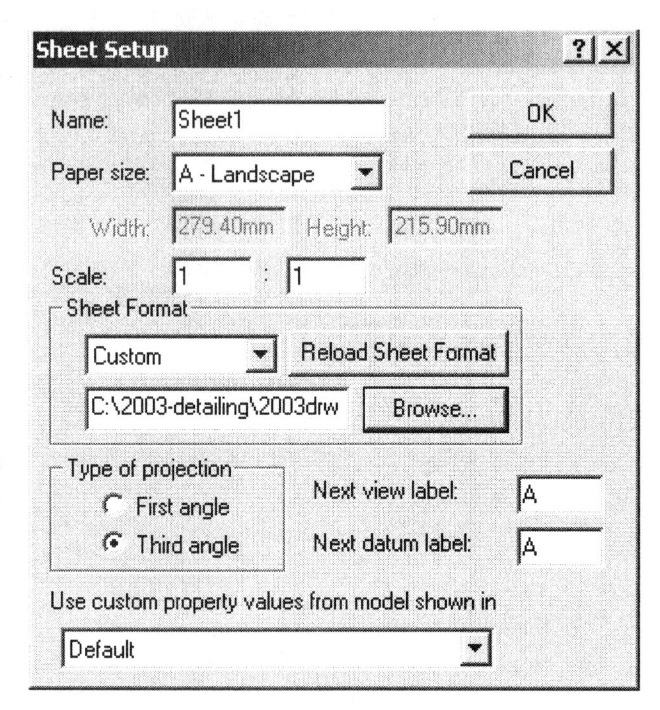

Note: The A-FORMAT is created in inches. The A-SIZE-ANSI-MM-EMPTY Drawing Template is created in millimeters.

The Drawing Template controls the units.

The A-FORMAT geometry, text and dimensions are created on separate layers. The None option is the current Layer. A-FORMAT is displayed in Edit Sheet mode.

Create a new Drawing Template: A-ANSI-MM.

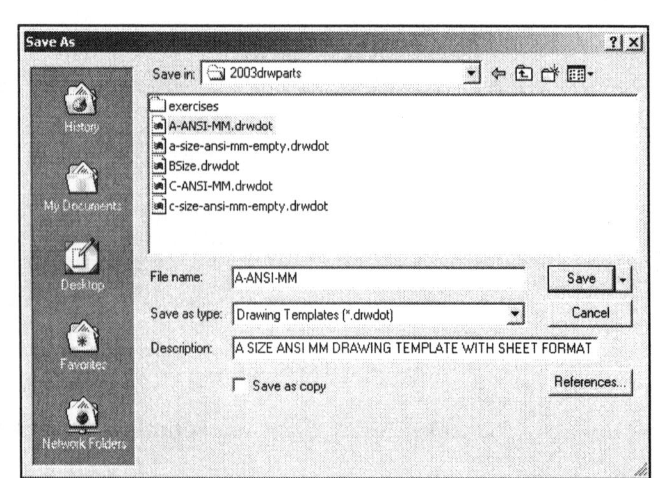

Combine the Sheet Format and the empty Drawing Template.

Save the new Drawing template.
259) Click **File**, **SaveAs**.

260) Select **Drawing Templates(*.drwdot)** for Save as type.

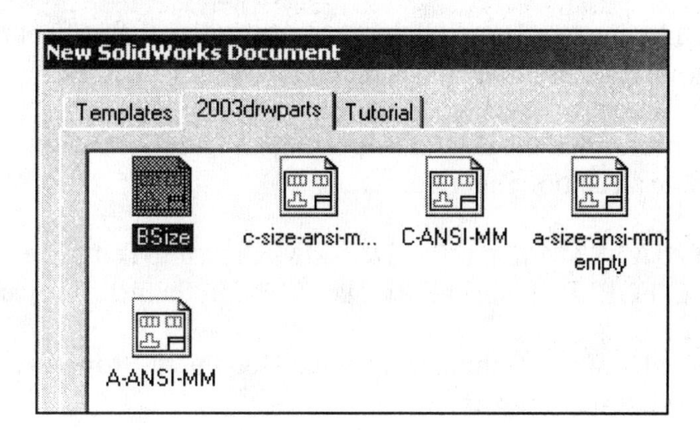

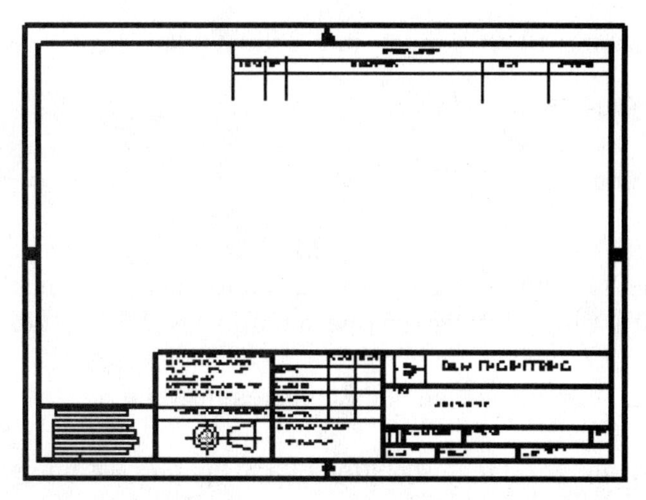

261) Select the **2003drwparts** file folder.

262) Enter **A-ANSI-MM**.

263) Close all documents. Click **Windows**, **Close All**.

264) Verify the template. Click **New**.

265) Click the **2003drwparts** tab.

266) Click the **A-ANSI-MM** template.

267) Click **OK**.

268) Close all documents. Click **windows**, **Close All**.

Project Summary

In this project you created a custom C-size and A-size Drawing Template and Sheet Format. The Drawing Template and Sheet Format contained global drawing and detailing standards.

You obtained and applied drawing properties that reflect the ASME Y14 Engineering Drawing and Related Drawing Practices.

You performed the task of importing an AutoCAD drawing to create and modify a custom Sheet Format.

The A-ANSI-MM and C-ANSI-MM Drawing Templates and A-FORMAT and C-FORMAT Sheet Formats are use in the next Project.

Create Drawing Templates for inch Document Properties in the Exercises at the end of this Project.

Import other Sheet Formats into SolidWorks.

Project Terminology

ASME – American Society of Mechanical Engineers. ASME is the publisher of the Y14 Engineering Drawing and Related Documentation Practices. ASME Y14.5M-1994 is a revision of ANSI Y14.5-1982.

ANSI – American National Standards Institute.

Drawing: A 2D representation of a 3D part or assembly. The extension for a SolidWorks drawing filename is .SLDDRW.

Drawing Template: A document that is the foundation of a new drawing. It contains document properties and user-defined parameters such as sheet format. The extension for Drawing Template filename is .DRWDOT.

Feature Manager: An outline view of the active part, assembly or drawing displayed on the left side of the SolidWorks window.

Sheet: A page in a drawing document.

Hidden Lines Removed (HLR): A view mode. All edges of the model that are not visible from the current view angle are removed from the display.

Hidden Lines Visible (HLV): A view mode. All edges of the model that are not visible from the current view angle are shown gray or dashed.

Import: The ability to open files from other software applications into a SolidWorks document. The A-size sheet format was created as an AutoCAD file and imported into SolidWorks.

Layers: Simplifies a drawing by combining dimensions, annotations, geometry, and components. Properties such as display, line style and thickness are assigned to a named layer.

OLE (Object Linking and Embedding): A Windows file format. A company logo or excel spread sheet placed inside a SolidWorks document are examples of OLE files.

Part: A 3D object made up of features. A part inserted into an assembly is called a component. A part's views and feature dimensions and annotations are inserted into 2D drawing. The extension for a SolidWorks part filename is .SLDPRT.

Sheet Format: A document that contains page size and orientation, standard text, borders, logos and title block information. Customize sheet format to save time. The extension for the Sheet Format filename is .SLDDRT.

Plane: To create a sketch choose a plane. Planes are flat and infinite. They are represented on the screen with visible edges. The default reference plane for this project is Front.

Menus: Menus provide access to the commands that the SolidWorks software offers.

Toolbars: The toolbar menus provide shortcuts enabling you to quickly access the most frequently used commands.

Mouse Buttons: The left and right mouse buttons have distinct meanings in SolidWorks.

System Feedback: Feedback is provided by a symbol attached to the cursor arrow indicating your selection. As the cursor floats across the model, feedback is provided in the form of symbols, riding next to the cursor.

Copy and Paste: Simple sketched features and some applied features can be copied and then pasted onto a planar face. Multi-sketch features such as sweeps and lofts cannot be copied.

Questions

1. Name the drawing options that are defined in the Drawing Template.
2. Name five drawing items that are contained in the Sheet Format.
3. Identify the paper dimensions for an A-size horizontal drawing.
4. Identify the paper dimensions for an A4 horizontal drawing.
5. The SolidWorks format Landscape corresponds to a_____________ drawing format and Portrait corresponds to a____________________ drawing format.
6. What Paper Size option do you select in order to define a custom paper width and height?
7. Identify the primary type of projection utilized on a drawing in the United States.
8. Describe the steps to display and modify the properties on a drawing sheet.
9. Identify the location of the stored System Options.
10. Name the four display modes for drawing views using SolidWorks 2003.
11. True or False. SolidWorks Line Font Types define all ASME Y14.2 type and style of lines.
12. Identify all Dimensioning standards options supported by SolidWorks.
13. Identify 10 drawing items that are contained in a title block.
14. SolidWorks Properties are saved with the _________________ Format.

Exercises

Create Drawing Templates for both inch units and Metric units. ASME Y14.5M has different rules for Metric and English unit decimal display.

English decimal display:

A dimension value is less than 1 inch. No leading zero is displayed before the decimal point. See Table 1 for details.

Metric decimal display:

A dimension value is less than 1mm. A leading zero is displayed before the decimal point. See Table 1 for details.

General Tolerances are specified in the Title Block. Specify tolerances are applied to an individual dimension. A dimension is displayed to the same number of decimal places as its tolerance for inch Unilateral Tolerance. Select ANSI for the SolidWorks Dimensioning Standard. Select inch or metric for Drawing units.

TABLE 1

TOLERANCE DISPLAY FOR INCH AND METRIC DIMENSIONS (ASME Y14.5M)

DISPLAY	INCH	METRIC
Dimensions less than 1	.5	0.5
Unilateral Tolerance	$1.417^{+.005}_{-.000}$	$36^{0}_{-0.5}$
Bilateral Tolerance	$1.417^{+.010}_{-.020}$	$36^{+0.25}_{-0.50}$
Limit Tolerance	.571 .463	14.50 11.50

Exercise 1.1:

a) Create an A-size ANSI Drawing Template using inch units. Use an A-FORMAT Sheet Format.

b) Create a C-size ANSI Drawing Template using inch units. Use a C-FORMAT Sheet Format.

The ASME Y14.2M, Minimum letter height for Title Block is as shown in Table 2.

c) Create three New Layers named DETAILS, HIDE DIMS and CNST DIMS (Construction Dimensions). Create New Layers to display CHAIN, PHANTOM and STITCH lines.

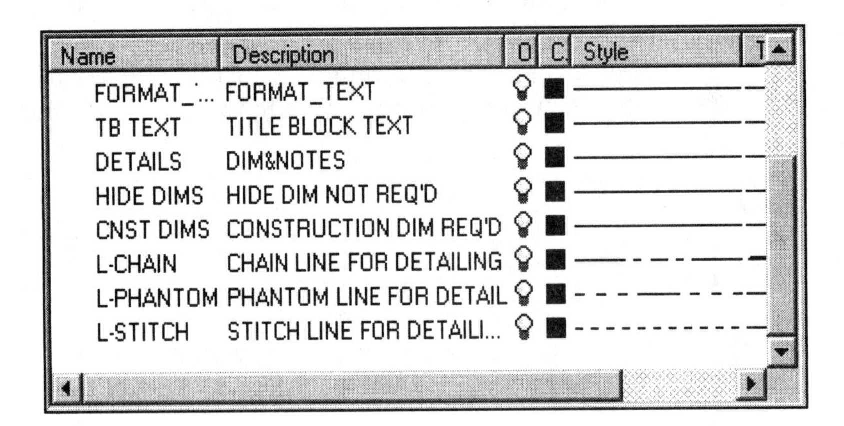

TABLE 2 MINIMUM LETTER HEIGHT FOR TITLE BLOCK (ASME Y14.2M)	
Title Block Text	Letter Height (inches) for A, B, C Drawing Size
Drawing Title, Drawing Size, Cage Code, Drawing Number, Revision Letter	.12
Section and view letters	.24
Drawing block letters	.10
All other characters	.10

Exercise 1.2:

Create an A4(horizontal) ISO Drawing Template. Use Document Properties to set the ISO dimension standard and millimeter units.

Exercise 1.3:

Modify the SolidWorks Drawing Template A4-ISO. Edit Sheet Format to include a new Sheet Metal & Weldment Tolerances box on the left hand side of the Sheet Format, Figure EX1.3.

Display sketched end points to create new lines for the Tolerance box. Click Tools, Options, System Options, Sketch. Check Display entity points. The endpoints are displayed for Sketch lines.

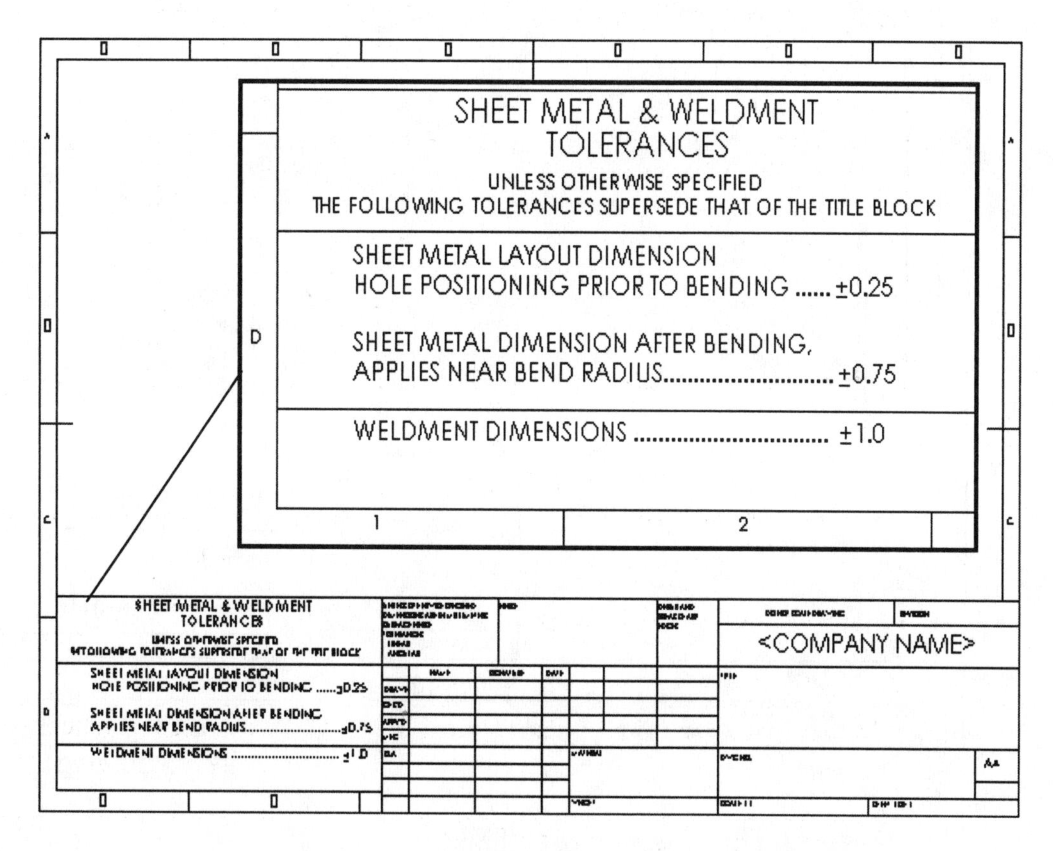

Figure EX1.3

SHEET METAL & WELDMENT TOLERANCES box courtesy of Ismeca, USA Inc. Vista, CA.

Exercise 1.4:

Your company uses SolidWorks and Pro/ENGINEER to manufacture Sheet Metal parts, Figure EX1.4. Import the empty A-size drawing format, FORMAT-A-PRO-E.DWG located in the 2003drwparts file folder. This document was exported from Pro/E as a DWG file. Save the PRO/E drawing format as a SolidWorks Sheet Format.

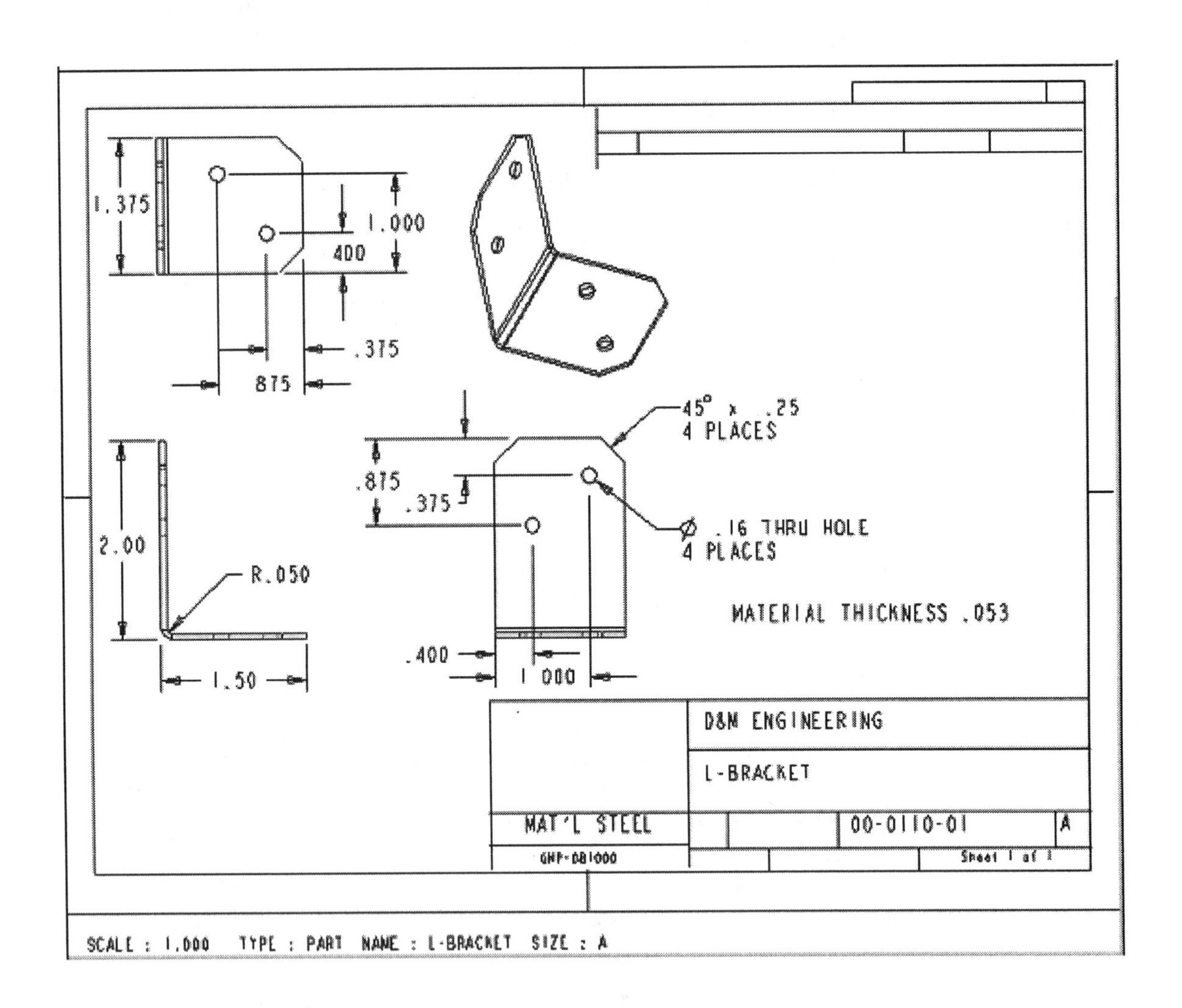

Figure EX1.4

Sheet Metal Strong Tie Reinforcing Bracket, courtesy of Simpson Strong Tie Corporation, CA, USA.

Exercise 1.5:

You require AutoCAD to perform Exercise 1.5. Your company uses SolidWorks and AutoCAD. Open an A-size drawing template from AutoCAD. Review the Dimension Variables (DIMVARS) in AutoCAD. Record the DIMSTATUS for the following variables:

DIMTXSTY	Dimensioning Text Style
DIMASZ	Arrow size
DIMCEN	Center Mark size
DIMDEC	Decimal Places
DIMTDEC	Tolerance Decimal Places
DIMTXT	Text Height
DIMDLI	Space between dimension lines for Baseline dimensioning

Identify the corresponding values in SolidWorks Document Properties to contain the AutoCAD dimension variables.

Favorite dimension style settings are defined for a particular dimension. Favorite dimension styles are applied to other dimensions on the drawing, part and assembly documents. The styles are accessed through the Dimension PropertyManager.

Note: Early AutoCAD drawing formats contain fonts not supported in a Windows NT/2000 environment. These fonts imported into SolidWorks will be misaligned in the Sheet Format. Modify older AutoCAD formats to a True Type Font in SolidWorks.

Notes:

Project 2

Drawing Views

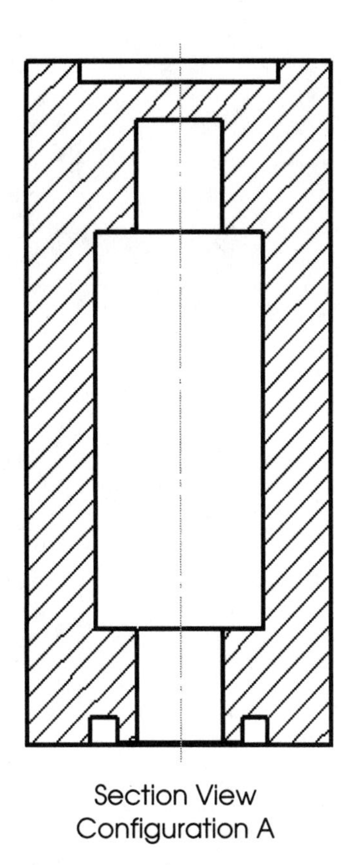

Section View
Configuration A

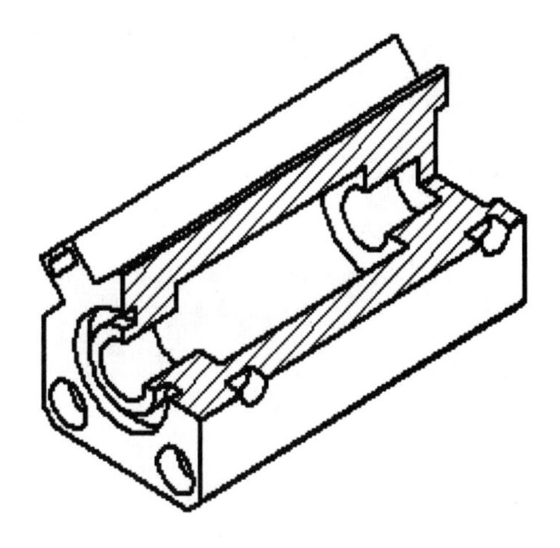

Half Section View
Configuration B

Below are the desired outcomes and usage competencies based on the completion of this Project.

Desired Outcomes:	**Usage Competencies:**
TUBE drawing. ROD drawing. COVERPLATE drawing.	An understanding of displaying and creating Standard, Isometric, Auxiliary, Section, Broken Section, Detail and Half Section (Cut-away) views.
	Ability to create multi-sheet drawings from various part configurations.

Notes

Project 2 – Drawing Views

Project Objective

Obtain a comprehensive understanding of displaying and creating Standard, Isometric, Auxiliary, Section, Broken Section, Detail and Half Section (Cut-away) views.

Achieve the ability to use SolidWorks drawing tools, other related view commands and the Fundamentals of Orthographic projection.

Create multi-sheet drawings from various part configurations.

Create three drawings:

- TUBE drawing.
- ROD drawing.
- COVERPLATE drawing.

The ROD drawing consists of three sheets. The TUBE drawing consists of one sheet. The COVERPLATE drawing consists of two sheets.

On the completion of this project, you will be able to:

- Create a single sheet drawing.
- Create a multi-sheet drawing.
- Insert 3 Standards views.
- Insert Named view.
- Insert Auxiliary view, Detail view and Section view.
- Insert Broken Section, Half Section, Offset Section and Aligned Section views.
- Add a part configuration.
- Utilize multiple configurations in a drawing.
- Insert Area Hatch.

Project Situation

A customer approaches the engineering department to address the current air cylinder for a new product application.

In the new application, there is an interference concern with the positions of the current air cylinder switches.

The engineering team proposes a new design that re-positions the switches in a 45°- grooved track.

The design incorporates three individual parts: TUBE, ROD and COVERPLATE.

The parts are mated to create the CYLINDER assembly.

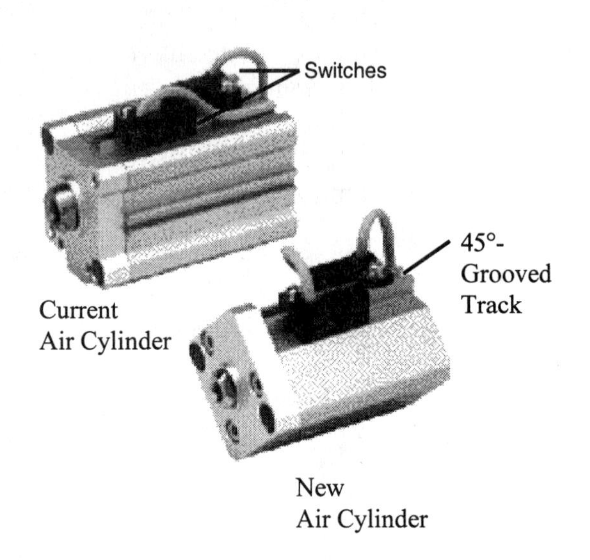

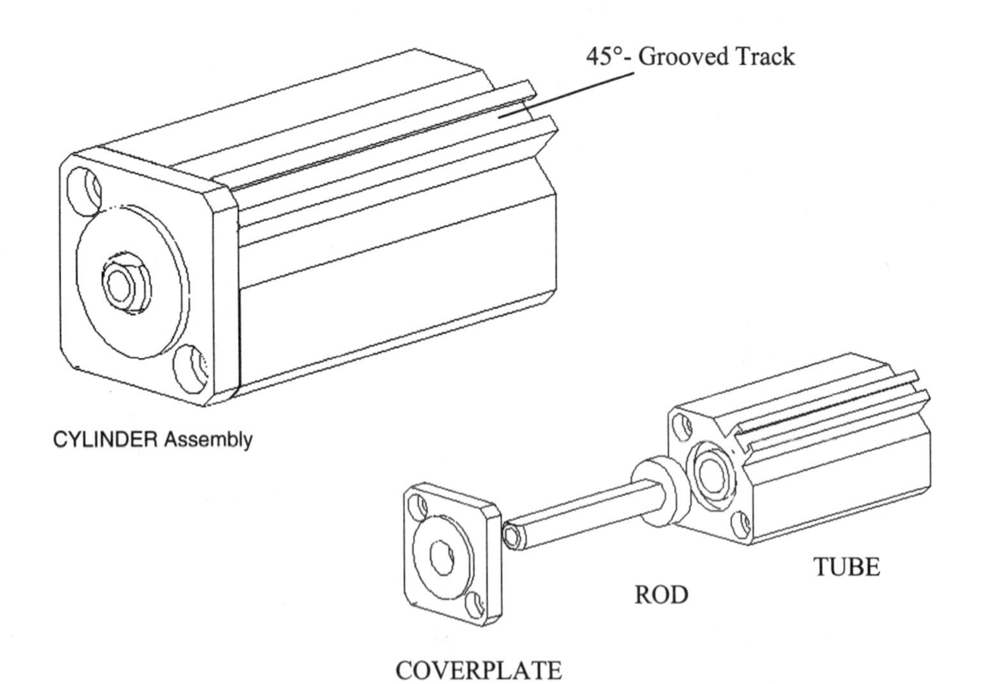

Courtesy of
SMC Corporation of America
The feature dimensions for components utilized in this assembly have been modified for educational purposes.

The marketing manager for the air cylinder product line views the new proposed assembly in SolidWorks. Numerous customers have voiced similar switch location requirements.

The design team decides to incorporate the new design in its standard product line. Note: The original designer that developed the current air cylinder was transferred to a different company division.

You are part of the CYLINDER project development team. All design drawings must meet the company's drawing standards.

What is the next step? You are required to create drawings for various internal departments, namely: production, purchasing, engineering, inspection and manufacturing.

Each drawing may contain unique information and specific footnotes. Example: A manufacturing drawing would require information on assembly, Bill of Materials, fabrication techniques and references to other relative information.

Project Overview

First, review and discuss the features used to create the three parts. The three parts are:

- ROD.
- TUBE.
- COVERPLATE.

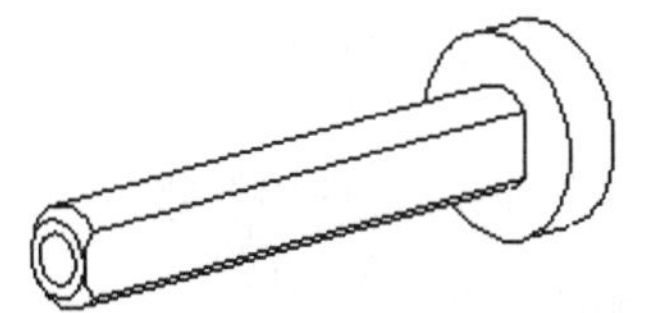

ROD

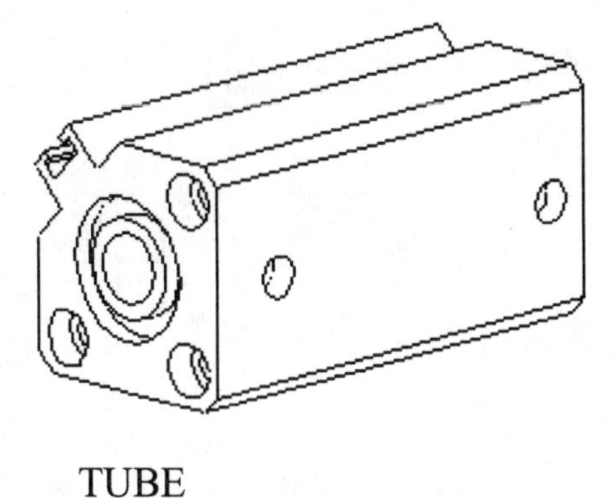

TUBE

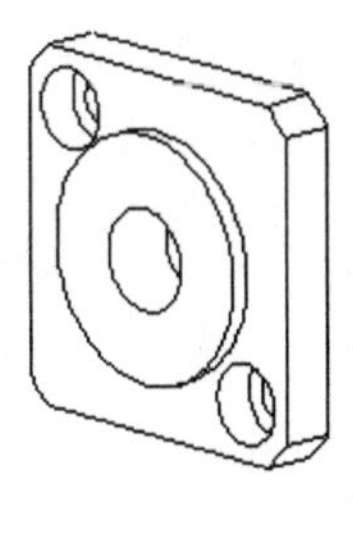

COVERPLATE

Second, create three drawings:

- ROD.
- TUBE.
- COVERPLATE.

The ROD drawing consists of three sheets.

The TUBE drawing consists of one sheet.

The COVERPLATE drawing consists of two sheets.

ROD drawing:

Use three separate drawing sheets to display the required information for the ROD drawing.

The first drawing sheet contains the Short Rod configuration and is comprised of three Standard views, (Principle views) and an Isometric view.

Do you remember what the three Standard views are? They are:

1. Top side.
2. Front side.
3. Right side.

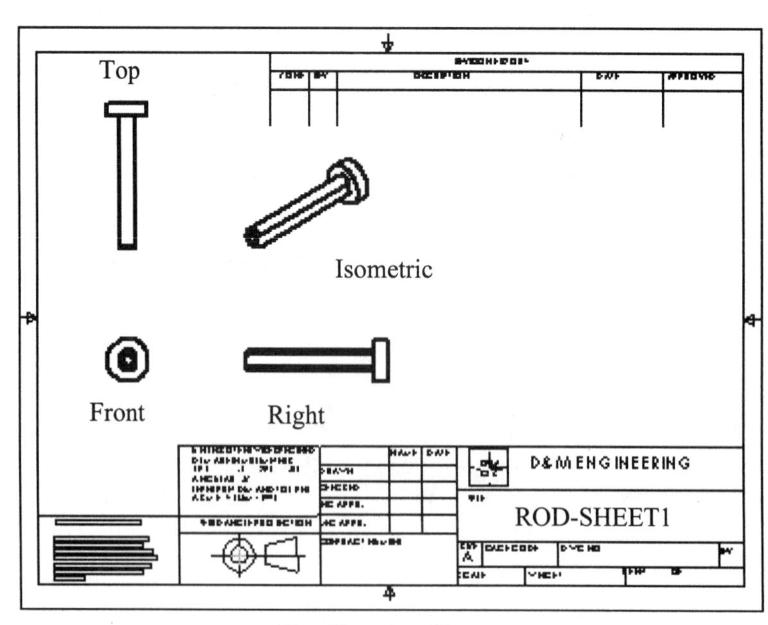

First Drawing Sheet:
Short Rod Configuration

The second drawing sheet contains the Long Rod configuration.

Use a drawing sheet scale of 2:1 to display the Front detail.

Using a 2:1 scale, the Right view is too long for the drawing sheet.

Use a Vertical Break to represent the Long Rod configuration with a constant cross section.

Add a Revolved Section to represent the cross section of the ROD.

Position the Revolved Section between the vertical break.

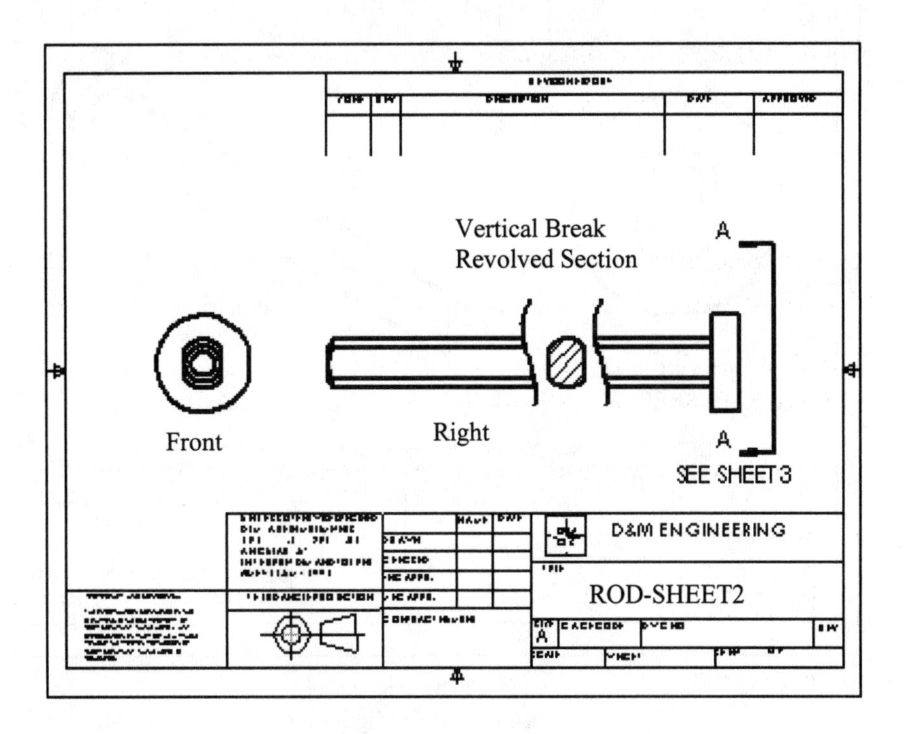

Second Drawing Sheet:
Long Rod Configuration

The third drawing sheet contains the Long Rod configuration.

Create the Removed view from the Right view in Sheet2.

The Removed view is at a 4:1 scale.

Combine two Detail views to construct the Isometric view.

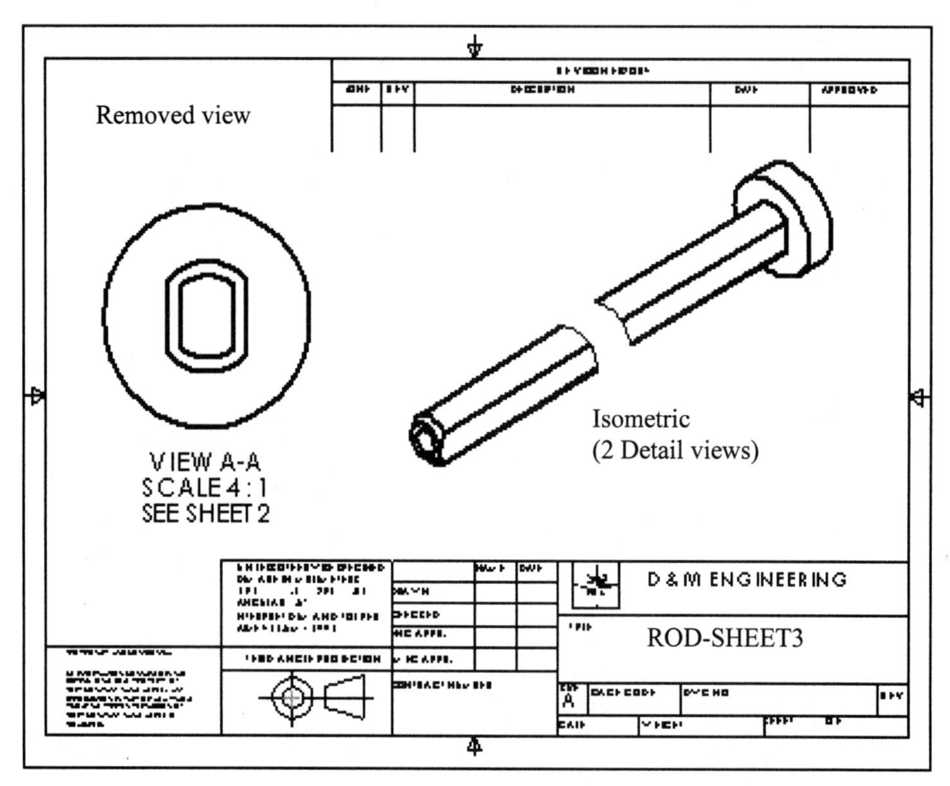

Third Drawing Sheet:
Long Rod Configuration

TUBE drawing:

The TUBE drawing sheet contains:

- Three Standard views.
- Projected Back view.
- Section view.
- Detail view.
- Auxiliary view.
- Half Section Isometric (Cut away) view.

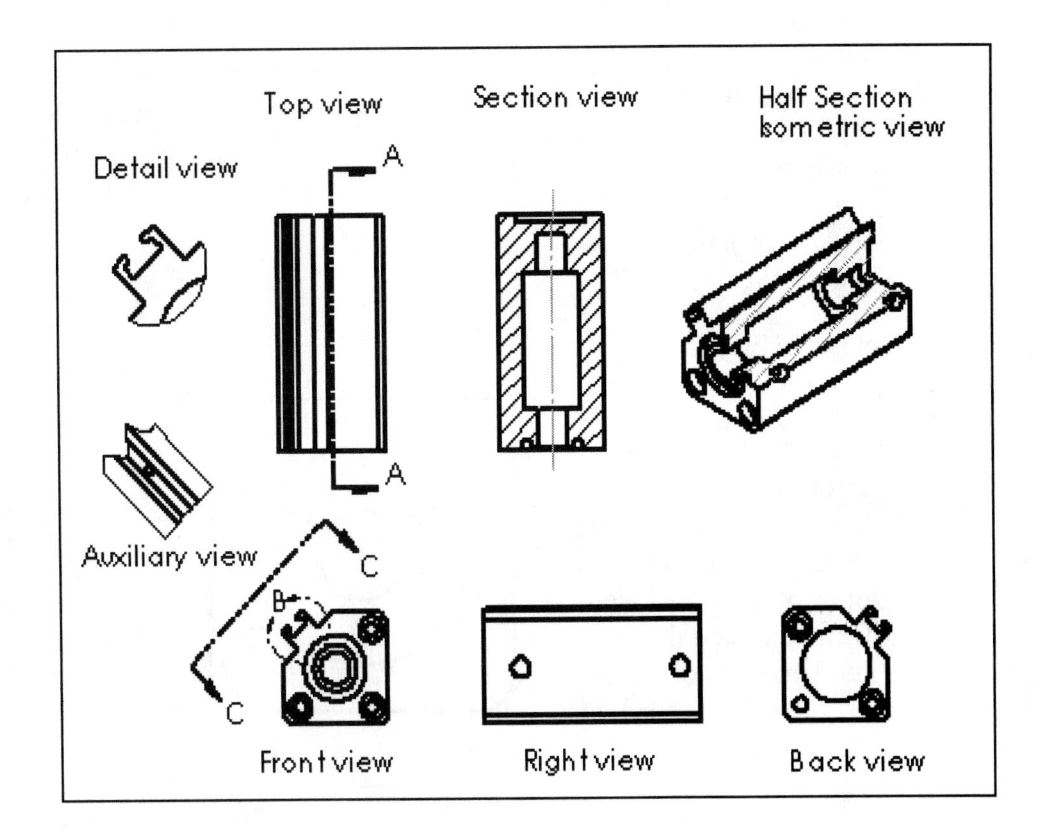

COVERPLATE drawing:

Use two separate drawing sheets to display the required information for the COVERPLATE drawing.

The first drawing sheet contains the Without Nose Holes configuration and is comprised of a Front view, Right view and an Offset Section view.

The second drawing sheet contains the With Nose Holes configuration and is comprised of a Front view and Aligned Section view.

Note: In Project 3 you will insert dimensions and create notes.

Before you create the drawings, let's review: SolidWorks drawing tools, other related view commands and the Fundamentals of Orthographic projection.

Offset Section

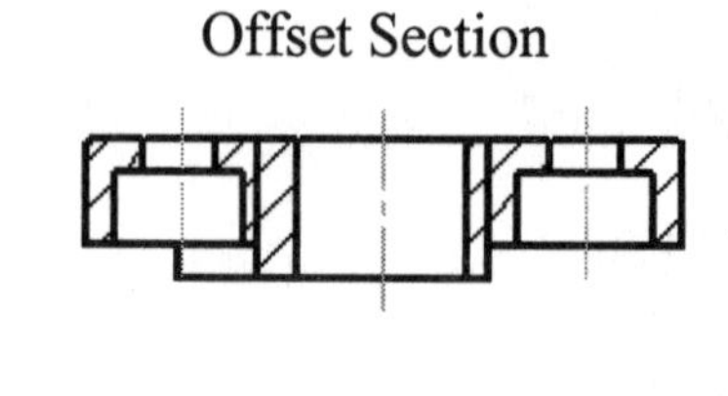

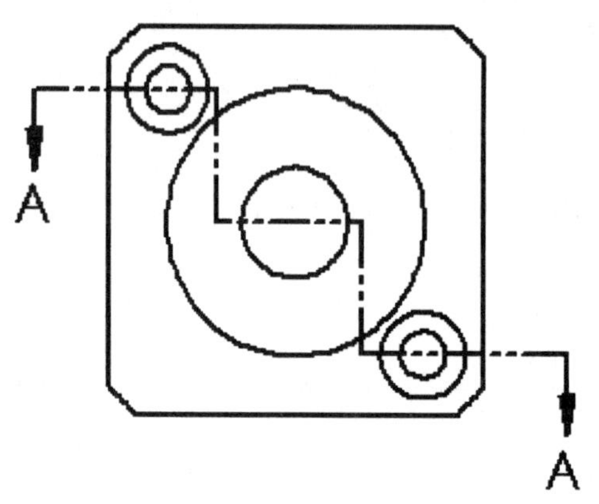

Front

Right

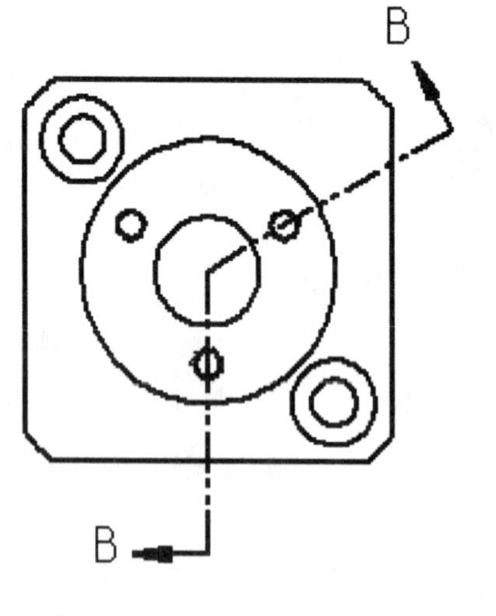

Front

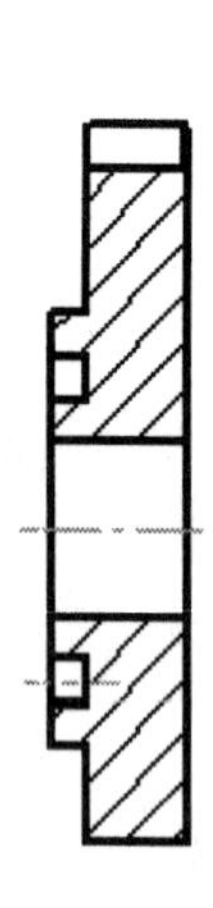

Aligned Section

SolidWorks Tools and Commands

The following Drawing tools are utilized in this Project:

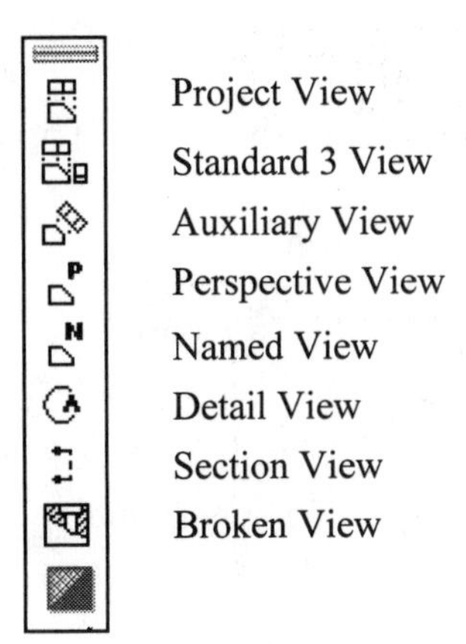

Other tools and commands included in this Project:

SolidWorks Tools and Commands:		
View Properties	Line, Spline, Rectangle	Broken View
Sheet Properties	Inference	Un Break View
Move View	Area Hatch	Layers
Note	Design Tables	Layer Properties
Linked Note	Configurations	Align Views
Copy/Paste	Suppress/UnSuppress	Partial Section
Hide View	Add Sheet	Shade View
Aligned Section View	Update View	Crop View

Additional information on Drawing tools and other commands are found in the SolidWorks On-line help.

Reference Planes and Orthographic Projection

The three default ⊥ reference planes, Front, Top and Right represent infinite 2D planes in 3D space. Planes have no thickness or mass.

Orthographic projection is the process of projecting views onto parallel planes with ⊥ projectors. The default reference planes are the Front, Top and Right side viewing planes.

In geometric tolerancing, the default reference planes are the Primary, Secondary and Tertiary ⊥ datum planes.

These are the planes used in manufacturing.

The Primary datum plane contacts the part at a minimum of three points.

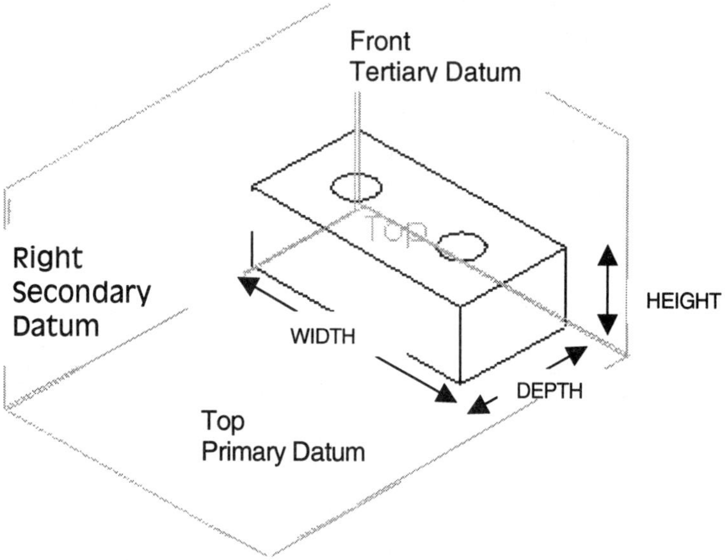

Front
Tertiary Datum
Right
Secondary
Datum
Top
WIDTH
HEIGHT
DEPTH
Top
Primary Datum

The Secondary datum plane contacts the part at a minimum of two points. The Tertiary datum plane contacts the part at a minimum of one point.

The six principle views of Orthographic projection listed in the ASME Y14.3M standard are: Top, Front, Right side, Bottom, Rear & Left side. SolidWorks Standard view names correspond to these Orthographic projection view names. The Standard 3 View  Drawing tool inserts the Front, Top and Right view into a drawing.

ASME Y14.3M Principle View Name:	SolidWorks Standard View:
Front	Front
Top	Top
Right side	Right
Bottom	Bottom
Rear	Back
Left side	Left

Orthographic Projection

In the third angle Orthographic projection example below, the standard drawing views are; Front, Top, Right and Isometric.

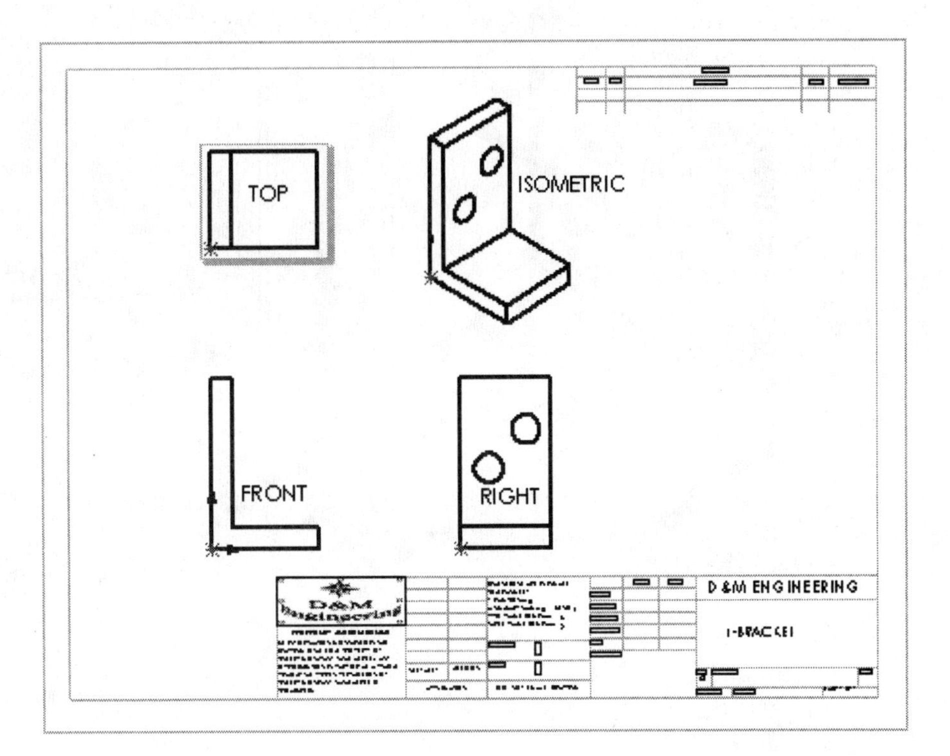

There are two Orthographic projection drawing systems.

The first Orthographic projection system is called the third angle projection.

The second Orthographic projection system is called the first angle projection.

The systems are derived from positioning a 3D object in the third or first quadrant.

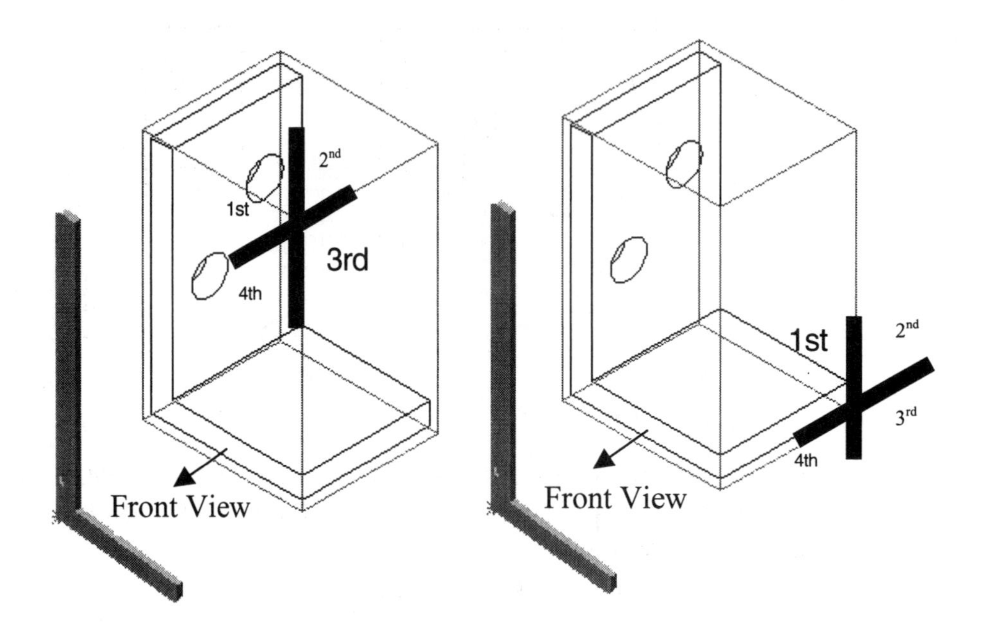

In third angle projection, the part is positioned in the third quadrant. The 2D projection planes are located between the viewer and the part.

The projected views are placed on a drawing.

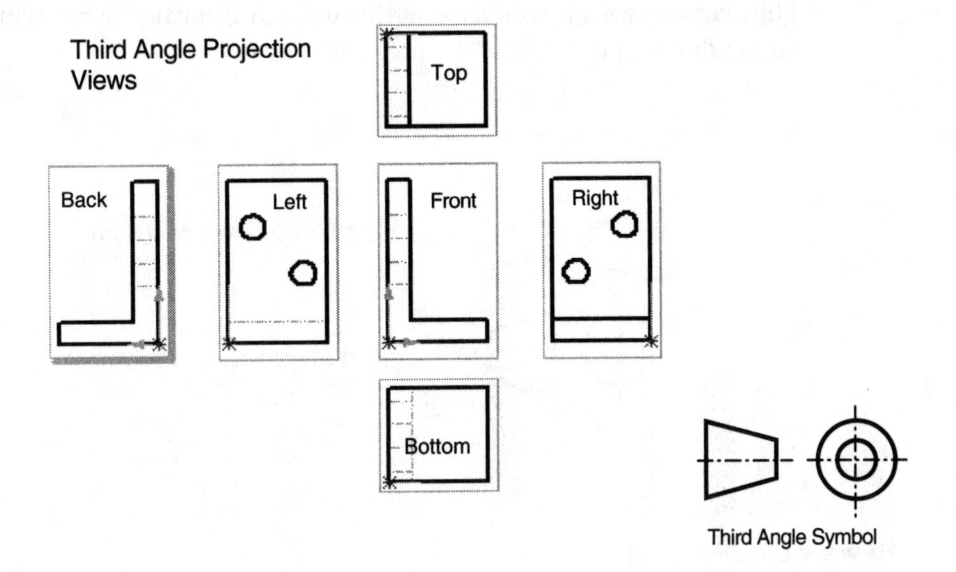

In first angle projection, the part is positioned in the first quadrant. Views are projected onto the planes located behind the part. The projected views are placed on a drawing.

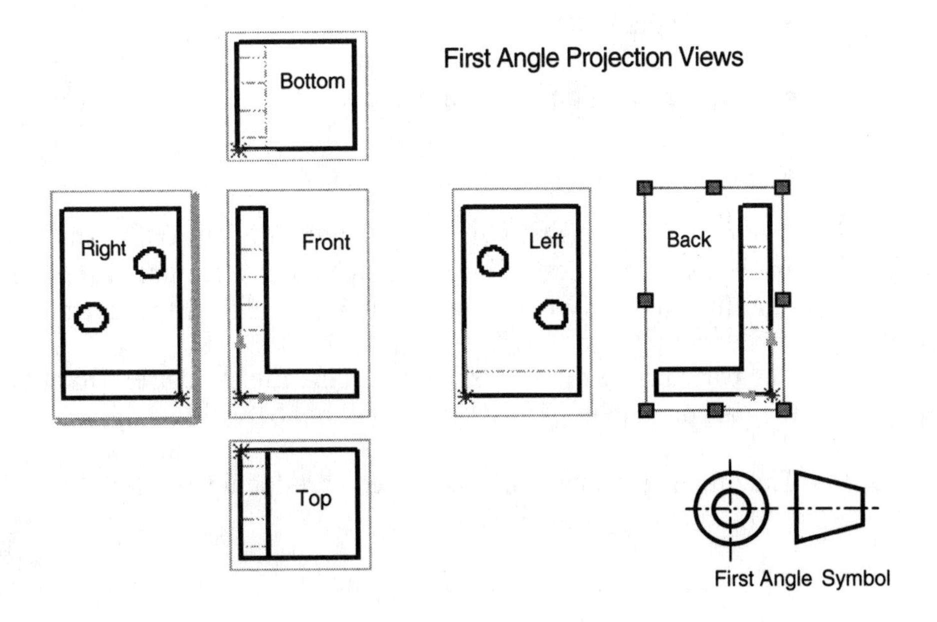

First angle projection is primarily used in Europe and Asia. Third angle projection is primarily used in the U.S. & Canada and is based upon the ASME Y14.3M Multi and Sectional View Drawings standard. Designers should have knowledge and understanding of both systems.

There are numerous multi-national companies. Example: A part is designed in the U.S., manufactured in Japan and destined for a European market.

Third angle projection is used in this text. A truncated cone symbol appears on the drawing to indicate the projection system:

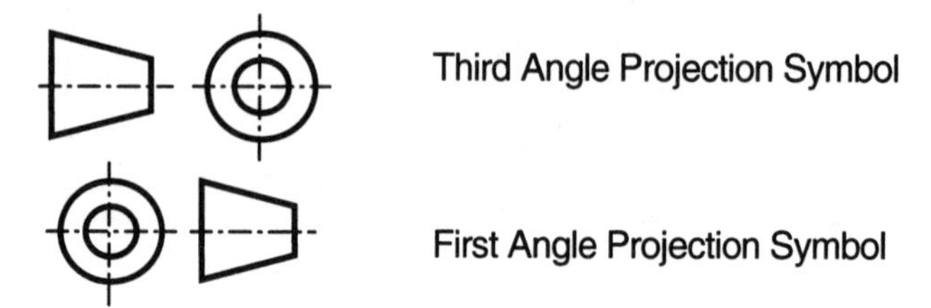

Review the ROD part

You are part of the new CYLINDER project development team. You were not part of the originally team that created the stock CYLINDER or its parts.

Perform the following recommended tasks before starting the ROD drawing:

- Verify the ROD part.
- View dimensions in each feature.

A drawing contains part views, geometric dimensioning and tolerances, notes and other related design information. When a feature or dimension of a part is modified in SolidWorks, the drawing automatically updates.

The part and the drawing share the same file structure. Do not delete or move the part document. The drawing will not be valid.

The drawing requires the associated part document.

Start a SolidWorks session.

1) Click **Start** on the Windows Taskbar, Start. Click **Programs**.

2) Click the **SolidWorks** SolidWorks folder.

3) Click the **SolidWorks** . SolidWorks application. The SolidWorks program window opens.

Review the ROD part.

4) Open the part. Click **Open** from the Standard toolbar.

5) Select the **2003drwparts** file folder for Look in.

6) Select **ROD**.

7) Click the **Open** button. The ROD is displayed in the Graphics window.

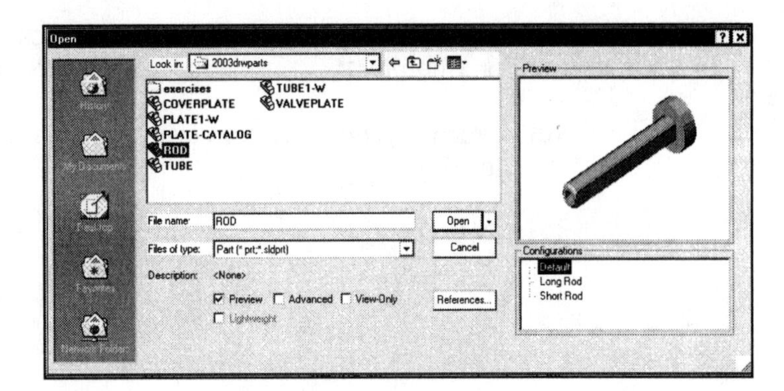

8) Expand the SolidWorks window to full screen. Click the **Maximize** icon in the top right hand corner of the Graphics window.

9) The FeatureManager list 6 individual features. There are too many dimensions to display all at once. Display **Hidden Lines Removed** .

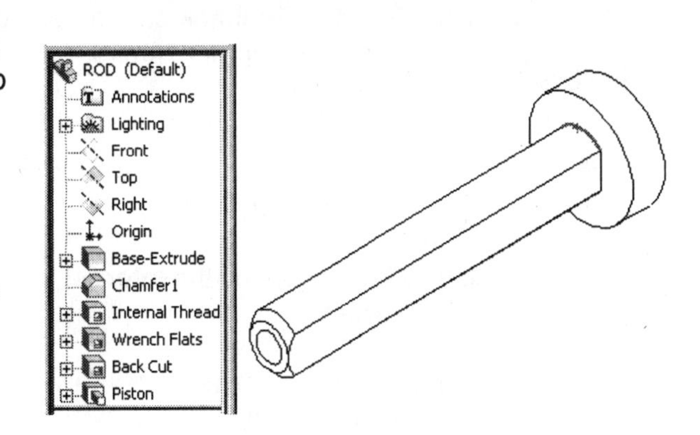

Review the ROD part features.

10) Position the Rollback bar. Place the **mouse pointer** over the yellow Rollback bar at the bottom of the FeatureManager design tree. The mouse pointer displays a symbol of a hand.

11) Drag the **Rollback** bar upward below the Base- Extrude feature.

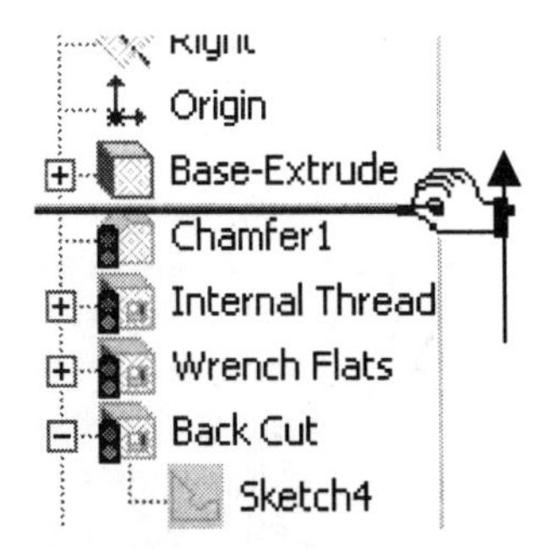

12) Display the feature dimensions. Double-click **Base-Extrude**.

13) Drag the **dimension text** away from the feature. Click **OK**. Note: Position dimension text off the feature before inserting the part into the drawing.

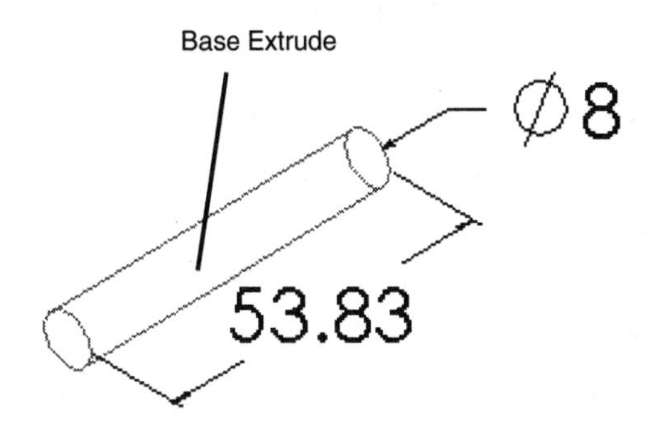

14) Drag the **Rollback** bar downward below Chamfer1.

15) Display feature dimensions. Double-click on **Chamfer1**.

16) Drag the **dimension text** away from the feature.

17) Click **OK**.

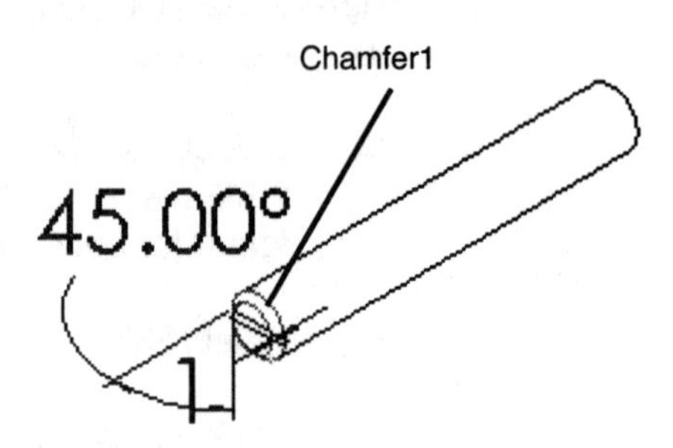

18) Drag the **Rollback** bar downward below the Internal Thread.

19) Click **Hidden Lines Removed** to view the feature. Display the feature dimensions.

20) Double click on the **Internal Thread**.

21) Drag the **dimension text** away from the feature.

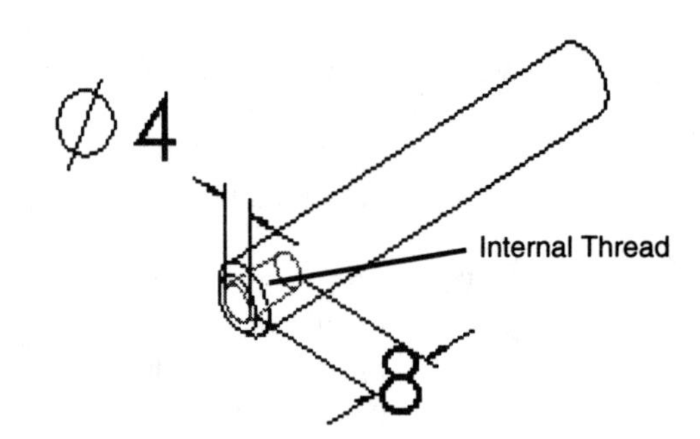

22) Click **OK**. The Internal Thread was created with an Extruded Cut sketched on the front plane. Use a Hole Callout in the Drawing to annotate the Internal Thread in Project 3.

The profile for the Wrench Flats is sketched on the Back plane. The design engineer uses a symmetric relation with the Right plane. You will add a centerline in the drawing to represent the Wrench Flat symmetry in Project 3.

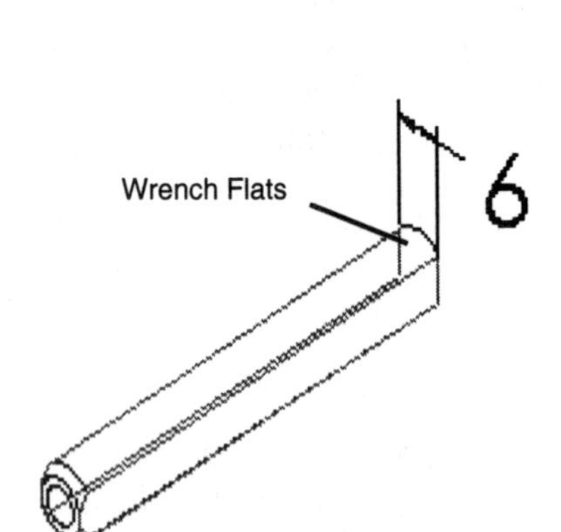

23) Drag the **Rollback** bar downward below the Back Cut. Display the feature dimensions.

24) Double click on the **Back Cut**.

25) Drag the **dimension text** away from the feature.

26) Click **OK**.

27) Drag the **Rollback** bar downward below the Piston.

28) Display the feature dimensions. Double click on the **Piston**.

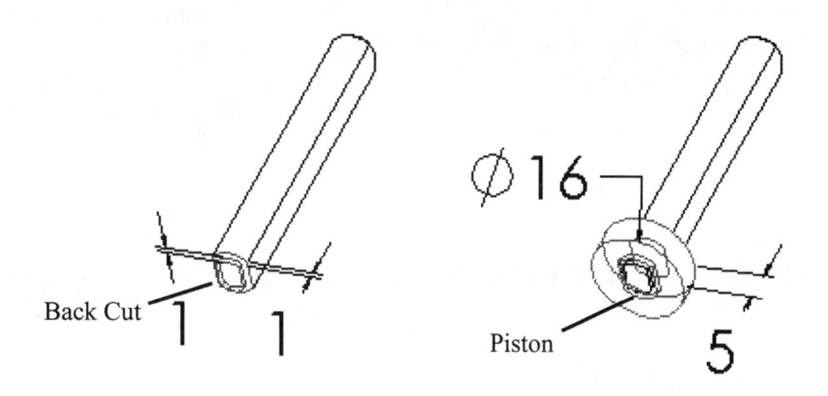

29) Drag the **dimension text** away from the feature.

30) Click **OK**.

Dimension schemes defined in the part require changes in the drawing.

Check to insure that all sketches are fully defined. A minus sign (-) displayed the FeatureManager indicates an under defined Sketch. Sketch1 through Sketch 5 are fully defined. This will address faster rebuild times and fewer configuration problems.

Utilize symmetry in the part whenever possible. This will save rebuild time. Use Symmetric relations in the sketch. Use Mirror All and the Mirror Feature in the part.

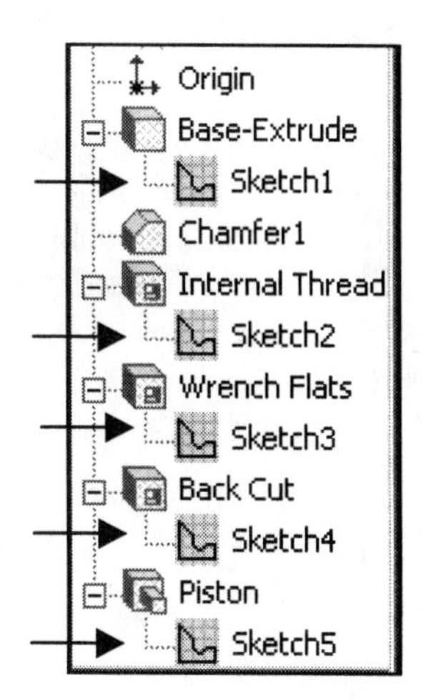

The ROD part consists of two configurations:

- Short Rod configuration.
- Long Rod configuration.

The first ROD drawing sheet contains the Short Rod configuration. The second and third ROD drawing sheets contains the Long Rod configuration.

Display the ROD part configurations.

31) Click **Configuration** at the bottom of the FeatureManager.

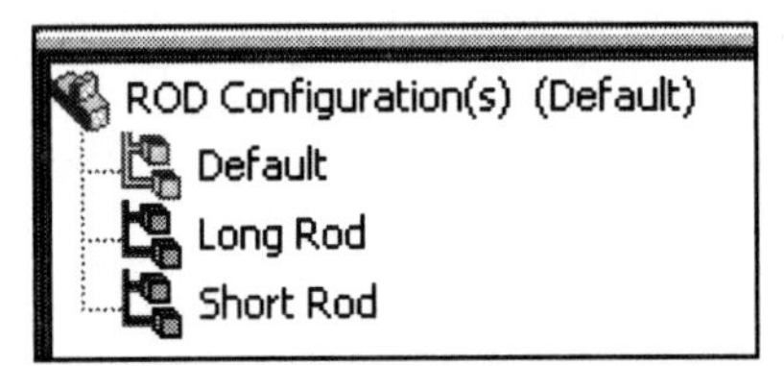

32) Double-click the **Long Rod** configuration.

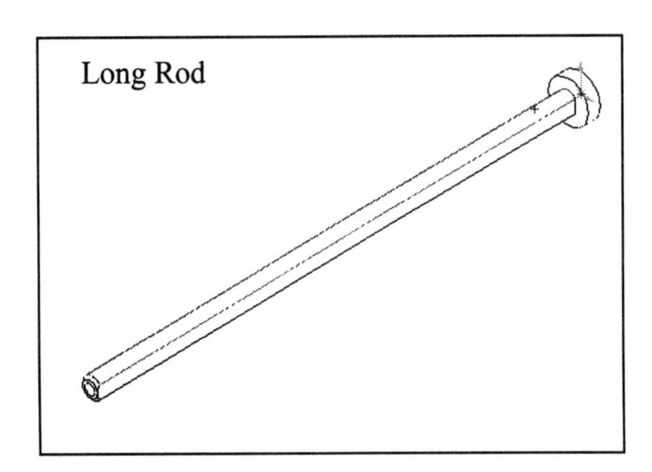

33) Double-click the **Short Rod** configuration.

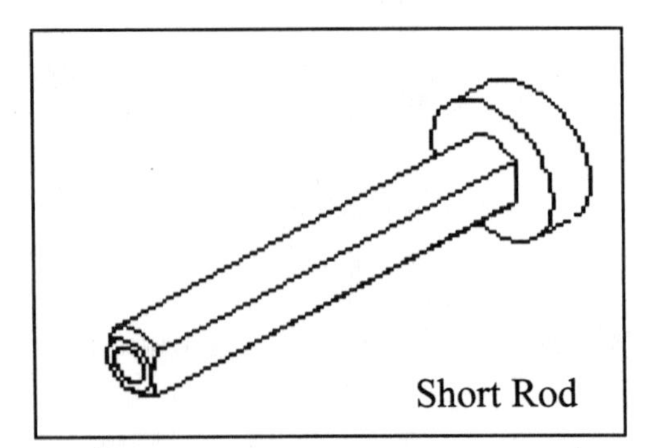

34) Return to the ROD FeatureManager. Click **Feature Manager** . The Short Rod is the current configuration.

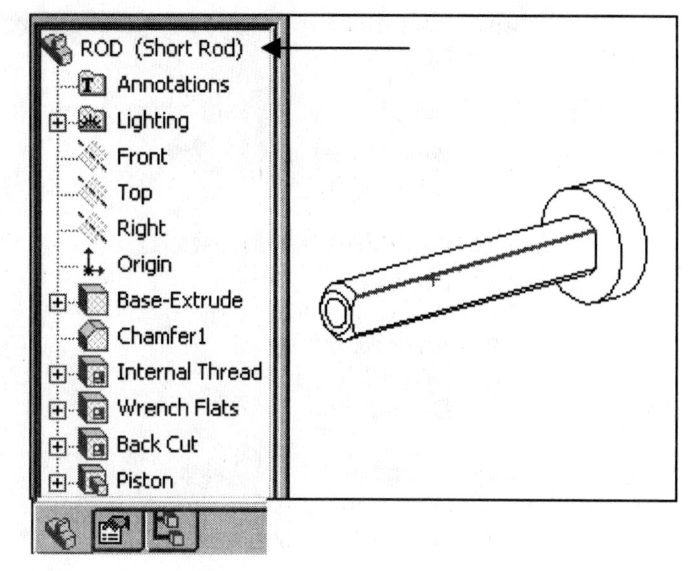

Create the ROD drawing: Sheet1 Short Rod Configuration

The ROD drawing consists of three drawing sheets. Sheet1 contains the Short Rod configuration and is comprised of three Standard views, (Principle views) and an Isometric view.

Create ROD-Sheet1. Use the A-ANSI-MM drawing template created in Project 1.

Select the A-ANSI-MM Drawing Template.

35) Click **New** . Click the **2003drwparts** tab.

36) Click **A-ANSI-MM** from the SolidWorks Document dialog box.

37) Click **OK**.

38) Save the empty drawing.

39) Click **File**, **Save As**.

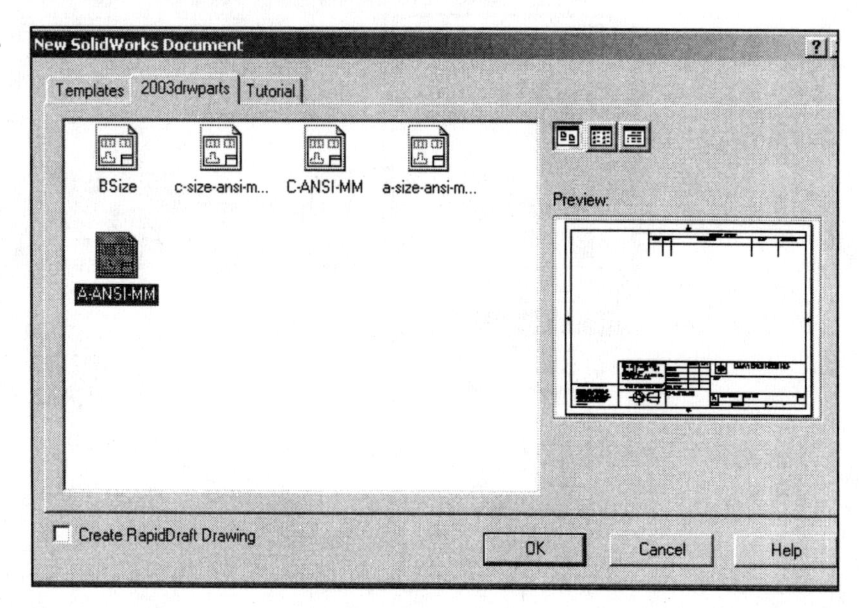

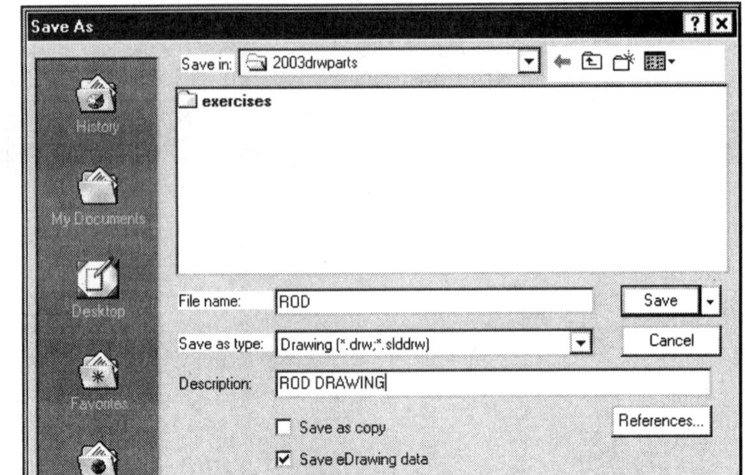

40) Enter **ROD**. Click **Save**.

41) Select **2003drwparts** for Save in.

42) Enter **ROD** for File name.

43) Enter **ROD DRAWING** for Description.

44) Click **Save**.

45) Display the Layer toolbar. Click **View**, **Toolbars**, **Layer**.

46) Set Layer Properties to **None**.

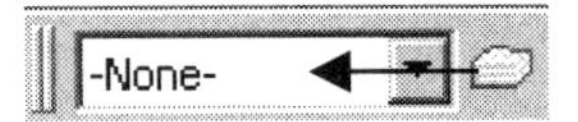

Insert the three Standard Views.

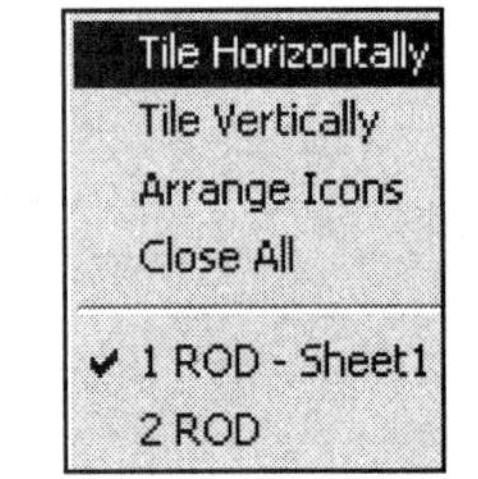

47) Display the ROD drawing and ROD part. Click **Window** from the Main menu. Click **ROD-Sheet1**.

48) Click **Tile Horizontal**. The ROD-Sheet1 drawing and ROD part are displayed.

49) Drag the ROD (Short Rod) **icon** from the Part FeatureManager into the drawing Graphics window. The Front, Top and Right views are displayed in the drawing.

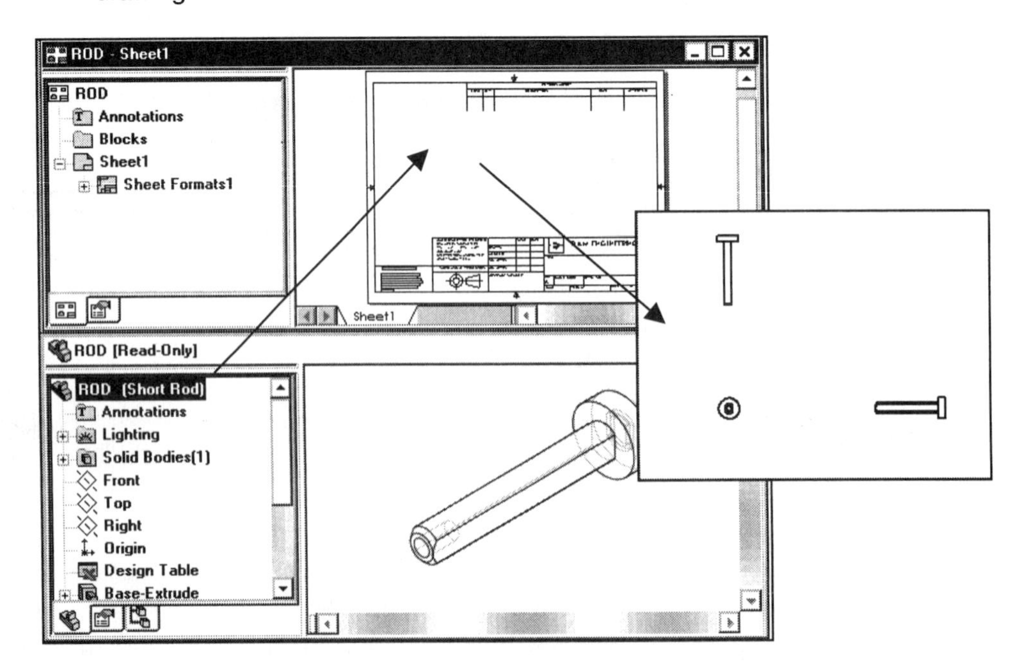

50) Maximize the drawing.

Standard views are displayed in the Graphics window. It may take a few seconds to create the standard views depending upon the configuration of your computer. The drawing views may be positioned too close to the title block. Views can be added, deleted and moved.

Drawing view names are displayed in the FeatureManager. Named views and Projected views are given a sequential numbered drawing view name.

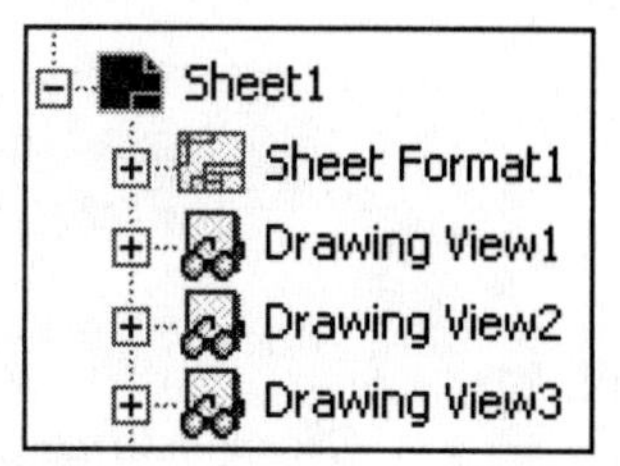

Example: The next view created is named Drawing View4.

Detail views and Section views are label with their view name followed by a letter for number. Example: Section View A-A.

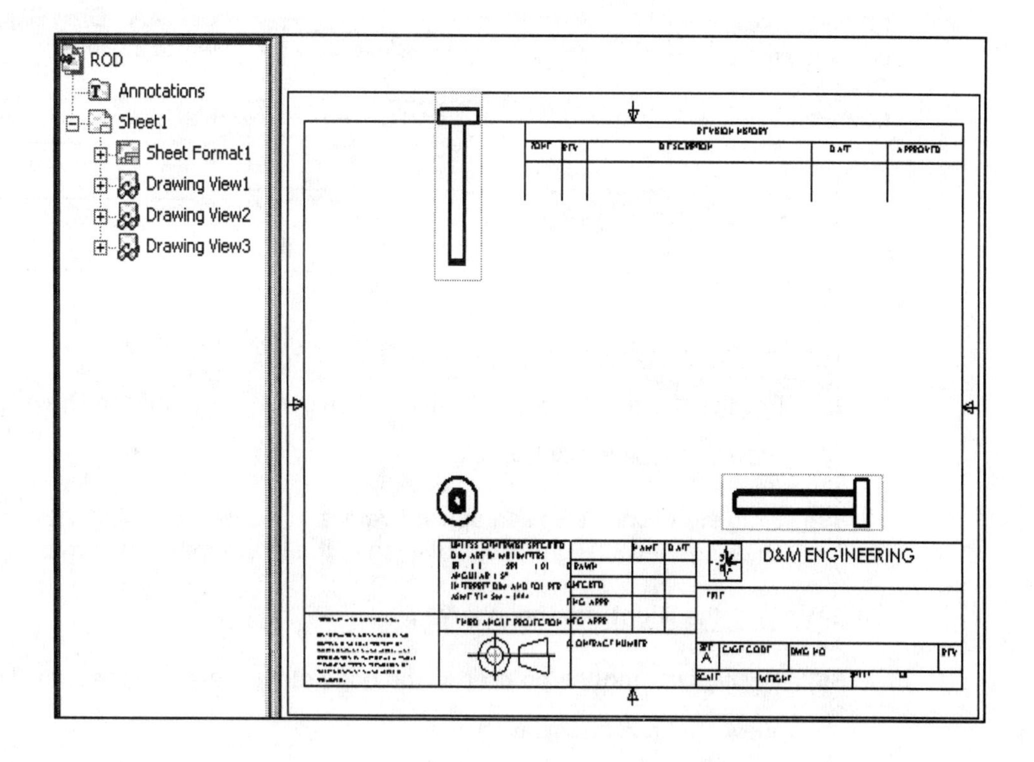

Reposition the view on a drawing. Provide approximately 1in. – 2in., (25mm – 50mm) between each view for dimension placement.

Move drawing views.

51) Click the view boundary of **Drawing View1** (Front). The mouse pointer displays the Drawing View icon. The view boundary is displayed in green.

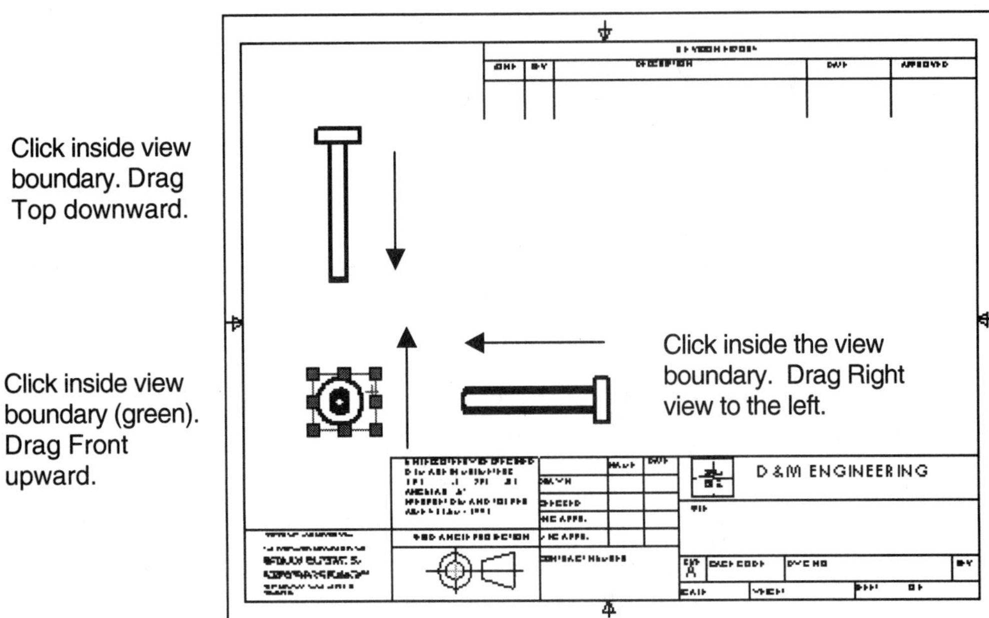

52) Position the **mouse pointer** on the edge of the view until the Drawing Move View icon is displayed.

53) Drag the **Front** view in an upward vertical direction. Drawing View2 (Top) and Drawing View3 (Right) move aligned to the Drawing View1 (Front).

54) Click the **Right** view boundary.

55) Position the **mouse pointer** on the edge of the view until the Drawing Move View icon is displayed.

56) Drag the **Right view** in a right to left direction towards the Drawing Front view.

57) Move Top view in a downward vertical direction. Click the **view boundary**.

58) Drag the **Top view** in a downward direction towards the Front view.

Note: You can add additional views to the drawing at anytime. Use Named View tool to create an Isometric view.

Property commands for the Sheet, Views and Geometry are accessible through the Graphics window.

Display the Properties of the Sheet, Views and Geometry.

59) Position the **mouse pointer** in the sheet area. The mouse pointer displays the Sheet icon.

60) Right-click in the **sheet area** (large white space) to display properties of the sheet. The sheet boundary turns green.

61) Position the **mouse pointer** in the view area. The mouse pointer displays the Drawing View icon.

62) Display the properties of the view. Right-click in the **view area**. The view boundary turns green.

63) Display the line properties. Right-click on the **vertical line** in the Top view. The vertical line turns green. The mouse pointer displays an Edge icon.

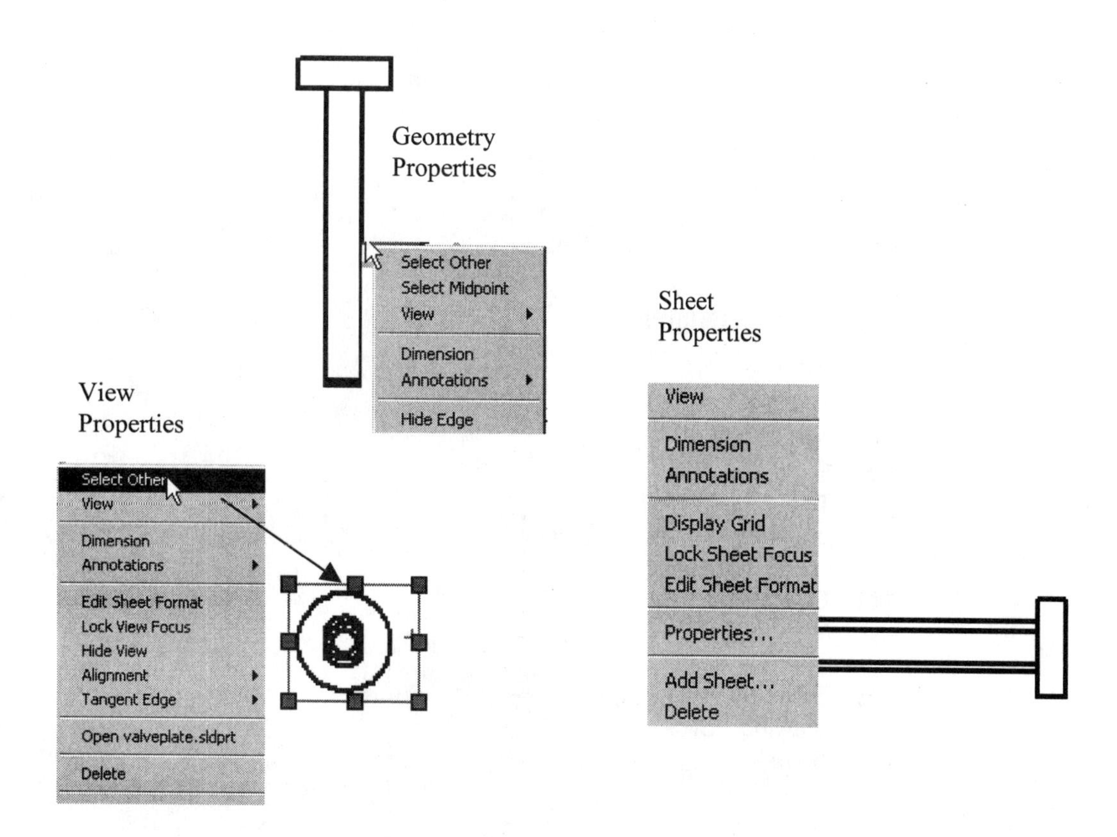

Add an Isometric view to the ROD-Sheet1 drawing.

64) Click **Drawing View1** (Front). The view boundary is displayed in green. Click **Named View** from the Drawing toolbar. The named views for the ROD are displayed. Isometric is the default view selected in the View Orientation text box. The Isometric view is placed on the mouse pointer

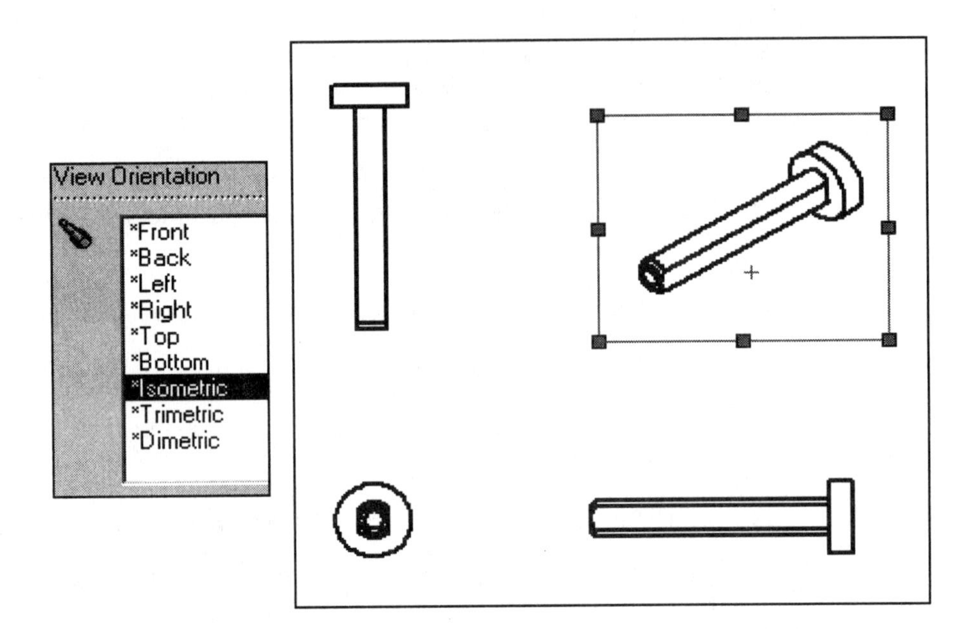

65) Click a **position** to the right of the Top view.

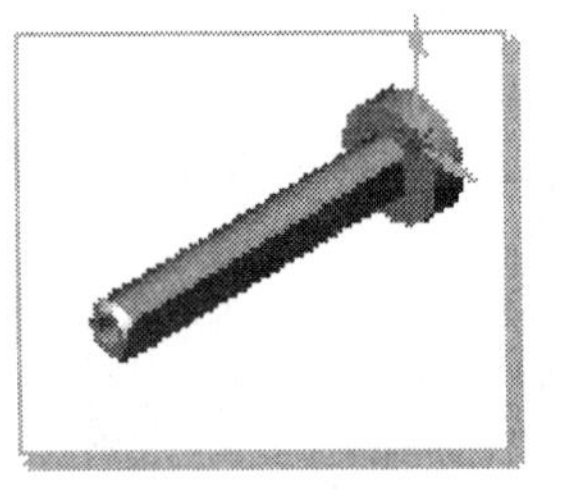

66) Display the Isometric view shaded. Click **Shaded** from the View toolbar.

67) Hide the Origin. Click **View**.

68) Uncheck **Origins**.

Hide the Top view.

69) Right-click the **Top** view boundary.

70) Click **Hide View**.

71) Save the ROD drawing. Click **Save**.

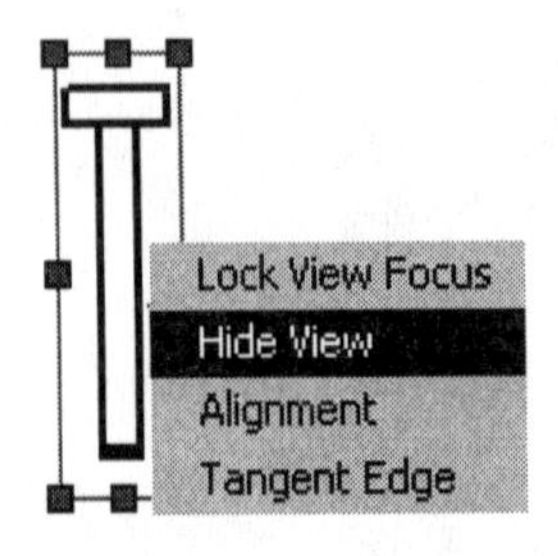

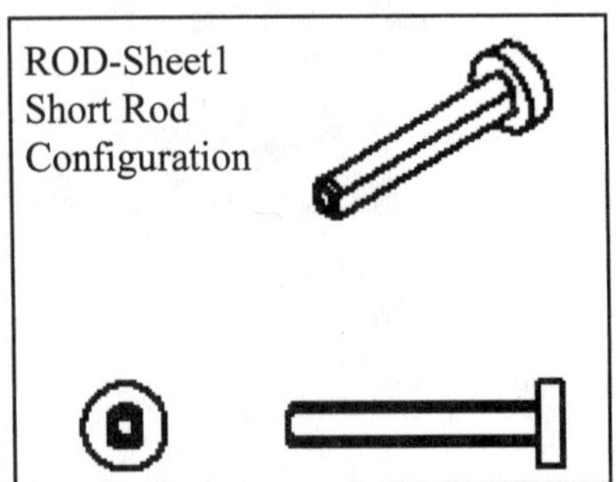

Create the ROD drawing: Sheet2 & Sheet3 Long Rod Configuration

You created the 3 standard views for the drawing ROD-Sheet1 with the Short Rod configuration.

Add ROD-Sheet2 and ROD-Sheet3 with the Long Rod configuration.

Add ROD-Sheet2. Create the Front view as a Named view. Select the Long Rod configuration. Create a projected Right view. Use a sheet scale of 2:1 to display the Front details. The Right view is too long for the sheet using the scale of 2:1.

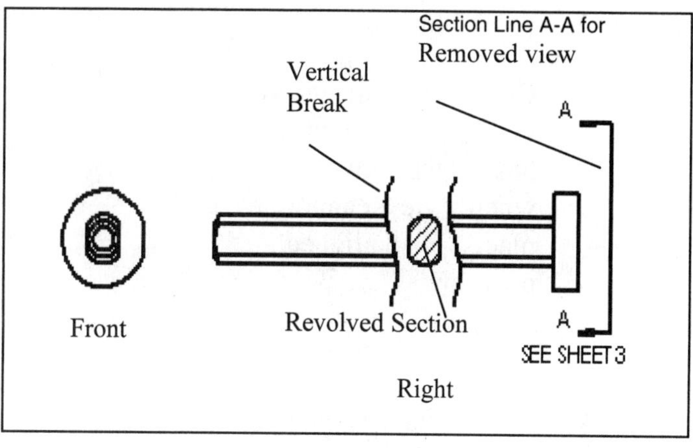

ROD-SHEET2
Long Rod Configuration

Create a Broken (or interrupted) view with a vertical break. This represents the Long Rod with a constant cross section. Dimensions associated with the Broken view reflect the actual model values.

Add a Revolved Section to represent the cross section of the ROD.

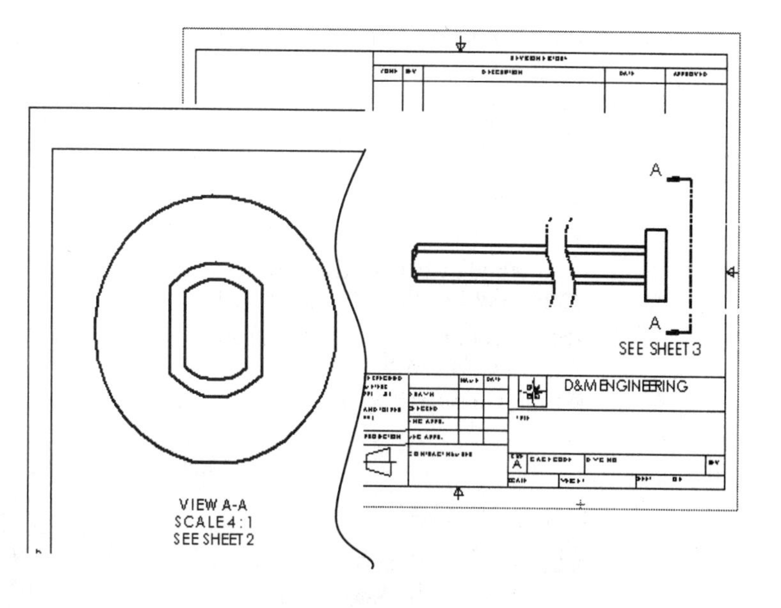

ROD-Sheet3 ROD-Sheet2

Add ROD-Sheet3. Create the Back view by using a Removed view on ROD-Sheet2. Copy the Back view from ROD-Sheet2 to ROD-Sheet3.

Create view arrows to indicate that the view has been removed when a view cannot be placed in an aligned position.

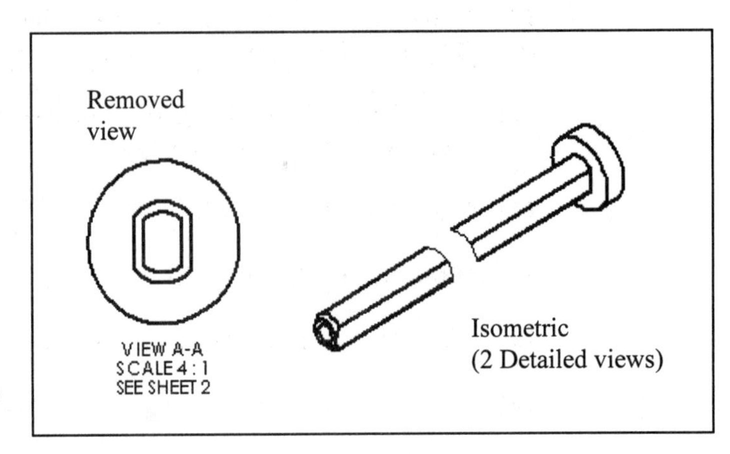

ROD-SHEET3
Long Rod Configuration

Use capital letters, Example: VIEW A-A to relate the viewing plane to the Removed view.

Note: Place Removed views on the same sheet as the parent view when room exists on the drawing sheet.

Indicate sheet name below the view name when a Removed view does not fit on the same sheet.

Combine the two Detail views to construct the Isometric view.

Add Sheet2 to the ROD drawing.

72) Right-click on the **Sheet1** tab.

73) Click **Add Sheet**.

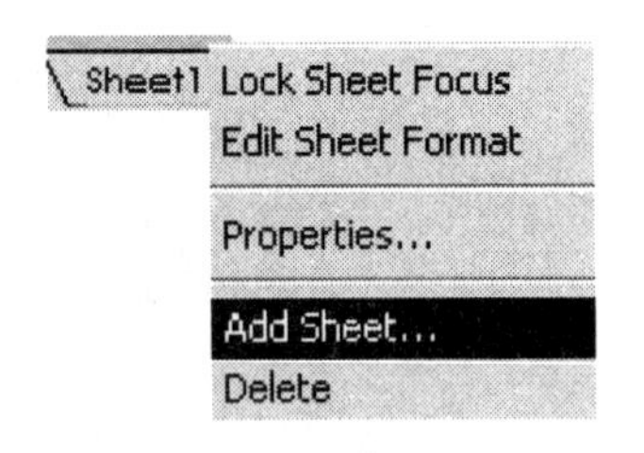

74) The Custom A-FORMAT is the default Sheet Format. Click **OK**. Sheet2 is displayed.

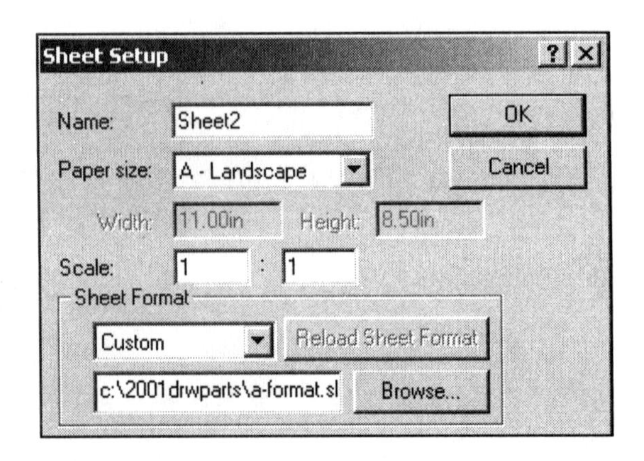

Add Views to Sheet2.

75) Copy the Front view from Sheet1 to Sheet2. Click the **Sheet1** tab.

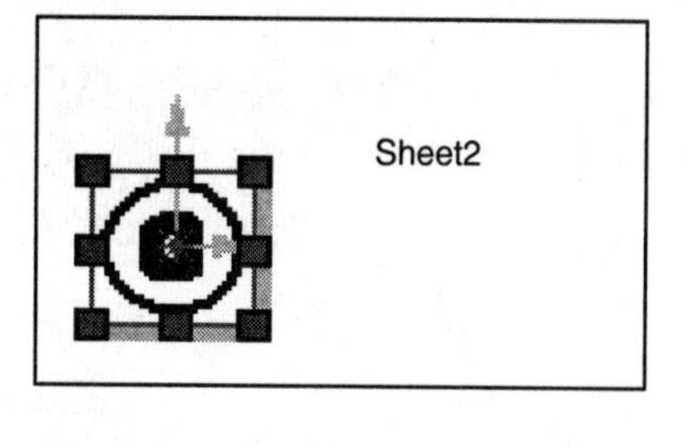

76) Click **Drawing View1** (Front). The view boundary is displayed in green.

77) Copy the view. Press **Ctrl C**.

78) Click the **Sheet2** tab.

79) Click the view **position** in the lower left corner of the Graphic window.

80) Paste the view. Press **Ctrl V**.

81) Click the copied **view boundary**.

82) Check **Custom Scale**.

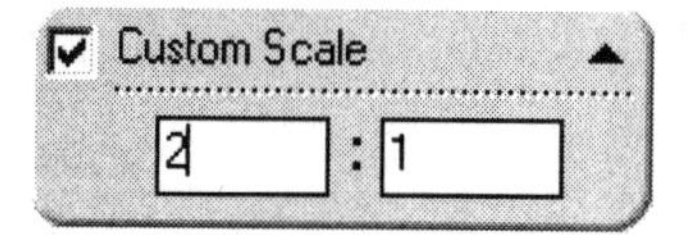

83) Enter **2:1**.

84) Modify the ROD configuration. Right-click **Properties** in the Drawing View5(Front), Sheet2.

85) Select **Long Rod** from the Use Named Configuration drop down list.

86) Click **OK**.

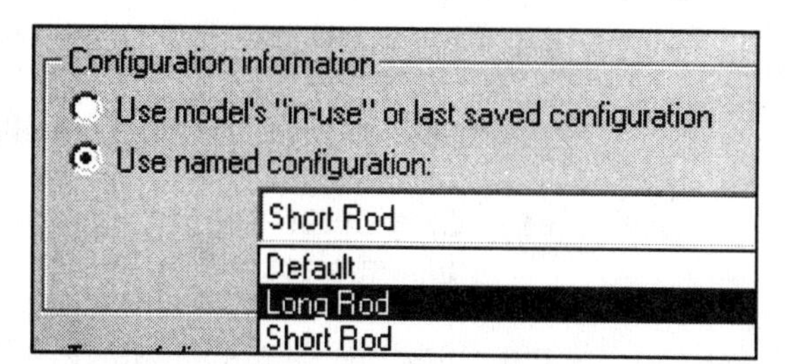

87) Create a Projected view. Click **Projected View** from the Drawing toolbar.

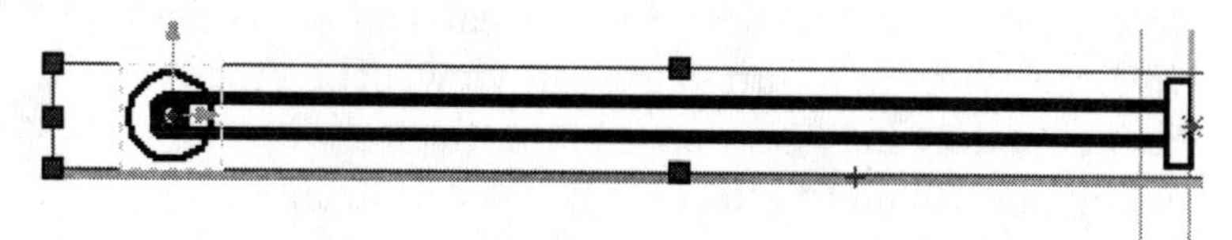

88) Drag the **mouse pointer** to the right of Front view.

89) Click a **position** to the right of the Front view.

90) Press the **z** key until the entire Right view is displayed.

91) Create a Vertical Break for the Right view. Click the Right view **boundary**.

92) Click **Insert, Vertical Break** from the Main toolbar. Two vertical break lines are displayed.

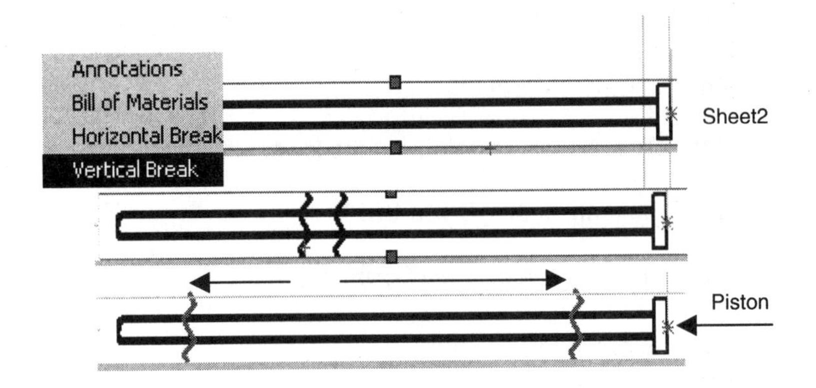

93) Drag the **left vertical break line** towards the Internal Thread.

94) Drag **the right vertical break line** towards the Piston.

95) Right-click inside the Right view **boundary**.

96) Display the Break view. Click **Break View**.

97) Right-click on the left **Break line**.

98) Click **Curve Cut** to display the curved break line.

99) Drag the **Break line** to the left to adjust the break.

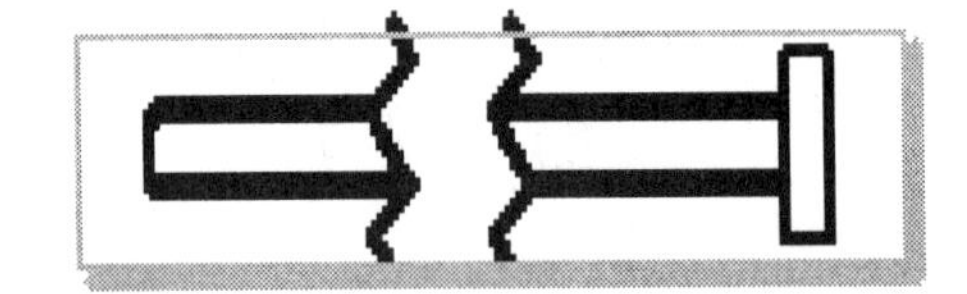

100) Drag the view **boundary** to the right of the Front view.

Note: Right-click on the Vertical Break and select the Un-Break view to return to the full view. The green view boundary may overlap other projected views making it difficult to select a projected view in the Graphics window.

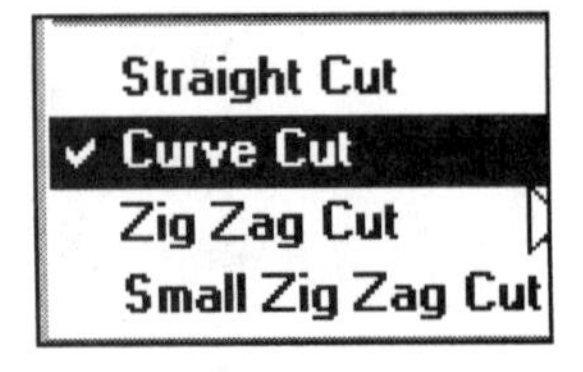

Select a projected view from the FeatureManager.

Modify Break Lines display. Right-click on the Break line. Select Straight Cut, Curve Cut, Zig Zag Cut or Small Zig Zag.

Utilize Tools, Options, Document Properties, Line Font to modify the Break Lines Font.

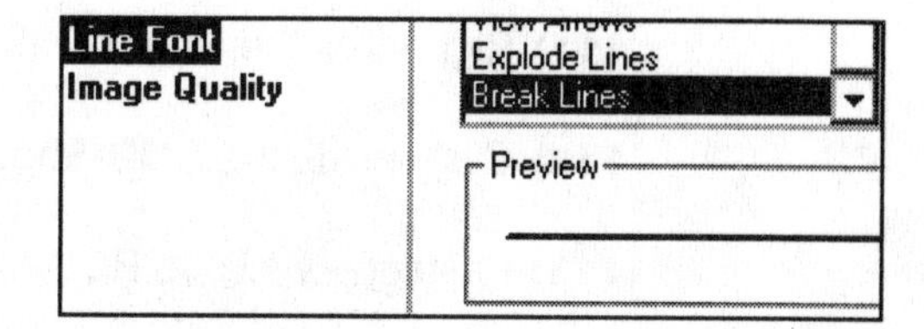

101) Save the ROD drawing. Click **Save**.

Create a Removed view.

102) Click the Right view boundary.

Click **Section** 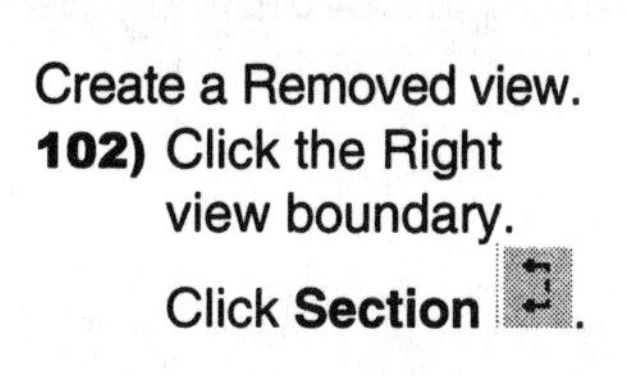.

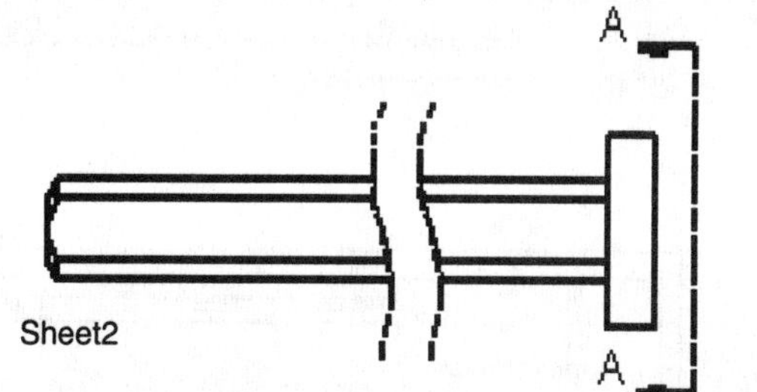

Sketch Section line outside the Right view boundary.

103) Sketch a vertical section **line** to the right of the Vertical Break Right view.

104) Flip the **arrows** if required.

105) Hold the **Ctrl** down key. Click a **position** above the Front view. Release the **Ctrl** key.

106) Click **Custom Scale**.

107) Enter **4:1**.

The Ctrl key prevents the Removed view from being aligned to the Vertical Break. Place the Removed view on ROD-Sheet3.

The Removed view depends upon the Vertical Break.

Copy both views to ROD-Sheet3. Place the Vertical Break to the right of the ROD-Sheet3 boundary.

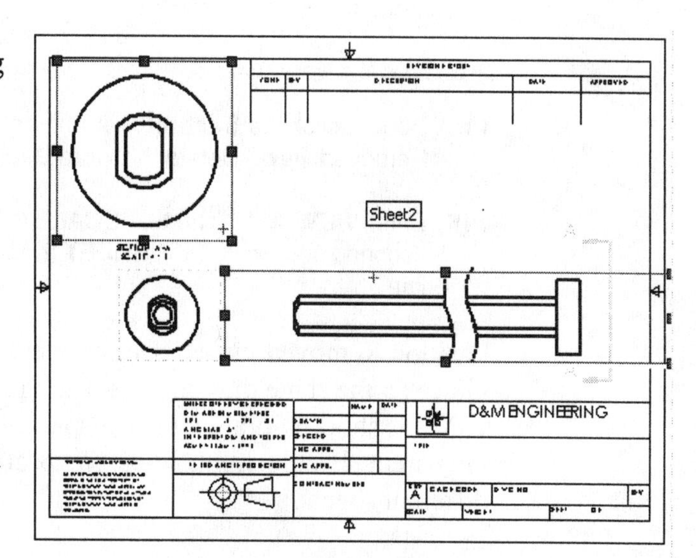

Switch Removed view from ROD-Sheet2 to ROD-Sheet3.

108) Press the **Ctrl** key. Select the **Vertical Break Right view**.

109) Select the **Removed view**.

110) Copy the views. Press **Ctrl C**.

111) Right-click the **Sheet2 tab**. Click **Add Sheet**. Click **OK.**

112) Click inside the **ROD-Sheet3 boundary**.

113) Paste the views to ROD-Sheet3. Press **Ctrl V**.

Note: Click inside the Sheet boundary to paste a view.

114) Zoom out. Press the **z** key 4 times. Drag the **Vertical Break Right view** off ROD-Sheet3 to the right of the title block. Drag the **Removed view** to the left of the Sheet boundary.

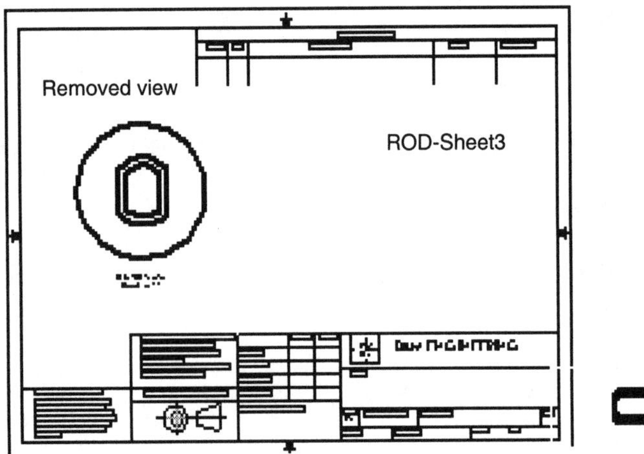

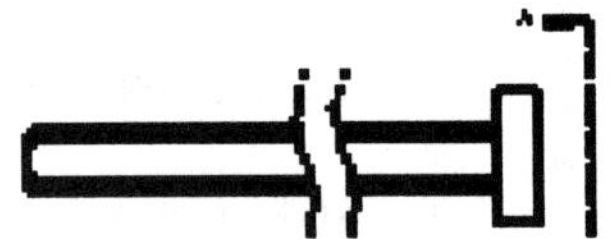

115) Double-click the **Section A-A** text below the Removed view. Delete the word **Section**.

116) Enter **VIEW A-A**. The text SCALE 4:1 is on the second line. Enter **SEE SHEET 2** on the third line.

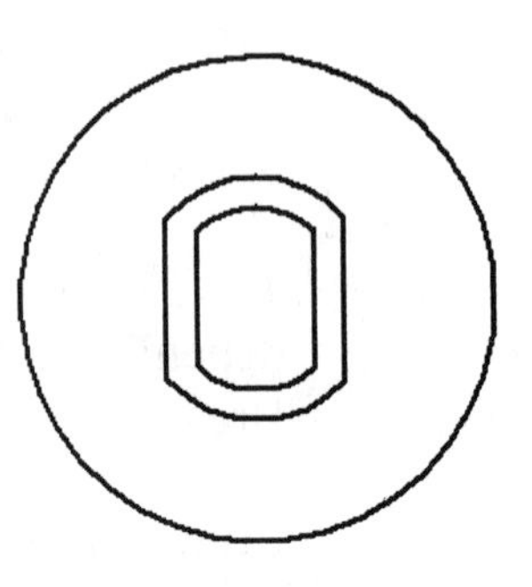

Position Removed views, Section views and Detail views on the same drawing sheet as the parent view. If short on drawing sheet space, document the parent sheet location when views are placed on different sheets.

Sheets larger than B size contain Zone letters and numbers along with their margins. The Zone location is placed below the “SEE SHEET #” text for drawing sizes larger than B.

Example: SEE SHEET 2
ZONE A2

117) Click the **Sheet2 tab**. Right-click the Removed view **boundary**.

118) Click **Delete**.

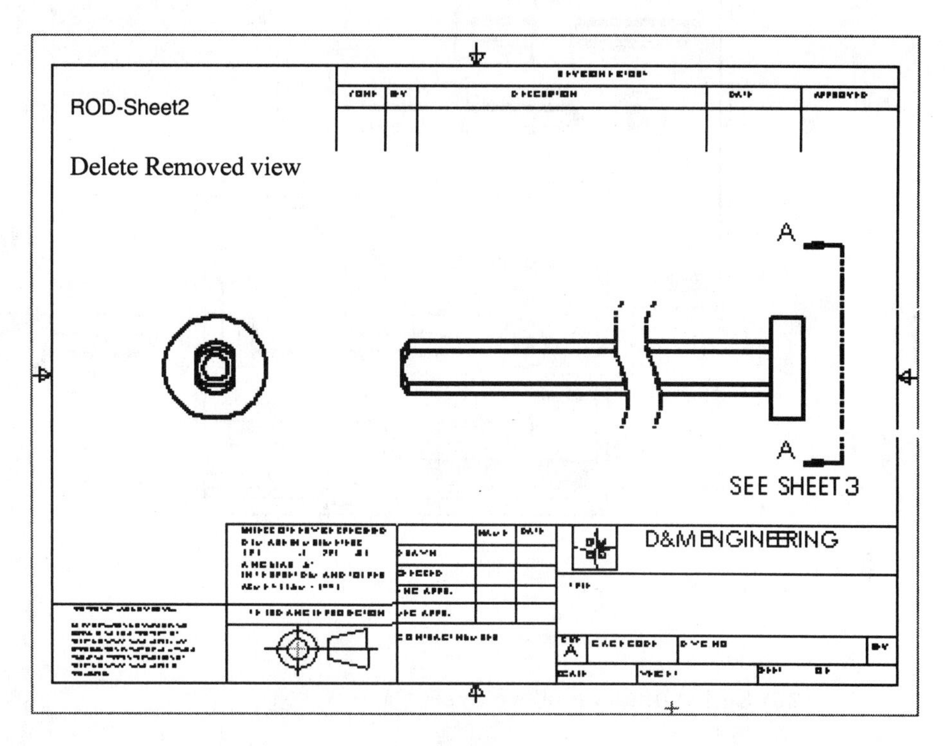

119) Click **Yes**. DO NOT DELETE ABSORBED FEATURES. The Section Line A-A remains on Sheet 2.

120) Click **Note** A.

121) Click a **position** below the View A-A arrows.

122) Enter SEE SHEET3.

Create a Revolved Section.

123) Increase the Break view gap to place the Revolved section. Click **Tools**, **Options**, **Document Properties**.

124) Enter **25** for Break line gap.

125) Click **Rebuild**.

126) Zoom in on the Graphics window. Press the **z key** until Sheet2 is approximately ½ its original size.

127) Copy the Vertical Break Right view. Click the **Vertical Break view boundary**.

128) Click **Ctrl C**.

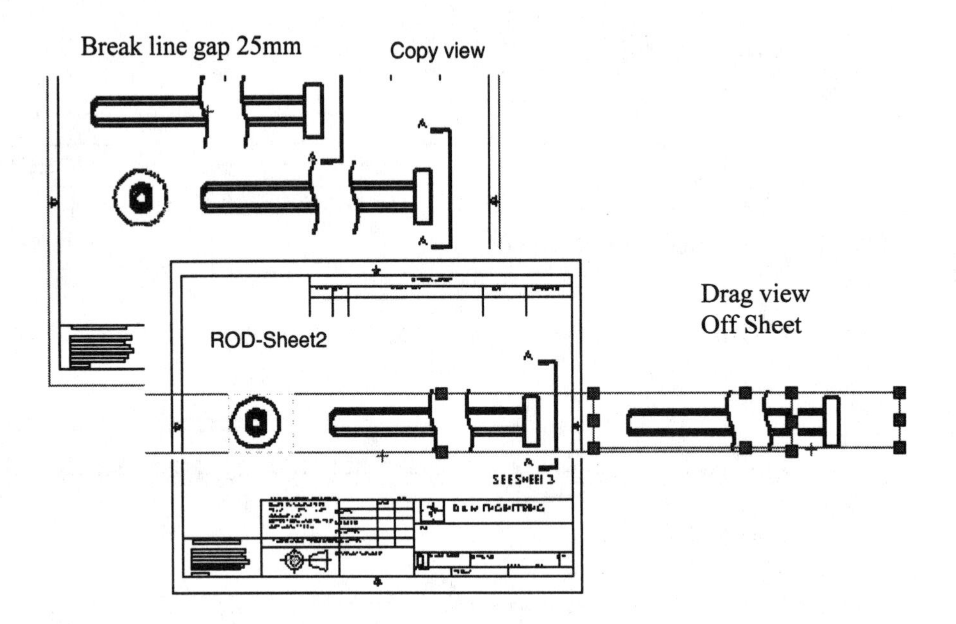

129) Click a **position** above the Right view.

130) Paste the view. Click **Ctrl V**.

131) Drag the copied **view** off the sheet boundary to the right of the Vertical Break Right view.

132) Resize the view boundaries. Drag the green view **boundaries** 1 - 2mm from the visible lines of the Right view.

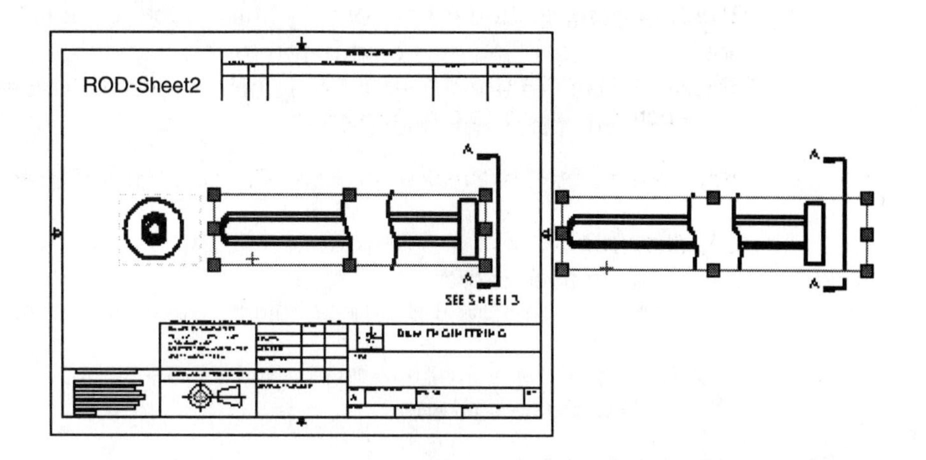

Align Horizontal by Origin
Align Vertical by Origin
Align Horizontal by Center
Align Vertical by Center

133) Align the views. Right-click the **copied Right view**.

134) Click **Alignment**.

135) Click **Align Horizontal by Origin**. The mouse pointer displays the Alignment icon .

136) Click the **Right view**. The copied Right view aligns to the Right view.

137) **Zoom** area to display the 2 Broken views.

138) Create a Revolved Section with a Section view. Click the **copied Right view boundary**.

139) Click **Section** . Sketch a **vertical line** to the left of the copied Vertical Break.

140) **Flip** the arrows to the left if required.

141) Place the **cross section** between the Vertical Break Lines.

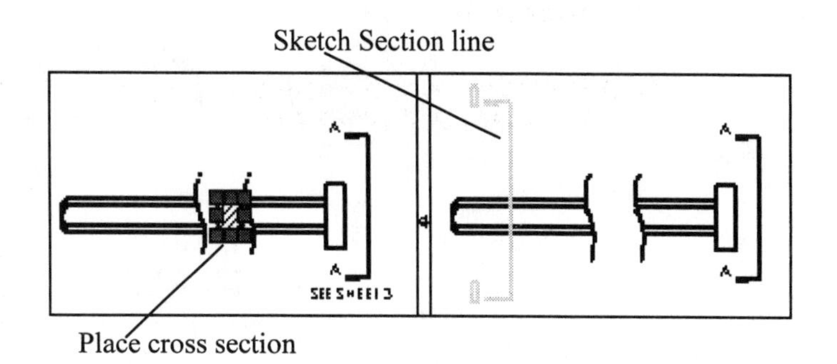

The Revolved Section is horizontally aligned to the Right view. The view boundary is small and may be difficult to select and move. Select the Section view from the FeatureManager. Lengthen the view boundary. Center the view between the Break line gap.

Center the Revolved Section.

142) Click **Section View B-B Sheet2** from the FeatureManager.

143) Drag the green **view boundary** downward, 5 – 10mm below the Vertical Break Right view.

144) Drag the bottom line of the **green view boundary** to center the Revolved Section.

145) **Delete** the Section B-B text.

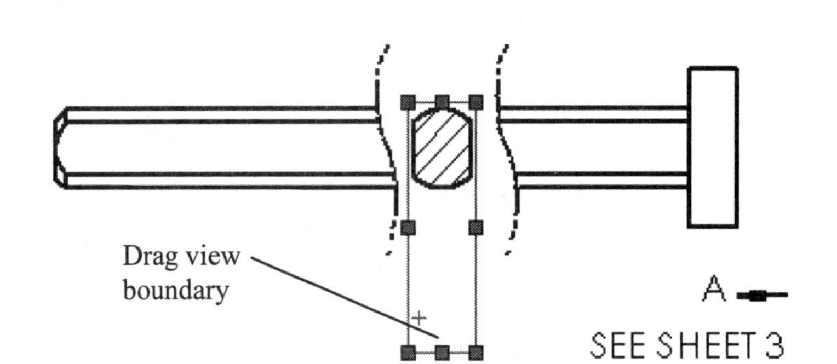

146) Drag the **SECTION B-B** text off the sheet boundary.

147) Save the ROD drawing. Click **Save**.

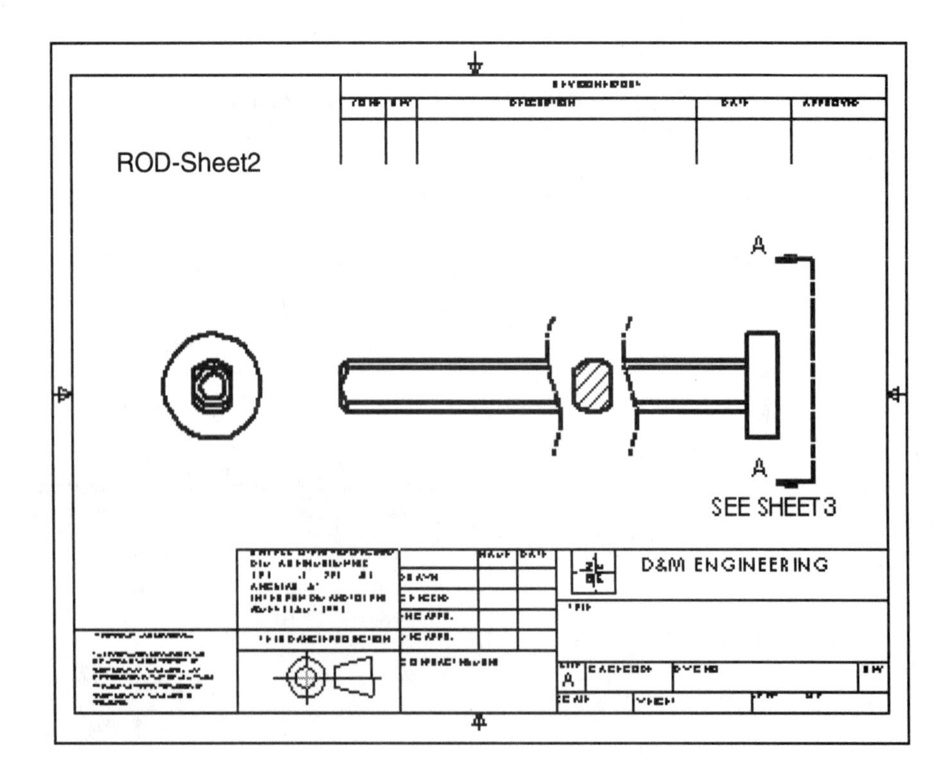

Create a broken Isometric view. Create a front Detail view and then a back Detail view.

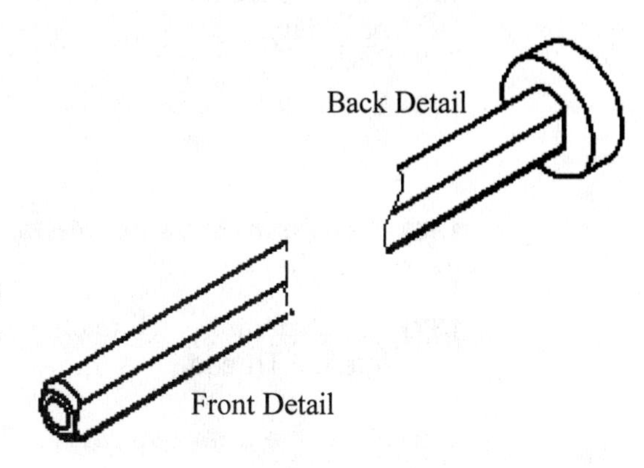

Align the two Detail views with a Sketched Line. The Sketched Line is created on a Drawing Layer.

Select the Detail views and Sketched Line together in order to move the geometry together.

Create Layers in a drawing document for construction geometry and dimensions.

Specify line color, thickness and line style. New entities are added to the active layer. Turn layers on/off to simplify drawings. Shut layers off when not in use.

Create a Broken Isometric view.

148) Click the **SHEET3 tab**.

149) Click **Zoom in** on the Sheet boundary.

150) Press the **z key** until ROD-Sheet3 is approximately ½ it's original size.

151) Create a Named Isometric view. Click **Removed view boundary** in ROD-Sheet3.

152) Click **Named View** from the Drawing toolbar.

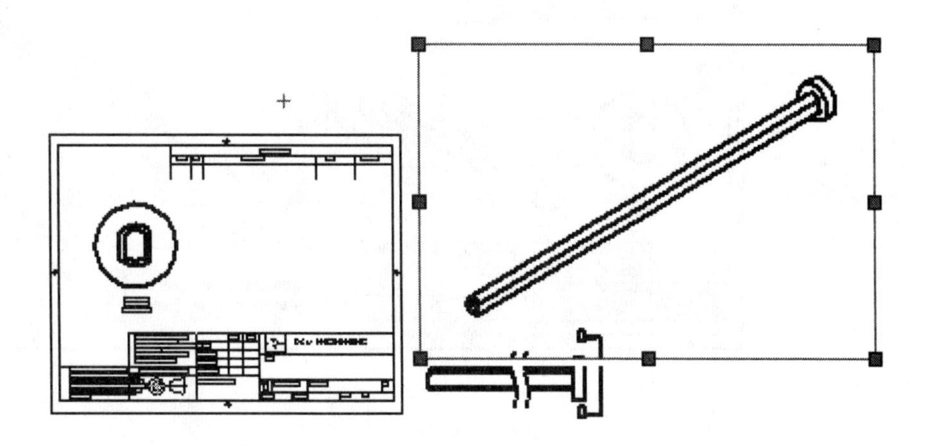

153) Click **Isometric** from the View Orientation text box.

154) Click the Isometric view **position** above the Right view, off the ROD-Sheet3 boundary.

155) Create the first Detail view. The Isometric view boundary is displayed in green. Enlarge the view.

156) Click **Zoom to Selection**.

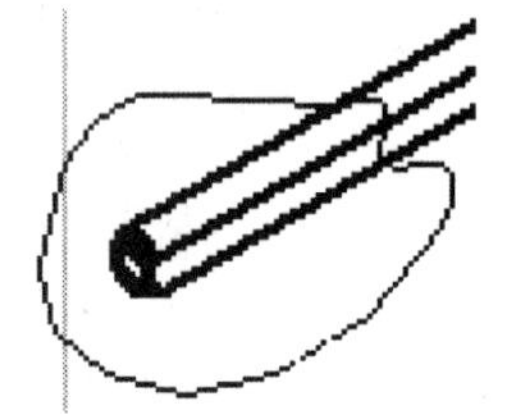

157) Click **Spline**. Sketch a closed **Spline** around the Internal Thread.

158) Press the **z** key two times.

159) Click **Detail View** from the Drawing toolbar. Drag the mouse pointer into ROD-Sheet3. Click a **position** to the right of the Removed view.

160) Create the second Detail view. Click **Spline**. Sketch a closed **Spline** around the Internal Thread.

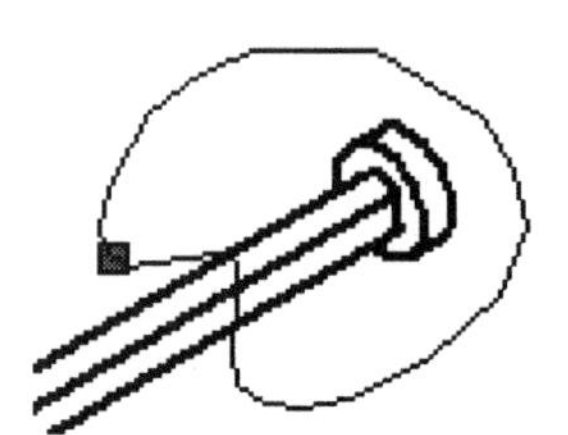

161) Click **Detail View** from the Drawing toolbar. Drag the mouse pointer into ROD-Sheet3.

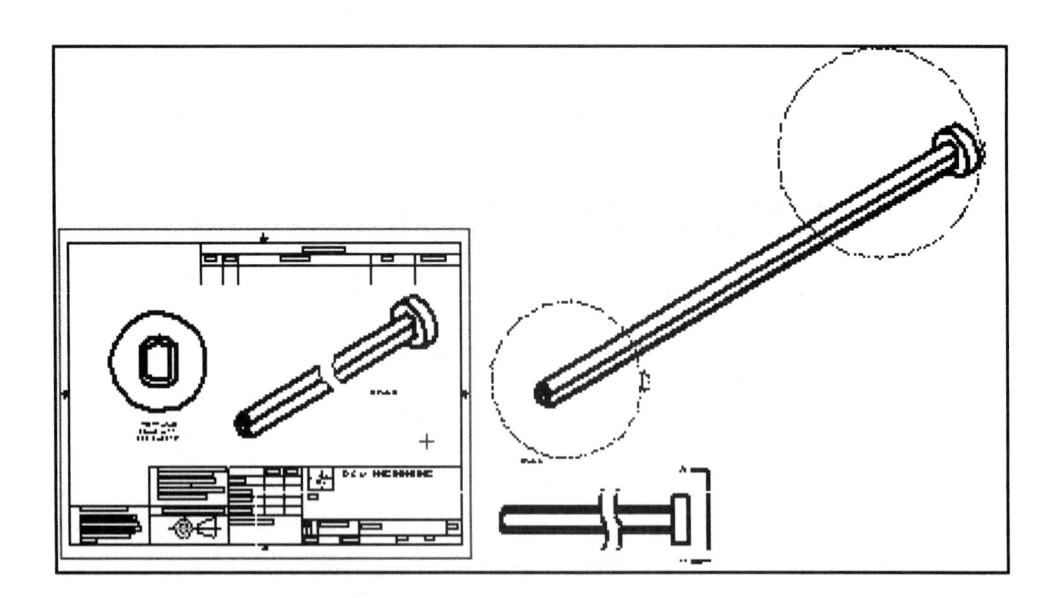

162) Click a **position** to the upper right of the first Detail view.

163) Drag the **Detail A** text and the **Detail B** text from the sheet boundary to the Isometric view.

Note: Detail A and Detail B text depend upon the Isometric view. Do not delete the Isometric view. Sketch the Spline back portion in Detail A similar to the Spline front portion in Detail B.

Align Detail A and Detail B with a sketched construction line. The construction line is created with the Line sketch tool, on a new layer, with Dashed Style.

Create a New Layer.

164) Display the Layer toolbar. Click **View**, **Toolbars**, **Layer**.

165) Display the Layers dialog box. Click **Layer Properties** .

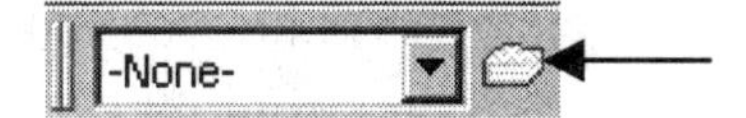

166) Create a new Layer. Click the **New** Button.

167) Enter **Construction** for Name.

168) Enter **Construction View Lines** for Description. Note: The Layer is on when the Light Bulb is yellow. Enter **Red** for Color. Select **Dashed** for Style. Click **OK**.

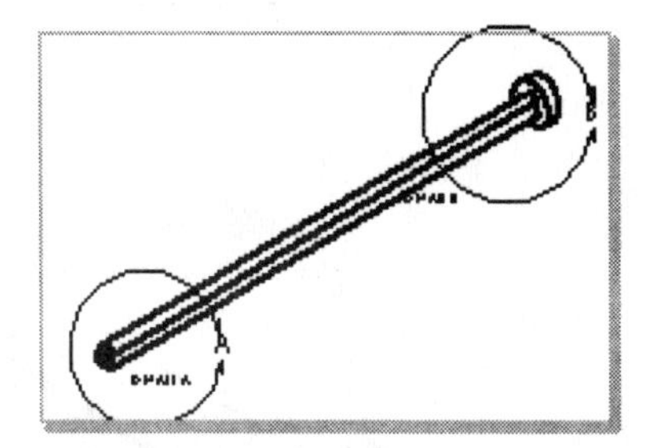

Note: The other Layers names were created from the AutoCAD Sheet Format in Project 1.

1

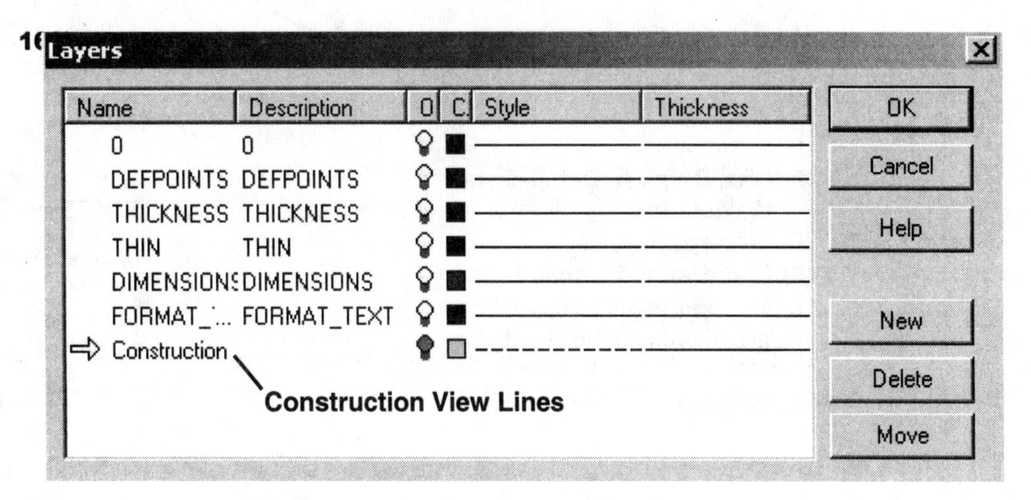

Click **Zoom in** on Detail view A and B.

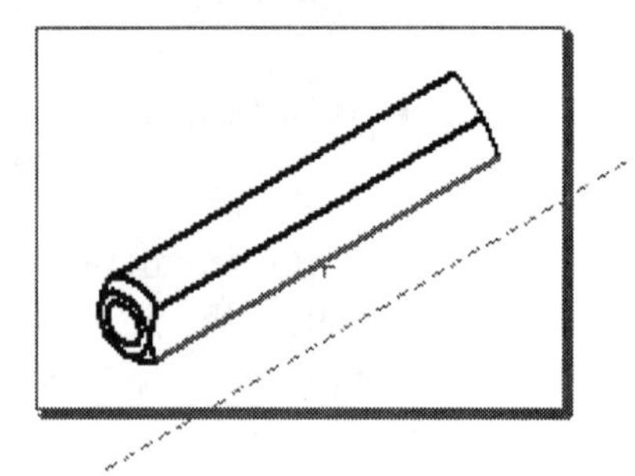

170) Sketch a **Line** along the lower profile line on Detail A. The line is displayed in red.

Add a Relation. Hold the **Ctrl** key down. Click the **right edge** of Detail view B. Click **Collinear**. Release the **Ctrl** key.

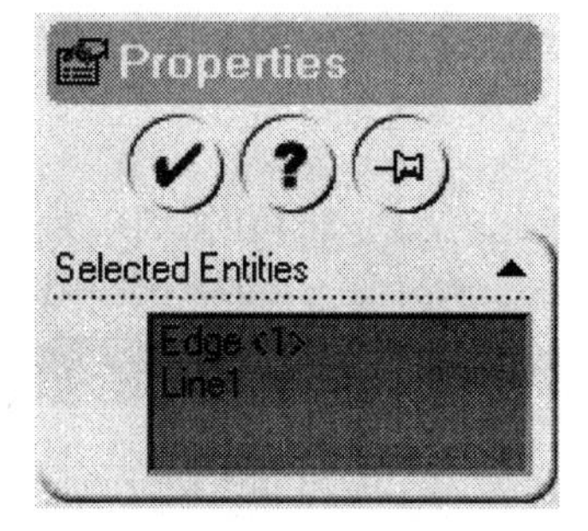

171) Move the Broken Isometric view. Hold the **Ctrl** key down.

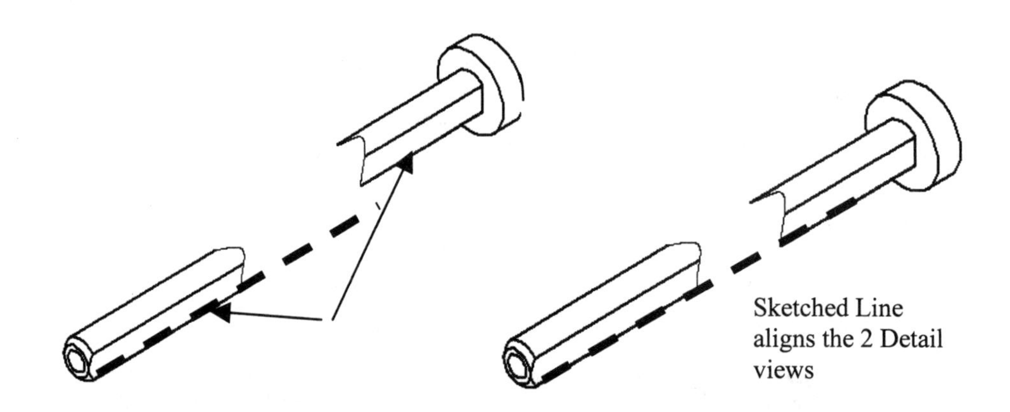

172) Click **Detail A**, **Detail B** and the red **Line**.

173) Drag **Detail A**, **Detail B** and the red **Line** to a new location directly above the Right view.

174) Release the **Ctrl** key.

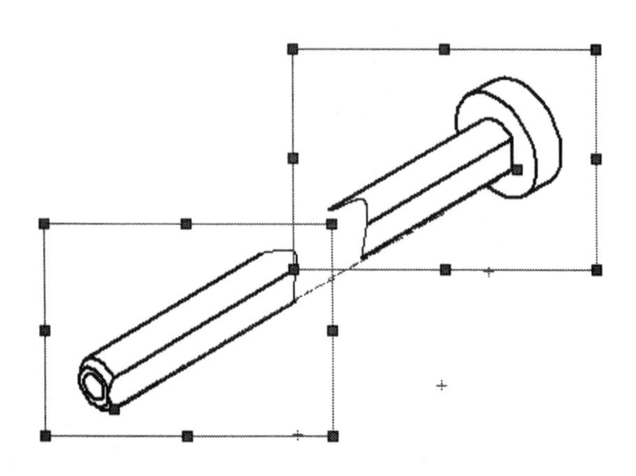

175) Hide the red Line. Click **Layer Properties**.

176) Turn off the Construction Layer. Click **Light Bulb** .

177) Click **OK**.

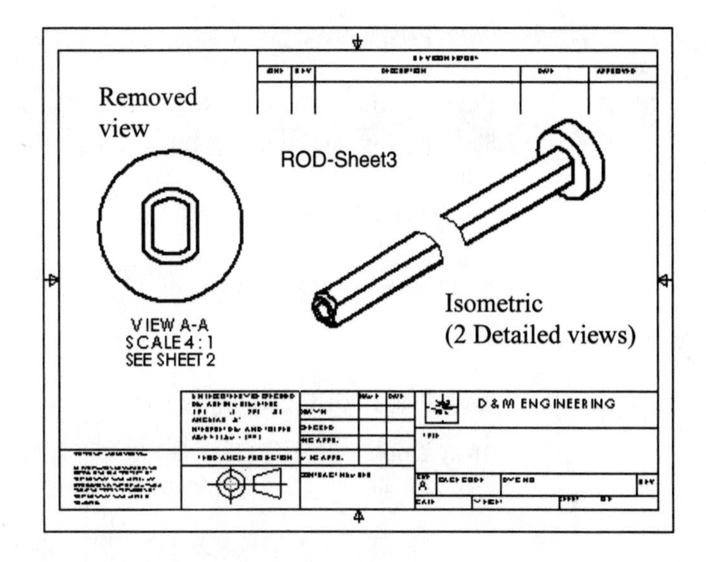

178) Return to the default Layer. Click **None** from the Layer toolbar.

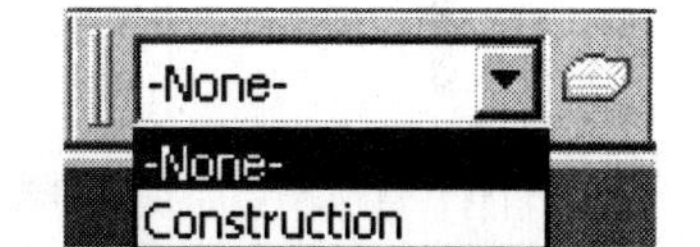

179) Save the ROD drawing. Click **Save**.

180) Close all parts and drawings. Click **Windows**.

181) Click **Close All**.

Note: Broken views can be copied and unbroken.

The exercise is developed at the end of the Project.

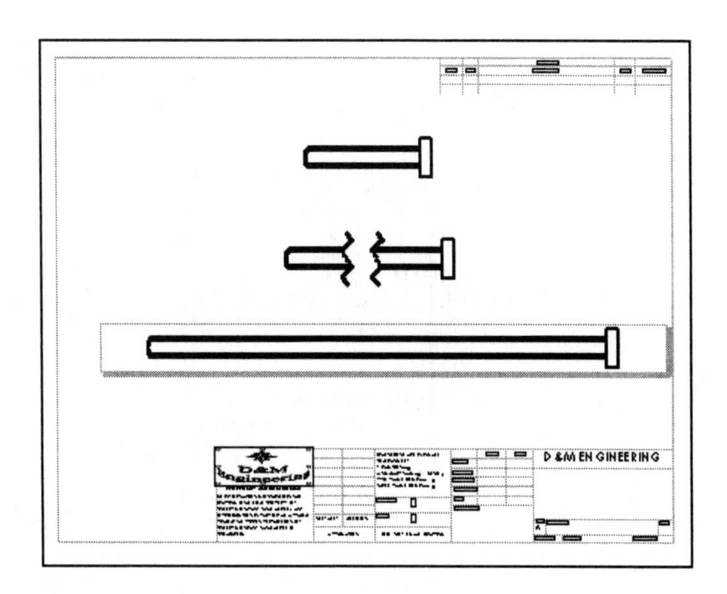

Review the TUBE part

Perform the following recommended tasks before starting the TUBE drawing:

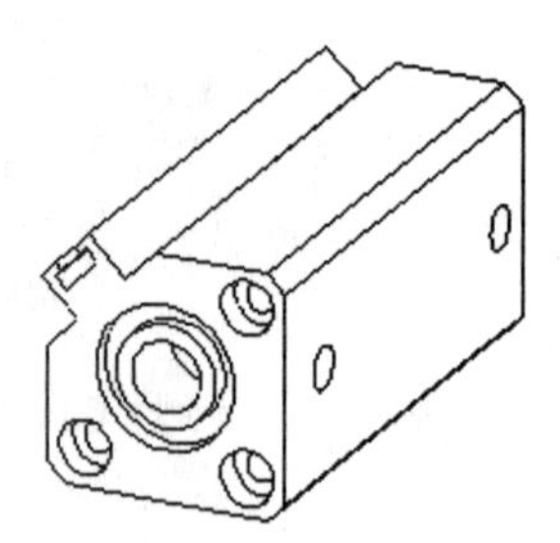

- Verify the TUBE part.
- View dimensions in each feature.

Review the TUBE part.

182) Open the part. Click **Open** from the Standard toolbar. The TUBE part is located in the 2003drwparts folder.

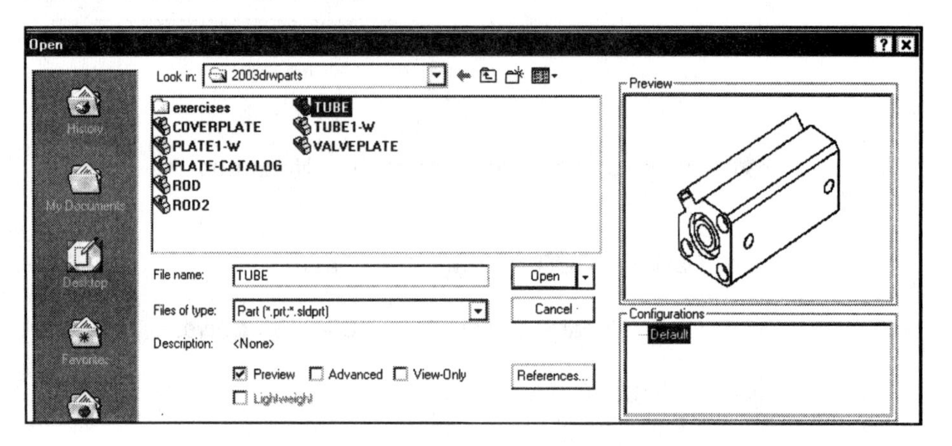

183) Click the **Open** button. The TUBE is displayed in the Graphics window.

184) Expand the SolidWorks window to full screen. Click the **Maximize** icon in the top right hand corner of the Graphics window. The FeatureManager lists 12 individual features.

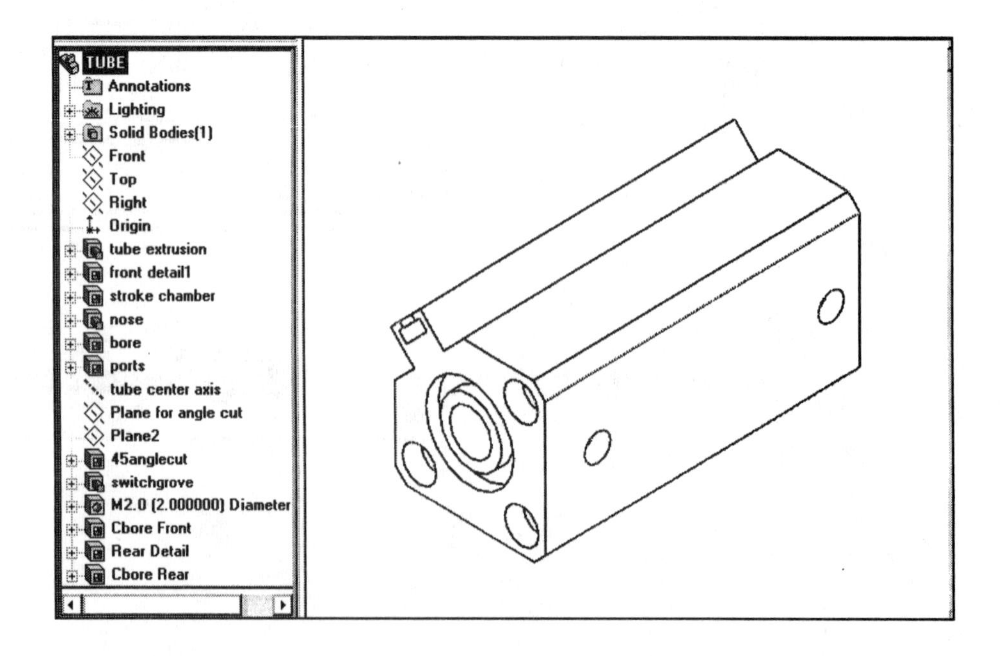

185) Click **Hidden Line Visible** . Internal features are displayed.

186) Display **Hidden Lines Removed** . Internal features are hidden.

Review the TUBE part features.

187) Position the Rollback bar. Place the **mouse pointer** over the yellow Rollback bar at the bottom of the FeatureManager design tree. The mouse pointer displays a symbol of a hand.

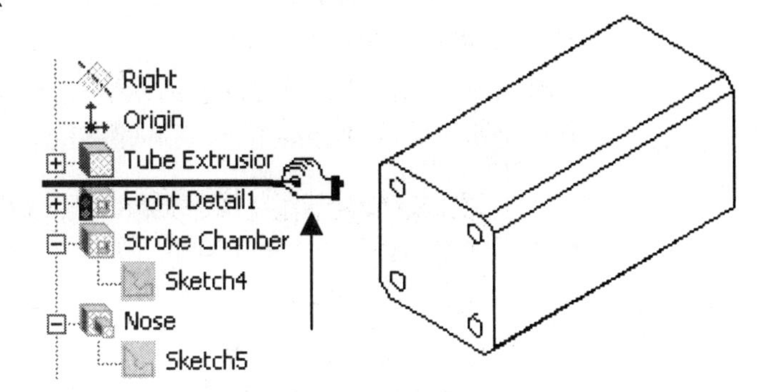

188) Drag the **Rollback** bar upward to below the Tube Extrusion.

189) Double-click on the **Tube Extrusion** to display feature dimensions.

190) Drag the **dimension text** away from the feature.

191) Flip the **dimension arrows** to clearly view the text.

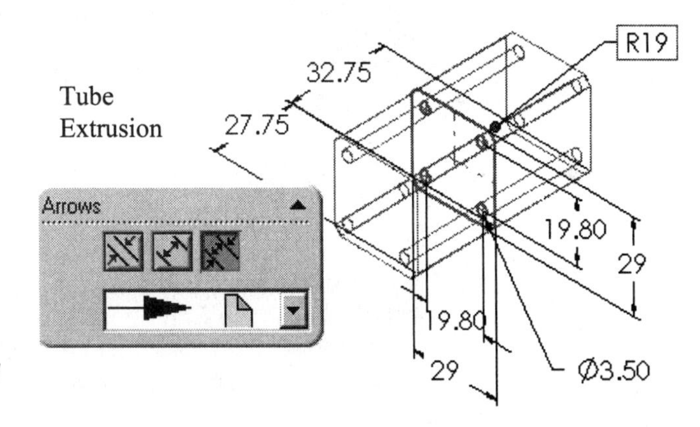

Position dimension text off the feature before inserting the part into the drawing.

Dimension schemes defined in the part require changes in the drawing. The profile is sketched on the Front plane.

Design engineers can use different dimension schemes than those required by manufacturing engineers.

Example, the design engineer referenced the depth dimensions 27.75mm and 32.75mm to the Front plane. The Front plane is referenced for analysis calculations.

The manufacturing engineer requires an overall depth of 60.50mm (27.75mm + 32.75mm) referenced from the front face. Create the overall depth dimension of 60.50mm in the drawing.

192) Drag the **Rollback** bar downward below the Front Detail1 feature.

193) Display feature dimensions. Double-click on **Front Detail1**.

194) Drag the **dimension text** away from the feature. Front Detail1 was created with an Extruded Cut. A 17mm circle was sketched at the coincident to the Origin on the front face.

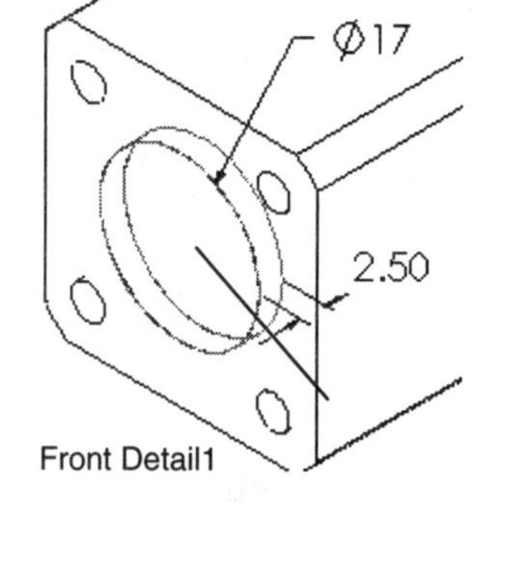

195) Drag the **Rollback** bar downward below the Stroke Chamber feature.

196) View the feature. Click **Hidden Line Visible**.

197) Display the feature dimensions. Double click on the **Stroke Chamber**.

198) Drag the **dimension text** away from the feature. The Stroke Chamber feature was created with an Extruded Cut sketched on the front plane and extruded in two directions.

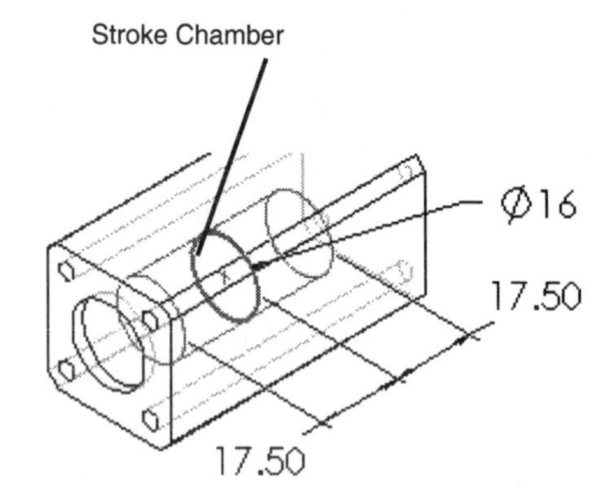

The design engineer referenced the depth dimensions to the Front plane for the Stoke Chamber (Depth1 = 17.50mm and Depth2 = 17.50mm).

The manufacturing engineering requires an overall depth of 35.00mm.

The Stroke Chamber is an internal feature; a Section view is required in the drawing. Stroke Chamber dimensions are displayed in the drawing Section view.

199) View the Boss. Drag the **Rollback** bar downward below the Nose feature.

200) Display the feature dimensions. Double-click on the **Nose**. The Nose depth is 2mm. The Nose feature requires a Detail view.

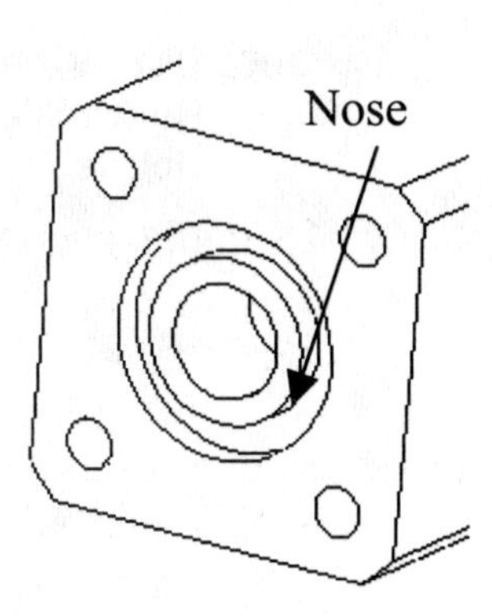

201) Review the Internal Cut. Drag the **Rollback** bar downward below the Bore feature. Display the feature dimensions.

202) Double-click on the **Bore**. The Cut requires a Section view. The circular sketch profile is extruded on both sides of the front plane with two different Depth options.

203) Drag the **dimension text** away from the feature.

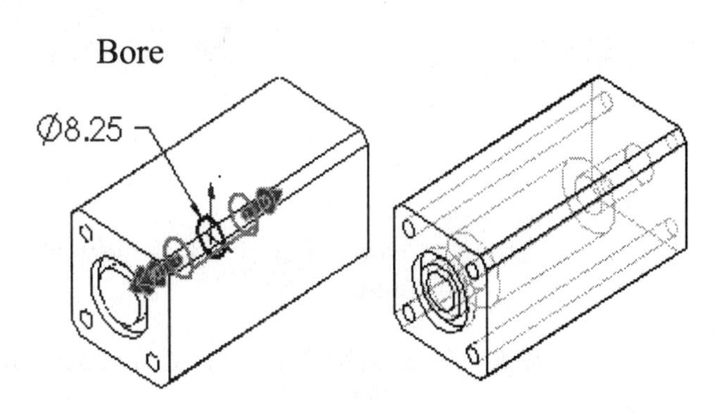

204) Display the Port location. Drag the **Rollback** bar downward below the Port feature. Display the feature dimensions.

205) Double click on the **Port**. The design engineer required the port distance from the Front Plane (17.75mm + 27.25mm). These dimensions will be redefined in the drawing for manufacturing and customer requirements (45.00mm).

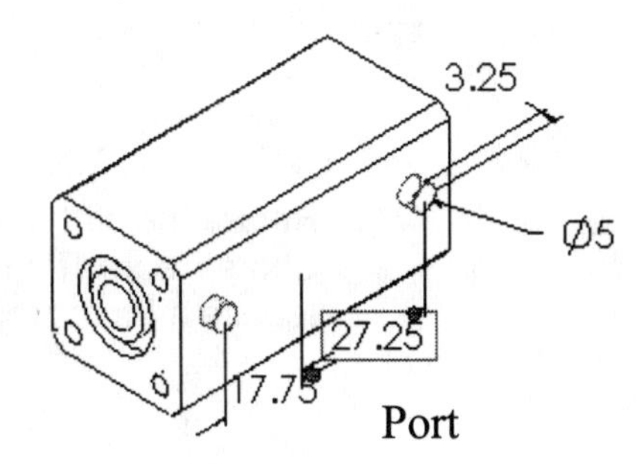

206) Display the 45 AngleCut, SwitchGrove and M2 Hole respectively. Drag the **Rollback bar** downward. The Switch Groove requires a Detail view. The M2 Hole requires an Auxiliary view to display these feature dimensions.

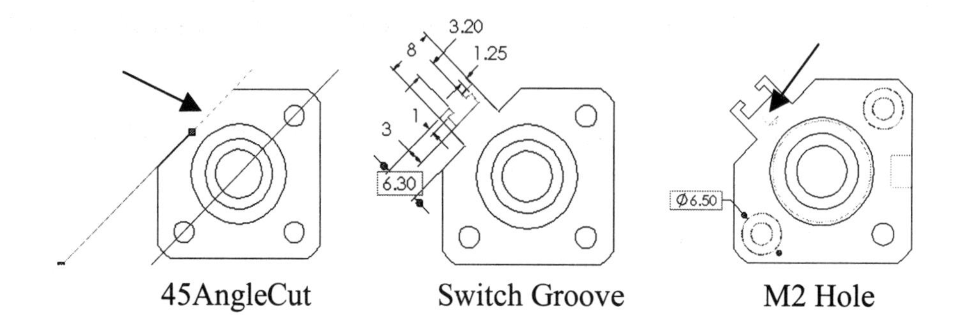

207) Drag the **dimension text** away from the features.

208) Display the Cbore Front. Drag the **Rollback bar** downward.

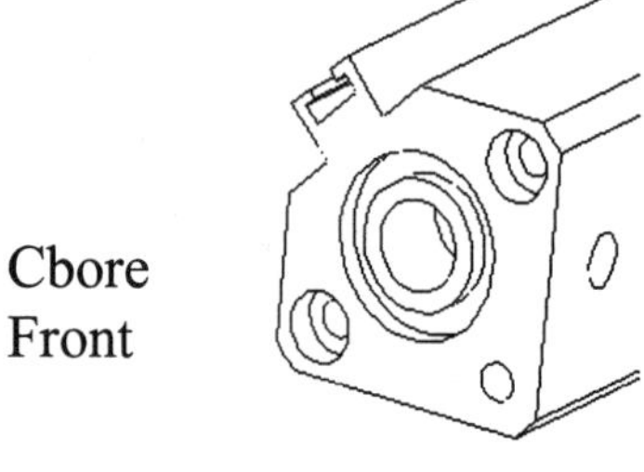

209) **Rotate** the part to view the Rear Detail and Cbore Rear. The Rear Detail and Cbore Rear require a Back projected view in the drawing.

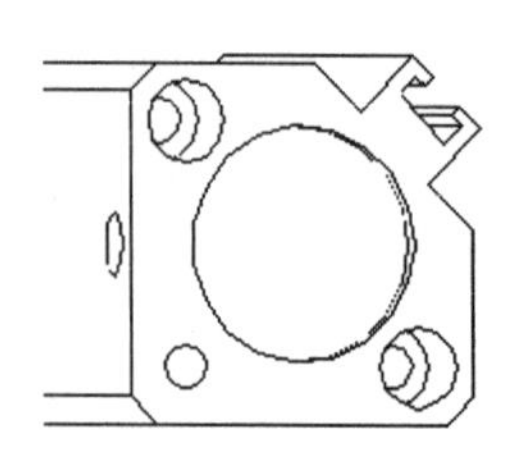

Cbore Rear
Rear Detail

Create the TUBE drawing

Create the TUBE drawing. The TUBE drawing consists of a single drawing sheet with eight views.

The eight views are: three Standard views, a Projected Back view, Section view, Detail view, Auxiliary view and Half Section Isometric (Cut away) view.

The Half Section Isometric view requires two part configurations named: Entire Part and Section Cutaway.

Use an A-ANSI-MM drawing template.

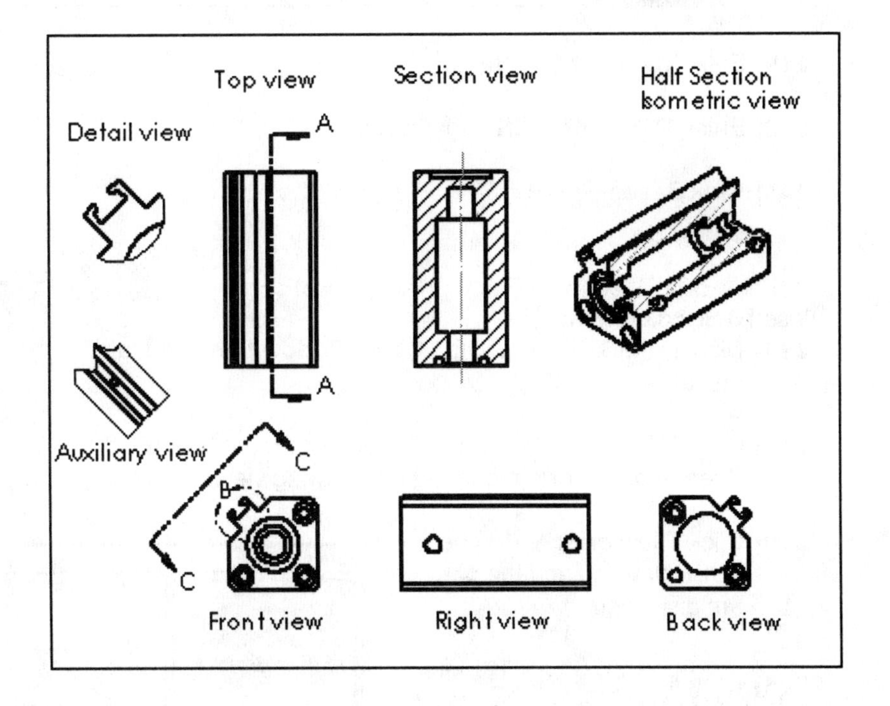

Create the TUBE drawing.

210) Click **New** .

211) Click the **2003drwparts** tab.

212) Click **A-ANSI-MM** from the SolidWorks Document dialog box.

213) Click **OK**.

214) Save the empty drawing. Click **File**, **Save As**.

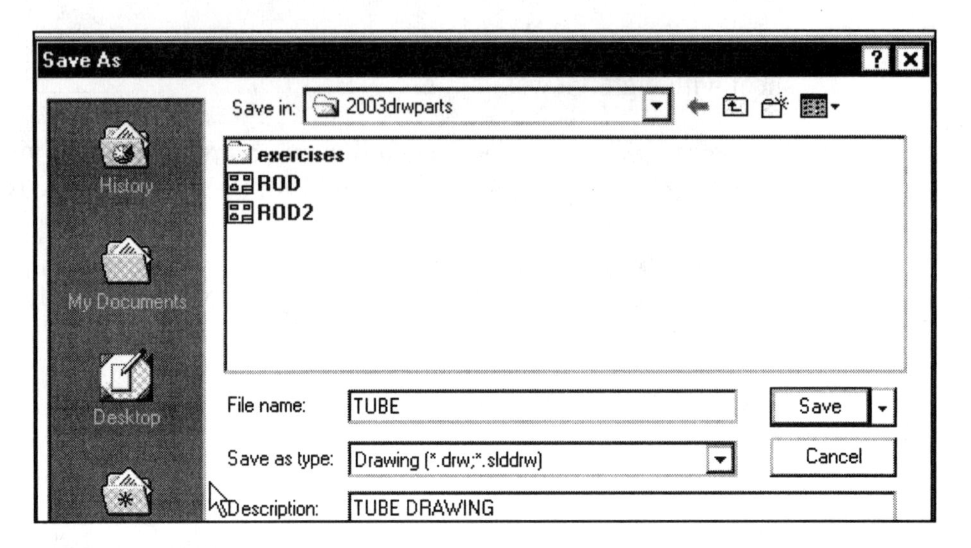

215) Enter **TUBE** for File name.

216) Enter **TUBE DRAWING** for Description..

217) Click **Save**.

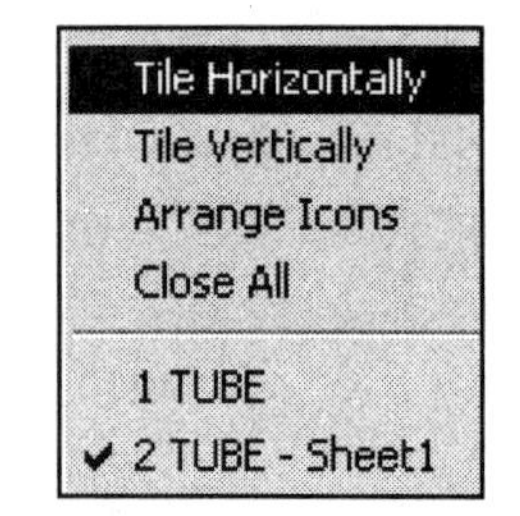

Insert 3 Standard views.

218) Display the TUBE drawing. Click **Window** from the Main menu.

219) Click **TUBE-Sheet1**. Note: The drawing must be in Edit Sheet mode in order to insert the drawing views.

220) Click **Tile Horizontal**. The TUBE drawing and the part are displayed.

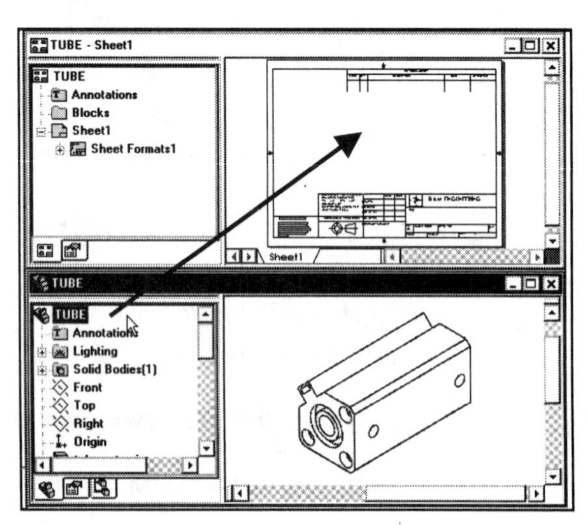

221) Drag the **TUBE Part** tube icon from the Part Feature Manager into the drawing Graphics window. The Front, Top and Right views are displayed in the drawing.

222) **Maximize** the drawing.

Observe the 3 Standard views on the A size drawing. Five other views are required to document the TUBE part.

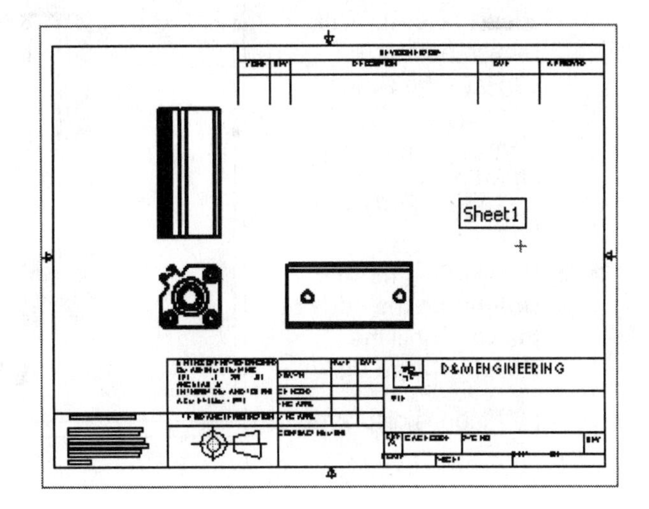

There is not enough space on an A size drawing.

You have two options: enlarge the sheet size or move additional views to multiple sheets. How do you increase the sheet size? Use Sheet Properties.

Modify Sheet Properties.

223) Right-click in the **Graphics window**.

224) Click **Properties**. The Sheet Setup dialog box is displayed.

225) Select **C-Landscape** for Paper size.

226) Enter **2:1** for sheet Scale.

227) Select **Custom** from the Sheet Format drop down list.

228) Click the **Browse** button.

229) Select the **C-FORMAT** for Sheet Format name from the 2003drwparts file folder.

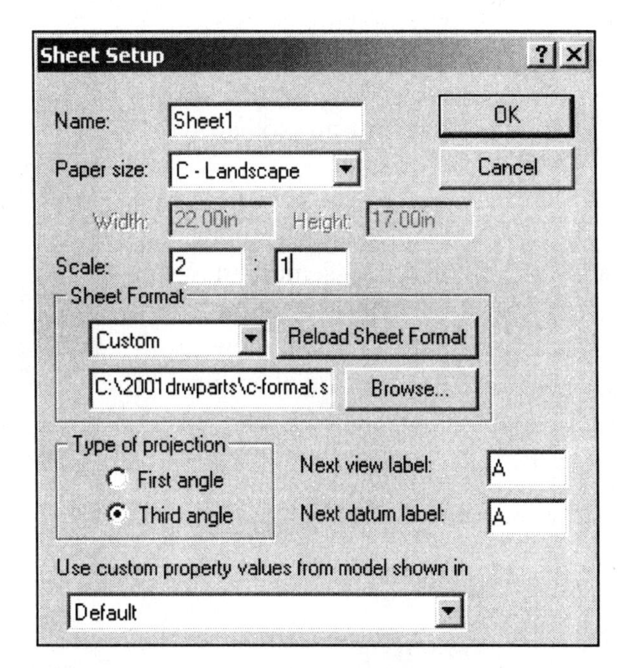

230) Click **OPEN**. Click **OK**. The new C-size sheet format is displayed.

Move the TUBE Drawing views.

231) Click the view boundary of **Drawing View1** (Front). The mouse pointer displays the Drawing View icon. The view boundary is displayed in green.

232) Position the **mouse pointer** on the edge of the view until the Drawing Move View icon is displayed.

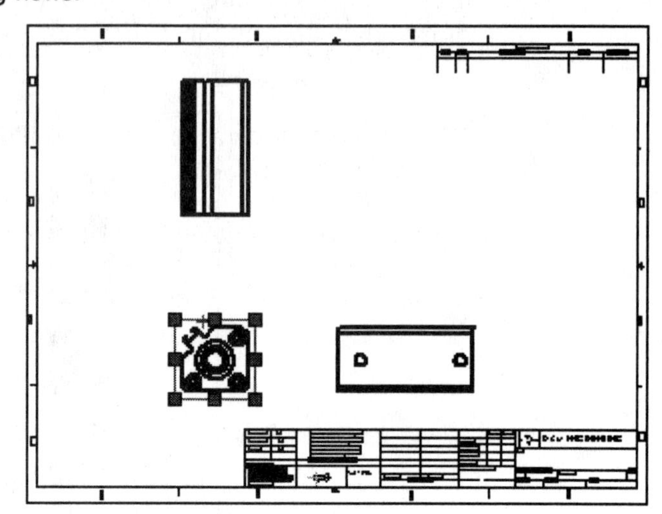

233) Drag the **Front** view in an upward vertical direction, above the title block.

234) Drag the **Right** view towards the left. Provide approximately 1in. – 2in., (25mm - 50mm) between views.

The default three standard views for the TUBE partially document the part. Display the part or assembly in various orientations with Named views, Projection views, Section views, Detail views and Auxiliary views. Views can be added to the drawing at anytime.

Use Projected view to create the Back view.

Position the Back view to the right of the Right view. Use Named view to create an Isometric view.

Named view displays the part or assembly in various orientations. Add a Named view to the drawing at anytime. The Named view is displayed in the View Orientation text box.

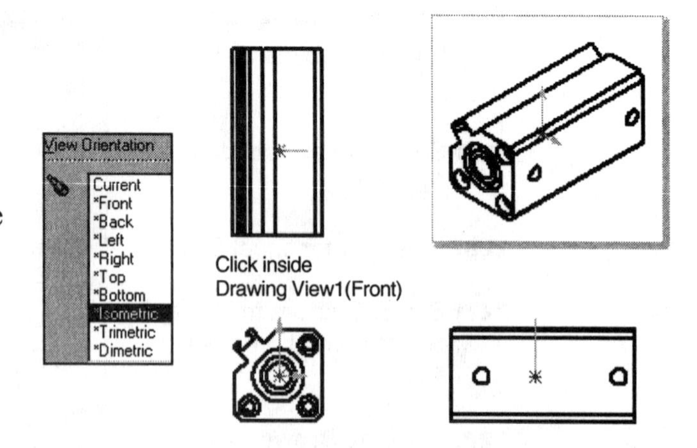

Add an Isometric view to the TUBE drawing.

235) Click **Drawing View1** (**Front**). The view boundary is displayed in green.

236) Click **Named View** from the Drawing toolbar. The named views for the TUBE are displayed.

237) Isometric is the default view. Click a **position** above the Right view.

238) Save the TUBE Drawing. Click **Save**.

Projected views display the part or assembly by projecting an Orthographic view using First angle or Third angle projection. Recall that Third angle projection is set in Sheet Properties in the drawing template.

Add a Projected Back view to the TUBE drawing.

239) Click **Drawing View3** (**Right**). The view boundary is displayed in green.

240) Click **Projected View** from the Drawing toolbar.

241) Drag the **mouse pointer** to the right of the Right view. Click the **position** for the Projected Back view.

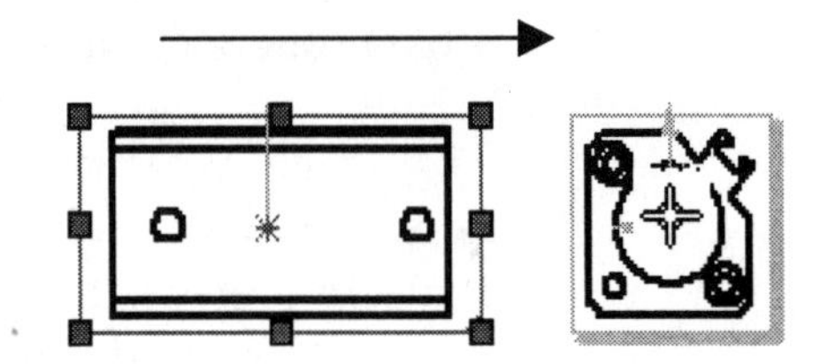

How many views should you use in a drawing? The number of views in a drawing depends upon how many views are required to define the true shape and size of the part.

Additional drawing views are required to display and enlarge features of the TUBE part.

Section View:

Section views are used to display the interior TUBE part features.

A Section view defines a cutting plane with a sketched line in a view perpendicular to the view. Create a full Section view by sketching a section line in the Top view.

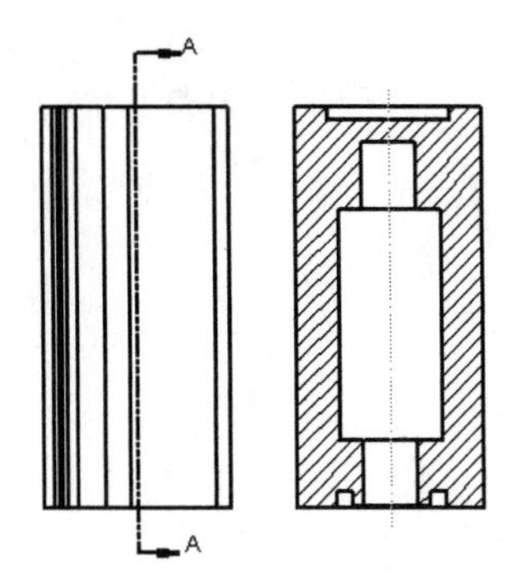

Detail View:

A Detail views enlarge an area of an existing view. Specify location, shape and scale. Create a Detail view from a Section view at a 4:1 scale.

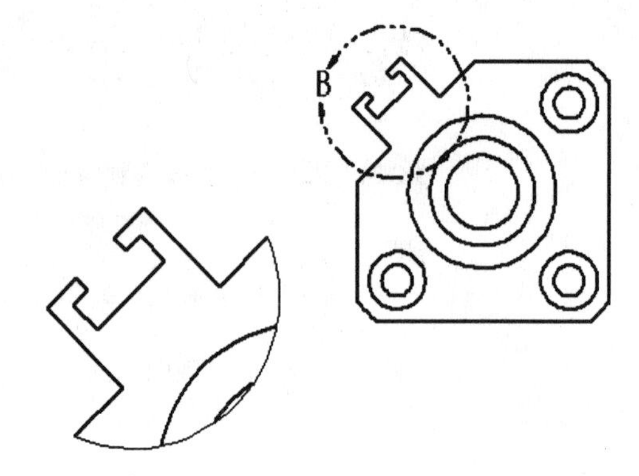

Broken-out Section:

A Broken-out Section removes material to a specified depth to expose the inner details of an existing view. A closed profile defines a Broken-out Section.

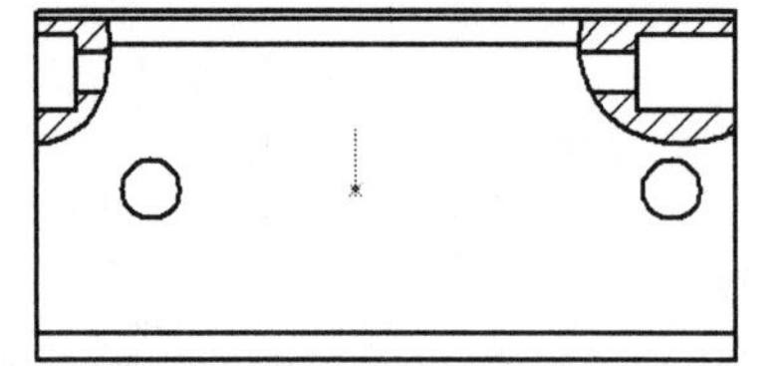

Auxiliary view:

An Auxiliary view displays a plane parallel to an angled plane with true dimensions. A primary Auxiliary view is hinged to one of the six principle views. Create a primary Auxiliary view that references the Front view. Use the Crop view to create a partial Auxiliary view.

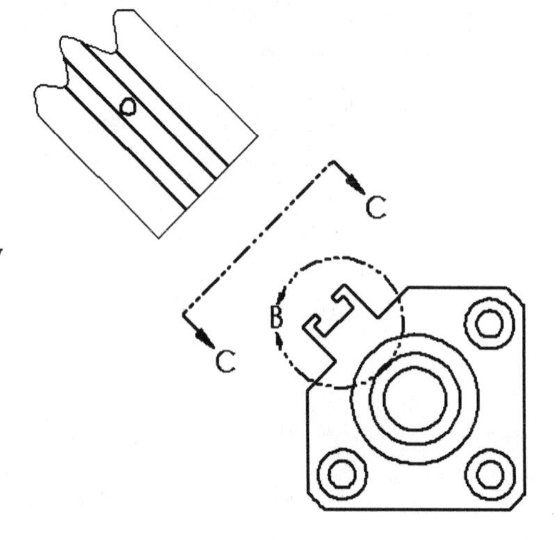

Add a Section view to the TUBE drawing.

242) Click **Drawing View2** (Top). The view boundary is displayed in green.

243) Click **Section View** from the Drawing toolbar.

244) Sketch a vertical section **Line** coincident with the Right plane, through the Origin. The Line must extend beyond the profile lines.

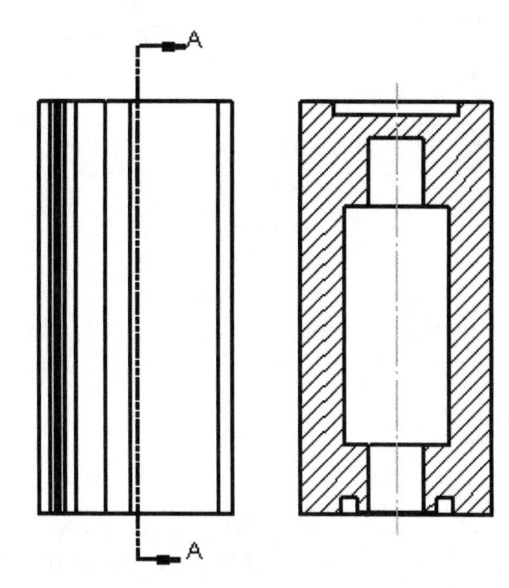

245) Click a **position** to the right of the Drawing View2 (Top) view. The section arrows point to the right.

246) If required, click **Flip direction**.

247) Click **OK** from the PropertyManager.

Add a Detail view to the TUBE drawing.

248) Click **Zoom to Area** on the lower left corner of the drawing.

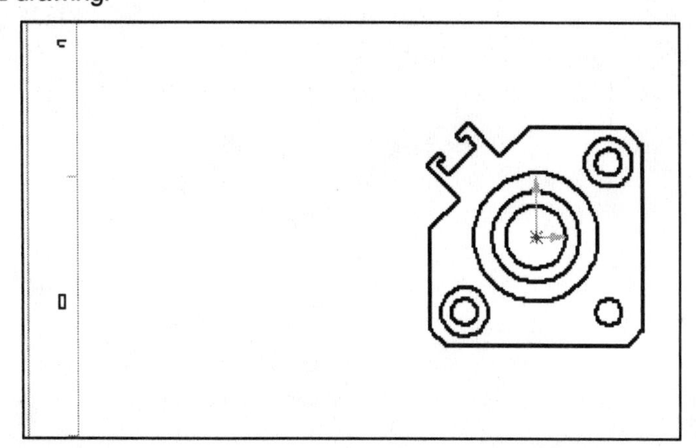

Click **Detail View** from the Drawing toolbar. The Circle Sketch tool is activated. Click the **middle** of the SwitchGrove in Front view.

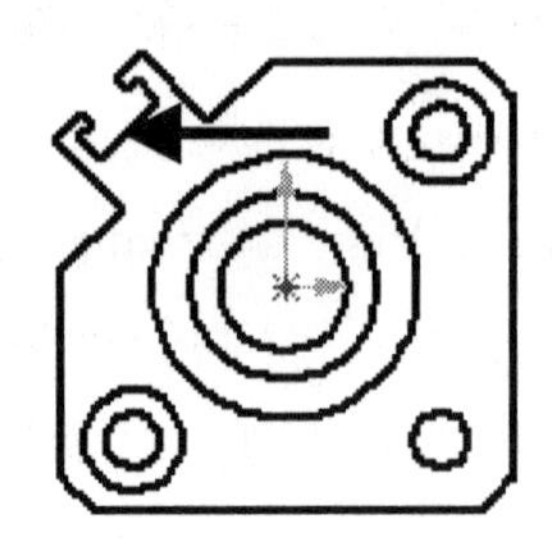

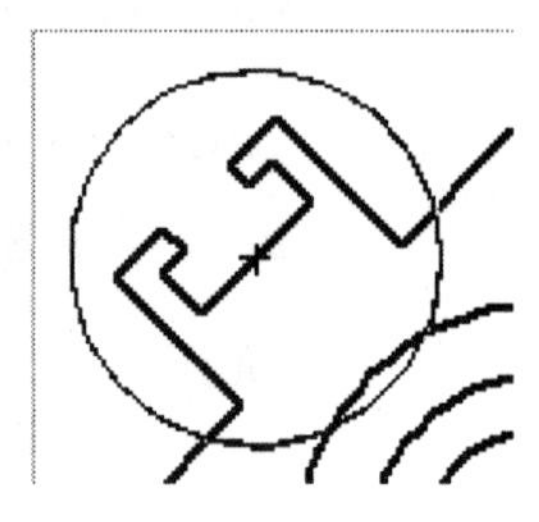

249) Drag the **mouse pointer** outward.

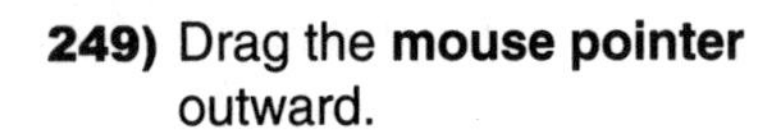

250) Click a **position** just below the large circle to create the sketched circle.

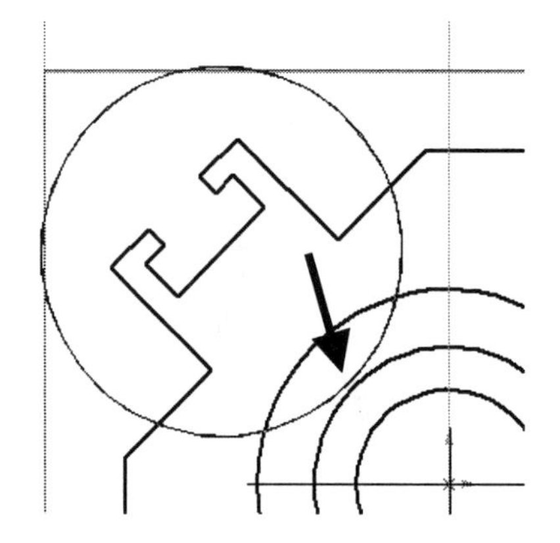

251) Click the **position** to the bottom left of DrawingView1 (Front).

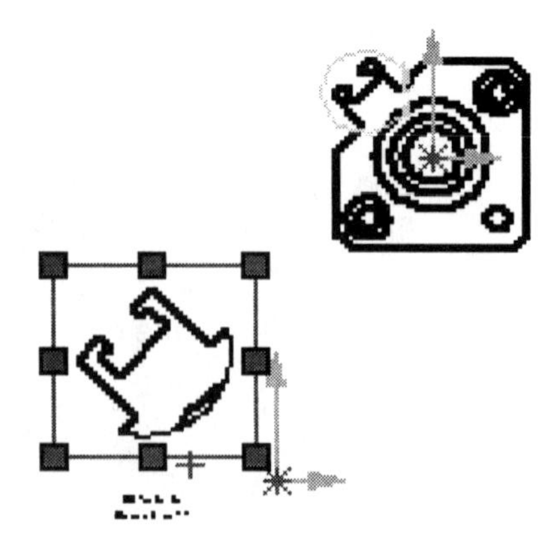

252) The Detail View name is B. Click the **down arrow** to expand the Custom Scale text box.

253) Check the **Custom Scale** box. Change the view Scale.

254) Enter **4:1** in the Custom Scale text box.

255) Click **OK** from the PropertyManager.

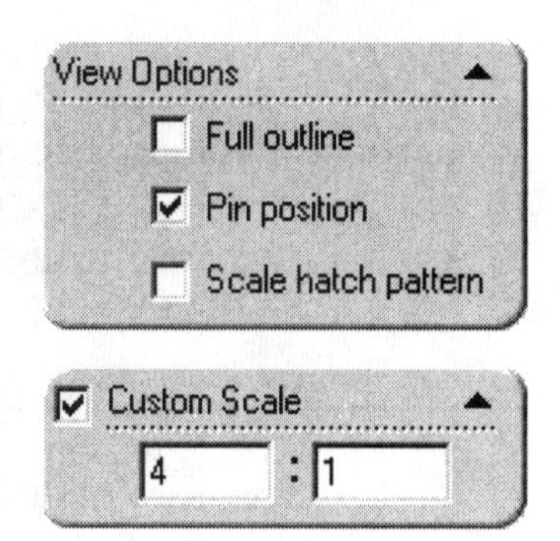

256) The B Detail circle is created in the Front view. Drag the text **B** off the profile lines.

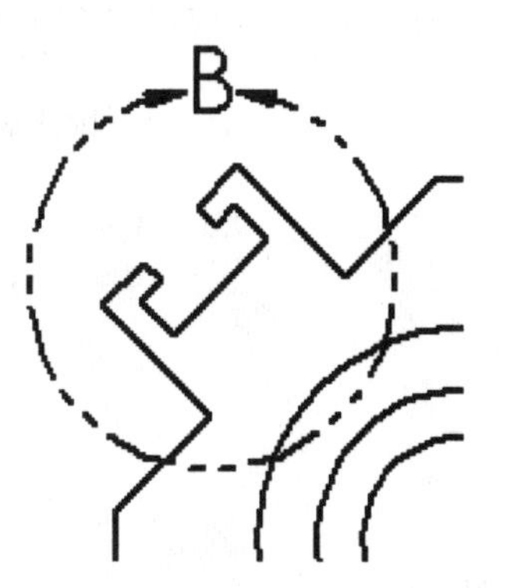

Note: After the Detail view is created, the Detail view size can be modified.

To modify the size of the Detail view, position the mouse pointer on the Detail circle.

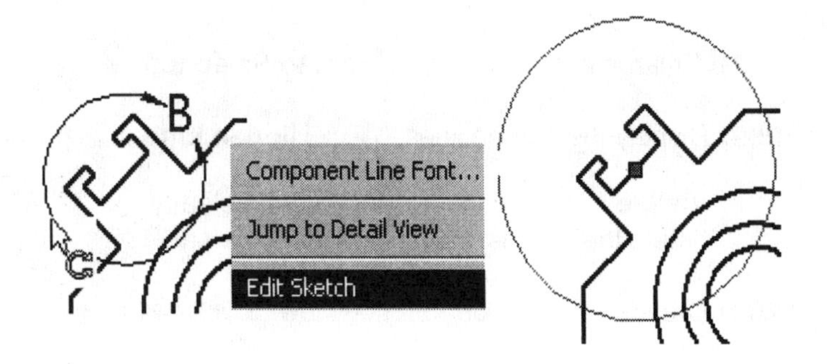

The mouse pointer displays . Right-click and select Edit Sketch. Drag the circumference of the sketch circle. Click Rebuild to update the Detail view.

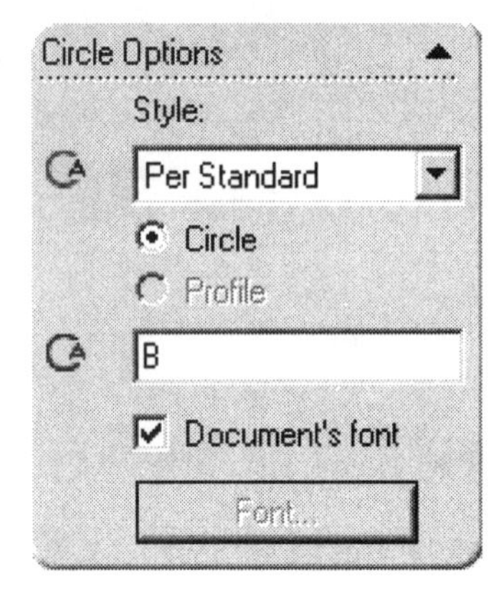

The Detail view profile is a circle. When a non-circular view is required, sketch the closed profile first. Then select the Detail View .

The view name; A, B, C, increments one letter at a time for Section views, Detail views and Auxiliary views. Note: If you delete the view, the view name still increments by a letter.

Add the first Broken-out Section view to the TUBE drawing.

257) Click **Right view boundary** from the FeatureManager. The view boundary is displayed in green.

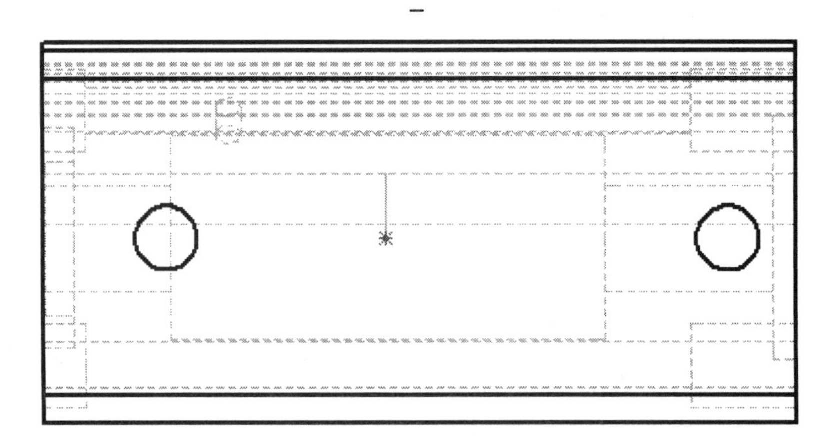

258) Enlarge the view. Click **Zoom to Selection** .

259) Display the hidden lines. Click **Hidden Line Visible** . The hidden lines do not clearly define the internal front features of the part.

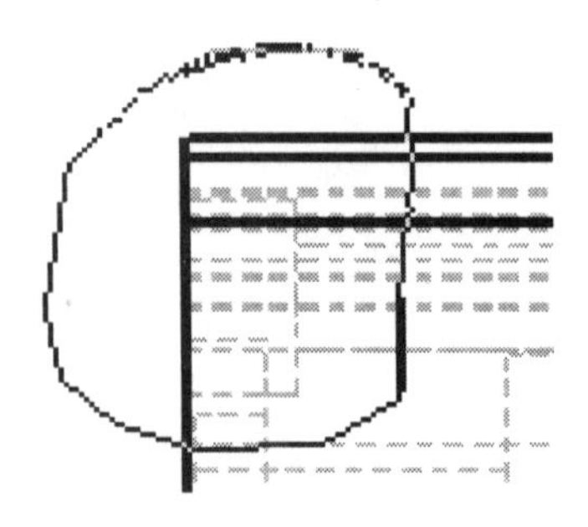

260) Create a Broken-out Section view: Click **Spline** .

261) Sketch a closed **Spline** in the top left corner. The Spline contains the Cbore Front feature.

262) Click **Broken-out Section** .

263) Enter **5** for Depth.

264) Click **Preview** to insure that the Cbore Front is displayed. Display the Broken-out feature.

265) Click **OK**.

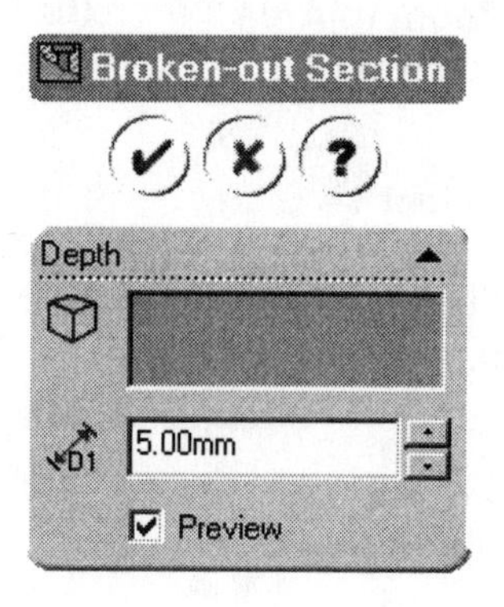

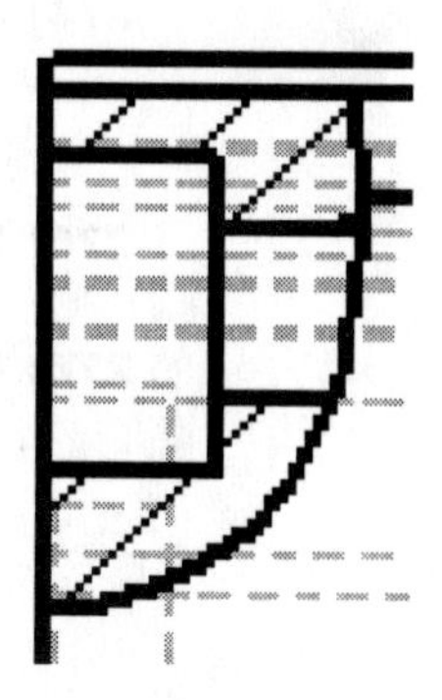

Add a second Broken-out Section view to the TUBE drawing.

266) Click **Spline** .

267) Sketch a closed **Spline** in the top right corner. The Spline contains the Cbore Rear feature.

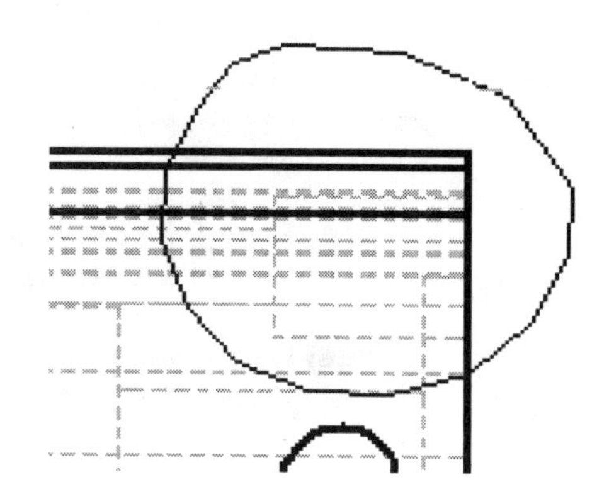

268) Click **Broken-out Section** .

269) Enter **5** for Depth.

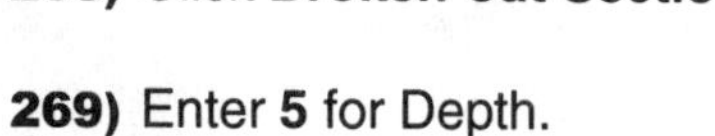

270) Click **Preview**. Display the Broken-out feature.

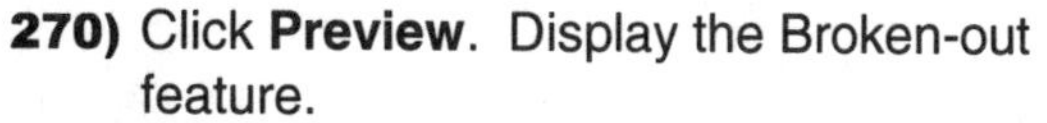

271) Click **OK**.

272) Display no hidden lines. Click **Hidden Lines Removed** .

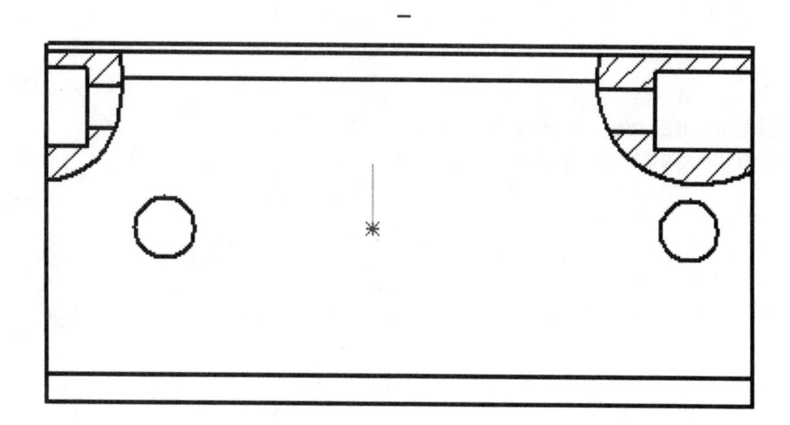

273) Fit the drawing to the screen. Press the **f** key.

Add an Auxiliary view to the TUBE drawing.

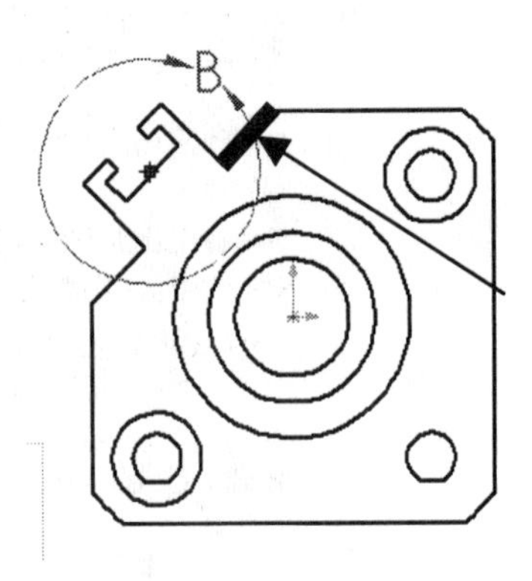

274) Click **Zoom to Area** on the left side of the TUBE drawing.

275) Click **Select**.

276) Click the **left angled edge** in Drawing View1 (Front).

277) Click **Auxiliary View** from the Drawing toolbar. Position the Auxiliary view.

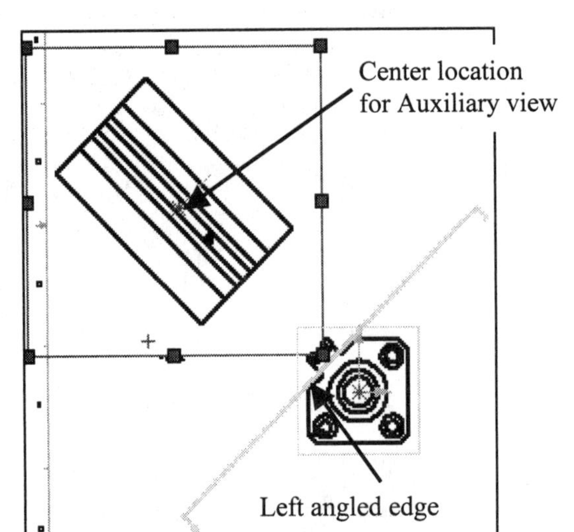

278) Click a **position** to the upper left of Drawing View1 (Front). The location selected is the center of the view.

279) Enter **C** for the View Name. Display a full Auxiliary view.

280) Click **OK** from the PropertyManager

281) Position the view arrows. Click **Line C-C**.

282) Drag the **midpoint** and position it between the Auxiliary view and Front view.

283) Click each **endpoint** and drag them towards the midpoint.

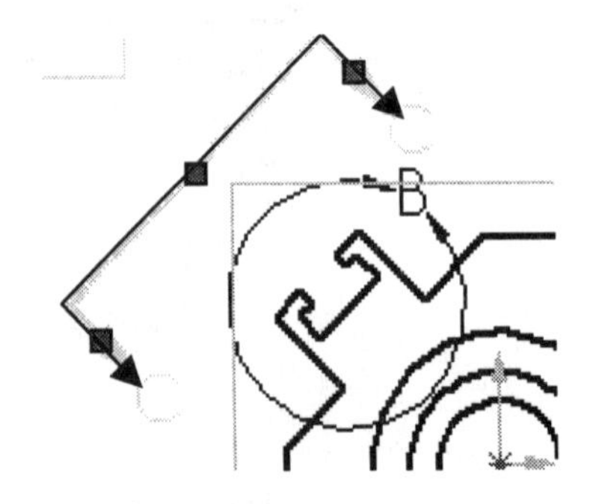

Display the M2 Hole information. Create a Partial Auxiliary view from the Full Auxiliary view. Sketch a closed profile in an active Auxiliary view. Create the Profile with a Spline and 3 line segments.

Watch the mouse pointer for inferencing (feedback). The endpoints of the Spline are collinear with the Switch Grove profile lines. The three line segments are collinear with the Switch Grove profile lines.

Create a Partial Auxiliary view. Crop the view.

284) Sketch a closed profile in the active Auxiliary view. An active view displays the green view boundary. Sketch a closed Spline.

Click **Spline**. The first point is coincident with the left line of the Switch Grove. The mouse pointer displays On line inference for the first point.

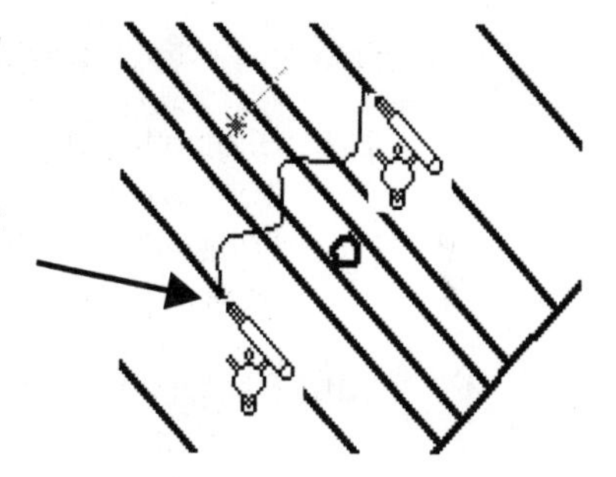

285) Sketch **6 Points** to create the Spline. The last point is coincident with the right line of the Switch Grove. The mouse pointer displays On Line inference at the line right of the Switch Grove.

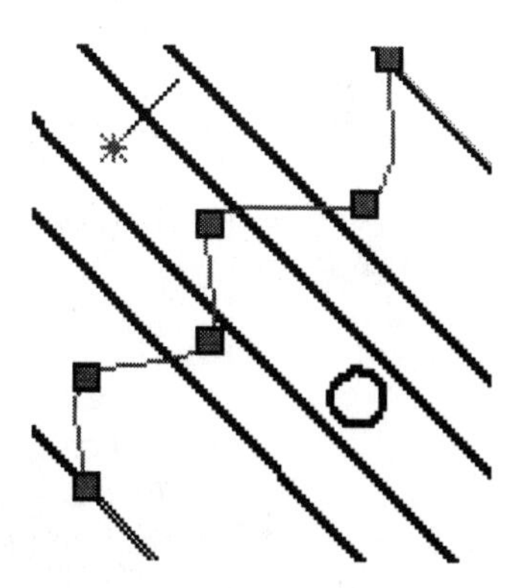

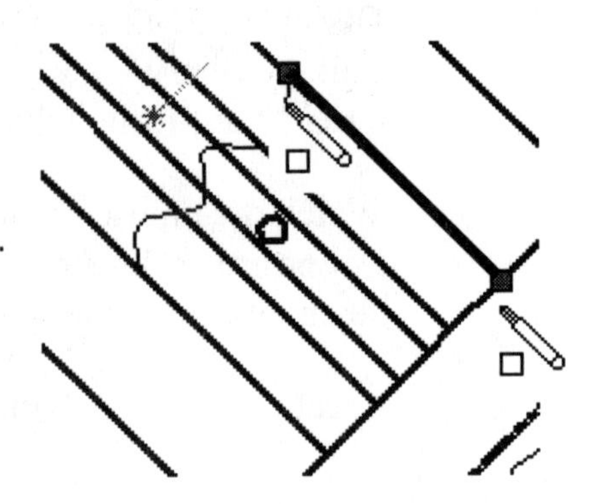

286) Sketch three lines. Click **Line**. **Sketch** the first line.

287) For the first point of the Line, the mouse pointer displays **Endpoint inference**.

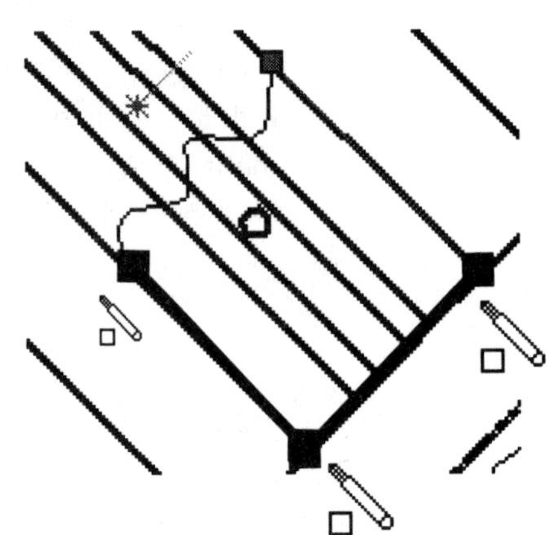

288) Sketch the second **Line** collinear with the bottom edge of the Auxiliary view. The first point and second point display Endpoint inference.

289) Sketch the third **Line**. The last point must display Endpoint interference with the first point of the Spline.

Note: Utilize a closed and continuous Sketch profile to create a Crop View.

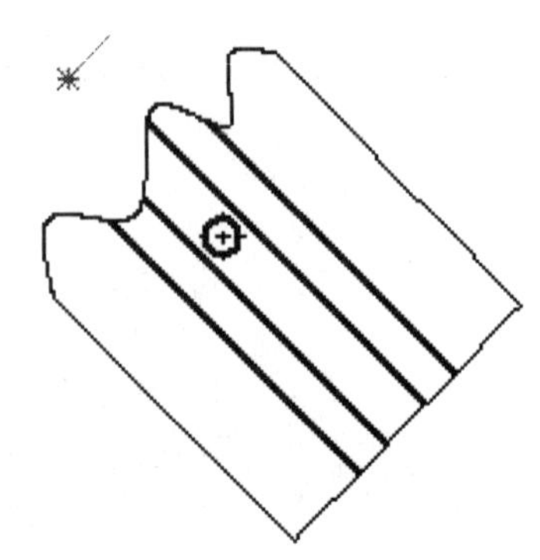

290) The Crop View icon turns bold when a closed profile is sketched and is currently selected. Crop the view. Display the Partial Auxiliary view. Click **Tools**, **Crop View**, **Crop** from the Main toolbar.

291) Fit the drawing to the Graphics window. Press the **f** key.

292) Drag the **C-C view arrow** between the Auxiliary view and the Front view.

293) Drag the **VIEW C** text below the Auxiliary view.

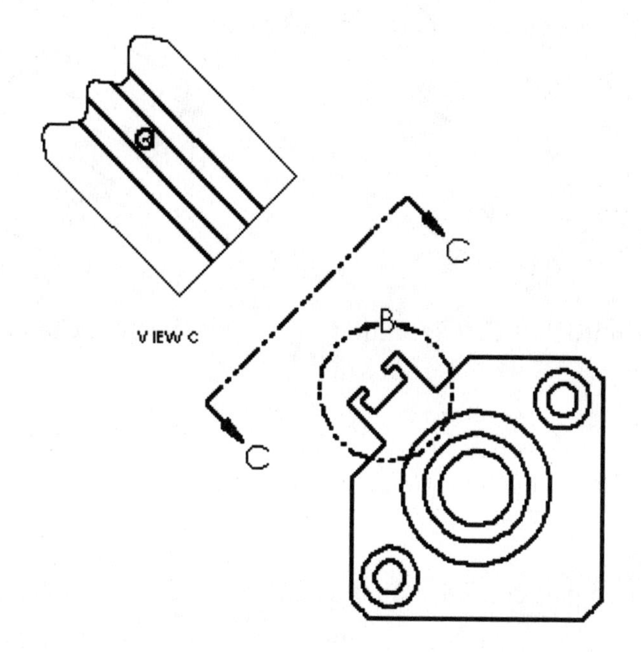

Note: Auxiliary and Section views are aligned to their parent view. These views are positioned in other locations on the sheet when space is limited. Press the Ctrl key before selecting the view tool in order to position the Auxiliary and Section views anywhere on the sheet.

294) Save the TUBE drawing. Click **Save**.

Half Section Isometric (Cut – Away)

A Half Section Isometric view requires a Cut feature created in the part. The Cut feature removes ¼ of the TUBE part. Create the Cut feature. Create a Design Table.

The Design Table will represent two configurations of the TUBE part: Entire Part and Section Cut.

Design Table for: tube-project1	
	$STATE@Cut-Extrude1
Entire Part	SUPPRESSED
Section Cut	UNSUPPRESSED

The Design Table contains configuration names, parameters to control and assigned values for each parameter.

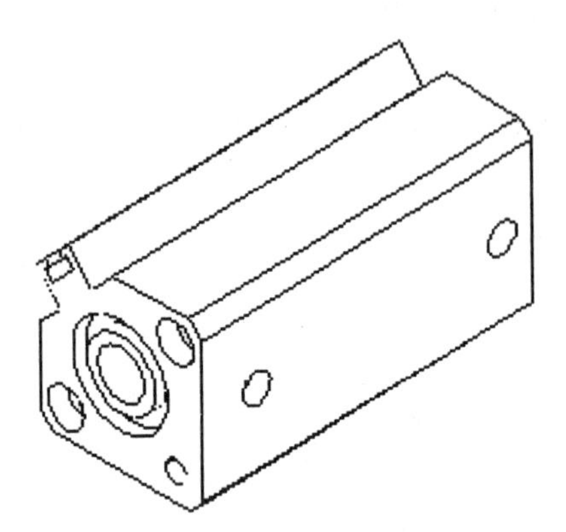

Entire Part Configuration

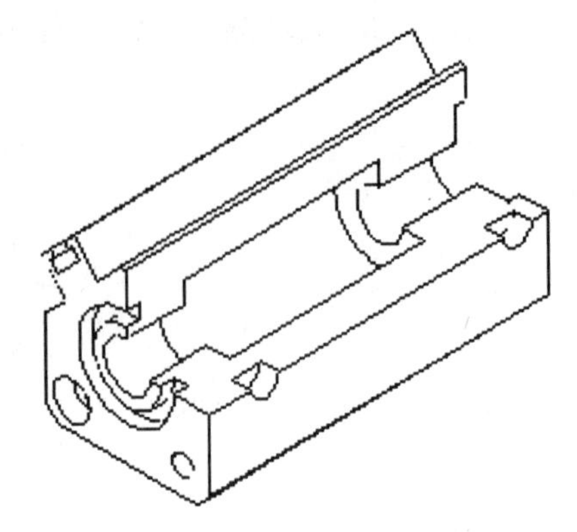

Section Cut Configuration

The two part configurations are referenced on the same TUBE drawing sheet.

Add the Section Cut Configuration as an Isometric view.

Insert an Area Hatch pattern in the Isometric view.

A Hatch Pattern (section lining or cross sectioning) represent an exposed cut surface based upon the material.

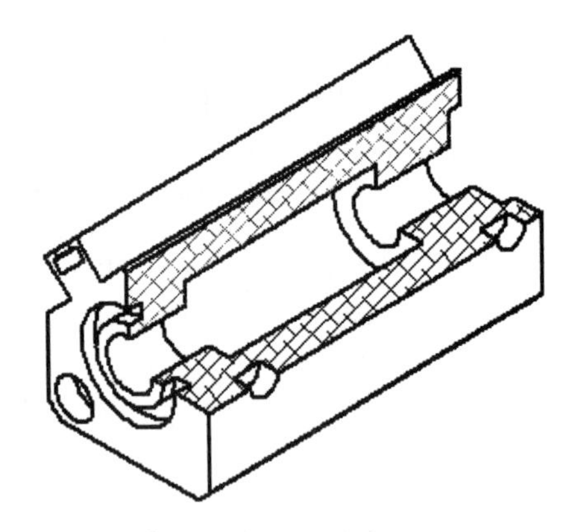

Create the Cut feature.

295) Open the TUBE part. Right-click in the **Front** view in the drawing Graphics window. Click Open **TUBE.sldprt**.

296) Create a Cut. Click the **front face**. Click **Sketch**.

297) Sketch a **Rectangle** through the Origin.

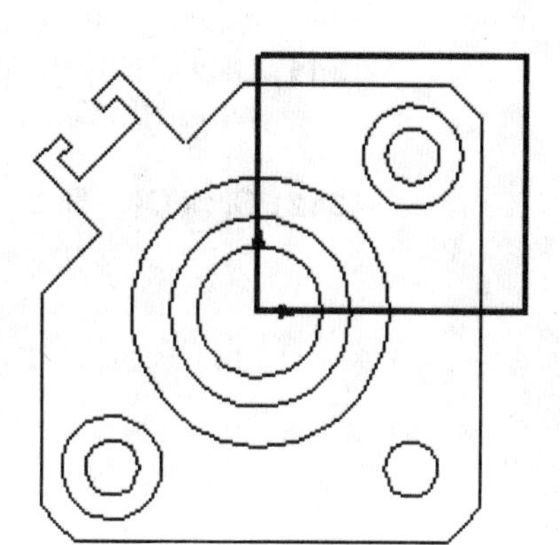

298) Click **Extruded Cut** from the Feature toolbar.

299) Click **Through All** for Depth.

300) Click **OK**.

301) Suppress the Extruded Cut. Right-click **Cut-Extrude1** from the FeatureManager.

302) Click **Suppress**.

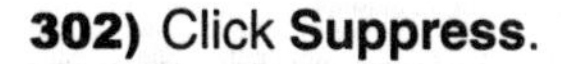

303) Save the TUBE part. Click **Save**.

Create a Design Table.

304) Click **Insert**, **Design Table**, **New** from the Main menu.

305) Check **Blank** for Source.

306) Click **OK**. The Design Table for the TUBE is displayed in the upper left corner.

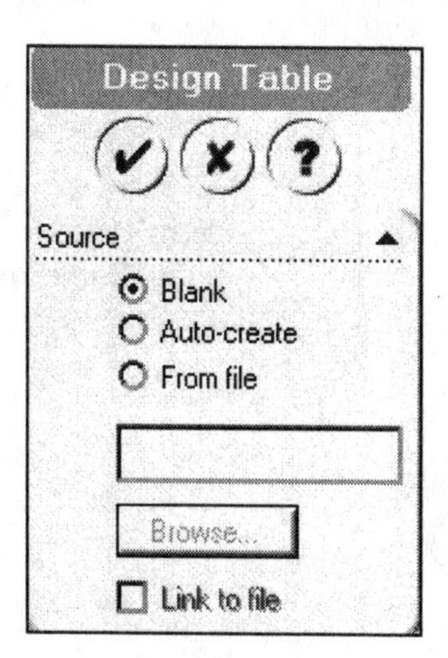

307) Click **Cut-Extrude1** from the FeatureManager. The variable $STATE@Cut-Extrude1 is displayed. The SUPPRESS state is displayed under the $STATE@Cut-Extrude1 column.

	$STATE@Cut-Extrude1
First Instance	SUPPRESSED

308) Rename the text **First Instance** to **Entire Part**. This is the first configuration.

309) Create the second configuration. Enter the text **Section Cut** under the Entire Part text.

310) Add **UNSUPPRESSED** under the SUPPRESSED text.

	$STATE@Cut-Extrude1
Entire Part	SUPPRESSED
Section Cut	UNSUPPRESSED

311) Click inside the part **Graphics window**. Close the Design Table.

312) Click **OK**. Both TUBE configurations are created: Entire Part and Section Cut.

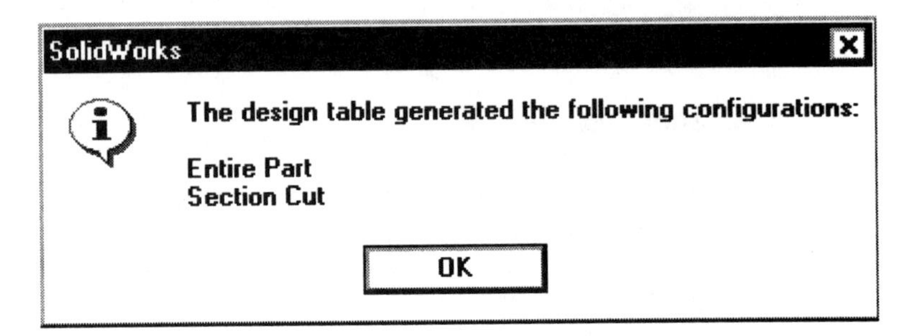

Note: Click Edit, Design Table from the Main toolbar to access an existing Design Table or Click the Design Table icon in the FeatureManager.

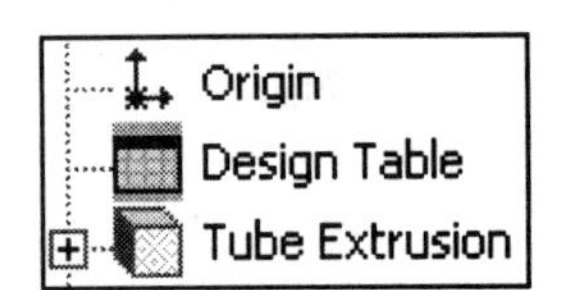

313) Display the TUBE part configurations. Click **Configuration** at the bottom of the FeatureManager.

314) Double-click **Entire Part** from the Configuration Manager. Double-click **Section Cut** from the Configuration Manager.

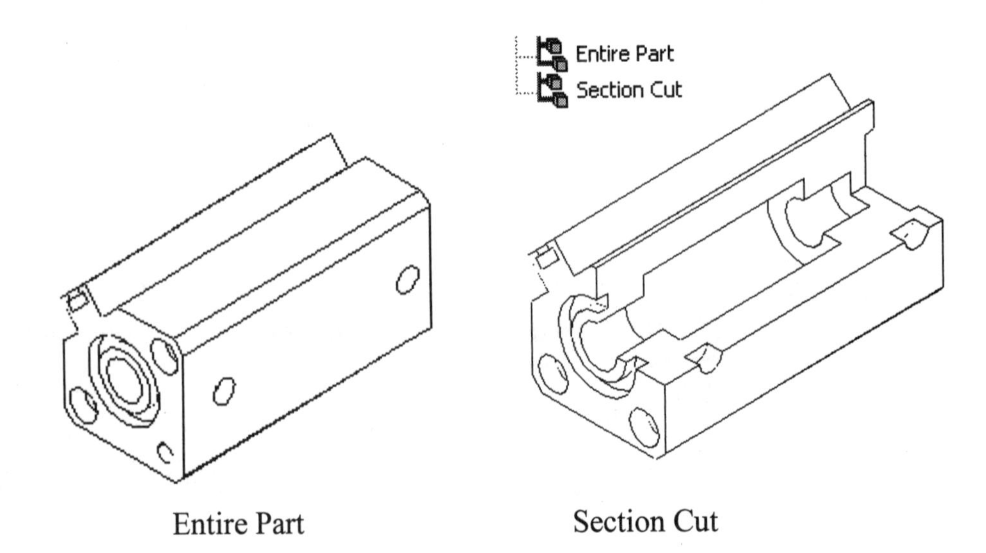

Entire Part Section Cut

315) Display the TUBE part default configuration. Click **Default** from the Configuration Manager.

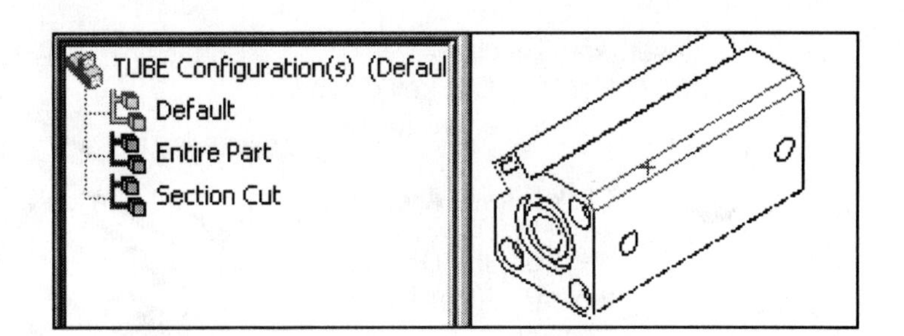

316) Return to the TUBE part FeatureManager. Click **Part** at the bottom of the FeatureManager.

317) Save the TUBE part. Click **Save**.

318) Open the TUBE drawing. Right-click **Tube(Default) Part** icon.

319) Click **Open Drawing**.

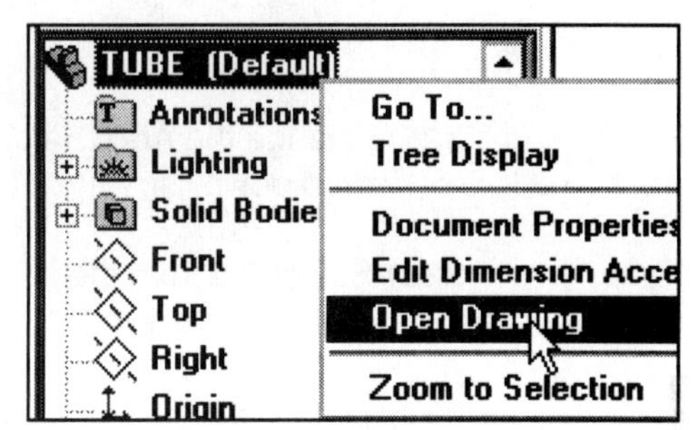

Select the TUBE configuration.

320) Right-click **Properties** on the Isometric view boundary.

321) Select the **Section Cut** from the Named Configuration drop down list. The Section Cut configuration is displayed.

322) Click **OK**.

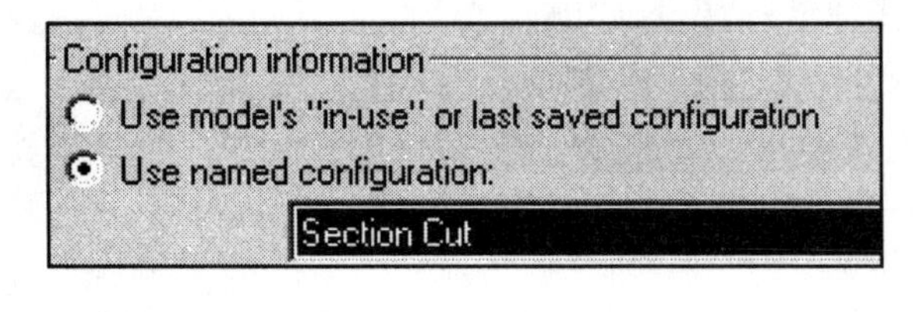

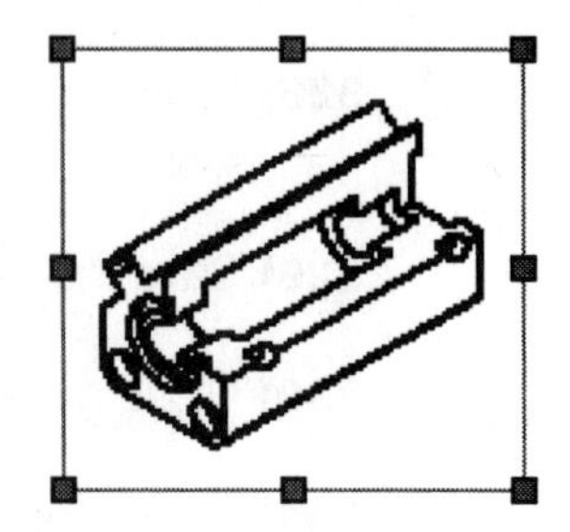

Insert Area Hatch.

323) Hold the **Ctrl** key down. Select the **two faces**.

324) Click **Insert**. **Area Hatch** ■ from the Drawing toolbar. Release the **Ctrl** key.

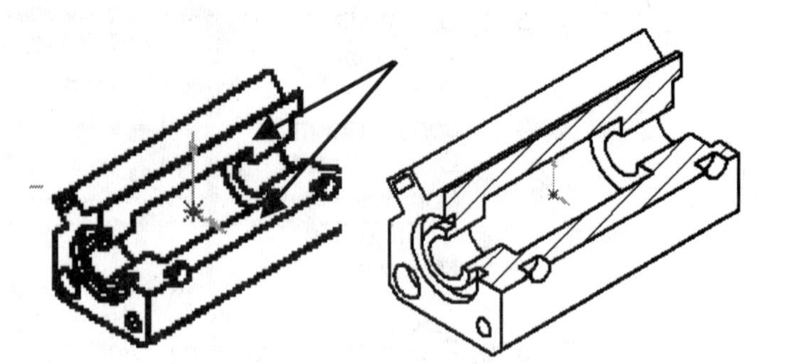

325) Change the Area Hatch type. Select **ANSI38(Aluminum)** from the Pattern drop down list.

326) Select **2.0** from the Scale drop down list.

327) Click **OK**.

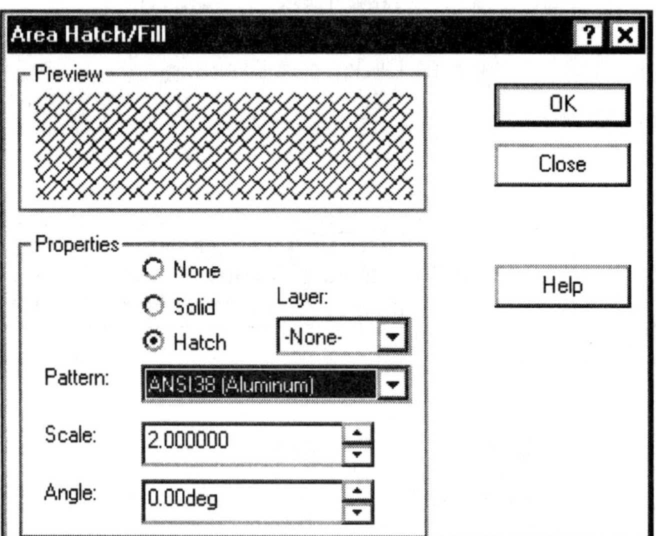

The Hatch type, ANSI138(Aluminum) represents the TUBE material.

328) Fit the drawing to the screen. Press the **f** key.

329) Save the TUBE drawing. Click **Save**.

330) Close all files. Click **Windows**.

331) Click **Close All**.

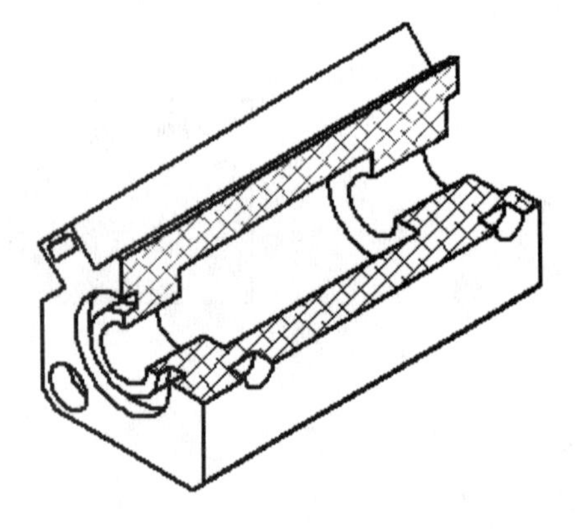

The views required for the TUBE drawing are complete. Create dimensions and notes for the TUBE drawing in Project 3.

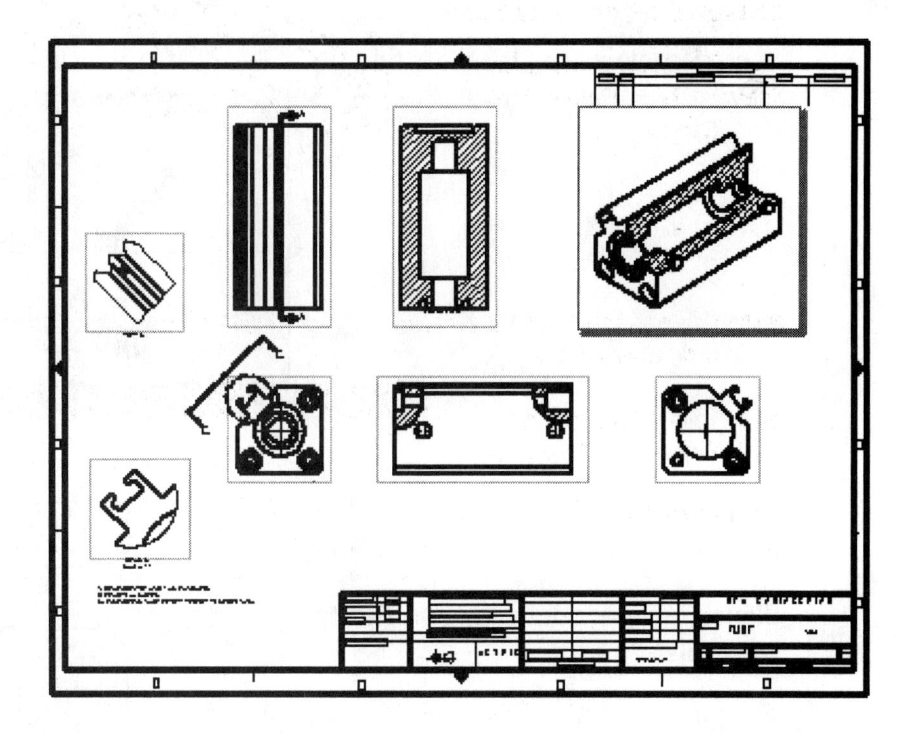

Create the COVERPLATE Drawing

Create the COVERPLATE drawing. The COVERPLATE drawing consists of two part configurations. The first part configuration contains Nose Holes. The second part configuration is without Nose Holes.

The COVERPLATE drawing requires an Offset Section view and an Aligned Section view for the two different part configurations.

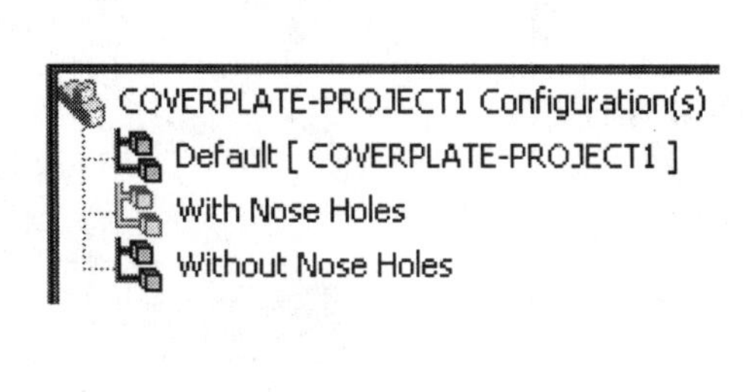

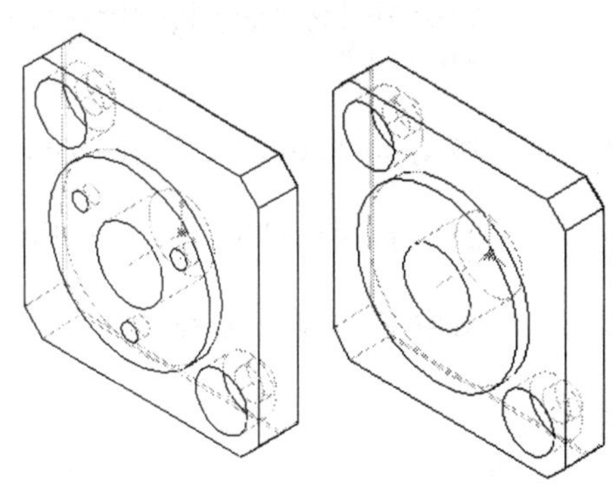

The COVERPLATE drawing consists of two sheets, one sheet for each part configuration.

Sheet 1 contains an Offset Section view.

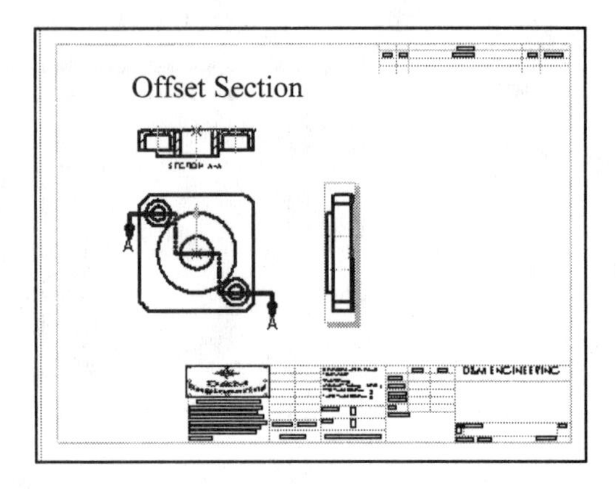

Create an Offset Section view by drawing a Sketched Line.

Sketch Lines are drawn in perpendicular segments.

Section A-A displays the offsets in a single plane.

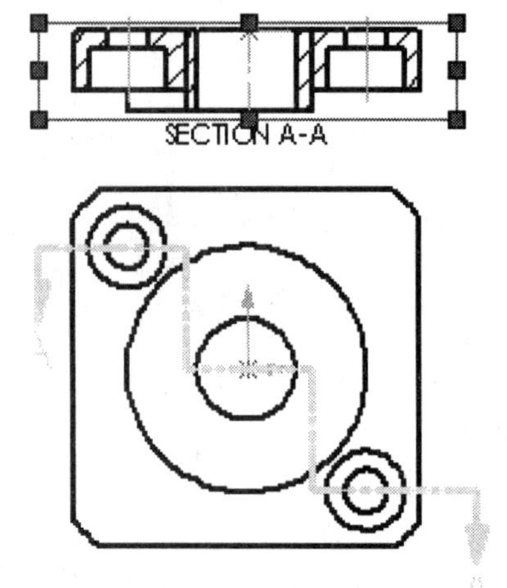

Sheet 2 contains an Aligned Section view.

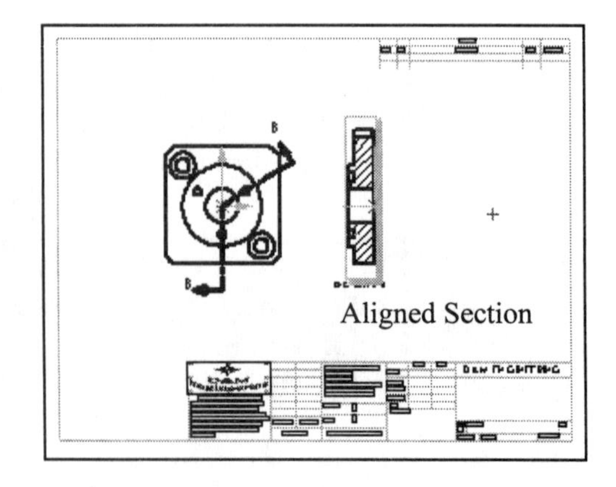

In the ASME Y14.3M standard, an Aligned Section occurs when features lend themselves to an angular change in the direction of the cutting plane. The bent cutting plane and features are rotated into a plane perpendicular to the line of sight of the sectional view.

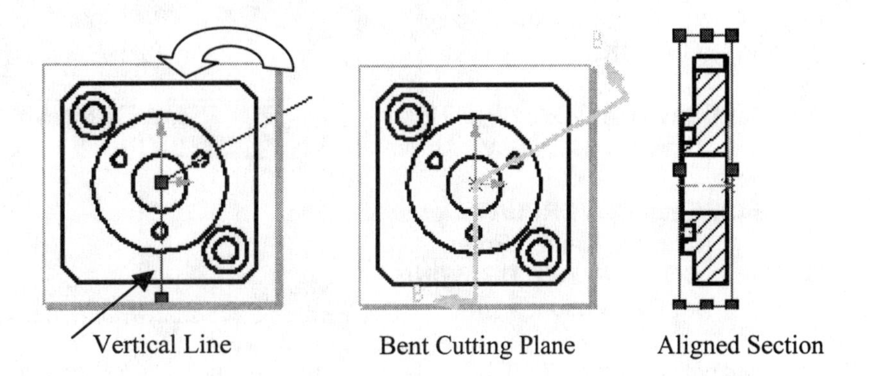

Review the COVERPLATE part.

332) Click **File**, **Open**.

333) Select the part, **COVERPLATE**.

334) Click **Open**.

335) Display the dimensions. Double-click on each feature **name**.

336) Drag the **Split Bar** downward to split the FeatureManager. Display the two COVERPLATE part configurations. Click **Configuration** at the bottom of the FeatureManager.

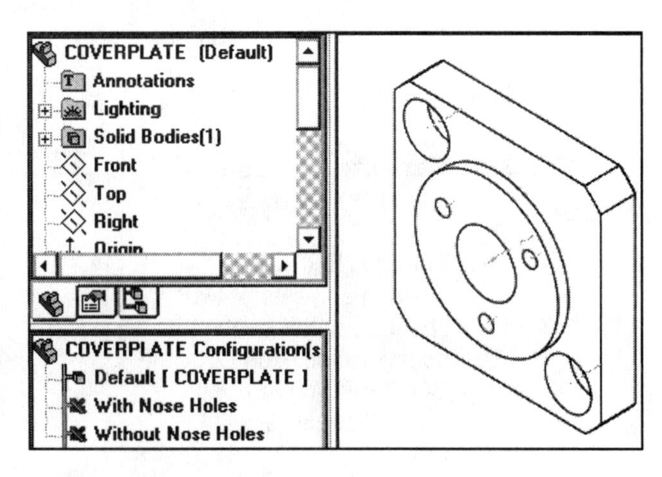

337) Double-click the **With Nose Holes** configuration. This is the first COVERPLATE part configuration.

338) Double-click the **Without Nose Holes** configuration. This is the second COVERPLATE part configuration.

Create the COVERPLATE drawing.

339) Click **New**.

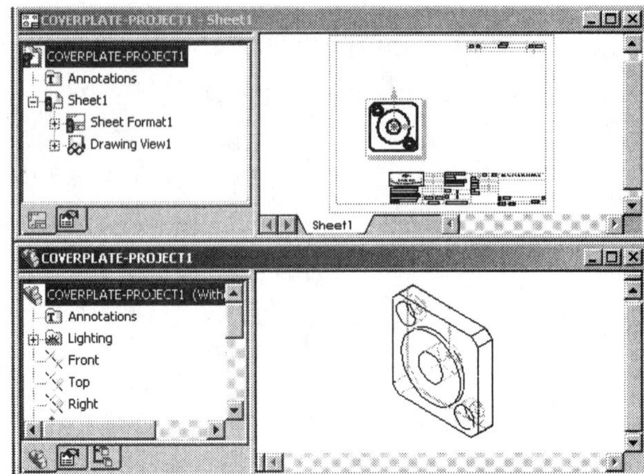

340) Select the **A-ANSI-MM** Drawing Template.

341) Click **OK**.

342) Save the COVERPLATE drawing. Click **Save As**.

343) Enter **COVERPLATE** for Drawing Name. Enter **COVERPLATE DRAWING** for Description.

344) Display the COVERPLATE part and drawing. Click **Windows**, **Tile Horizontal**.

345) Click inside the drawing **Graphics window**.

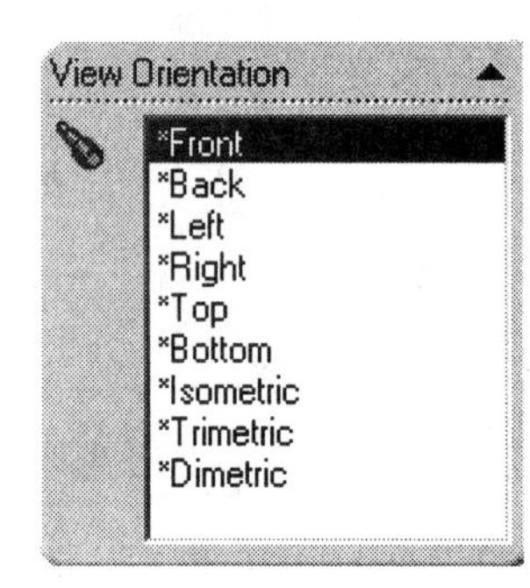

346) Add a Front view as a Named view. Click **Named View** from the Drawing toolbar. Click inside the Graphics window on the **COVERPLATE Without Nose Holes**.

347) Select **Front** from the Named view list box.

348) Enter **2:1** for Custom Scale.

349) Click a **position** in the lower left cover of the drawing Graphics window.

350) Click **OK**.

351) **Maximize** the drawing Graphics window.

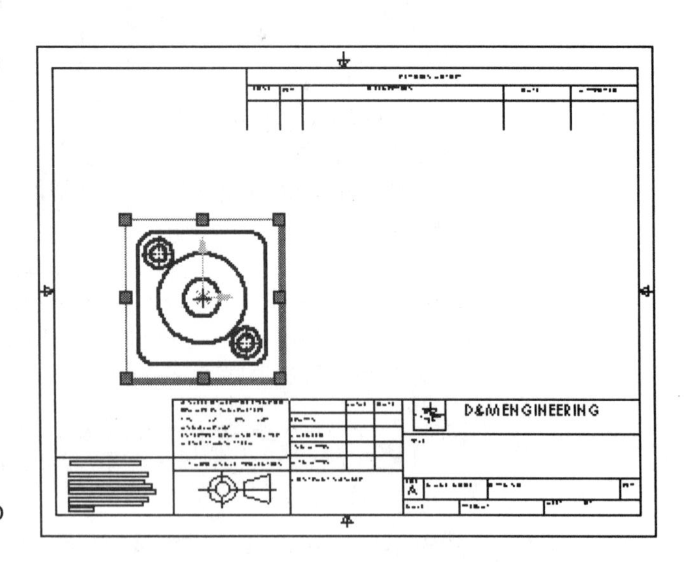

352) The COVERPLATE drawing consists of two sheets, one for each part configuration. Add COVERPLATE-Sheet2. Right-click the **Sheet1** tab.

353) Click **Add Sheet**.

354) Copy the Front view from COVERPLATE-Sheet1 to COVERPLATE-Sheet2. Click the **Sheet1** tab.

355) Click **Front view boundary**.

356) Copy the view. Click **Ctrl C**.

357) Click the **Sheet2** tab.

358) Click inside the **Graphics window**.

359) Paste the view. Click **Ctrl V**.

360) Modify the configuration. Right-click in **Front view COVERPLATE-Sheet2**. Click **Properties**.

361) Select **With Nose Holes** from the Configuration text box. A pattern of 3 Holes is displayed.

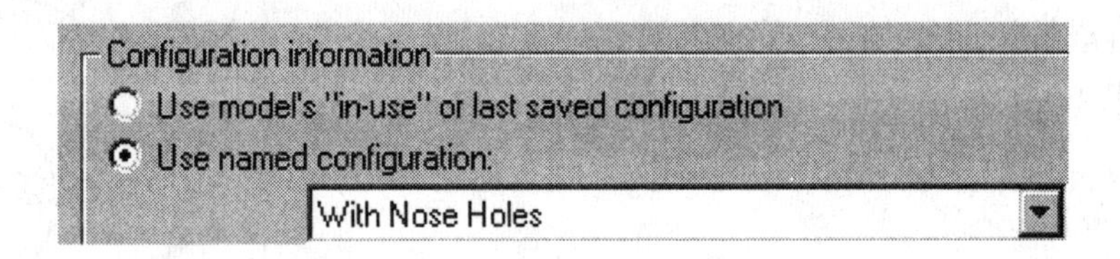

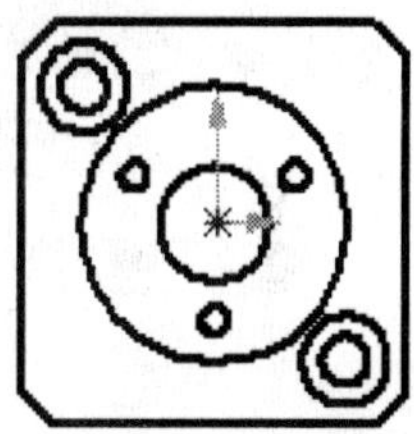

362) Return to the COVERPLATE-Sheet1. Click the **Sheet1** tab.

363) Select the **Front view**. The view boundary is displayed in green.

364) Enlarge the view. Click **Zoom to Selection**.

Create an Offset Section view.

365) Sketch an open contour with 5 connecting Center Line segments. Click **Center Line**.

366) Position the **mouse pointer** on the **circumference** of the upper left Cbore. The center point is displayed.

367) Drag the **mouse pointer** directly to the left. A blue dashed line, aligned with the center point is displayed.

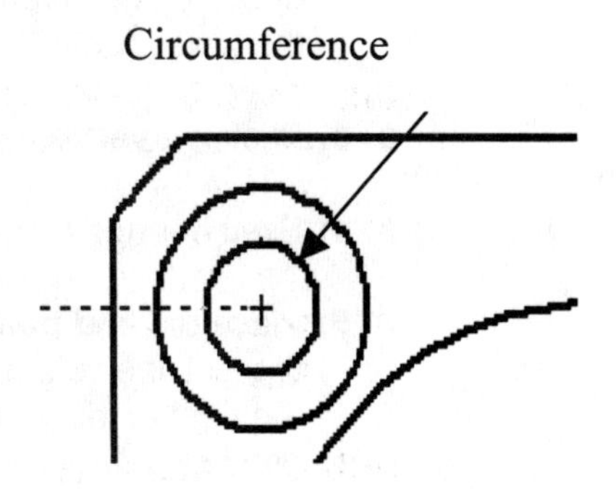

368) Click the **start point** to the left of the vertical profile line.

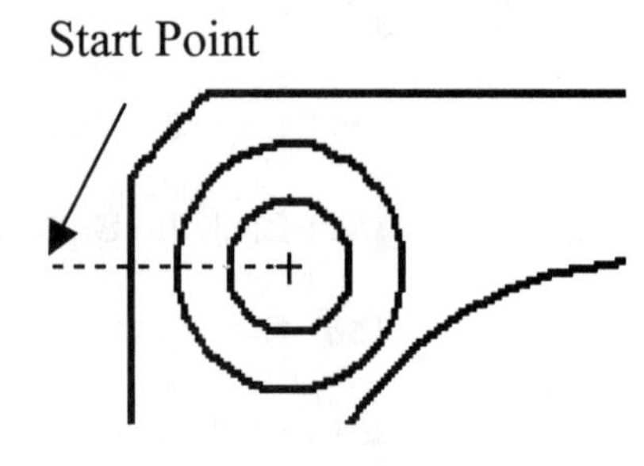

369) Drag the mouse pointer over the **circumference** of the Bore to display the center point.

370) Sketch a **vertical line** to the left of the Bore circle.

371) Sketch a **horizontal line** through the center of the Bore.

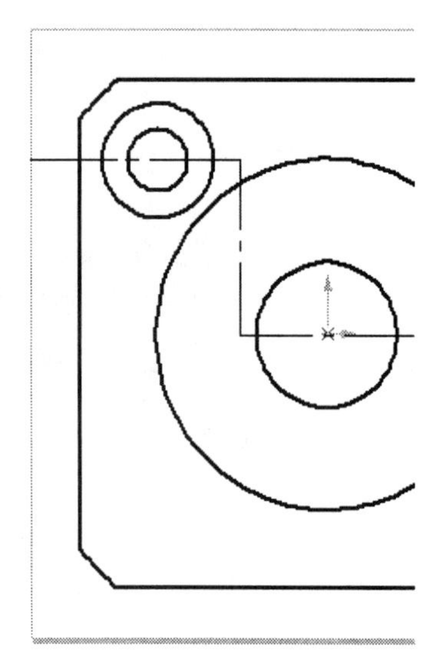

372) Display the center mark. Drag the **mouse pointer** over the circumference of the lower right Cbore.

373) Sketch a **vertical line**.

374) Sketch a **horizontal line**.

375) Place the **end point** of the last line segment to the right of the vertical profile line.

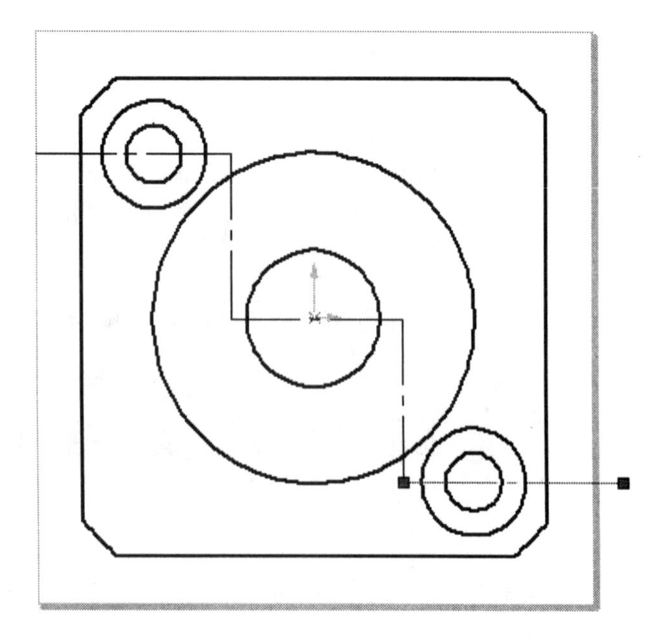

376) Create the Offset Section view. Click **Section View** from the Drawing toolbar.

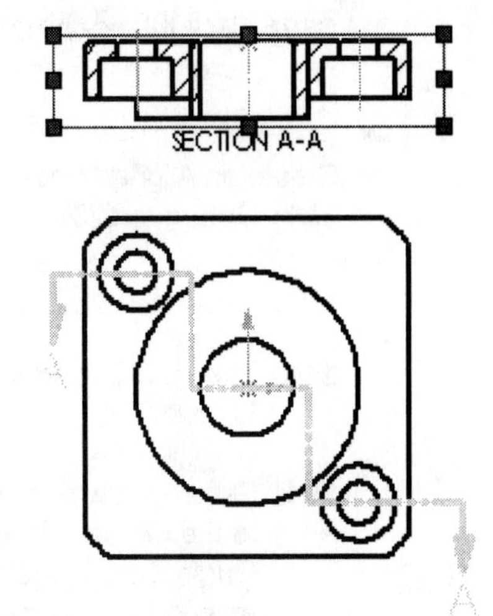

377) Click a **position** above the Front view. The section arrows point downward.

378) If required, click **Flip direction**. Create the Offset Section A-A.

379) Click **OK** from the PropertyManager.

380) Save the COVERPLATE drawing. Click **Save**.

Edit a Section Line.

381) Right-click on **Section Line 1** inside the Front view. Click **Edit Sketch**. The Centerline is displayed.

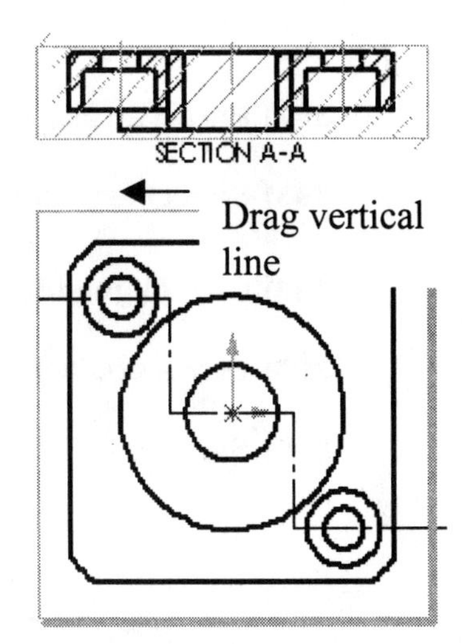

382) Drag the **left vertical line** towards the left Cbore.

383) Click **OK**.

384) Click the **Offset Section A-A boundary**. Update the view if light gray hatching appears across the view boundary.

Click **Update View**.

Note: You can use the Rebuild command to update the drawing.

385) Add a Project Right view. Click **Front view boundary**. Click **Projected View**.

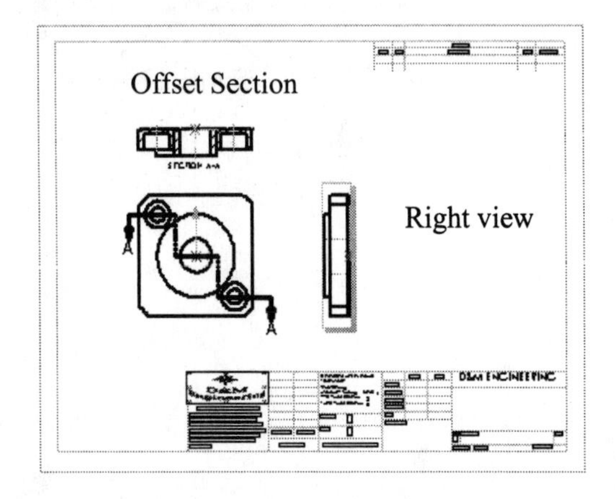

386) Click a **position** to the right of the Front view.

387) Display the Hidden lines. Click **Hidden Line Visible**.

388) Save the COVERPLATE drawing. Click **Save**.

Create an Aligned Section view.

389) Return to COVERPLATE-Sheet2. Click the **Sheet2** tab.

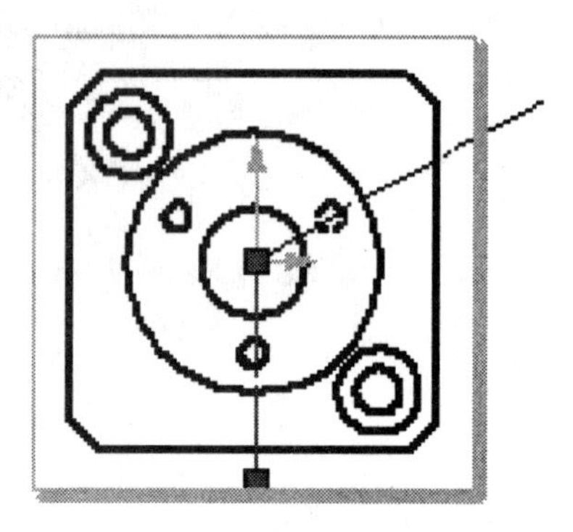

390) The view boundary is green. Click **Center Line**.

391) Sketch a **vertical line** through the bottom hole to the Origin. Sketch an angled **line** from the Origin though the left hole. The endpoint must extend beyond the right profile.

392) Project an aligned view from the vertical line. Click **Select** .

393) Select the **vertical line**.

394) Click **Aligned Section View** .

395) Click a **position** to the right of the Front view.

396) Save the COVERPLATE drawing. Click **Save**.

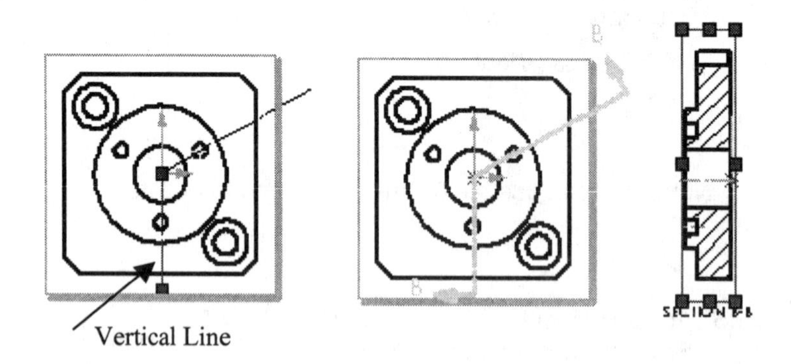

Note: Select a vertical or horizontal sketched line before you create the Aligned view. Otherwise, the aligned view is created from the bent sketched line.

Utilize 2 sketch line segments for Aligned Section view. Utilize 1, 3 or more sketch line segments and arc segments for Section view.

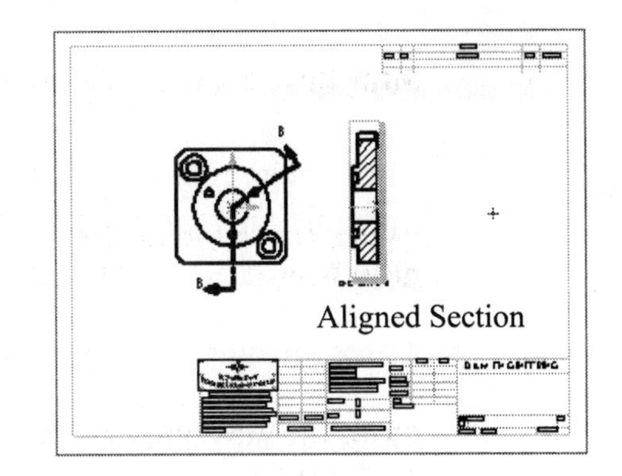
Aligned Section

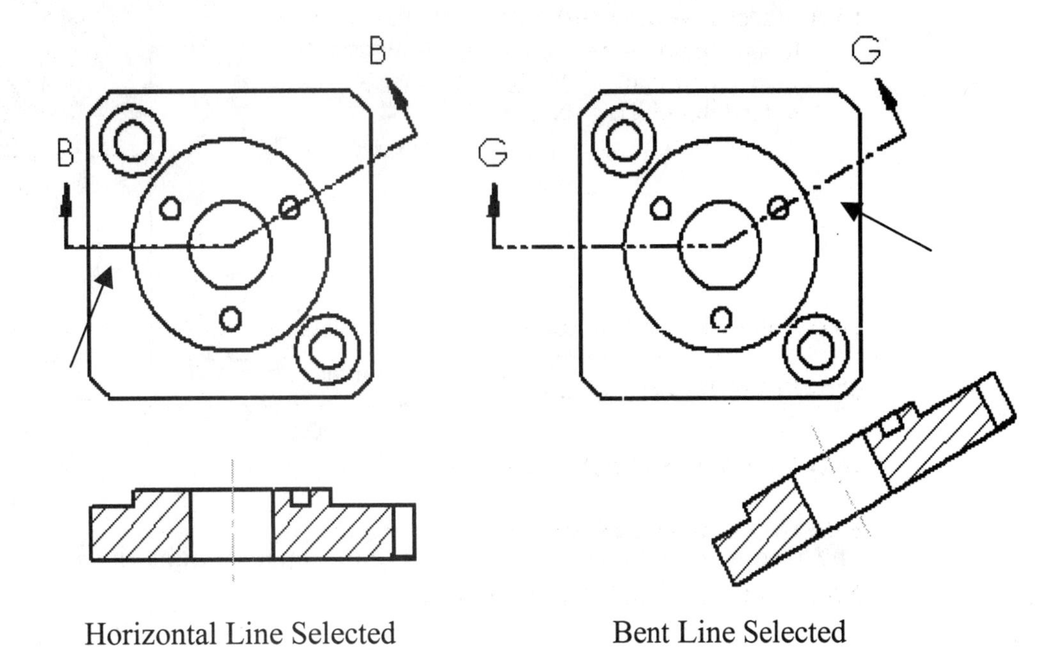

Horizontal Line Selected

Bent Line Selected

There are two additional section view options. The two additional section view options are: Partial section and Display only surface. A Half Section is created with a section line.

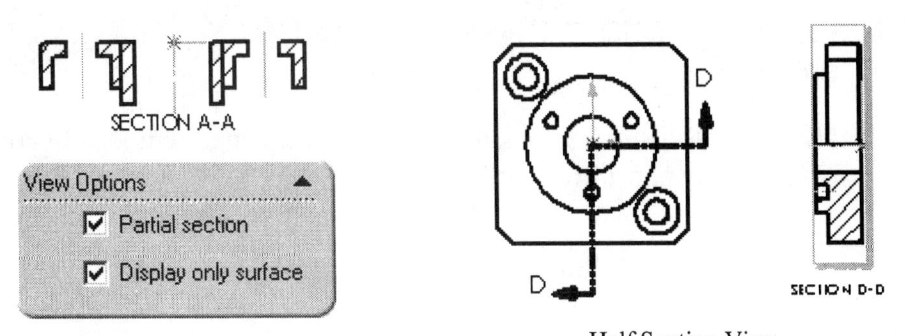

Partial Section View

Half Section View

Multi-view Drawings

The following section is not related to the CYLINDER assembly. The information is provided to you in order to create other types of multi-view drawings.

There are no step-by-step instructions in this section.

Part drawings can contain multiple views. Multiple views may be required to represent the true shape of the part.

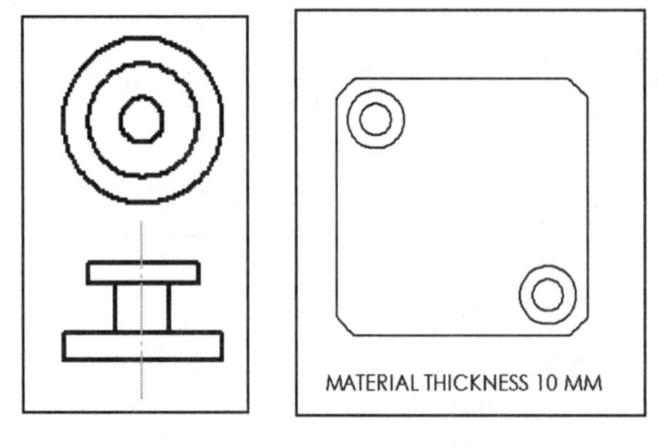

2 View Drawing 1 View Drawing

The Standard View displays the Front, Top and Right views.

Create the Left, Bottom and Rear views with a Projection or Named View .

A Section View , Detail View and or Auxiliary View may be required for part fabrication.

How do you create a two or one view drawing?

Let's first create a two view drawing. Create the Standard views: Front, Top and Right.

Hide one of the views. Another two view drawing method is to create a Named view and a Projected view.

Create a single view drawing. Create the drawing by using the Named view. Use a parametric note to represent material thickness.

Orient the part based upon its position in the assembly. Use Named view, Rotate view and Projection view to orient the part in a fabrication drawing. Select the view orientation that minimizes hidden lines and outlines the profile of the feature.

The ASME Y14.3M standard defines other view types not required in this Project. These views are applied to different type of drawings.

A Primary Auxiliary view: VIEW A is aligned and adjacent to the angled edge of the principle Front view.

A Secondary Auxiliary view: VIEW B is aligned and adjacent to a Primary Auxiliary view, VIEW A.

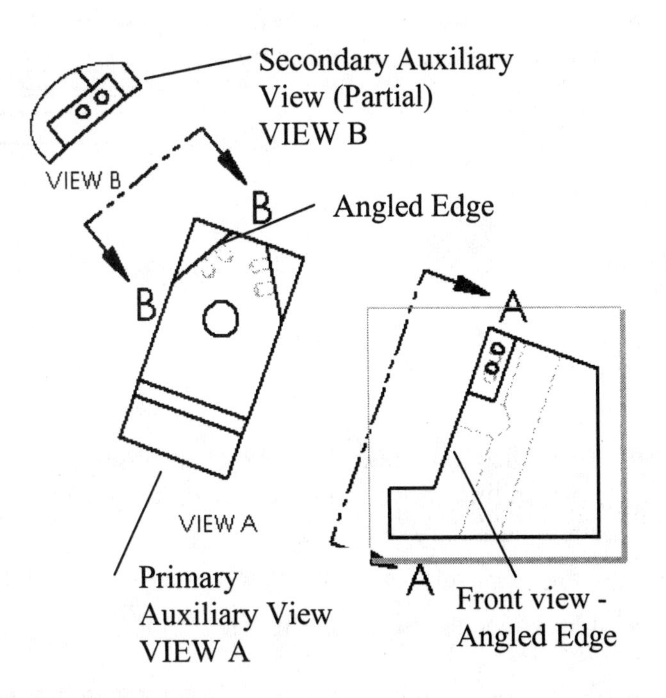

Select an edge in the Primary Auxiliary.

In SolidWorks, Secondary Auxiliary views are created from a Primary Auxiliary view. Use Auxiliary View to create the Secondary Auxiliary view.

Views can be rotated to fit within the sheet boundary. The angle and direction of rotation is placed below the view title.

Example: A Front view and Projected Left view are displayed in an A size drawing. Steps to Rotate the Front view:

- Select the Front view boundary.
- Click from the View toolbar.
- Enter Drawing view angle from the Rotate Drawing View dialog box. Example: 45°

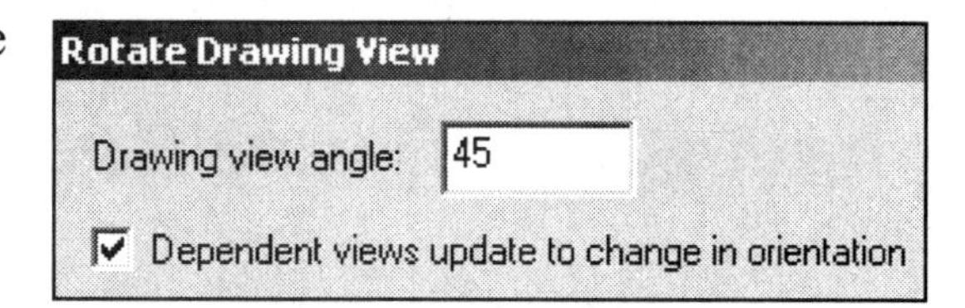

- The Left view depends upon the Front view and rotates by 45°.

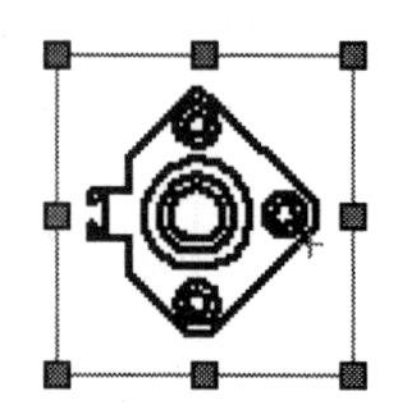

Steps to create a new Rotated view:

- Click Section View . Sketch a straight section line above the Front view. Click the view boundary to rotate.

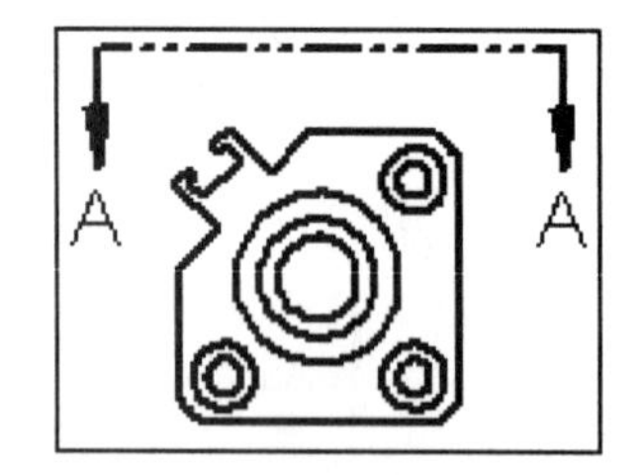

- Click Rotate from the View toolbar.

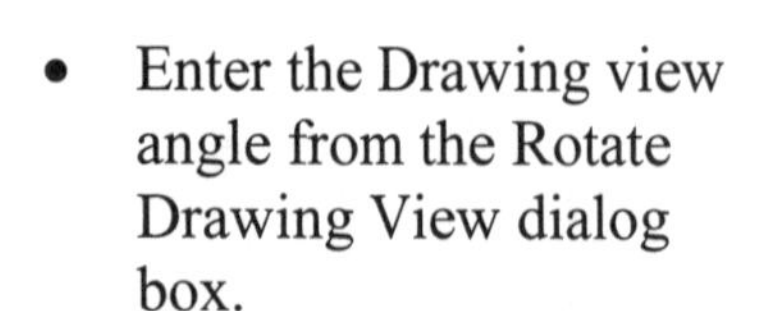

- Enter the Drawing view angle from the Rotate Drawing View dialog box.

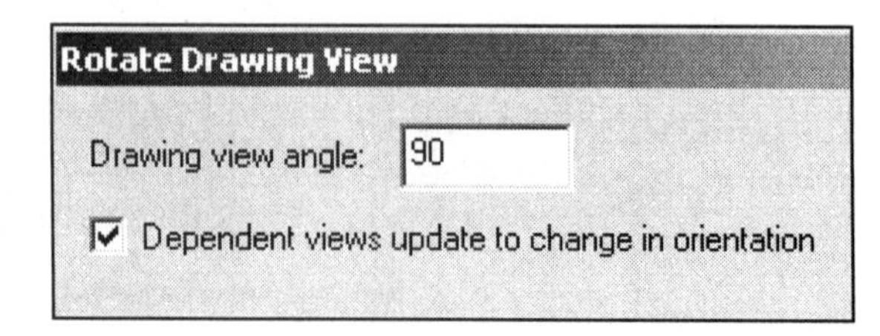

- Break alignment of the rotated view to position the view in a new location.

Break Alignment
Align Horizontal by Origin
Align Vertical by Origin
Align Horizontal by Center
Align Vertical by Center
Default Alignment

- Realign the view if required.

- Add a note with view name, rotated angle and direction. Example: VIEW A-A, ROTATED 90 CCW.

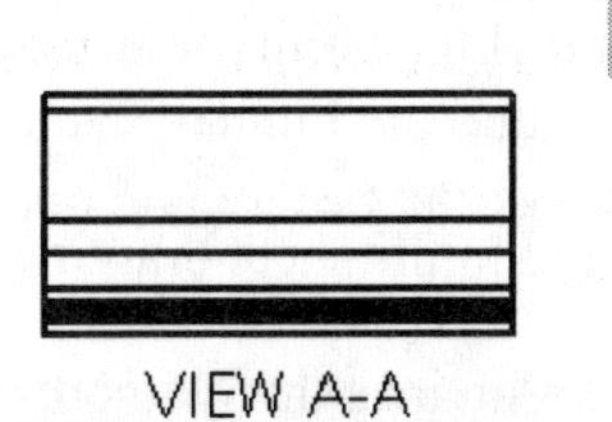

VIEW A-A
ROTATED 90° CCW

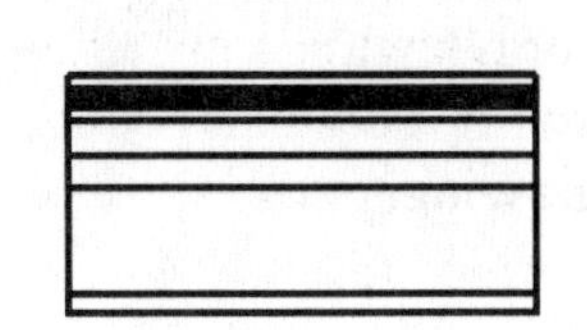

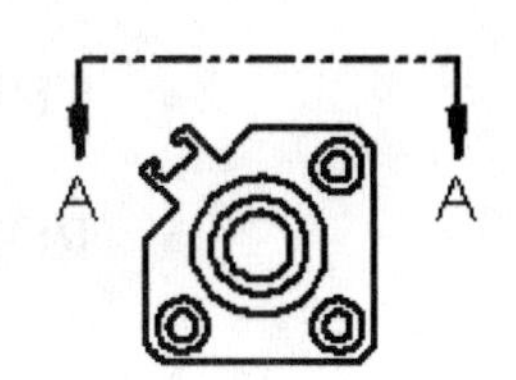

A Perspective view is the view normally seen by the human eye. Parallel lines recede into the distance to a vanishing point.

Utilize View, Display, Perspective in the Part.

Create a user defined Named view. Enter Perspective for View name.

Insert the Perspective named view into the drawing.

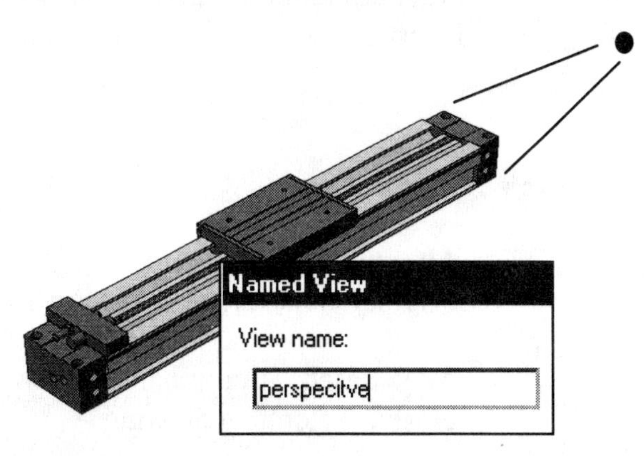

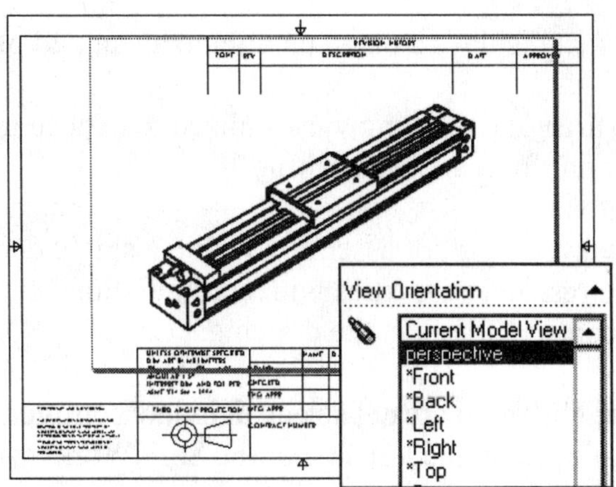

Project Summary

In this project you displayed and created Standard, Isometric, Auxiliary, Section, Broken Section, Detail and Half Section (Cut-away) views.

You obtained the ability to use SolidWorks drawing tools, other related view commands and the Fundamentals of Orthographic projection.

You created multi-sheet drawings from various part configurations.

The three drawings that you created were:

- TUBE drawing.
- ROD drawing.
- COVERPLATE drawing.

The ROD drawing consisted of three sheets. The TUBE drawing consisted of one sheet. The COVERPLATE drawing consisted of two sheets.

Practice creating drawings, views and more options in the project exercises. Insert dimensions from the part and create new annotations in Project 3.

Project Terminology

Alternate Section: A drawing view superimposed in phantom lines on the original view. Utilized to show range of motion of an assembly.

Area hatch: A crosshatch pattern or fill applied to a selected face or to a closed sketch in a drawing.

Auxiliary view: A view that displays a plane parallel to an angled plane.

Broken-out section: An area of a drawing view utilized to expose inner details by removing material from a closed profile.

Configuration: A variation of a part or assembly within a single document. Control dimensions, features, and properties with configurations in the part and assembly. Display configurations in the drawing.

ConfigurationManager: Utilize to create, select and view configurations of a part or assembly and displayed on the left side of the SolidWorks window.

Cosmetic thread: is an annotation that represents threads.

Detail view: A portion of a larger view, usually at a larger scale than the original view.

Driving dimensions: Sets the value for a sketch entity.

Extruded feature: Projects a sketch perpendicular to a sketch plane to add material to a part or remove material from a part.

Handle: An arrow, square, or circle that you drag to adjust the size or position of an entity such as a view or dimension.

Named view: A specific view of a part or assembly. Standard named views are listed in the view orientation dialog box such as isometric or front. Named view can be user-defined name for a specific view.

Origin: Represents the (0,0,0) coordinate of the model.

Section line: A line or centerline sketched in a drawing view to create a section view.

Section view: A drawing view created by cutting another drawing view with a section line. Section views show the interior of a model.

Spline: A sketched 2D or 3D curve defined by a set of control points.

Standard views: Are the three orthographic projection views, Front, Top and Right positioned on the drawing according to First angle or Third angle projection.

Mouse Buttons: The left and right mouse buttons have distinct meanings in SolidWorks.

System Feedback: Feedback is provided by a symbol attached to the cursor arrow indicating your selection. As the cursor floats across the sheet, view or geometry, feedback is provided in the form of symbols, riding next to the cursor.

Copy and Paste: Utilize copy/paste to copy views from on sheet to another sheet in a drawing.

Questions:

1. Name the three default reference planes: __________, ___________ and ______________.
2. Identify the six principle drawing views in Orthographic Projection: ______________, ______________, ______________, ______________, ______________, ______________.
3. Name the two Orthographic projection systems: ________________, ________________ .
4. A drawing contains multiple __________ of a part.
5. True or False. Delete the part when a drawing is complete.
6. True of False. All drawings contain a single part configuration.
7. A Design Table is inserted into two documents. Identify the two documents.
8. Describe the difference between view properties and sheet properties.
9. Identify the Drawing tool used to create an Isometric view.
10. Describe the procedure to copy a view from one sheet to another sheet in the same drawing.
11. True or False. Drawing layers exist in a SolidWorks drawing.
12. True or False. Use the Broken Isometric View Drawing Tool to create a Broken Isometric view.
13. Identify the command used to change the scale of a Detailed view.
14. Describe the procedure to display internal features of a part in an Isometric view of a drawing.
15. You create a multi-sheet drawing. On the first sheet, the correct Sheet Format is displayed. On the second sheet the incorrect Sheet Format is displayed. Identify the Sheet Properties to display the correct Sheet Format.

Exercises:

Exercise 2.1:

Add a second sheet to the TUBE Drawing. Copy/Paste the first Isometric view from the first sheet to the second sheet. Modify the part configuration to display the Entire Part. Modify the Sheet Scale from 2:1 to 4:1.

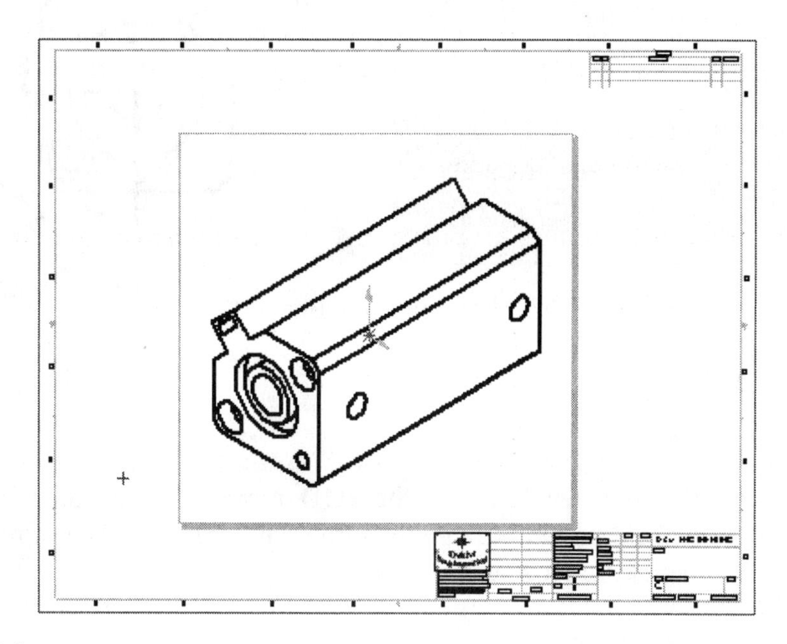

Exercise 2.2:

Modify the ROD Drawing. Add a fourth sheet. Use the Copy/Paste commands to place the Short Rod and Long Rod configuration on the third sheet; ROD-Sheet4. Copy the Broken Right view. Right-click in the view to remove the break.

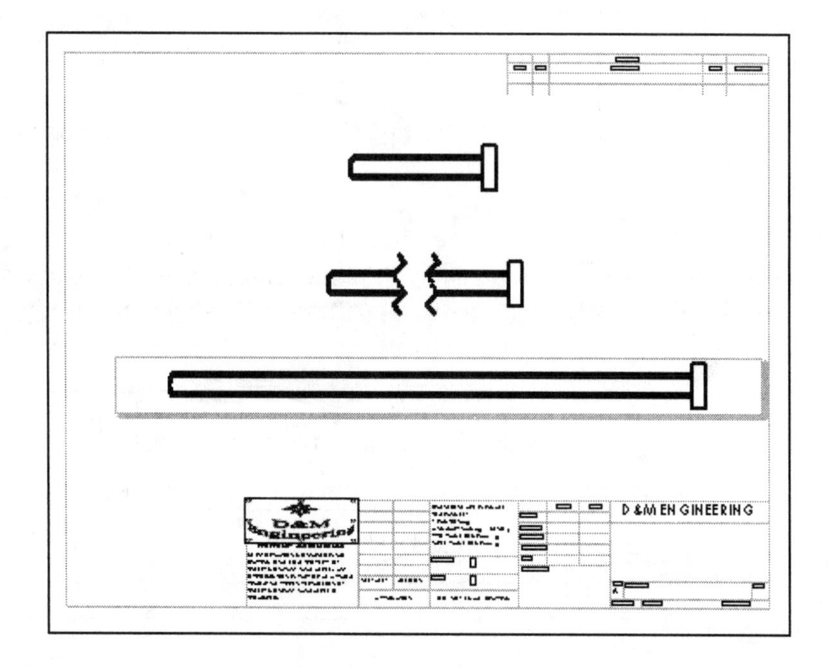

Exercise 2.3:

Modify the COVERPLATE drawing. Add a third sheet. Add a Partial Section view and a Half Section view.

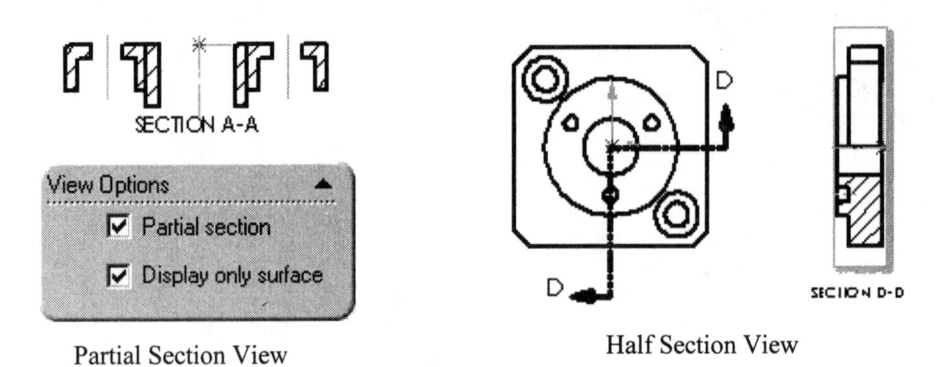

Partial Section View

Half Section View

Exercise 2.4:

Create a new Sheet for the ROD drawing. Select the SHORT ROD Configruation. Create the three standard views Top, Front , Right, an Isometric view and a Back view.

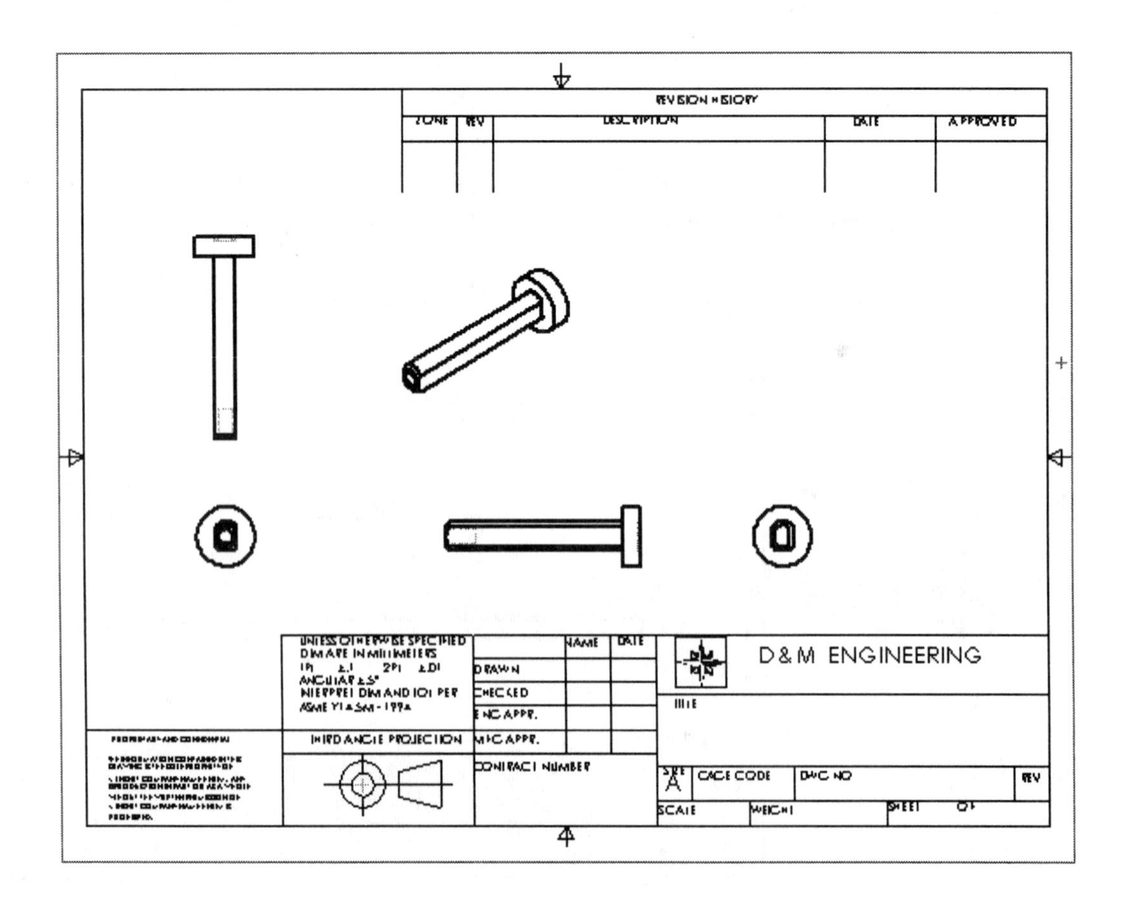

Exercise 2.5:

Create a new drawing for the part, AUXSEC located in the 2003drwparts file folder. Create a primary and secondary Auxiliary view.

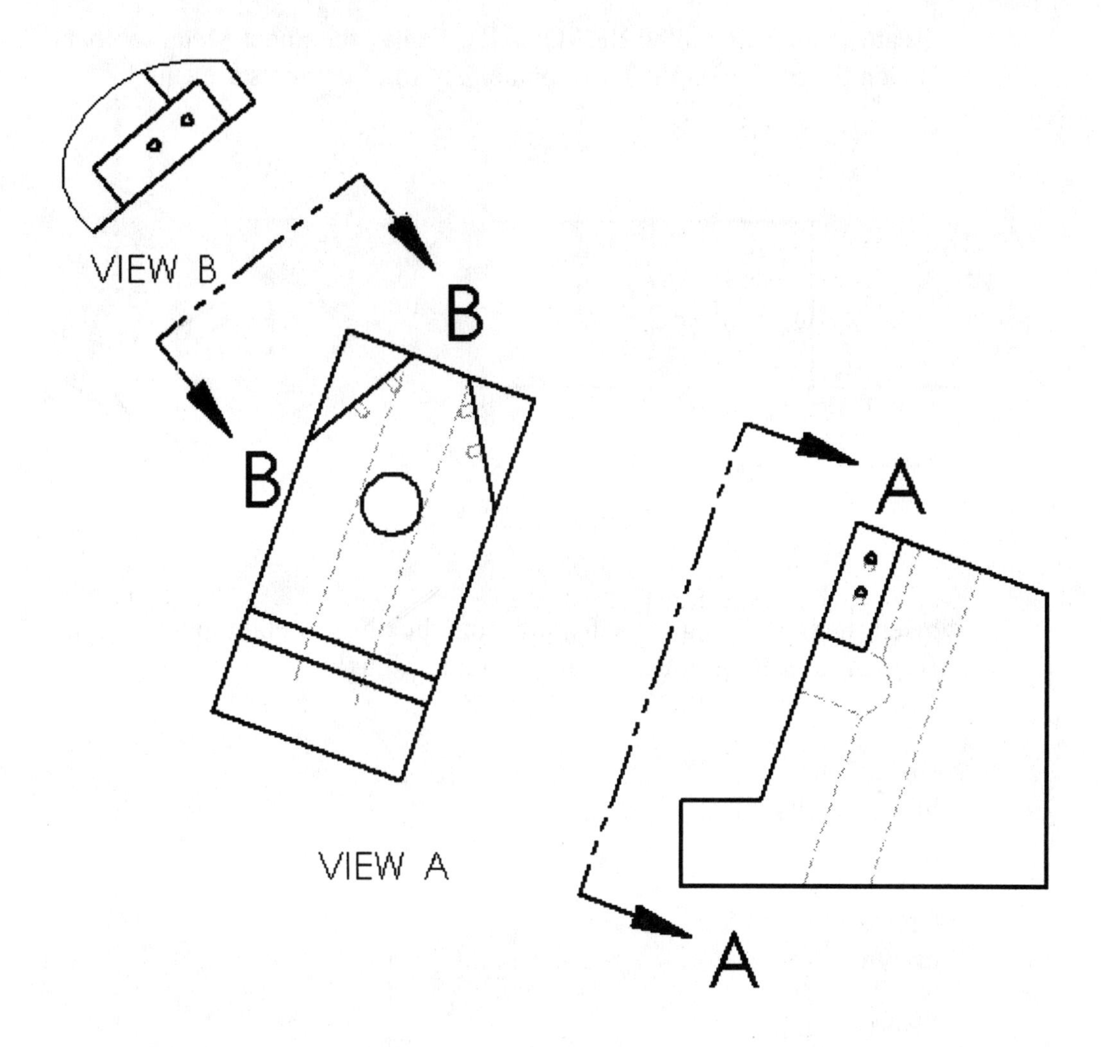

Exercise 2.6:

Sheet Metal Drawings utilized the form state (3D) a flat state (2D) on a drawing.

Create a new part, called BRACKET. Display the Sheet Metal toolbar. Sketch the profile on the Front plane. All line segments are equal.

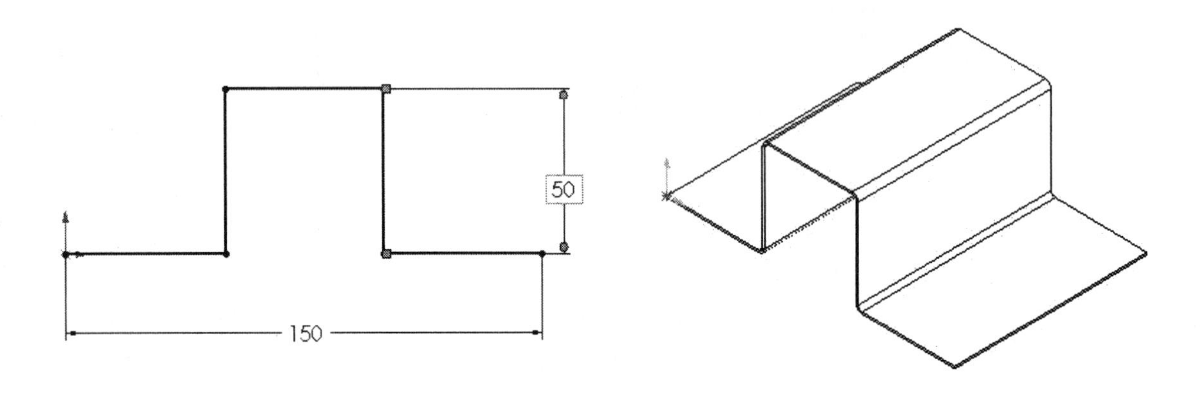

Insert the Base Flange feature from the Sheet Metal toolbar. Enter 100mm for Depth. Enter 2mm for Bend Radius.

Display the Flat State. Click Flatten .

Create a Sheet Metal drawing.

Insert a Named view, Isometric

Insert a Named view, Flat Pattern.

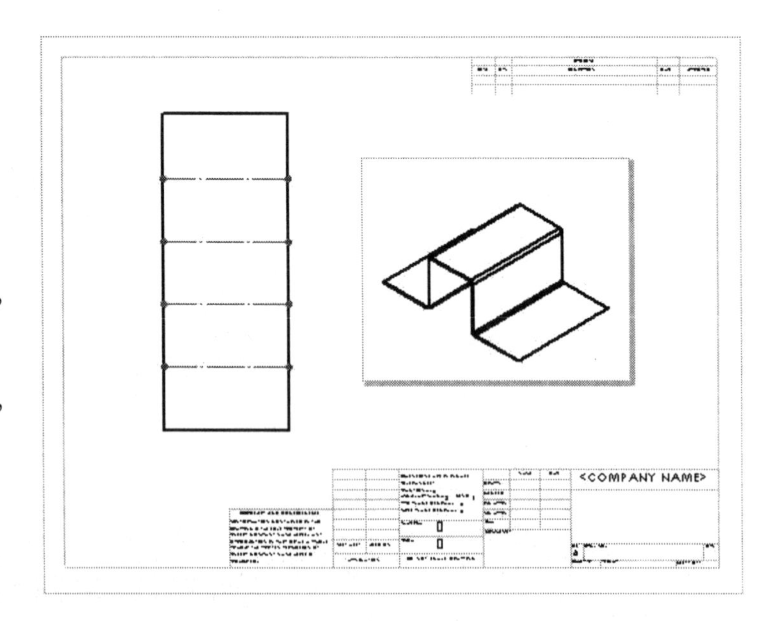

A sheet metal part contains two configurations. The Default configuration is in the formed state. The Default-SM-FLAT-PATTERN configuration is in the flatten state.

Exercise 2.7:

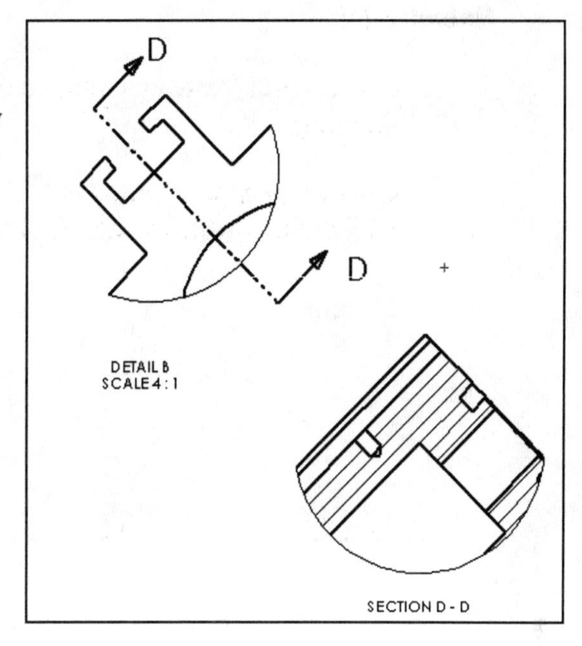

Create a Section view from the Detail B. A Section view cannot be created directly from a Detail view. Create Section D-D.

Hints:

- Copy the Front view to the bottom left corner of the Sheet1.
- Drag the Front view off the Sheet1 boundary.
- Create a Section view (SECTION E-E) to the left of the copied Front view.
- Create a circular Detail view.
- Drag the Detailed view back into the Sheet boundary. Enter SECTION D-D for text.
- Create two new layers. Layer Section Line contains a Thick Phatom Line Sytle. Layer Arrow contains a Solid Continuous line. Sketch a Line on the Section Layer through Detail B at 45°. Sketch two short ⊥ lines. Sketch a triangle on the end of each line to create an arrow head. Add Note D to the right of the arrow head.

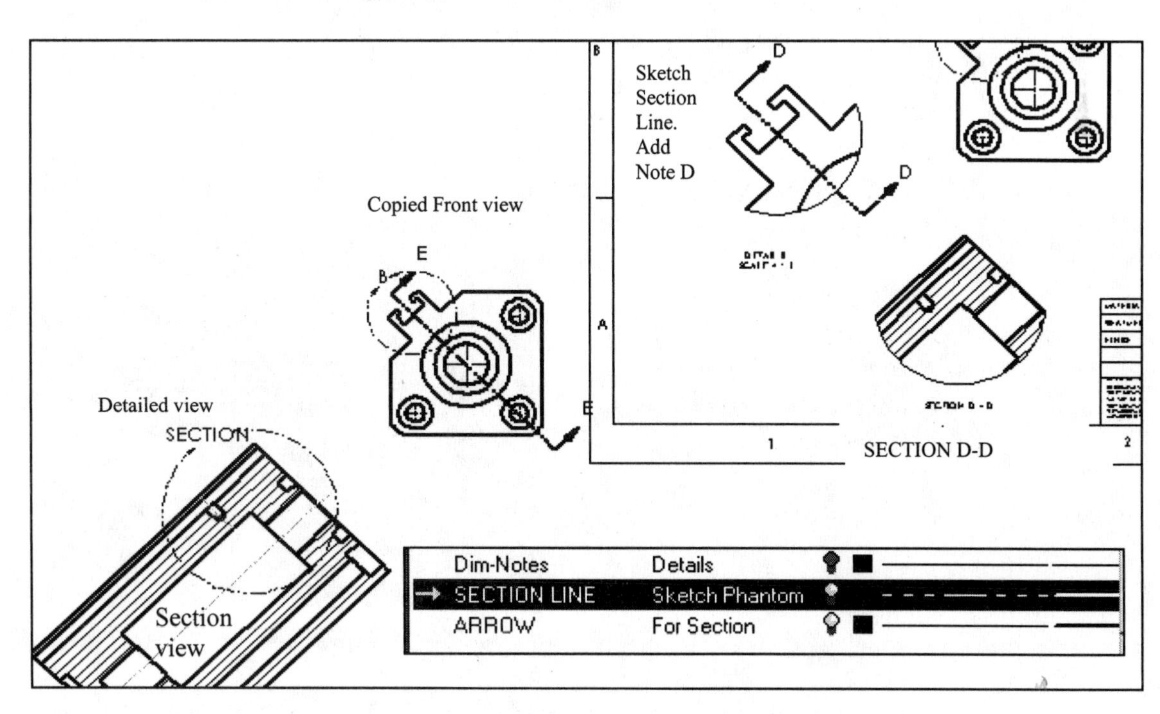

Notes:

Project 3

Fundamentals of Detailing

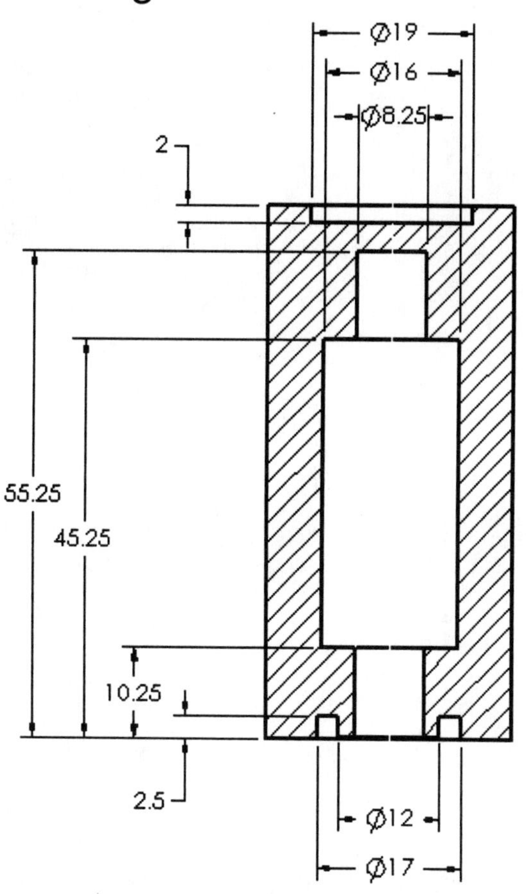

Below are the desired outcomes and usage competencies based on the completion of this Project. Note: Drawing refers to the SolidWorks module used to insert, add and modify views in an engineering drawing. Detailing refers to the SolidWorks module used to insert, add and modify dimensions and notes in an engineering drawing.

Project Desired Outcomes	Usage Competencies
TUBE drawing with detailing. ROD drawing with detailing. COVERPLATE drawing with detailing.	Ability to insert, add and modify dimensions.
	An understanding of inserting and adding notes.
	Knowledge of dimensioning standards.

Notes

Project 3 – Fundamentals of Detailing

Project Objective

Create three detailed drawings:

- Detailed TUBE drawing.
- Detailed ROD drawing.

Detailed COVERPLATE drawing.

Details are the drawing dimensions and notes required to document part features.

Achieve the ability to insert, add and modify dimensions along with obtaining an understanding of inserting and adding notes in a drawing.

Attain an exposure to Drawing dimensioning.

Explore methods to move, hide and recreate dimensions to adhere to a various drawing standard.

On the completion of this project, you will be able to:

- Add Dimensions
- Add Notes.
- Address Hole Call Outs.
- Add Center Marks and Centerlines.
- Use various methods to move, hide and recreate dimensions.
- Use the ASME Y14.5 standard for Types of Decimal Dimensions.
- Add Modifying Symbols and Hole Symbols.

Project Situation

You inserted and added views for the TUBE, ROD and COVERPLATE drawings in Project 2. In this Project, you will insert, add and modify dimensions along with obtaining an understanding of inserting and adding notes in a Drawing.

Note: Details are the drawing dimensions and notes required to document part features.

Insert dimensions are part dimensions. Insert dimensions are associative. If a dimension is modified in the drawing, the part will be modified.

Added dimensions are drawing dimensions. Added dimensions are reference dimensions. Reference dimensions are driven by part features. A reference dimension cannot be edited.

Annotations such as notes, hole call outs and center marks are added in the drawing document.

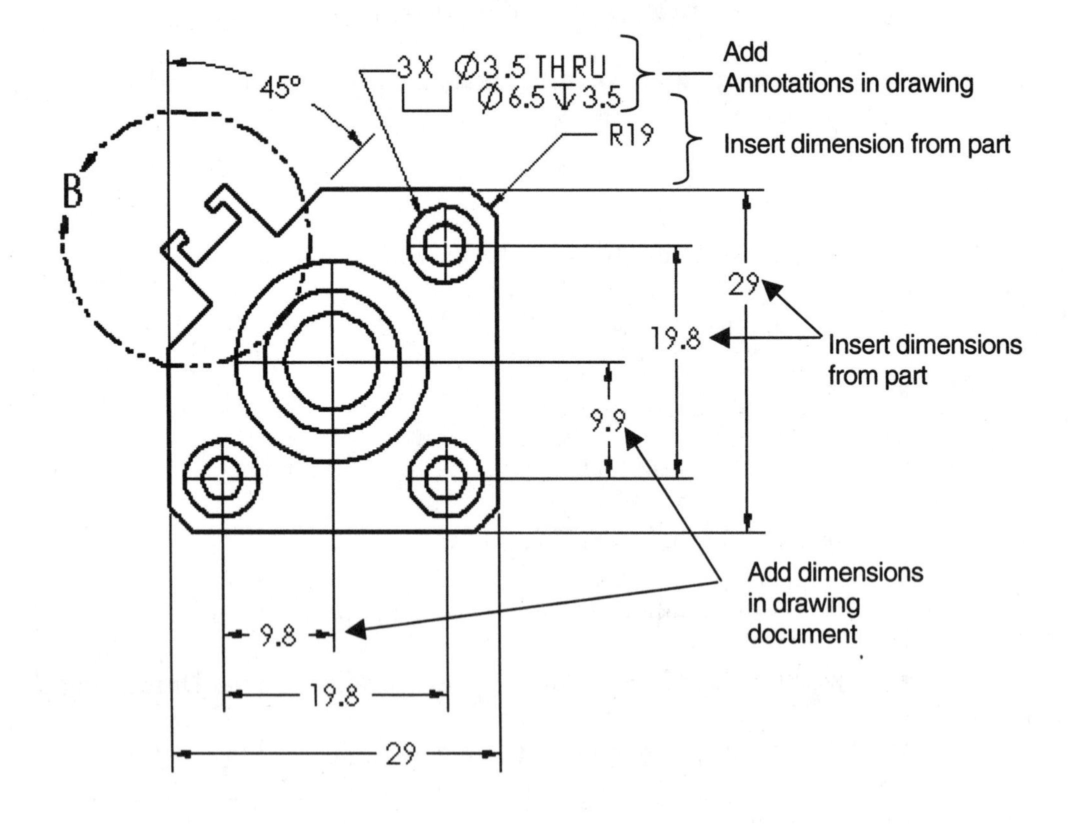

Project Overview

Create three detailed drawings:

- Detailed TUBE drawing.
- Detailed ROD drawing.
- Detailed COVERPLATE drawing.

The design intent of this project is to work with dimensions inserted from parts and to incorporate them into the drawings. Explore methods to move, hide and recreate dimensions to adhere to a drawing standard. Note: There are other solutions to the dimensioning schemes illustrated in this project. The TUBE, ROD and COVERPALTE drawings are sample drawings; they are not complete. A drawing requires tolerances, materials, Engineering Change Orders, etc. to release the part to manufacturing and other notes prior to production.

Review a hypothetical "worse case" drawing situation. You just inserted dimensions from a part into a drawing. The dimensions, extensions lines and arrows are not in the right locations. How can you improve the position of these details?

No	Situation
1	Extension line crosses dimension line. Dimensions not evenly spaced
2	Largest dimension placed closest to profile
3	Arrow heads overlapping
4	Extension line crossing close to arrowhead
5	Arrow gap too large
6	Dimension pointing to feature in another view. Missing dimension – inserted into Detail view (not shown)
7	Dimension text over centerline, too close to profile.
8	Dimension from other view – leader line too long
9	Dimension inside section lines
10	No visible gap
11	Arrows overlapping text
12	Incorrect decimal display with whole number (millimeter)

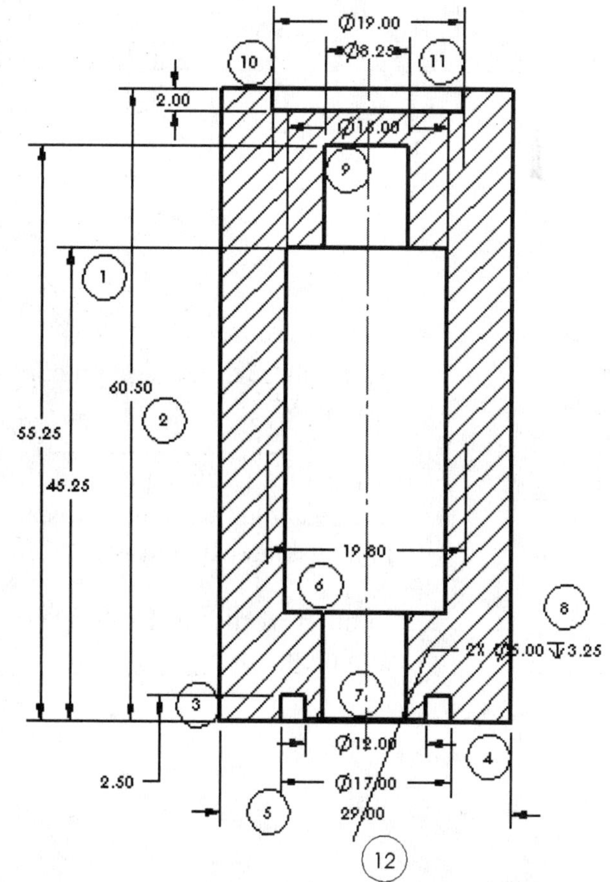

The ASME Y14.5M standard defines an engineering drawing standard. Review the eleven changes made to the drawing to meet the standard.

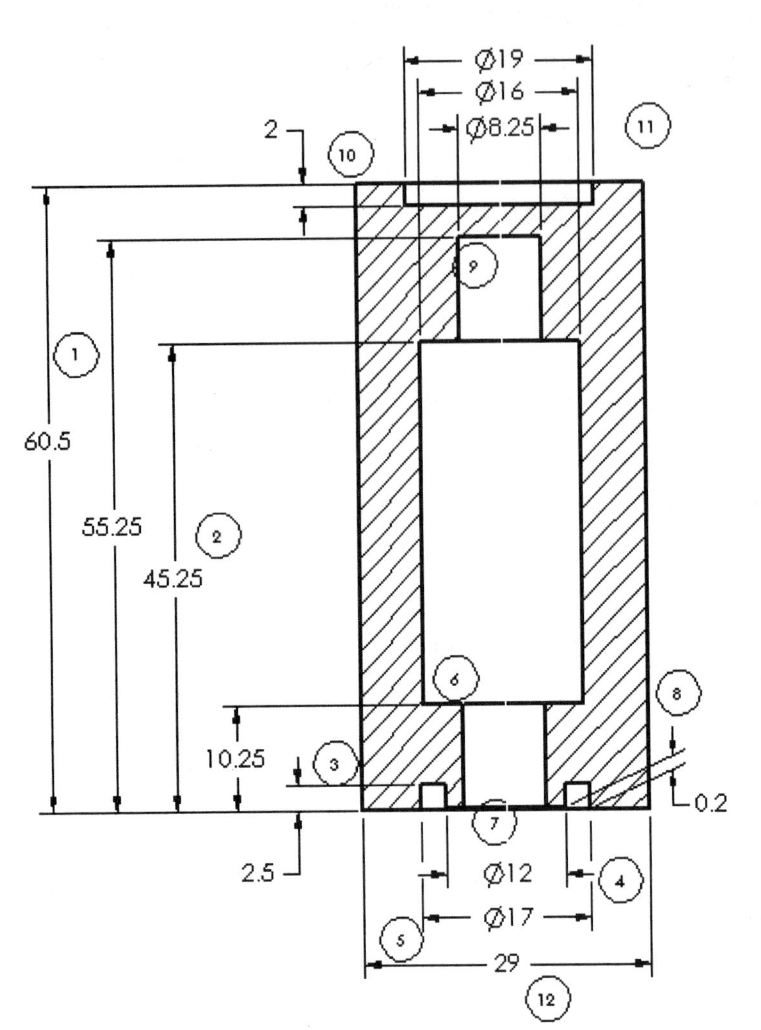

No:	Preferred Application of the Dimensions:
1	Extension lines do not cross unless situation is unavoidable. Stagger dimension text.
2	Largest dimension placed farthest from profile. Dimensions are evenly spaced and grouped.
3	Arrow heads do not overlap
4	Break extension lines that cross close to arrowhead.
5	Flip arrows.
6	Move dimensions to the view that displays the outline of the feature. Insure that all dimensions are accounted for.
7	Move text off of reference geometry (centerline).
8	Drag dimensions into their correct view boundary. Create reference dimensions if required. Slant extension lines to clearly illustrate feature.
9	Locate dimensions outside off section lines.
10	Create a visible gap between extension lines and profile lines.
11	Arrows do not overlap the text.
12	Whole numbers displayed with no zero and no decimal point (millimeter)

Apply these dimension practices to the TUBE, ROD and COVERPLATE drawings.

A Detailed Drawing is used to manufacture a part. A mistake on a drawing can cost your company substantial loss in revenue. The mistake could result in a customer liability lawsuit.

In other words, as the designer, dimension and annotate your parts clearly to avoid common problems.

Review the ASME Y14.5 standard for Types of Decimal Dimensions.

TYPES of DECIMAL DIMENSIONS (ASME Y14.5M)			
Description:	**Example: MM**	**Description:**	**Example: INCH**
Dimension is less than 1mm. Zero precedes the decimal point.	0.9 0.95	Dimension is less than 1 inch. Zero is not used before the decimal point.	.5 .56
Dimension is a whole number. Display no decimal point. Display no zero after decimal point.	19	Express dimension to the same number of decimal places as its tolerance. Add zeros to the right of the decimal point. If the tolerance is expressed to 3 places, then the dimension contains 3 places to the right of the decimal point.	1.750
Dimension exceeds a whole number by a decimal fraction of a millimeter. Display no zero to the right of the decimal.	11.5 11.51		

The SolidWorks dimensioning standard is set to ANSI. Trailing zeroes is set to Smart. The Primary units are Millimeters.

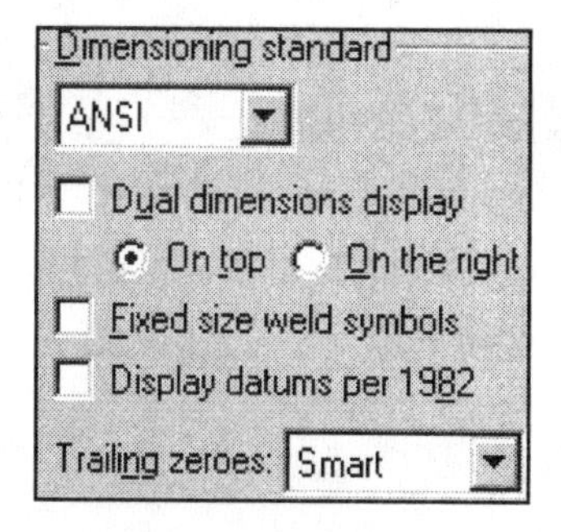

SolidWorks displays a leading zero for millimeter dimensions less than one. SolidWorks displays no decimal point and no zero after the decimal point for whole number dimensions.

The Dimension Precision option controls the number of decimal places displayed for dimension and tolerance values.

Modify individual millimeter dimensions if the decimal fraction ends in a zero.

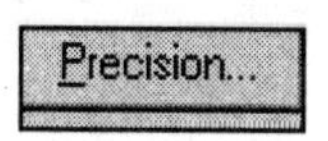

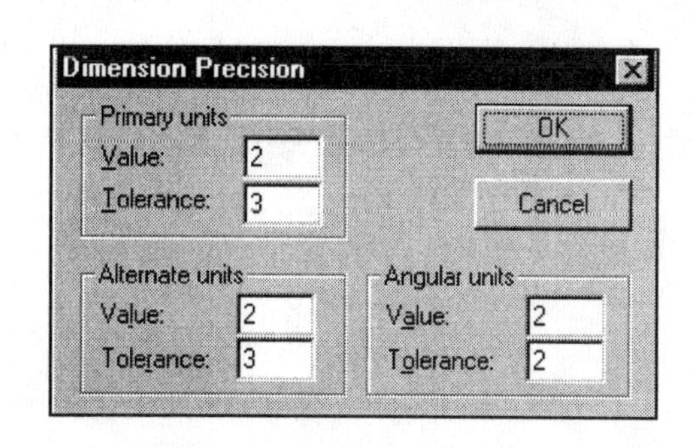

Example 1:

Set Precision Primary Units to 2 places.
The drawing dimension displays 0.95.
The number of decimal places is two.
No change is required.

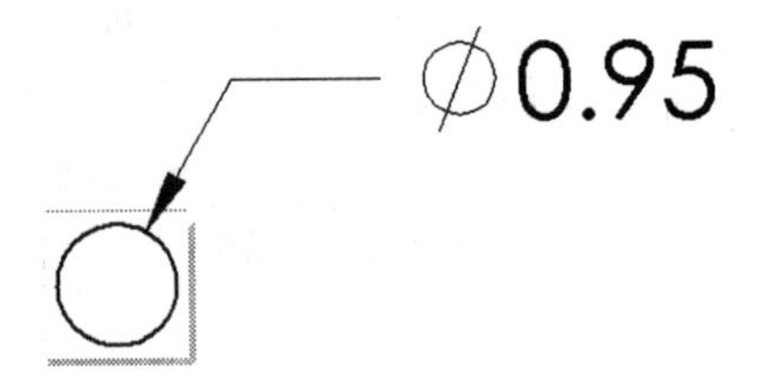

Example 2: The drawing dimension displays 0.90. Control individual dimension precision through the Dimension Properties Tolerance/Precision text box.

Modify the dimension Primary Units display to .X, (one decimal place). The drawing dimension displays 0.9.

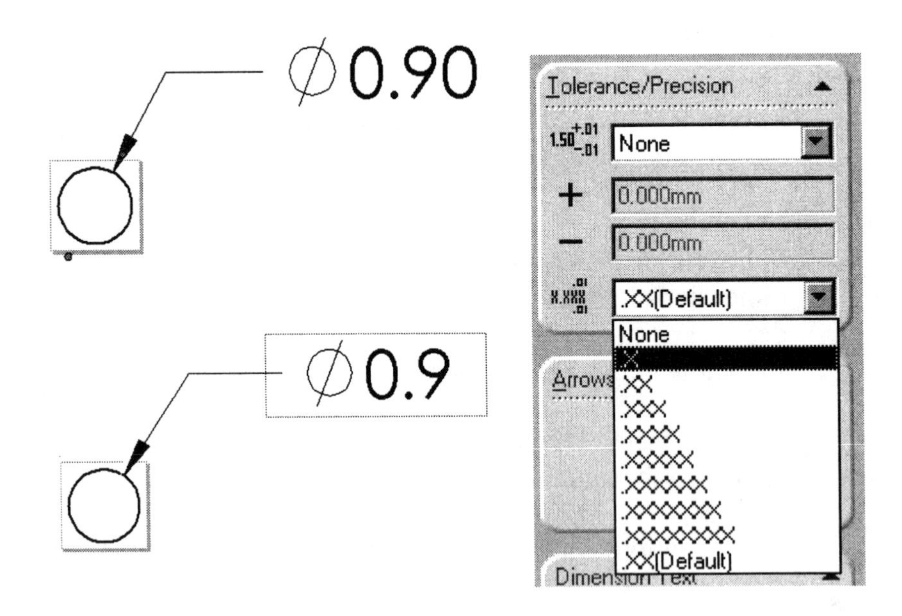

General Tolerance values apply to all dimensions on a drawing except reference dimensions and material stock sizes.

Tolerance values are displayed with 1, 2 and or 3 decimal places.

UNLESS OTHERWISE SPECIFIED
DIM ARE IN MILLIMETERS
1PL ±0.2 2PL ±0.05
ANGULAR ±.5°
INTERPRET DIM AND TOL PER
ASME Y14.5M - 1994

Values differ for machined parts, plastic parts, sheet metal parts, castings and other manufacturing processes.

Example:

1PL is ±0.2.

The dimension 0.9 has a tolerance value of ±0.2.

The feature dimension range is 0.7mm - 1.1mm. The tolerance equals 1.1mm - 0.7mm = 0.4mm.

Example:

2PL is ±0.05.

The dimension 0.95 has a tolerance value of ±0.05.

The feature dimension range is 0.90mm - 1.00mm. The tolerance equals 1.00mm - 0.90mm = 0.10mm.

The Document Property, Trailing Zeros has three options: Smart, Show and Remove. Smart removes all zeros to the right of the decimal point for whole numbers.

Show displays the number of zeros equal to the number of places specified in the Units option. Remove displays no trailing zeros to the right of the dimension value.

Set the Trailing Zeros to Smart. Control individual dimensions with the Primary Units Precision option.

Trailing Zeros do not affect tolerance display. The Tolerance display for the drawing is located in the Dimensioning Precision dialog box.

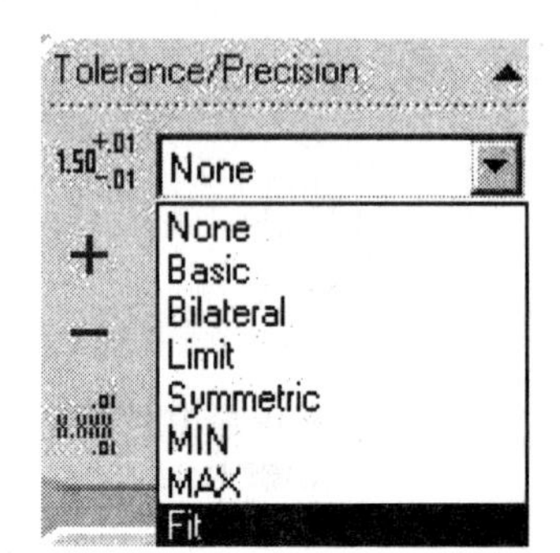

The Tolerance/Precision text box determines specific display for individual dimensions.

Specific Tolerance, such as None, Basic, Bilateral and Limit, are explored in Project 5.

Tolerance display for inch and metric dimensions is as follows:

Tolerance Display for Inch and Metric DIMENSIONS (ASME Y14.5M)		
Display:	**Metric:**	**Inch:**
Unilateral Tolerance	$36^{\;0}_{-0.5}$	$1.417^{+.005}_{-.000}$
Bilateral Tolerance	$36^{+0.25}_{-0.50}$	$1.417^{+.010}_{-.020}$
Limit Tolerance	14.50 11.50	.571 .463

SolidWorks Tools and Commands

The following Annotation tools are utilized in this Project:

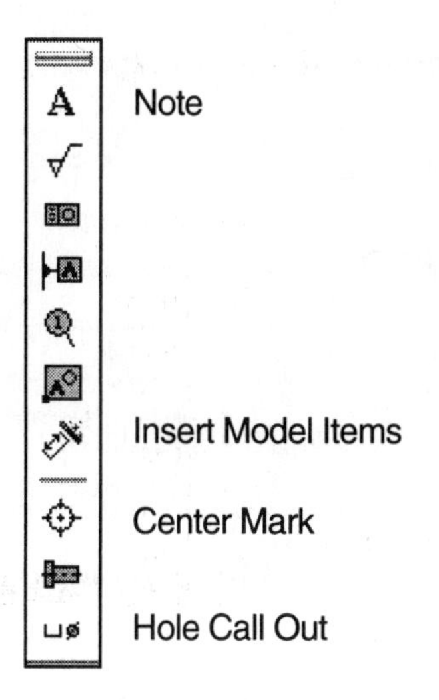

Other tools and command in this Project:

SolidWorks Tools and Commands:		
Layers	Gap	Cosmetic Thread
Hide/Show Annotations	Hide View/ Hide Edge	Break/Show Alignment
Align Parallel/Concentric	Display Options: Slant, Center Text, Show ( )	Break Dimension Witness/Leader Line
Grid	Foreshortened Radius	Min/Center/Max Arc Conditions
Configurations/Design Table	Baseline	Horizontal Ordinate/Vertical Ordinate

Additional information on Annotation tools and other commands are found in the SolidWorks On-line help.

Detailing the TUBE Drawing

Detailing the TUBE Drawing requires numerous steps. Example:

- Insert part dimensions into the Tube Drawing.
- Reposition dimension to the appropriate view.
- Add reference dimensions to the drawing.
- Add annotations.
- Review each view.
- Apply dimensions according to your company's standard.

There are three methods to import model dimensions from the part or assembly to the drawing. They are:

- Entire model.
- Selected Component (for assembly drawings).
- Selected Feature (for part and assembly drawings).

Review the view names, before you Insert Dimensions.

Open the TUBE drawing.

1) Click **File**, **Open**.

2) Select **Drawing Files** for file type.

3) Enter **TUBE**.

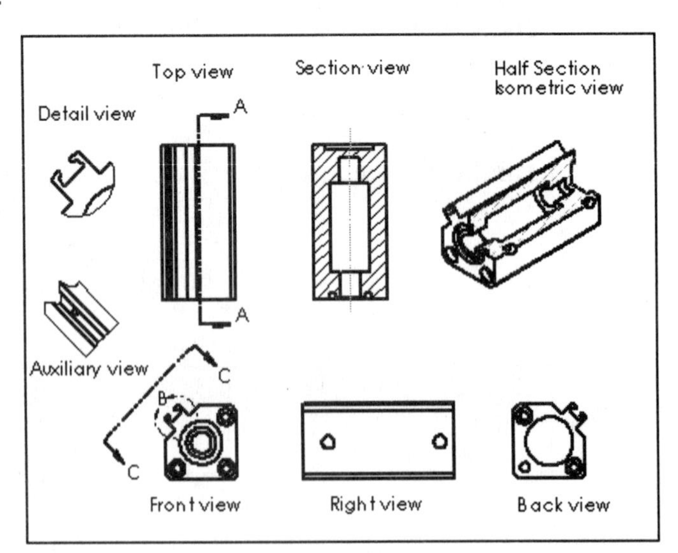

Hide the part Origins.

4) Click **View** from the Main menu. Uncheck **Origins**. The Origins in all the views are no longer displayed.

Create a New Layer.

5) Display the Layer toolbar. Click **View**, **Toolbars**, **Layer**.

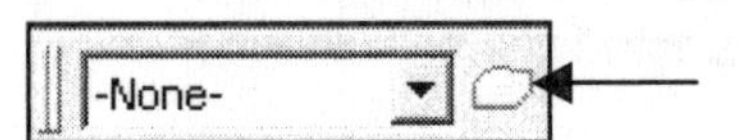

6) Display the Layers dialog box. Click **Layer Properties**.

7) Click **New** Button.

8) Enter **Details** for Name.

9) Enter **Dim & Notes** for Description. The Layer is On when the Light Bulb is yellow. Example: Enter **Cyan** for Color. Accept the default Style and Thickness.

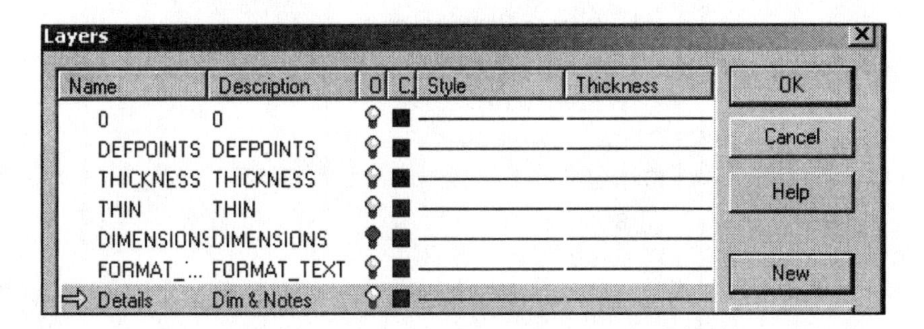

10) Click **OK**.

Layer Details is the active layer. Inserted dimensions from the part are displayed on layer Details.

Added dimensions from the Sketch toolbar and notes from the Annotation

toolbar are also displayed on layer Details.

Note: Layer names are truncated to 31 characters. Typed spaces in layer names are replaced with underscores.

Insert Dimensions for the Entire model.

11) Click a **position** inside the sheet boundary and outside any view boundary.

Note: No view boundaries are displayed in green. The dimensions will only be inserted into active views.

12) Click **Model Items** from the Annotations toolbar. The Dimensions option and Entire model option are selected.

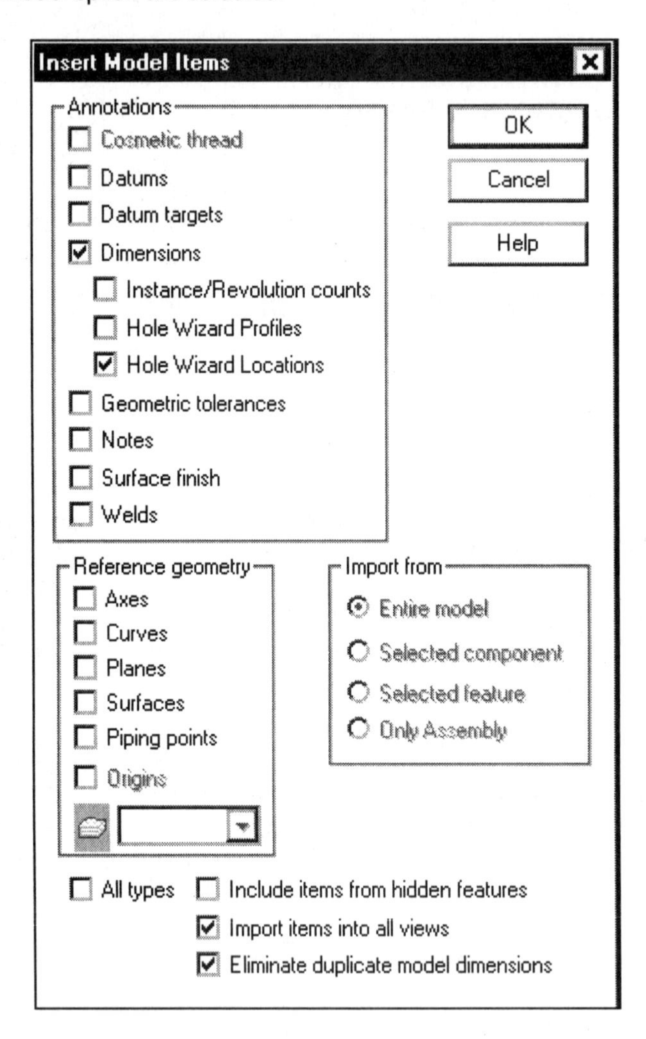

13) Display dimensions. Click **OK**.

Dimensions are displayed in a specific order. First, SolidWorks imports dimensions into all section views and detail views.

Next, the dimensions are positioned in the standard views.

How do you reposition numerous dimensions? Answer: One view at a time. Use the following tips:

- Hide views temporarily when not in use.
- Hide dimensions that are not longer required. Do not delete them. It takes less time to show a hidden dimension that to create one.
- Temporarily move views to see dimensions on top of other views.

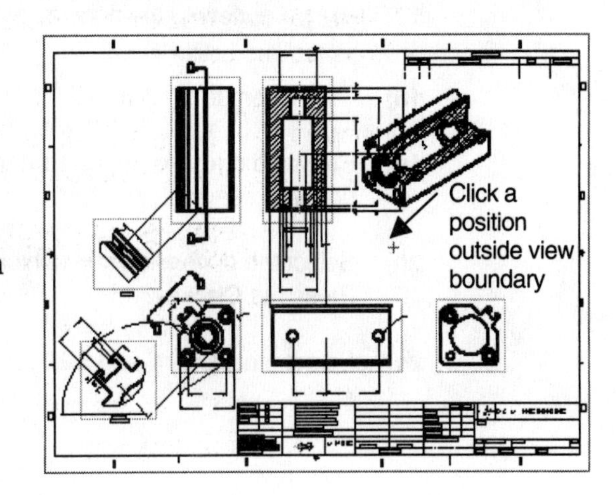

- Turn the default parenthesis display off when creating baseline dimensions.

There are different dimension techniques.

Example: To display minimum and maximum tolerance conditions, some companies create a single part configuration for the assembly and a single part configuration for the drawing.

Other companies define all part features from datum planes used in manufacturing.

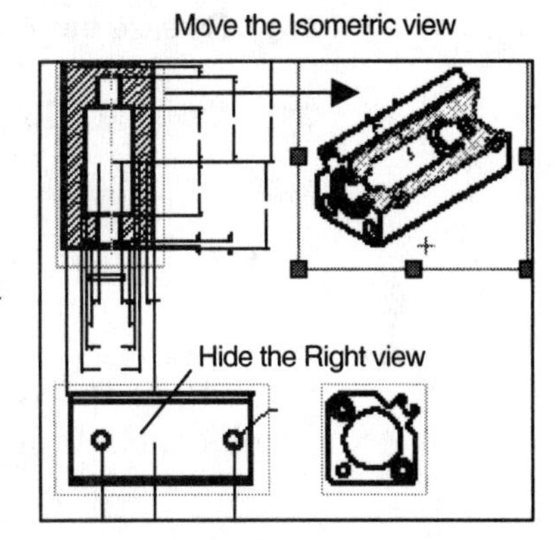

Temporarily hide views when not in use.

14) Hide the Right view. Right-click the **view boundary**.

15) Click Hide view.

Hide dimensions from the Half Section Isometric view.

16) Click the Cut-Away Section **view boundary**.

17) Drag the **view** to the right, away from the Section view dimensions.

18) Click Zoom to Selection .

19) Click and drag the **dimension text** until you see each number.

20) Select the dimension text to hide. Press the **Ctrl** key.

21) Select the **11**, **3**, **1**, and **6.30** text.

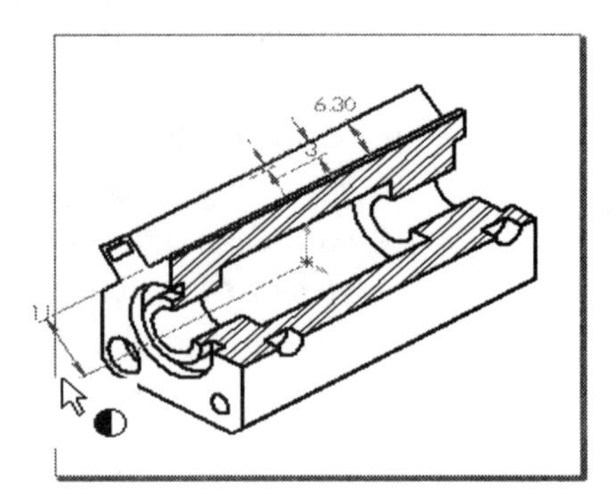

22) Right-click on the **dimension text**.

23) Click **Hide**.

24) Release the **Ctrl** key.

25) Save the TUBE drawing. Click **Save**.

Note: Hide and Delete commands that may not completely remove all of the graphic bits. If dimensions do not erase completely, click Rebuild.

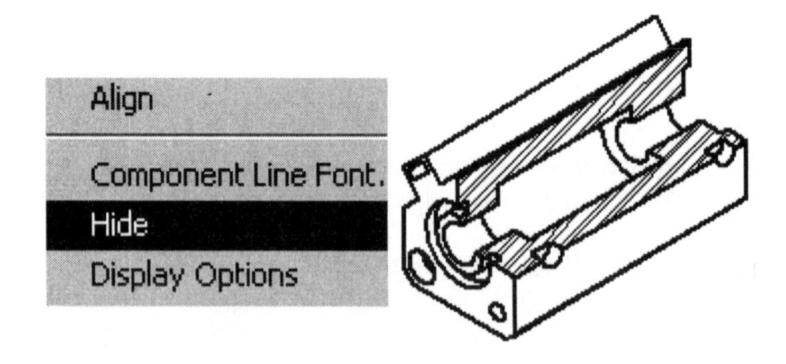

What command do you select when dimensions are no longer required? Answer: The Hide command. Hide dimensions versus delete dimensions.

Caution should be used when deleting a dimension. You may need it again.

If you hide a dimension you can get it back easily.

Utilize View, Hide/Show Annotations from the Main toolbar. The hidden dimensions are displayed in gray. Position the mouse pointer over the dimension to Show. The mouse pointer displays ◐. Click on the dimension text to display.

The dimensions added to a drawing are called Reference dimensions. Model dimensions drive Reference dimensions.

Model dimensions are created in a part or an assembly. You cannot change a Reference dimension. The dimension for overall length and Stroke Chamber are defined from the Front reference plane.

Detailing
Dimensions
Add parentheses by default
Snap text to grid

The part dimension scheme was the engineer's intent. Now as the detailer you define the dimensions to a base line. Hide the dimensions to avoid superfluous dimensions.

Reference dimensions may be displayed with parenthesis.

The command, Add parentheses by default, is unchecked to conserve time.

The ASME Y14.5M-1994 standard uses parenthesis to represent a Reference dimension for an Overall Reference dimension and an Intermediate Reference dimension.

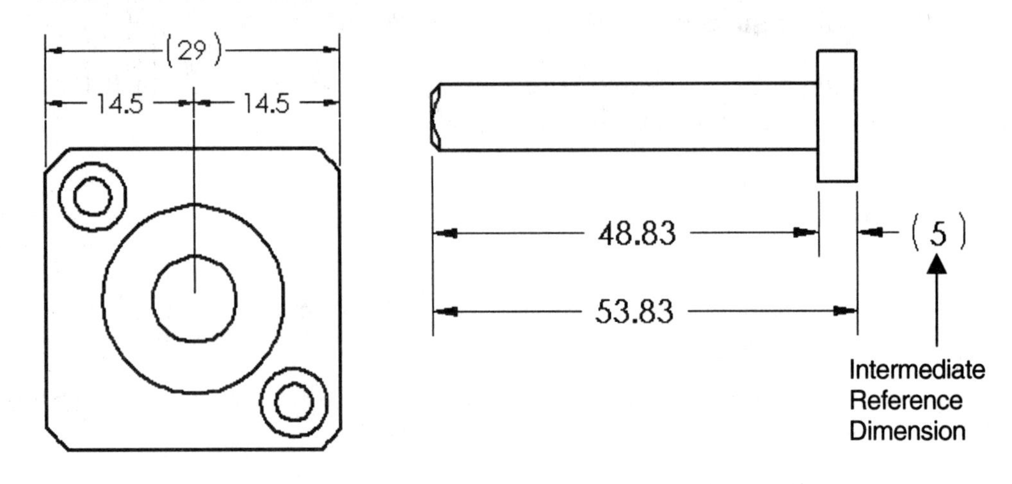

Control the dimensions that contain a parenthesis to adhere to your company's drawing standard.

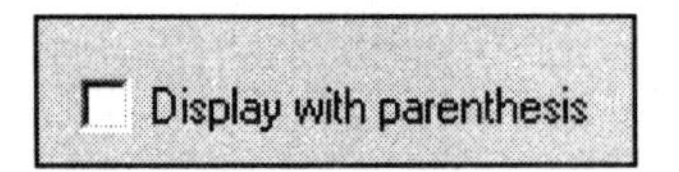

Select Properties on the dimension text.

Uncheck the Display with Parenthesis option to control the individual Reference dimensions.

Temporarily move the Section view.

26) Drag the Section view **boundary** to the right until the dimension text is off the Top view.

Create the overall depth dimension.

27) Click the **Top** view. Click **Zoom to Selection**. Click the **Dimension** .

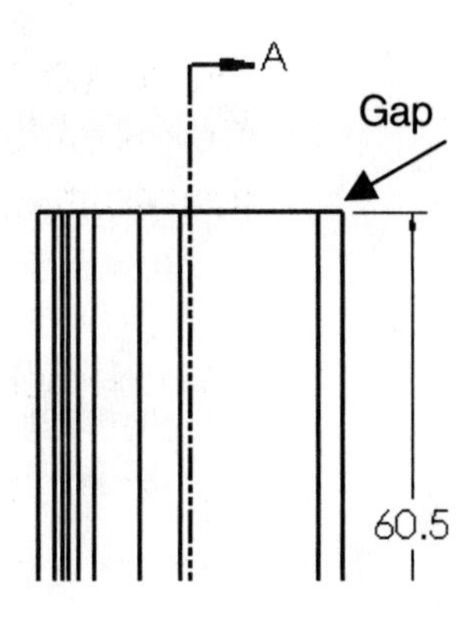

28) Click the **right vertical line** in the Top view.

29) Position the **60.50** dimension text to the right of the Top view. A visible gap exists between the extension lines and the right vertical profile lines.

30) Click **.X** from the Primary Unit Precision text box. The 60.5 dimension text is displayed.

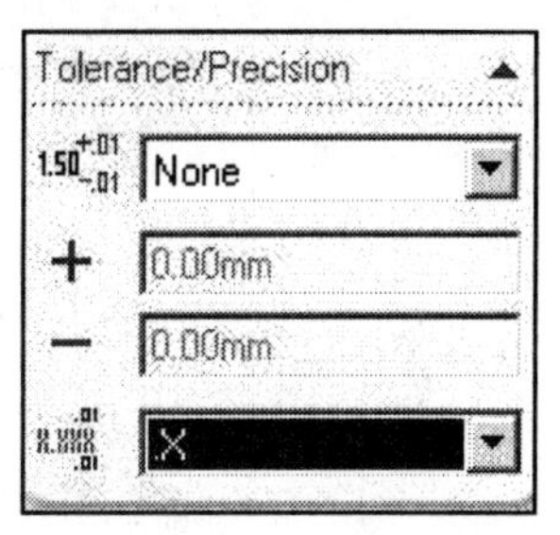

Note: Dimension text should be positioned between views when possible.

Hide vertical dimensions in the Section view.

31) Click **Select** . Hold the **Ctrl** key down.

32) Click the **32.75**, **27.75**, **27.75**, **17.50**, **17.50** and **2.30** dimension text.

33) Right-click **Hide**.

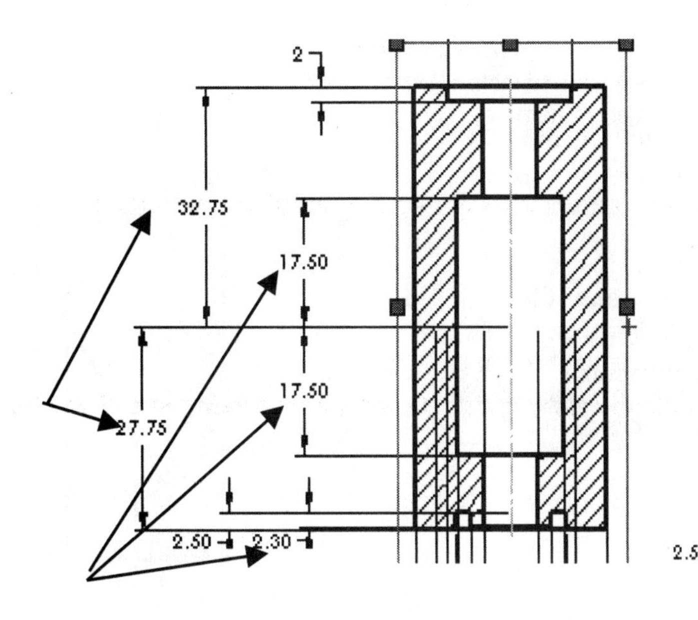

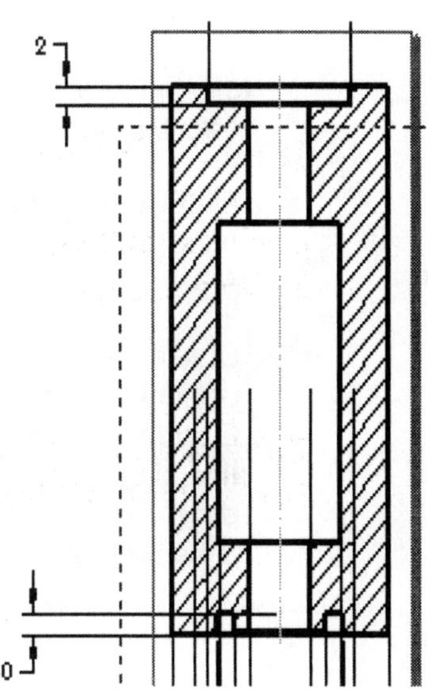

34) Release the **Ctrl** key. The vertical dimensions 2 and 2.50 are displayed.

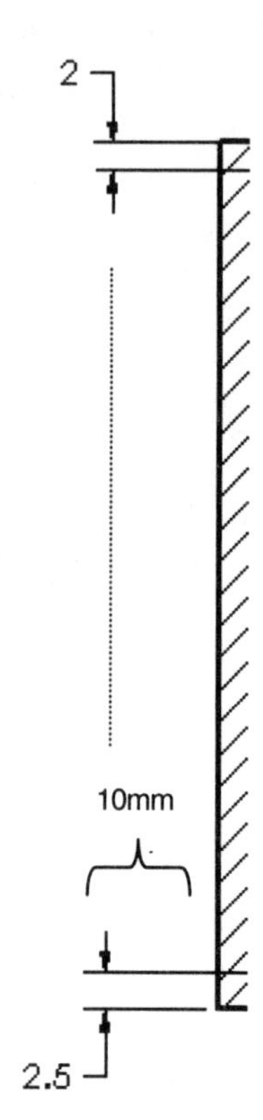

35) Click the **2.50** vertical dimension text.

36) Click **.X** from the Primary Unit Precision text box. The 2.5 dimension text is displayed.

37) Align the vertical dimensions. Drag the **2.5** vertical dimension text approximately 10mm away from the left vertical profile line.

38) Drag the **2.5** vertical dimension text below the horizontal extension line. Note: If required, flip the arrows to the outside.

39) Drag the **2** vertical dimension text until it is aligned with the 2.5 vertical dimension text. A dotted line is temporarily displayed when the 2 and 2.5 dimension text are aligned.

40) Create the vertical dimensions for the Stroke Chamber. Create the first dimension. Click **Dimension** . Click the **lower left horizontal line** of the Stroke Chamber. Click the **lower left horizontal line** of the Tube Extrusion. Drag the **10.25** dimension text to the left of the 2.5 dimension text.

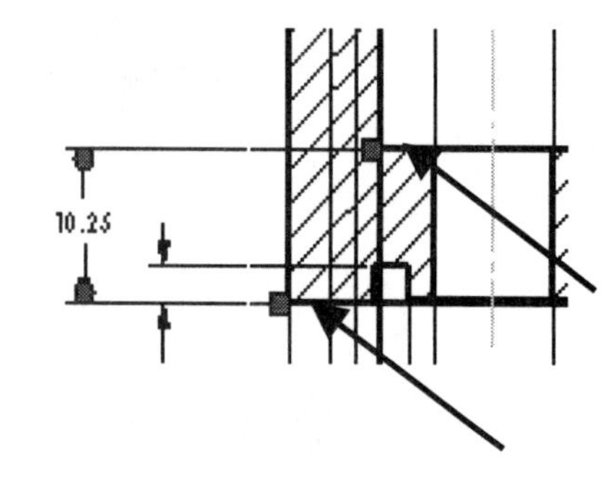

Note: Select a line segment not a point to create a dimension. Fillet and Chamfer features remove points.

41) Create the second dimension. Click the **upper left horizontal line** of the Stroke Chamber. Click the **lower left horizontal line** of the Tube Extrusion. Drag the **45.25** dimension text to the left of the 10.25 dimension text.

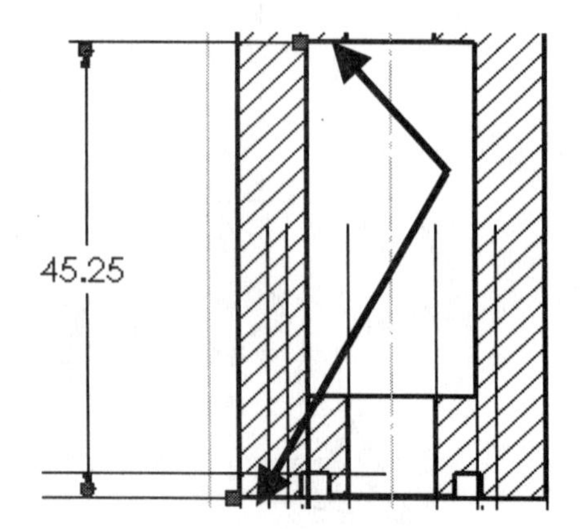

42) Create the third dimension. Click the **upper left horizontal line** of the Bore. Click the **lower left horizontal line** of the Tube Extrusion. Drag the **55.25** dimension text to the left of the 45.25 dimension text.

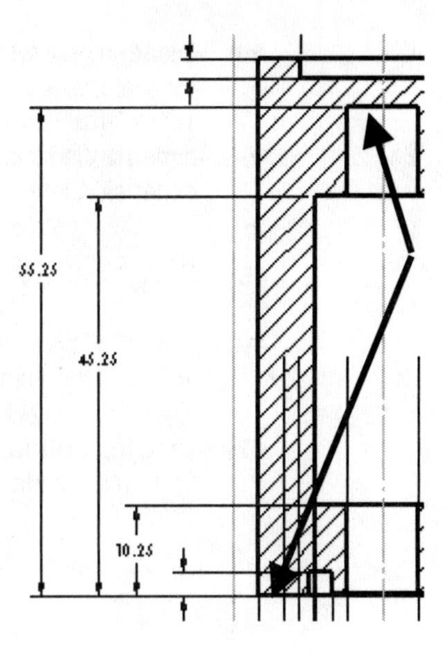

Section views require internal Gaps between visible lines and extension lines. A Gap exists between the internal visible lines of the Stroke Chamber and Bore features and the extension lines of the 45.25 and 55.25 dimension text.

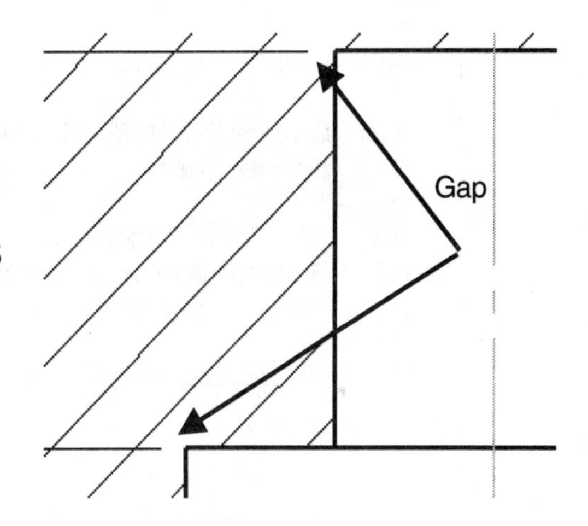

43) Move extension lines to create a gap for Front Detail1. Click the vertical **2.5** dimension text. Enlarge the text. **Zoom to Selection**. Drag the **top extension line** to the vertex of the FrontDetail1. Drag the **bottom extension line** to the bottom left vertex.

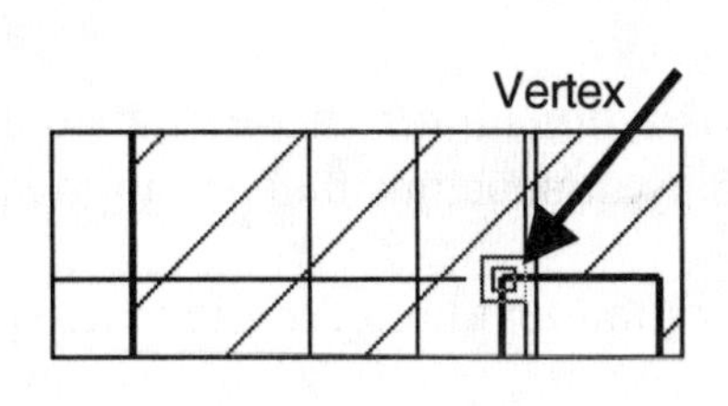

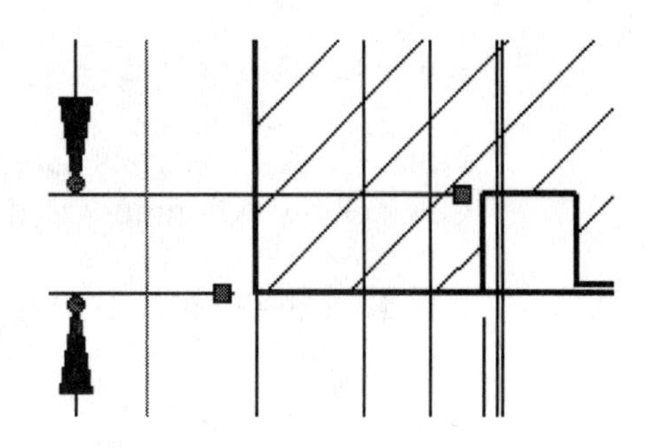

44) Create a gap for Rear Detail1. Move the extension lines. Drag the right **Scroll Bar** in the Graphics window until the 2 dimension text is visible. Click the **2** dimension text.

45) Click Zoom to Selection.

46) Drag the **top extension line** to the left vertex of the RearDetail1.

47) Drag the **bottom extension line** to the bottom left vertex.

Align the vertical dimensions.

48) Click **Select**.

49) Hold the **Ctrl** key down.

50) Click the **2.5**, **10.25**, **45.25** and **55.25** dimension text.

51) Click **Tools**, **Dimensions**, **Align Parallel/Concentric** from the Main Toolbar. Release the **Ctrl** key. The dimensions are set 6mm apart. The 4 vertical dimensions move as one entity.

Note: Do not select the 2 dimension text.

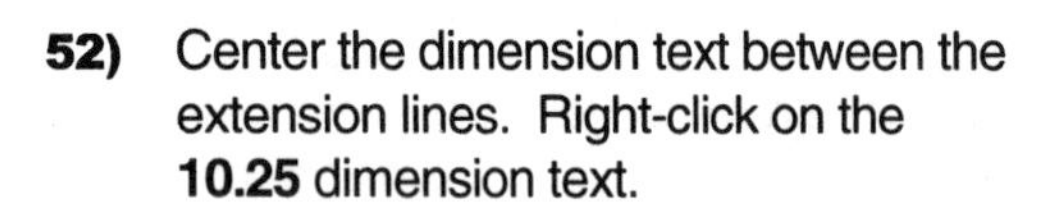

52) Center the dimension text between the extension lines. Right-click on the **10.25** dimension text.

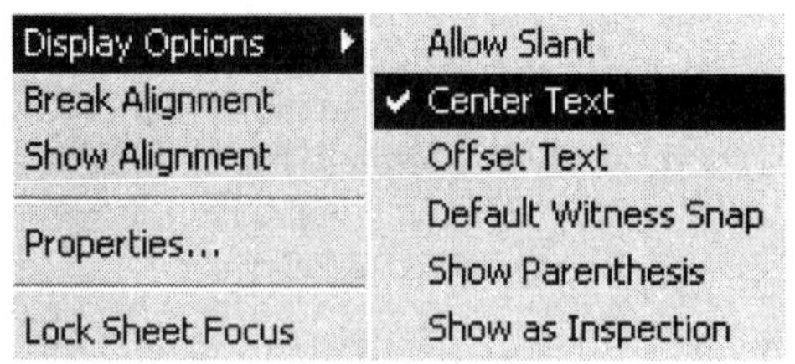

53) Click Display Options.

54) Click Center Text.

Note: Click Properties, Break Alignment to remove aligned dimensions. Click Show Alignment to display dimensions that are aligned.

Uncheck the Center Text to position text between the extension lines.

55) Review the dimensions and view position. If required, click and Drag the **Section view**, **dimensions** and the **extension lines**.

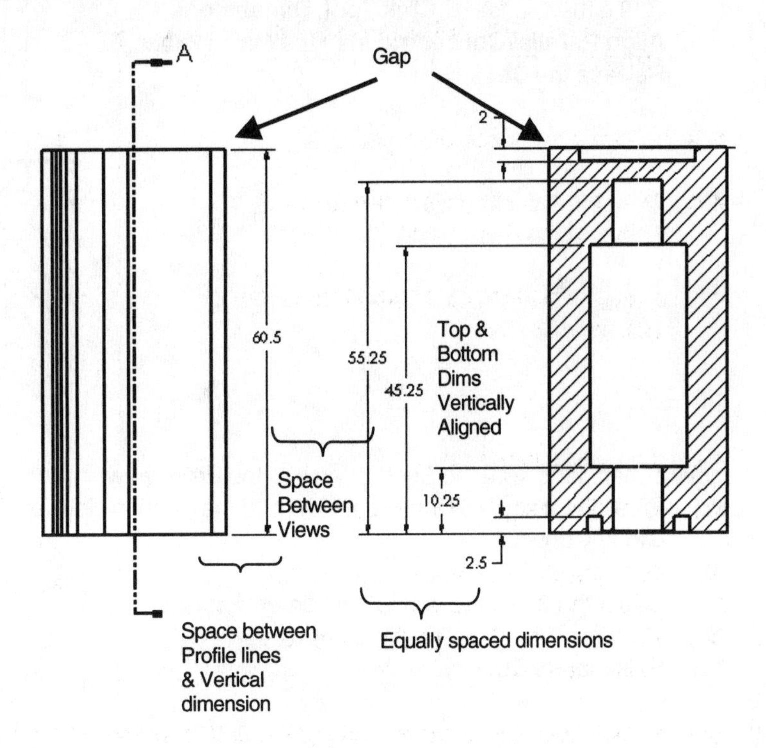

The vertical dimensions are equally spaced and positioned off the profiles. The 2 and 2.5 dimension text are vertically aligned.

The Top view and Section view are adequately spaced. The text and arrows are visible. There is a gap between the profile lines and the extension lines.

Move the horizontal dimensions.

56) Click the **Ø8.25** text at the bottom of the Section view.

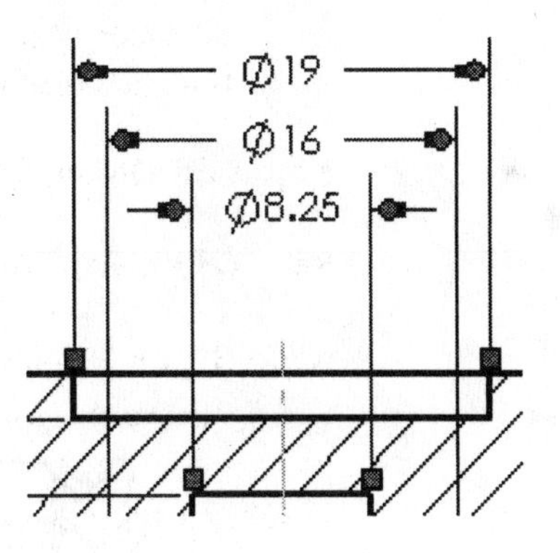

57) Drag the **Ø8.25** text upward approximately 10mm above the top horizontal profile. The text is positioned above the centerline.

58) Create a gap. Drag each **extension lines** to the top vertex of the Bore.

59) Click the **Ø16** text at the bottom of the Section view.

60) Drag the **Ø16** text upward above the Ø8.25 text.

61) Create a gap. Drag the extension lines to the **top vertex** of the Bore.

Align the top horizontal dimensions.

62) Hold the **Ctrl** key down. Click the **∅8.25**, **∅16** and **∅19** dimension text. Click **Tool**, **Dimensions**, **Align Parallel/Concentric** from the Main toolbar. Release the **Ctrl** key.

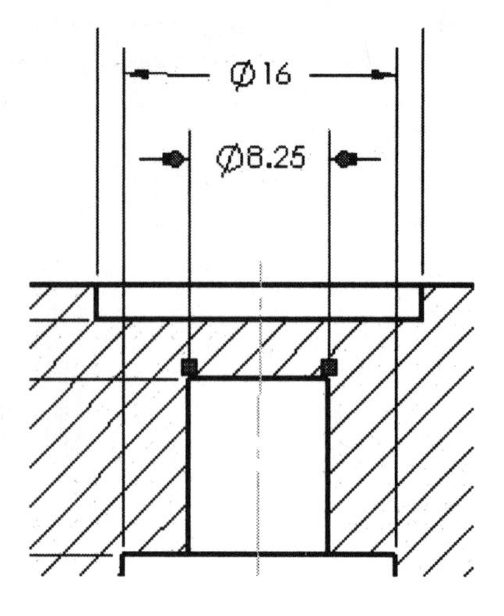

Move dimensions from the Section view to the Front view.

63) Press the **z** key until the Front view and the Section view are displayed. Hold the **Ctrl** key down.

64) Select the **29**, **19.80** and **∅3.50** dimension text.

65) Release the **Ctrl** key.

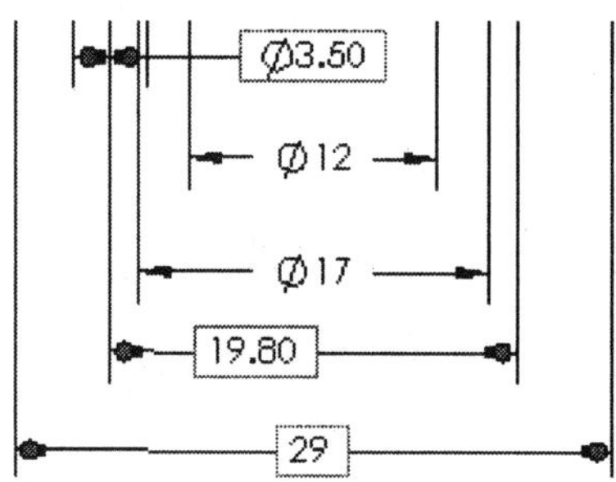

66) Hold down the **Shift** key. Drag the **29**, **19.80** and **∅3.50** dimension text to the inside of the Front view. Release the **Shift** key. Note: Move the text in the Front view in the next section.

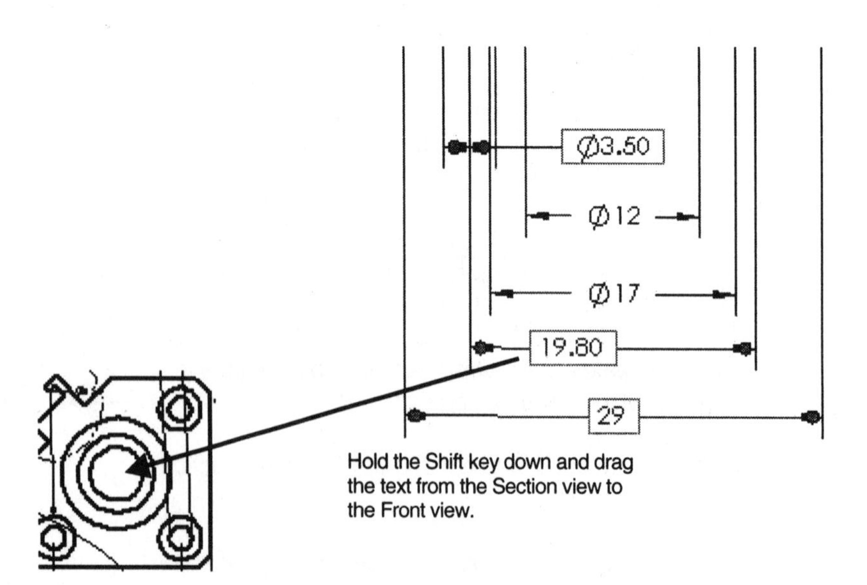

Move the horizontal dimension text.

67) Click the ∅**12** dimension text at the bottom of the Section view.

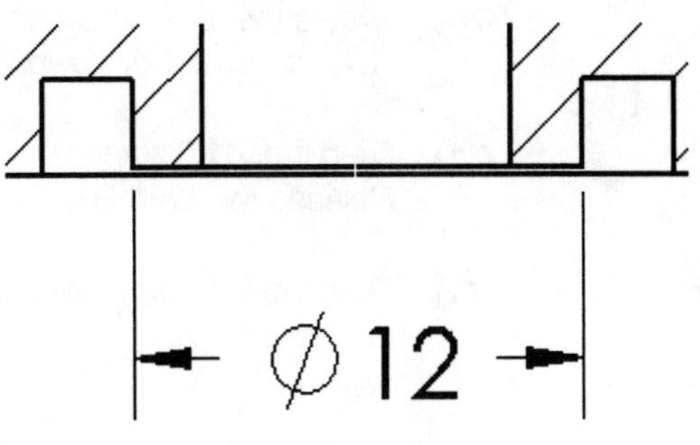

68) Drag the ∅**12** dimension text upward 10mm below the bottom horizontal profile line.

69) Drag each **extension lines** off the profile. Do not use the Nose vertex. The Nose feature is too close to the bottom horizontal line of the Tube to utilize the vertex.

There is a horizontal dimension required to describe Front-Detail1 in the Section view. The ∅17 dimension was created in the TUBE part.

Where is the dimension? Look for it in the Detail view if the ∅17 is not displayed in the Section view.

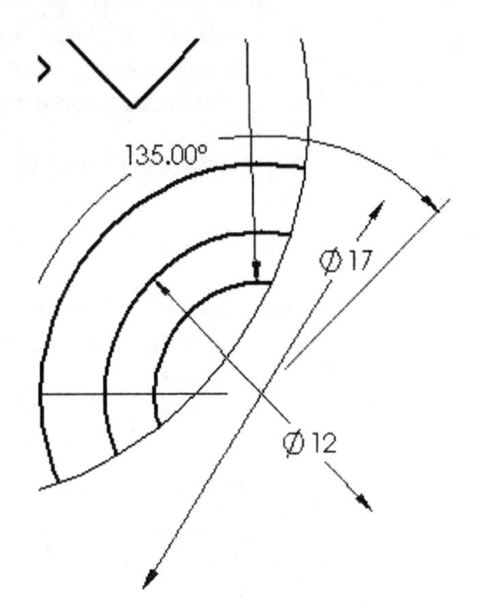

SolidWorks first inserts dimensions into the Section view and Detail view. The ∅17 text is located in the Detail view if the Detail view contains the Front-Detail1 feature.

A small part of the circle is displayed in the Detail view. The leader line is long and extends into the boundary of the Front view.

Move the ∅17 text from the Detail view to the Section view.

Detail view contains the ∅17 and ∅12 dimensions

Detail view contains the ∅17 dimension

Detail view contains no ∅17 dimension

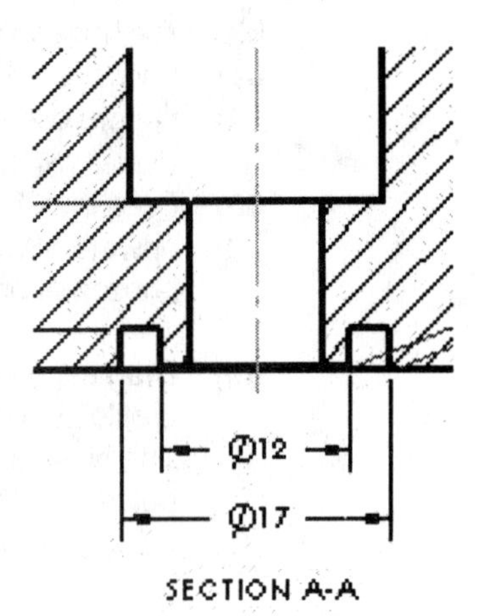

Position the dimension.

70) Locate the ∅17.00 dimension text in the Detail view. Hold the **Shift** key down.

71) Drag the **∅17** dimension text into the Section view. Release the **Shift** key.

72) Drag the **∅17** dimension text below the ∅12 dimension text.

Align the top horizontal dimensions.

73) Hold the **Ctrl** key down. Click the **∅12** and **∅17** dimension text. Click **Tools**, **Dimensions**, **Align Parallel/Concentric** from the Main Toolbar.

74) Release the **Ctrl** key.

75) Position the **Section A-A** text below the bottom horizontal dimensions.

76) Center the **Section A-A** text. A red dotted line displays when you position the text along the Tube Center Axis.

Extension lines are drawn at oblique angle to clearly illustrate a small feature. Slant the extension lines in small spaces to fit a dimension. The dimension lines are displayed in the direction of the feature.

Add a vertical dimension.

77) Redefine the Nose depth dimension Click **Dimension** . Click the **horizontal line** of the Nose. Click the **bottom horizontal line** of the Tube Extrusion. Click a **position** directly to the right. Drag the **dimension** to the right, off the Section view.

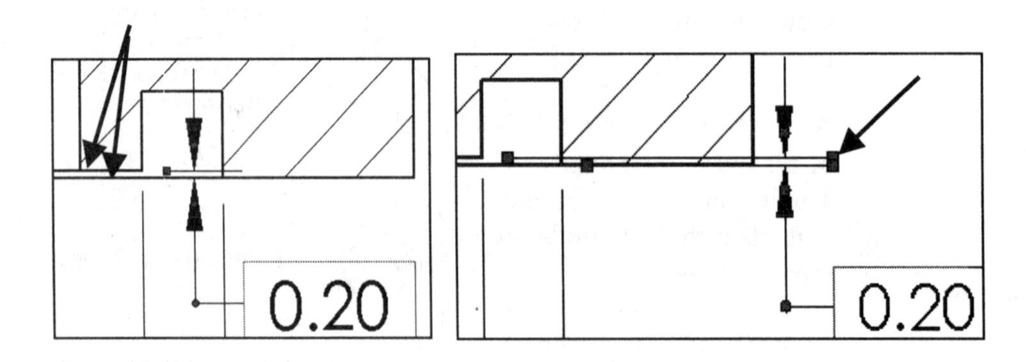

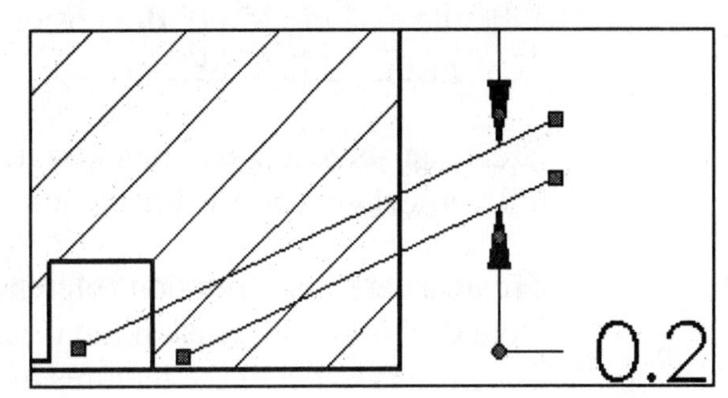

78) Right-click **Select**.

79) Right-click on the **0.20** dimension text.

80) Click Display Options, Allow Slant.

81) Drag the **endpoints** of the extension line upward to form a ~20° angle.

82) Click **.X** from the Primary Unit Precision text box. The 0.2 dimension is displayed.

83) Save the TUBE drawing. Click **Save**.

Review the current status of the Front view and Detail view. Dimensions are not clear or Dimensions are on top of each other.

Dimensions are too far or too close to the profile.

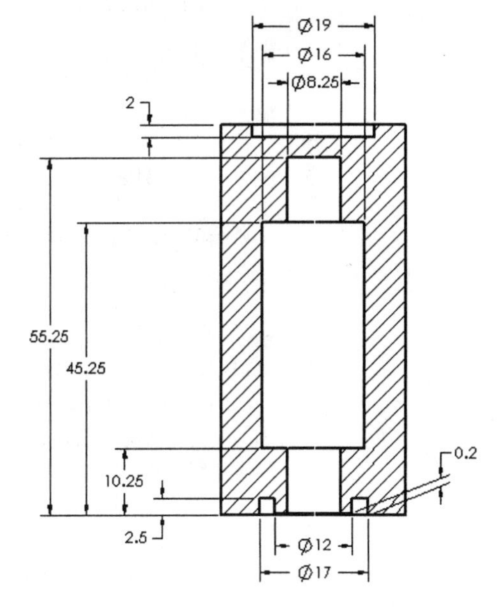

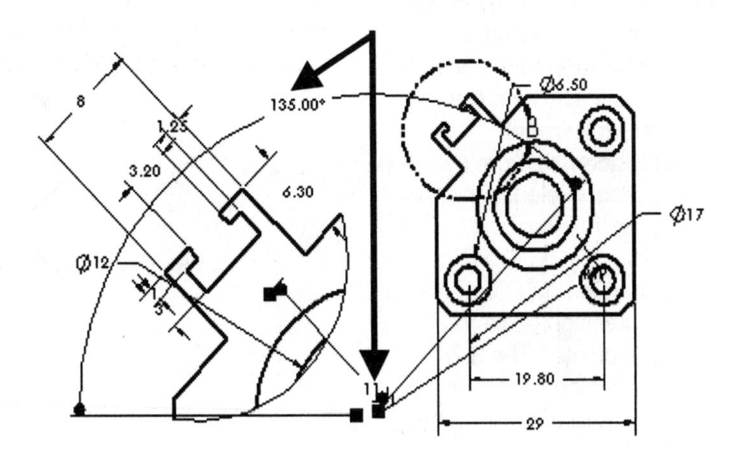

Hide the 11 and 135.00° dimensions in the Detail view or Front view. Replace the 135.00° obtuse angle.

Create an acute angle dimension from a construction line collinear with the left vertical edge in the Front view.

Create a vertical dimension referencing the bottom horizontal edge. Use the Hole Call out to dimension the Counter bores. Move dimensions and add dimensions to detail the features in Front view and Detail view.

Hide the dimensions in the Detail view or Front view.

84) Right-click the **11** dimension text.

85) Click **Hide**.

86) Right-click the **135.00°** angle text.

87) Click **Hide**.

The location of the 11 and 135.00° dimension depends upon the size of the Detail view.

88) Drag the **1.00** dimension text approximately 10mm away from the profile.

89) **Flip** the arrows if necessary. Drag the **3.00** text to the left of the 1.00 text.

90) Hold the **Ctrl** key down. Select the **1** and **3** dimension text. Click **Tools**, **Dimensions**, **Align Parallel/ Concentric** from the Main menu. Release the **Ctrl** key.

91) Drag the **8**, **3.20** and **1.25** dimension text away from the profile. The 3.20 extension lines overlap the 1.25 extension lines.

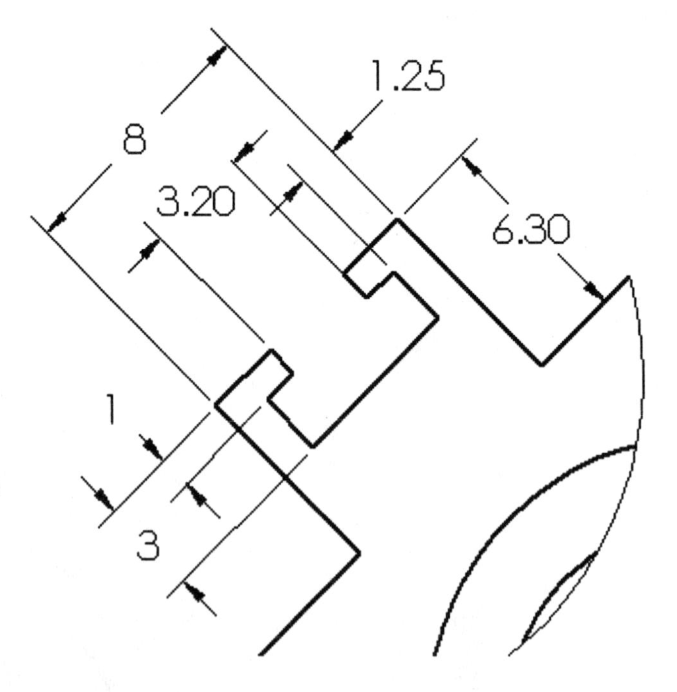

A gap is required when the extension lines cross the dimension lines.

92) Select the **3.20** horizontal dimension.

93) Click the **Break Dimension Lines** check box.

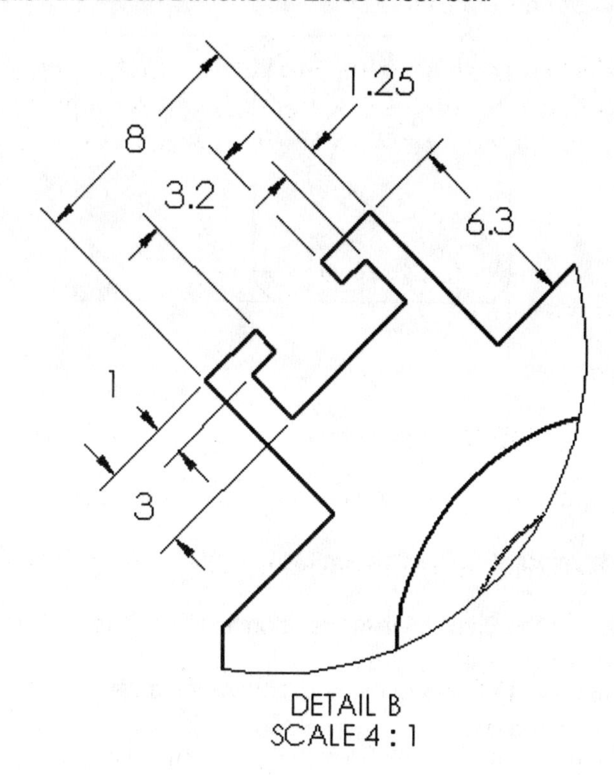

Display the Precision

94) Click the **3.20** dimension.

95) Click **.X** from the Primary Precision text box.

96) Repeat for **6.30**. The dimensions 3.2 and 6.3 are displayed.

Align the dimension.

97) Hold the **Ctrl** key down.

98) Select the **8**, **3.2** and **1.25** dimension text.

99) Click Tools, Dimensions, Align Parallel/Concentric from the Main menu.

100) Release the **Ctrl** key.

101) Position the **DETAIL B** text below the profile.

102) Save the TUBE drawing. Click **Save**.

Move and Hide dimensions in the Front view.

103) Click the **view boundary** of the Front view.

104) Enlarge the view. Click **Zoom to Selection**.

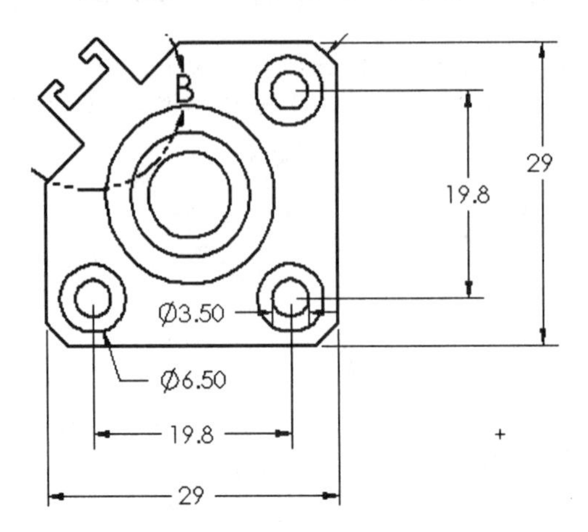

105) Drag the vertical **29** dimension text off the profile to the right.

106) Drag the vertical **19.80** dimension to the right.

107) Click **.X** from the Primary Units Precision text box. The 19.8 dimension is displayed.

108) Drag the horizontal **29** dimension text below the profile.

109) Drag the horizontal **19.80** dimension below the profile.

110) Click **.X** from the Primary Units Precision text box. The 19.8 dimension is displayed.

111) Drag the **R19** dimension upward. The arrow of the leader line is aligned to the centerpoint of the arc.

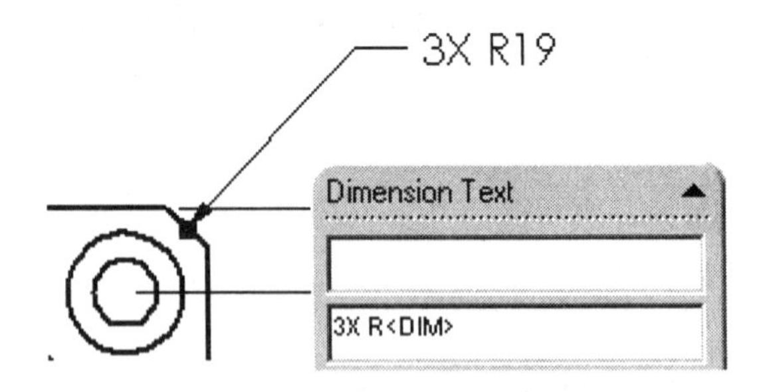

Edit Radius text.

112) Right-click on the **R19** text. Enter **3X** in the Dimension text box.

113) Hold the **Ctrl** key down.

114) Click the **∅3.50** and **∅6.50** dimension text.

115) Right-click **Hide**. Release the **Ctrl** key.

Dimension the angle cut.

116) Click **Dimension**.

117) Click the left vertical profile line and the angled edge.

118) Position the **dimension** inside the acute angle. Insure that there is a gap between both extension lines. The extension lines extend past the dimension arrows.

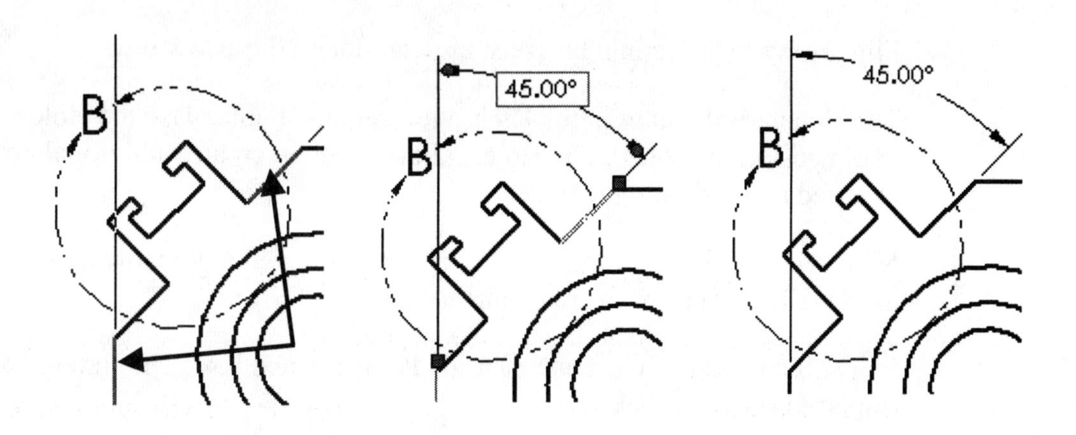

119) Display the angular dimension with no zeros. Click **None** from the Primary Units Precision check box.

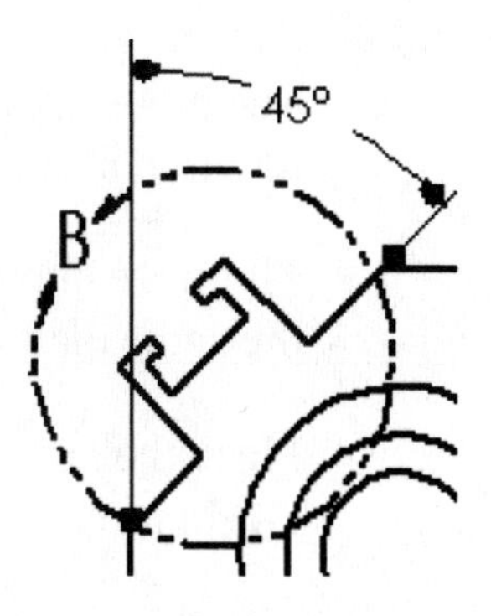

Simple Holes and other circular geometry can be dimensioned in three ways: Diameter, Radius and Linear (between two straight lines).

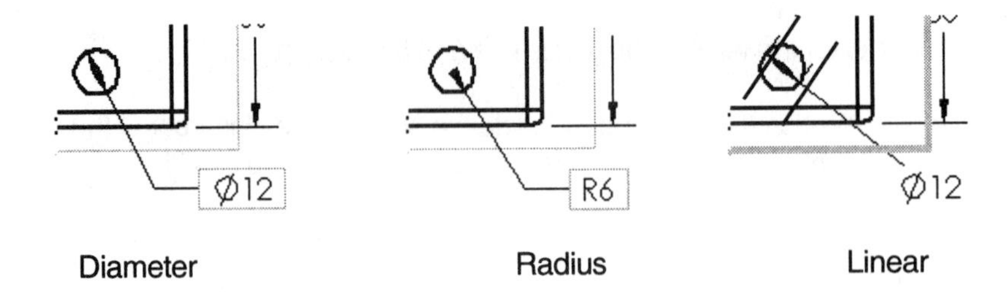

Diameter Radius Linear

Flip arrows by selecting the green circular dot on the arrowhead.

The Counterbore holes in the Right view require a note. Use the Hole Callout to dimension the holes. The Hole Callout function creates additional notes required to dimension the holes.

Note: The dimension standard symbols are displayed automatically if you used the Hole Wizard to create the holes.

Symbols are located o in the Dimension Properties text box. The current text is displayed in the text box.

- <MOD-DIAM>: Diameter symbol Ø.
- <HOLE-DEPTH>: Deep symbol.
- <HOLE-SPOT>: Counterbore symbol.
- <DIM>: Dimension value 3.

The mouse pointer displays the Hole Callout icon, when the Hole Callout function is active.

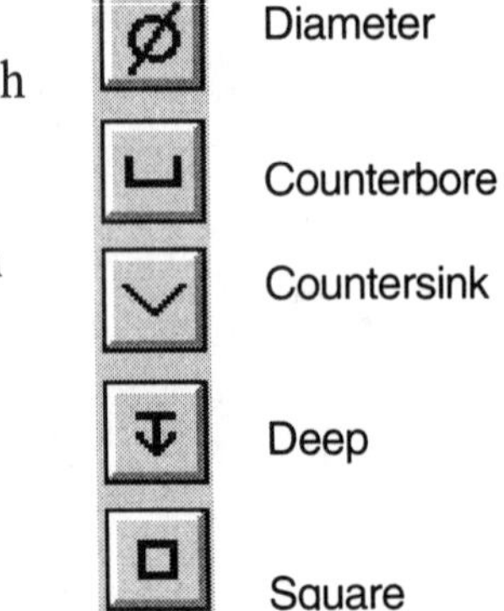

There are numerous ways to represent multiple holes with the same diameter.

Place the number of holes (2) and the multiplication sign (X) before the diameter dimension. Example:

3X <MOD-DIAM><DIM> is displayed on the drawing as: 3X Ø6.50.

120) Display the Right view. Right-click **DrawingView 3** in the FeatureManager.

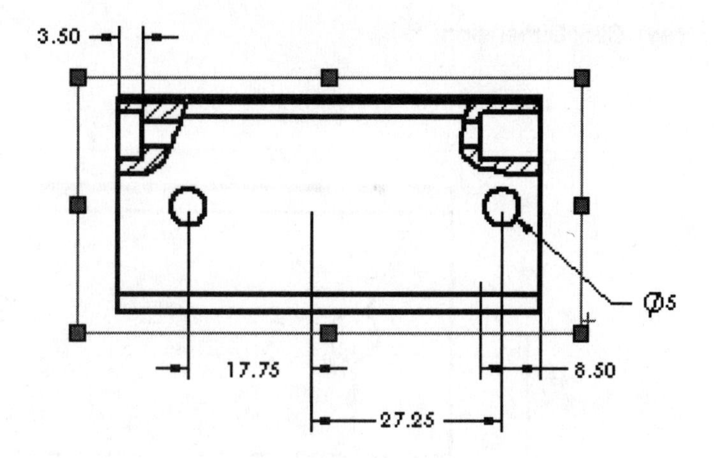

121) Click Show View.

122) Hide the dimensions. Press the **Ctrl** key.

123) Click the **∅5**, **3.50, 8.50**, **17.75** and **27.25** dimension text.

124) Right-click **Hide**.

125) Release the **Ctrl** key.

Dimension the Ports.

126) Click Hole Call Out.

127) Select the **circumference** of the left Port.

128) Enter **2X** before the **∅5** text.

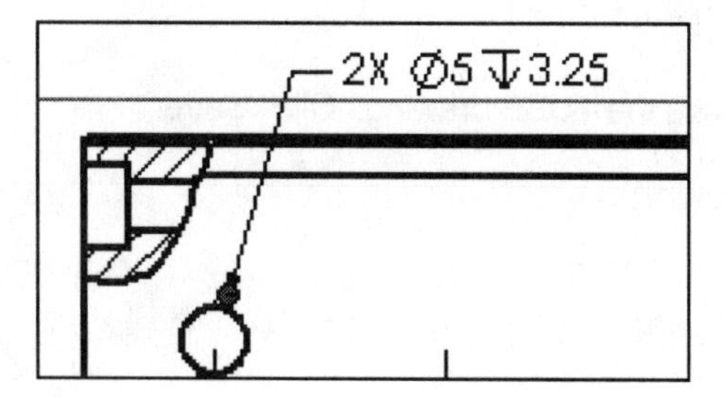

129) Click **OK**.

130) Turn off the Hole Call Out. Click **Hole Call Out**.

Add the vertical and horizontal dimensions.

131) Click **Dimension** .

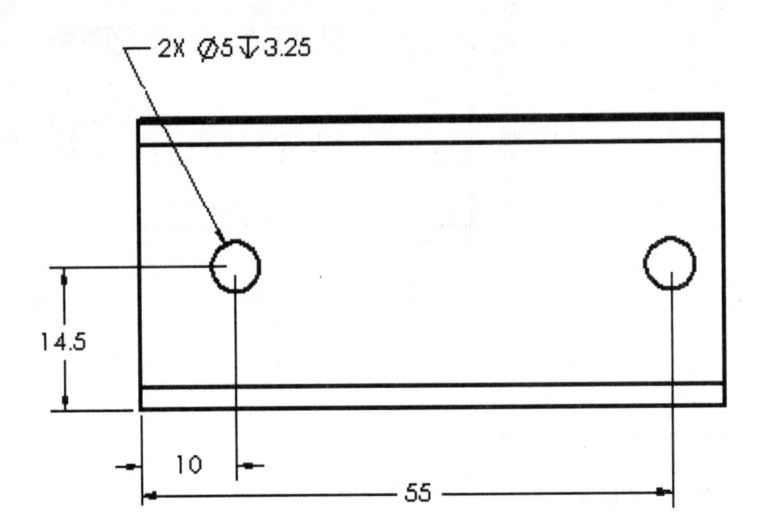

132) Click the left vertical edge.

133) Click the **circumference** of the first circle.

134) Click a **position** to the left of the vertical profile line. The 14.50 dimension text is displayed.

135) Click **.X** from the Primary Unit Precision text box. The 14.5 dimension text is displayed.

136) Drag the **left vertical extensions lines** downward to create a gap in the left corner.

137) **Flip** the arrows if required.

138) Repeat for the horizontal dimensions.

139) Save the TUBE drawing. Click **Save**.

The Counter Bore hole in the Front view was created with two Extruded Cut features.

A Counter Bore hole was added in the Back view with a third Extruded Cut feature. These Extruded Cut features do not produce the correct Counter Bore Hole Callout.

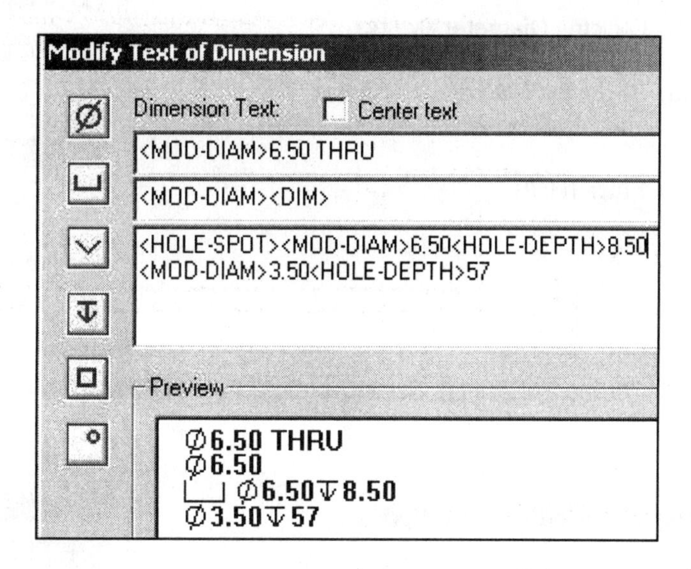

The dimension values displayed are directly related to the feature dimensions.

Break the relationship between the Hole Callout and the feature dimensions.

Create two parametric notes to represent the Counter Bore hole in the Front view and Back view.

Display the required Dimensions.

140) Click **View**, **Hide/Show Annotations** from the Main toolbar.

141) Drag the **mouse pointer** over the hidden diameter dimensions in the Front view.

142) Click the **∅3.50** and **∅6.50** dimension text.

143) Turn off Hide/Show. Click **View**, **Hide/Show Annotations**.

144) Click **.X** Primary Unit Precision for the **∅3.5** and **∅6.5** dimension text.

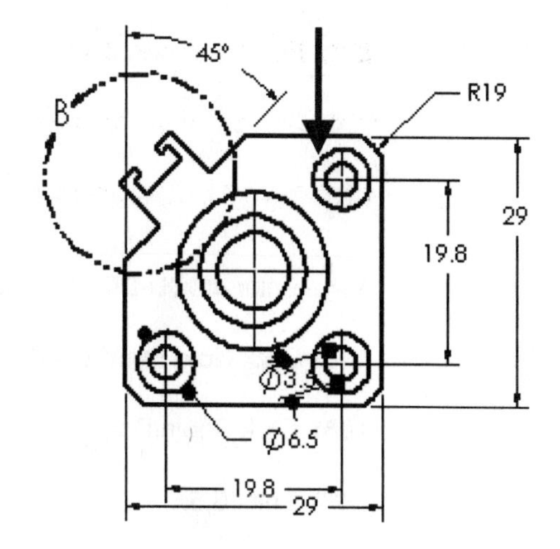

145) Click the **circumference** of the top right Counter Bore in the Front view.

146) Click **Note** A.

147) Click **Always Show Leader** for Leader Type.

148) For line 1: Enter **3X**. Click the **Symbol** button. Click **Modifying Symbols** for Symbol library.

149) Click the **Diameter** Symbol.

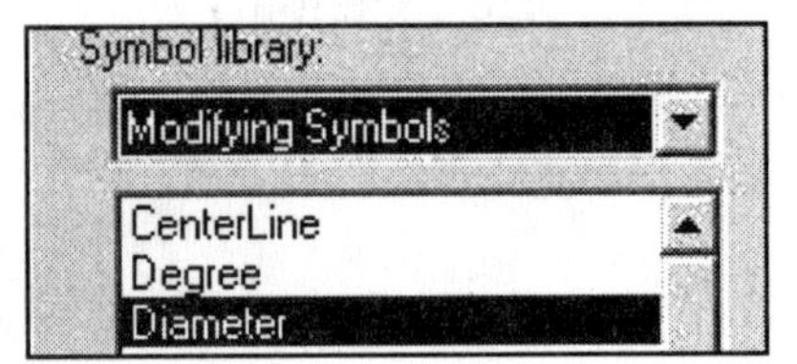

150) Select the ∅**3.5** dimension text. The variable name is entered into the text box.

151) Enter **THRU**.

152) For line 2: Click the **Symbol** button. Click **Hole Symbols** for Symbol library.

153) Click Counter bore (Spotface).

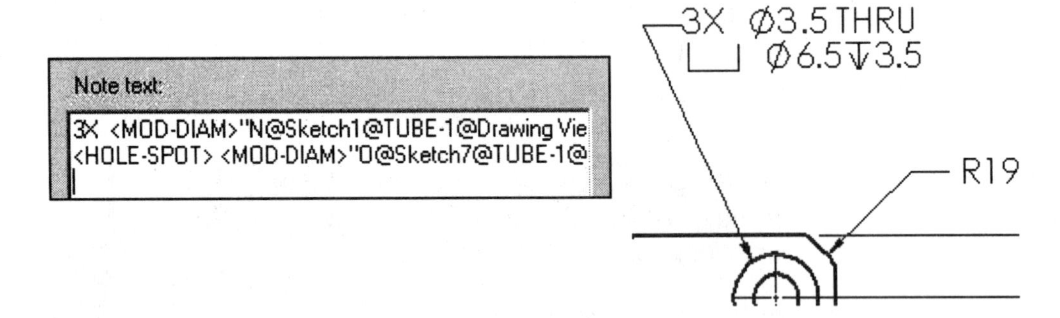

154) Enter **<MOD-DIAM>**. Select the ∅**6.5** dimension text.

155) Click **Hole Symbols** for Symbol library.

156) Click Depth/Deep.

157) Enter **3.5**.

158) Click **OK**.

159) Hold the **Ctrl** key down.

160) Click the **∅3.50** and **∅6.50** dimension text.

161) Right-click **Hide.**

162) Release the **Ctrl** key.

Note: Right-click the Dimension. Click Properties to modify the Note text.

The Counter bore Rear is detailed in the Back view for clarity. Notes and Hole Call outs may be modified at anytime.

Right-click Properties on the text to view the complete syntax of the dimension text.

The Counter bore depth dimension 8.5 is located either in the Section view or the Right view.

Add the depth dimension to the Note.

Add a Note to the Counterbore Rear.

163) Click **View, Hide/Show** Annotations.

164) Click the **8.50** depth dimension. The 8.50 depth dimension is displayed either in the Section view or Right view.

3X ∅3.5 THRU
⌴ 6.5 ↧ 3.5

165) Click **.X** for Primary Unit Precision. The dimension 8.5 is displayed.

166) Double-click on the **Cbore note** in the Front view. Copy the text.

<HOLE-SPOT> "O@Sketch7@TUBE-1@Drawing View1" <HOLE-DEPTH> 3.5

167) Select the **second line** of text.

168) Click **Ctrl C**.

169) Click **Note** A.

170) Click the **circumference** of the top left Counter Bore in the Back view. Click a position for the Note.

171) Paste the Note. Click **Ctrl V**.

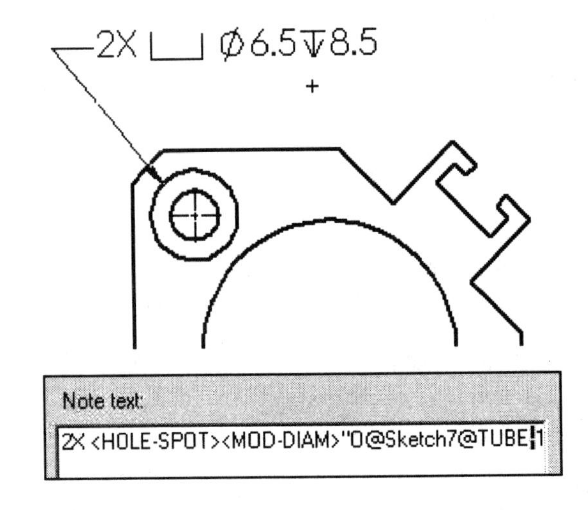

172) Enter **2X** before the <HOLE-SPOT>.

173) Delete the **3.5** Note text.

174) Click the **8.5** dimension text.

175) Click **OK**.

Hole centerlines are composed of alternating long and short dash lines. The lines identify the center of a circle, axes or cylindrical geometry.

Center Marks represents two perpendicular intersecting centerlines. Adjust adjacent extension lines after applying Center Marks.

Add Center Line.

176) Add a Center Line to the Section view. Click **Centerline** from the Annotation Toolbar. Click the **left most vertical edge** of the Section view.

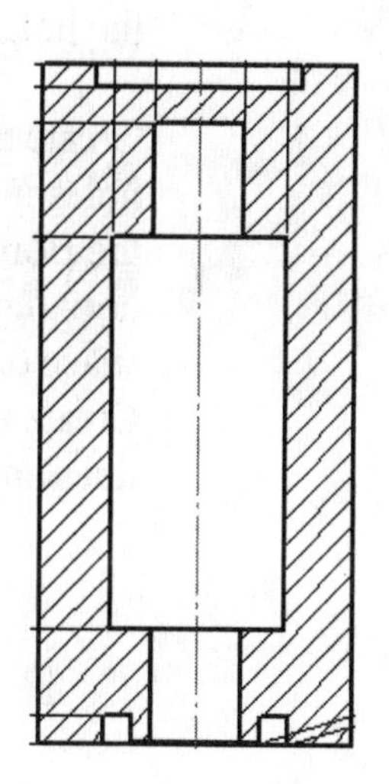

177) Click the **right most vertical edge** of the Section view.

178) Click **OK**. The Centerline is displayed.

179) Add a Center Line to the Right view. Click **Centerline** from the Annotation Toolbar.

180) Click the **top most horizontal edge** of the Right view.

181) Click the **bottom most horizontal edge** of the Right view.

182) Click **OK**. The Centerline is displayed.

183) Drag the **centerlines** and **extension** lines. Do not overlap the center marks.

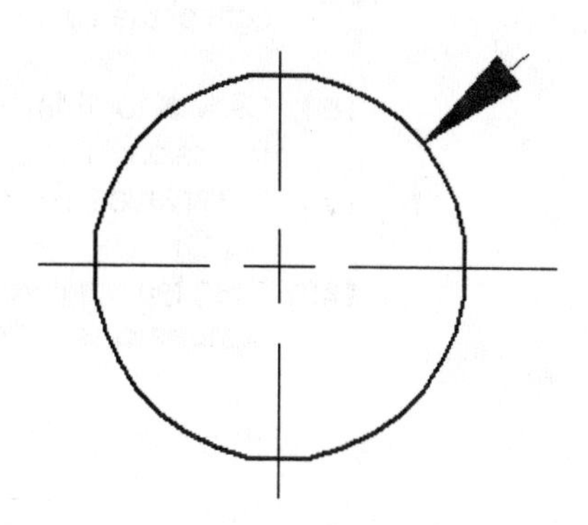

The TUBE holes utilized symmetry in the initial Base feature sketch.

Horizontal and vertical construction lines were created from the Origin to a midpoint. A symmetric relationship about vertical and horizontal construction lines created a fully defined sketch.

No additional dimension was required from the Origin to the center point of the hole.

Create new dimensions to locate these holes in relationship to the center of the Bore. Adjust all vertical and horizontal dimensions. Stagger and space dimension text for clarity. Create a gap between the extension lines and the center mark.

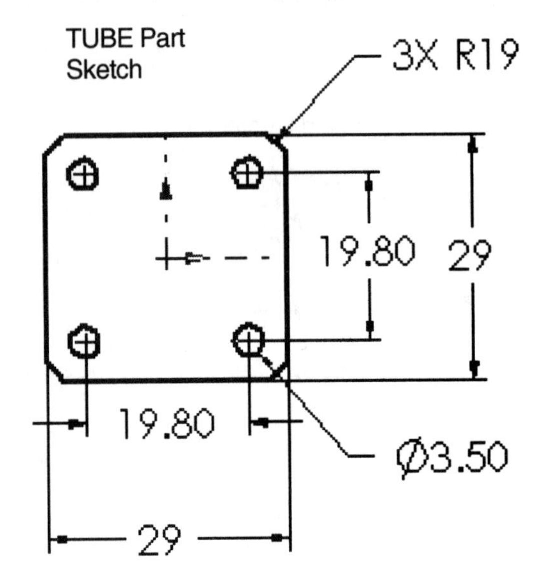

Create the linear Dimensions for the small holes.

184) Create a vertical dimension. Click **Dimension**.

185) Click the **circumference** of the small right bottom hole.

186) Click the **circumference** of the Bore.

187) Drag the **9.90** dimension to the right of the vertical profile line.

188) Click **.X** for Primary Unit Precision. The 9.9 dimension is displayed.

189) Drag the other vertical **dimensions** to the right.

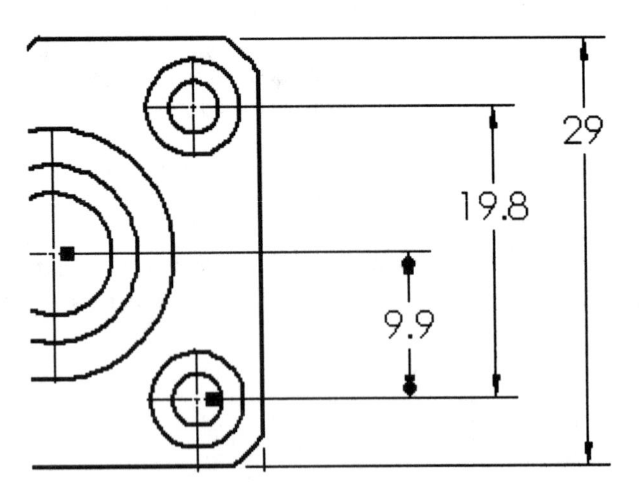

190) **Flip** the dimension arrows if required.

The largest dimension is on the outside. The smallest dimension is closest to the profile. Stagger the vertical dimension text.

191) Hold the **Ctrl** key down.

192) Select the **9.9**, **19.8** and **29** dimension text.

193) Click Tools, Dimensions, Align Parallel/Concentric from the Main menu.

194) Release the **Ctrl** key.

195) Create a horizontal dimension. Click **Dimension**. Click the **circumference** of the small left bottom hole.

196) Click the **circumference** of the right bottom hole.

197) Drag the **dimension** below the horizontal profile line.

198) Click **.X** for Primary Unit Precision. The 9.8 dimension is displayed.

199) Drag the other horizontal **dimensions** downward.

200) **Flip** the dimension arrows if required. The largest dimension is on the outside. The smallest dimension is closest to the profile. Stagger the dimension text.

201) Drag the **29** dimension text to the right of the 19.8 dimension text.

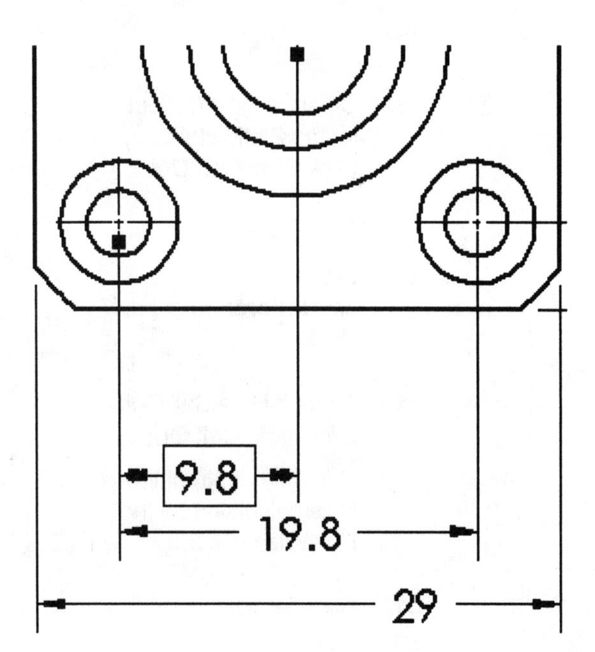

202) Hold the **Ctrl** key down.

203) Select the **9.8**, **19.8** and **29** dimension text.

204) Click Tools, Dimensions, Align Parallel/Concentric from the Main menu.

205) Release the **Ctrl** key.

The Auxiliary view is the last view to move and to add dimensions. Can you recognize what is wrong with the current dimension scheme? Answer: Dimensions are missing.

Use a Hole Callout to specify hole size and depth.

Add a dimension to locate the center off the hole. Move extension lines off the profile.

Hide, Move and Add dimensions to the Auxiliary view.

206) Click **Select** .

207) Right-click the **2** dimension text.

208) Click **Hide**.

209) Create a Hole Callout for the 2 MM Hole. Click **Hole Call Out** .

210) Click the **circumference** of the Hole.

211) Turn off Hole Call Out. Click **Hole Call Out** . The depth of the Hole calculated by the HoleWizard is 3.

212) Click Centerline.

213) Sketch a **Centerline** from the 2mm Hole parallel to the left edge of the Switch Groove.

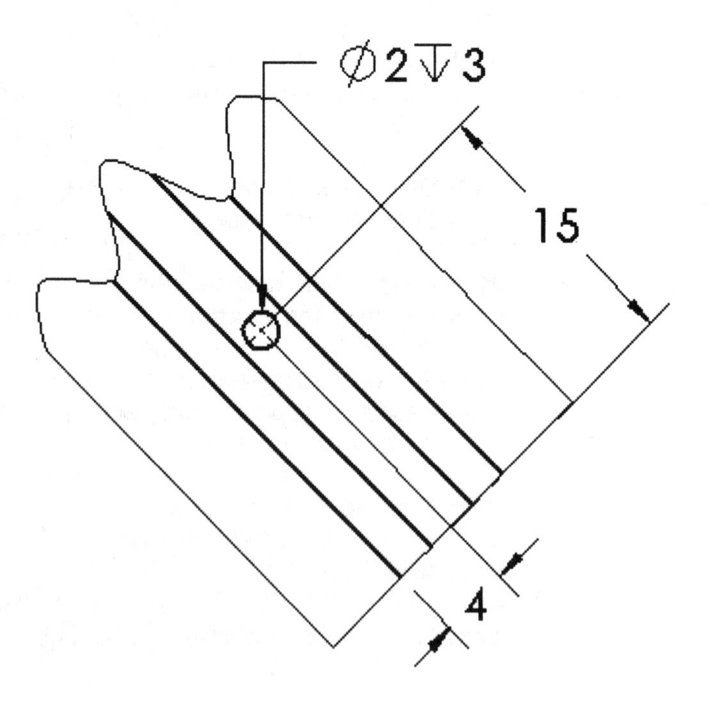

214) Click **Dimension**.

215) Click the **circumference** of the small hole.

216) Click the **left edge** of the Switch Groove.

217) Drag the **4** dimension text off the profile.

218) Drag the **extension lines** off the center mark to create a gap.

219) Save the TUBE drawing. Click **Save**.

The TUBE drawing is complete.

Where should dimensions be created? Do you return to the part and change your dimension scheme to accommodate the drawing? Answer: No.

Build parts with symmetric relationships. Use a line of symmetry in a sketch. Add geometric relationships.

Add dimensions for a symmetric feature. Use Mirror All when multiple features are symmetric about a planar face.

Modify part sketches and features to accommodate a drawing. Create multiple part configurations with a Design Table. These actions conserve time and provide a record of the changes in a single part file.

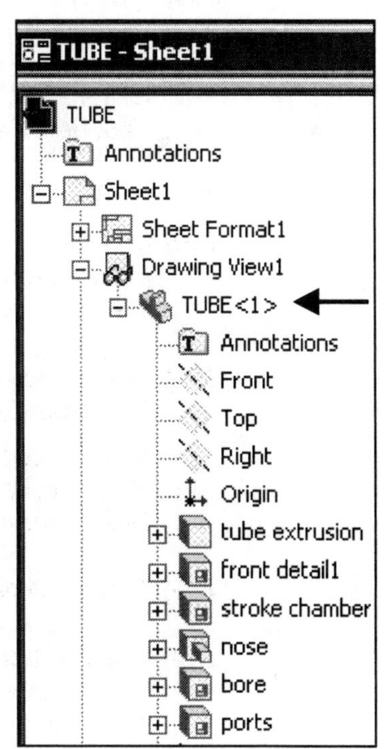

The TUBE drawing contained dimensions from 11 features. The option, Insert Model Items, Entire model, displayed numerous dimensions.

How do you display hundreds of feature dimensions for a part? Answer: Utilize the FeatureManager in the drawing and Drawing Layers.

Create a new Drawing Layer, named DETAILS. Expand the FeatureManager in the drawing. Select a Feature. Select Insert Model Items, Selected Feature option.

Position dimensions in the best view to document the feature.

Move dimensions with the Shift key. Add reference dimensions to document part Geometric Relations such as Equal and Symmetry.

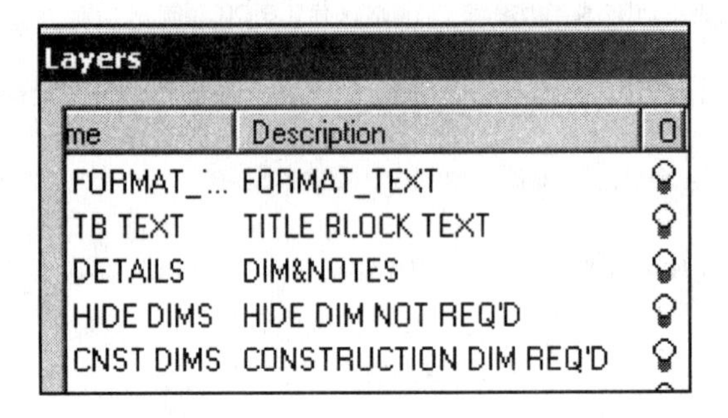

Create two new Drawing Layers named: HIDE DIMS and CNST DIMS. Place unwanted feature dimensions on the HIDE DIMS (HIDE DIM NOT REQ'D) Layer.

Place dimensions created in the drawing on the CNST DIMS (CONSTRUCTION DIM REQ'D) layer.

Add additional Layers to control Notes and different Line Fonts. Control the Layer display with the On/Off option.

Utilize views outside the sheet boundary to display interim dimensions and view types.

Detailing the ROD Drawing

The ROD requires dimensions for both the Long and Short part configuration. Dimension the Long configuration on Sheet2 and Sheet3 in this section.

The Short configuration is left as an exercise at the end of this Project.

You inserted dimensions from the entire TUBE part into the drawing. For the ROD drawing, insert dimensions by the individual view.

To insert dimensions by a view, select the Front view and Insert model items. The dimensions are displayed in the view. Position the dimensions.

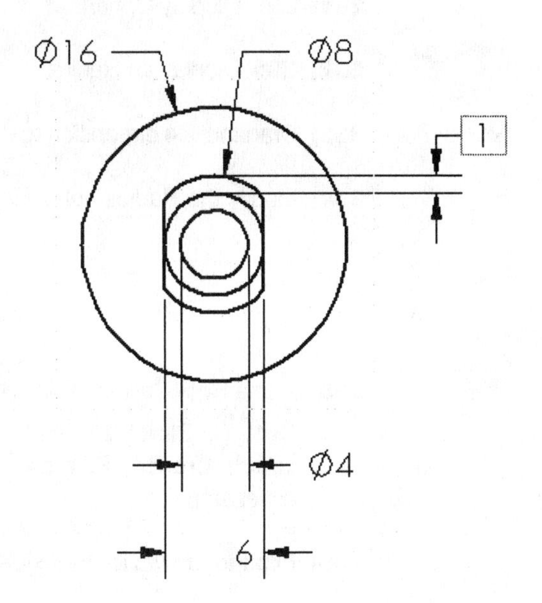

Hide the Ø4 and 1 dimension text. Create a new Hole Callout off the profile.

The Hole Callout indicates the depth of the Hole. Drag the horizontal dimension text closer to the profile. Modify the Ø8 dimension text to display extension lines as a Radial dimension.

Insert Dimensions into the Front view.

220) **Open** the ROD drawing.

221) Click ROD-Sheet2.

222) Click the **Front** view.

223) Click **Model Items** from the Annotations toolbar.

224) Click **OK**.

225) Drag the **Ø16** dimension text approximately 10mm off the profile.

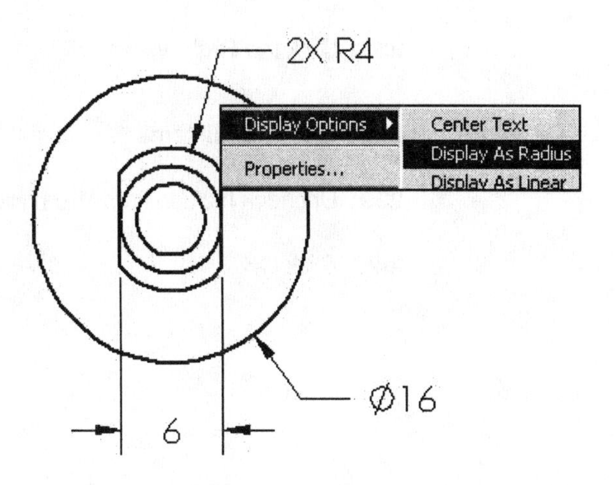

226) Drag the horizontal **6** dimension text 10mm off the profile.

227) Hide the ∅4.00 dimension text. Right-click the **∅4.00** dimension text.

228) Click **Hide**.

229) Repeat for the **1** dimension text.

230) Display the ∅8 dimension text as a Radius. Right-click the **∅8** dimension text.

231) Click Display Options.

232) Click Display as Radius.

233) Drag the **∅4** dimension text off the profile.

234) Modify the Radius note. Enter **2X** in the R4 text box. The note displays 2X R4.

235) Add a Hole Callout. Click **Hole Call Out** . Click the inside **circumference**. Position the dimension.

236) **Flip** the arrows to the inside.

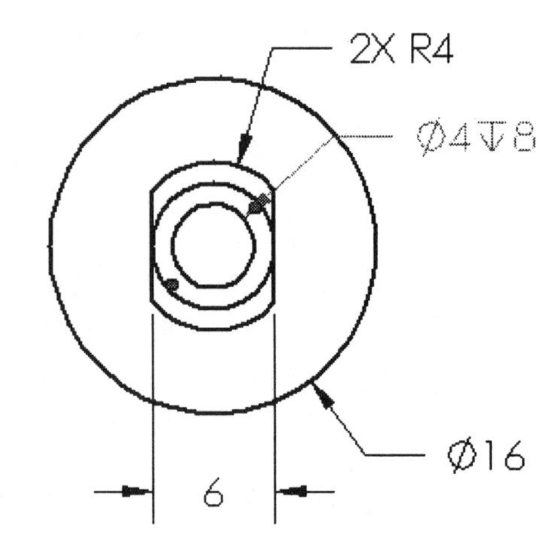

Insert Additional Dimensions.

237) Click the **Right** view.

238) Click **Model Items** from the Annotations toolbar.

239) Uncheck Include items from Hidden features.

240) Click **OK**.

241) Drag the **5** text and the **200** text approximately 10mm away from the profile line.

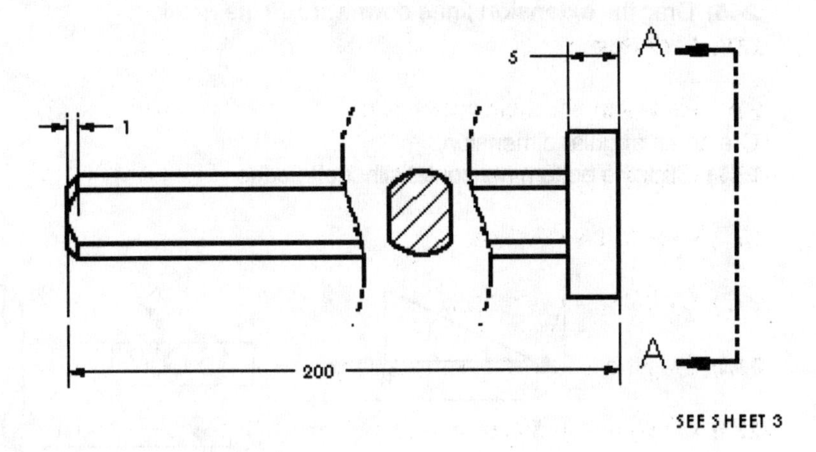

242) Create a gap. Drag their **extensions lines** to the vertex of the profile.

243) Drag the **A-A arrows** and **SEE SHEET 3** text to the right of the profile.

The ASME Y14.5M standard defines Chamfer dimension techniques in a variety of ways. Create a Chamfer with a note. The note can be written as distance/angle (1 X 45°) or distance/distance (1 X 1). The leader line arrow points to the midpoint of the Chamfer.

Create a Chamfer with a distance/angle (1 X 45°). Explore the distance/angle option to work with angle dimensions.

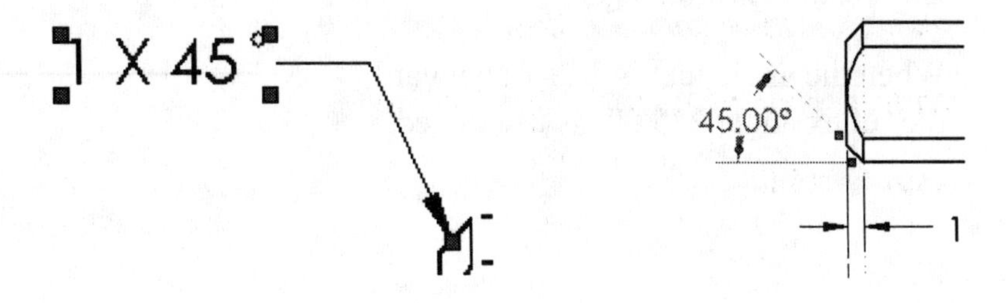

Create the Chamfer.

244) Create a gap. Drag the Chamfer **1** dimension line below the profile.

245) Drag the **extension lines** downward off the Profile.

Create an angular dimension.

246) Click the bottom horizontal silhouette edge.

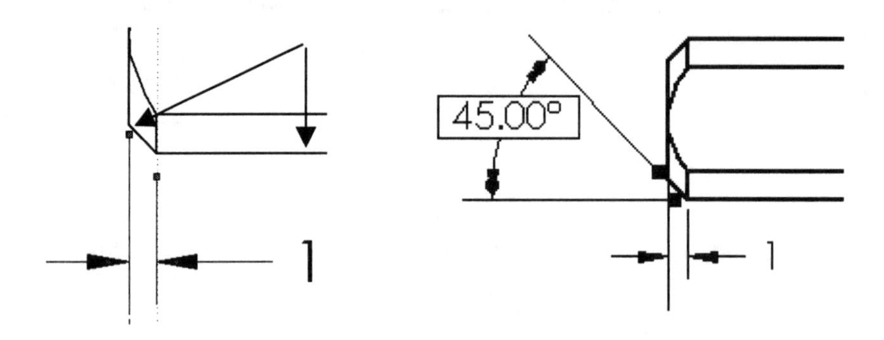

247) Click the **small silhouette edge** of the Chamfer.

248) Click a **position** to the left.

249) Drag the **45.00°** text between the two arrows.

250) Click **None** from the Primary Units Precision text box.

251) The **45°** angular dimension is displayed.

Silhouette edges create a variety of options for angular dimensions. Examples: When the top Chamfer edge is selected, the acute angle 45.00° is draw to the right.

When the angle text is drag downward, the reflex angle 225.00° is calculated.

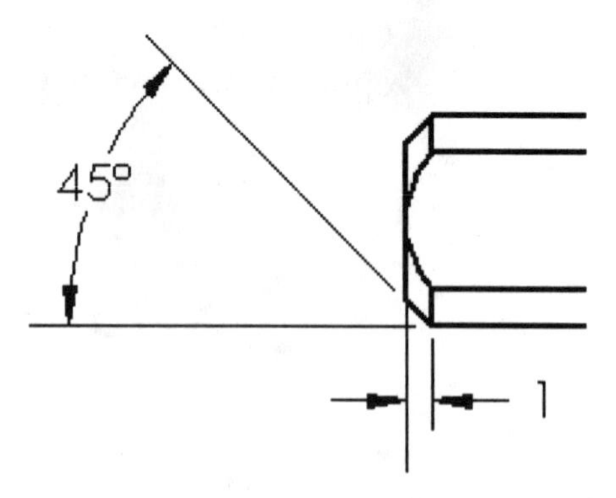

Create a construction line collinear to a silhouette edge when an angular dimension cannot be automatically determined from existing geometry.

The acute angle 45.00° is created on the left side of the top Chamfer edge.

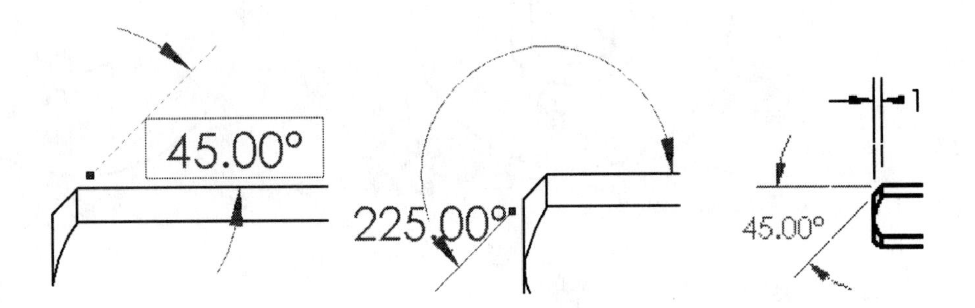

Add the Diameter dimension.

252) Click ROD-Sheet3.

253) Click the **Front** view.

254) Click **Model Items** from the Annotations toolbar.

255) Click **OK**.

256) Drag the dimension text into the Front view. Create a reference dimension. Right-click the **∅16** dimension text. Click **Properties**.

257) Click Display with parenthesis.

Driven
Display with parenthesis
Display as dual dimension

258) Drag the **1** vertical dimension off the profile. This dimension is difficult to interpret.

259) Right-click the **1** dimension text.

260) Click Display Option, Allow Slant.

261) Drag the **1** dimension text upward.

262) Use two diameter dimensions to document the feature. Right-click the **R4** dimension text.

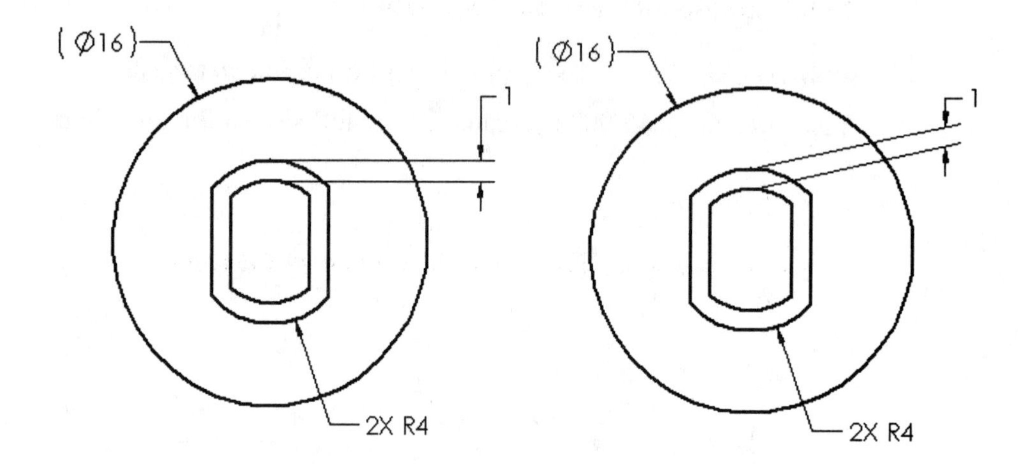

263) Click Display Options, Display as Diameter.

264) Drag the ∅**8** dimension text into the center until the dimension arrows are vertical.

265) Right-click the ∅**8** dimension text.

266) Click Display Options, Display as Linear.

267) Drag the ∅**8** dimension text to the right of the view boundary.

268) Create a reference dimension. Click **Dimension**.

269) Select the **R3 arc**.

270) Right-click **R3**.

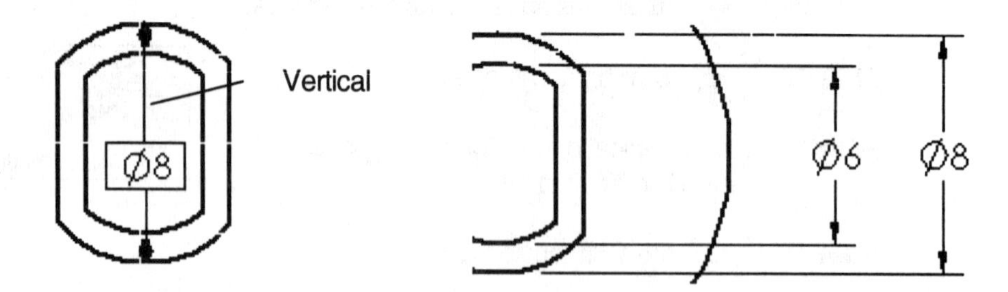

271) Click Display Options, Display as Diameter.

272) Drag the ∅**6** dimension text until the dimension arrows are vertical.

273) Right-click the ∅6 dimension text.

274) Click Display Options, Display as Linear.

275) Drag the ∅**6** dimension text to the right of the view boundary.

276) Save the ROD drawing.

277) Click **Save**.

278) Close all Parts and Drawings. Click **Windows, Close All.**

Detailing the COVERPLATE Drawing

The COVERPLATE utilized geometric relationships such as Symmetric to define the position of the features.

The ∅8.25 dimension requires a Precision value of 2 decimal places.

All the remaining dimensions require a Precision value of 1 or None. Conserve time.

Modify the Document Properties Precision value to 1 decimal place.

Insert and add dimensions to create the COVERPLATE drawing.

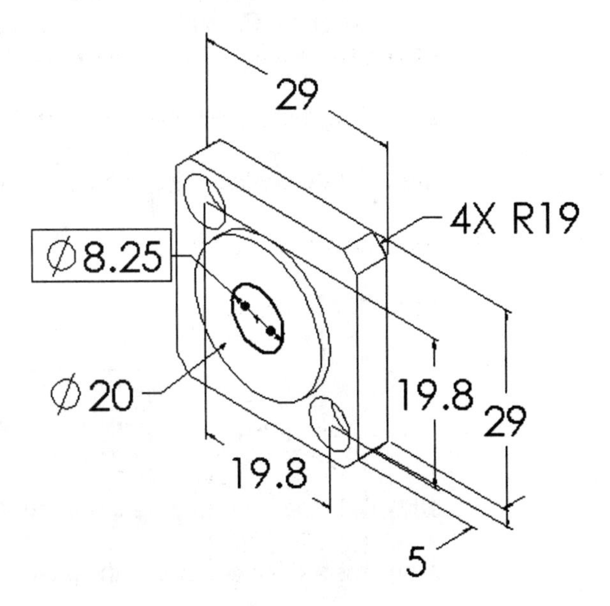

Set Document Precision to 1 decimal place.

279) **Open** the COVERPLATE drawing.

280) Click Tools, Options, Document Properties, Dimensions.

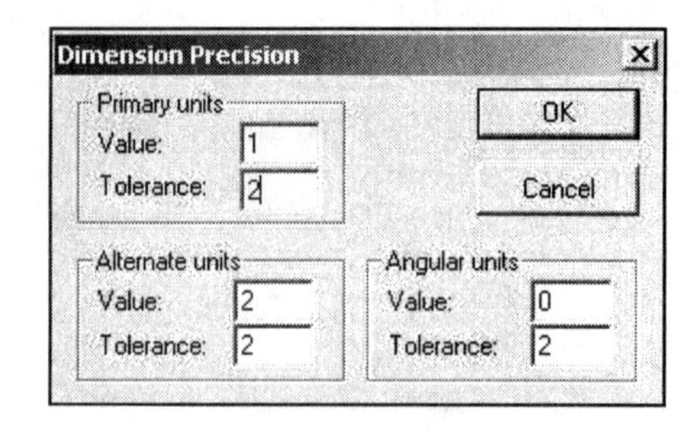

281) Click the **Precision** button.

282) Enter **1** for Primary units value.

283) Enter **2** for Tolerance.

284) Enter **0** for Angular units value.

285) Click **OK**.

Create Hole Callout.

286) Click COVERPLATE-Sheet1.

287) Click Hole Callout ⌴ø.

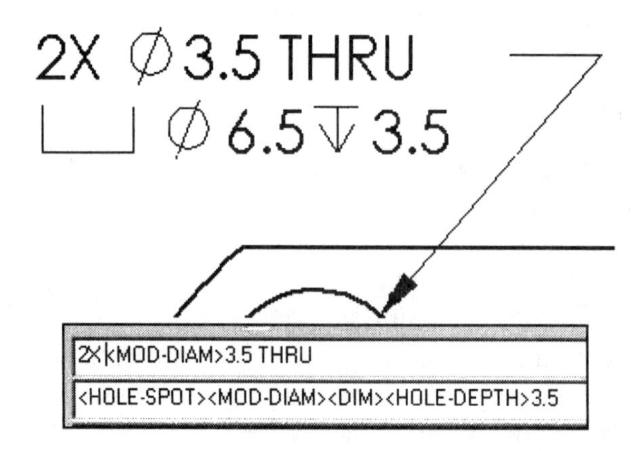

288) Click the Counter Bore **circumference**.

289) Position the **text** above the profile.

290) Turn off the Hole Callout. Click **Hole Callout** ⌴ø. There are 2 Counter Bores.

291) Enter **2X** in the Dimension Text box.

292) Click **OK**.

293) Deactivate the Hole Callout. Click **Hole Callout** ⌴ø.

Insert the Model Item for individual features.

294) Hold the **Ctrl** key down. Click the **small center circle** in the Front view.

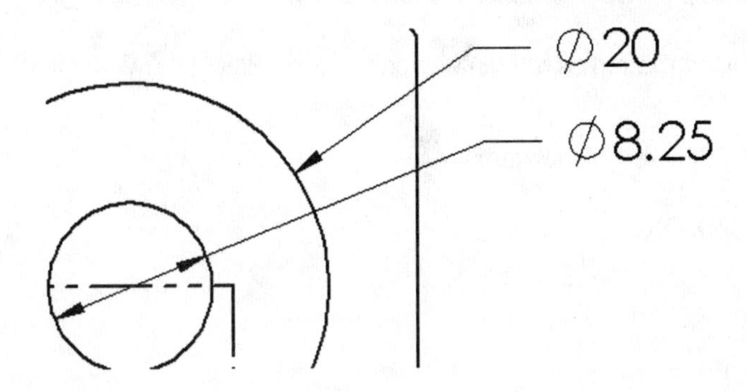

295) Click the large center circle.

296) Click **Insert Model items**. Release the **Ctrl** key. Insert Model Items by Feature.

297) Click **OK**.

298) Right-click the **∅8.3** text.

299) Click **.XX** from the Primary Unit Precision text box.

300) Click Display Options.

301) Click Display as Diameter.

302) Drag the **∅20** text and **∅8.25** text off the profile.

303) Position the **∅20** text and the **∅8.25** text off the Front profile.

304) **Flip** the dimension arrows if required.

Insert the Model Item for Section view.

305) Click the top **Section view**.

306) Click Insert Model items.

307) Hide the ∅**6.5**, ∅**3.5** and **3.5** dimension text.

308) Drag the horizontal dimensions **29** and **19.8** 10mm away from the Profile line. A gap exists between the extension lines and profile of the 29 dimension. No Gap exists between extension lines and the centerlines (light gray) of the 19.8 dimension.

309) Insert the remaining dimensions. Right-click in the **Graphics window**.

310) Click Insert Model items.

311) Click **OK**.

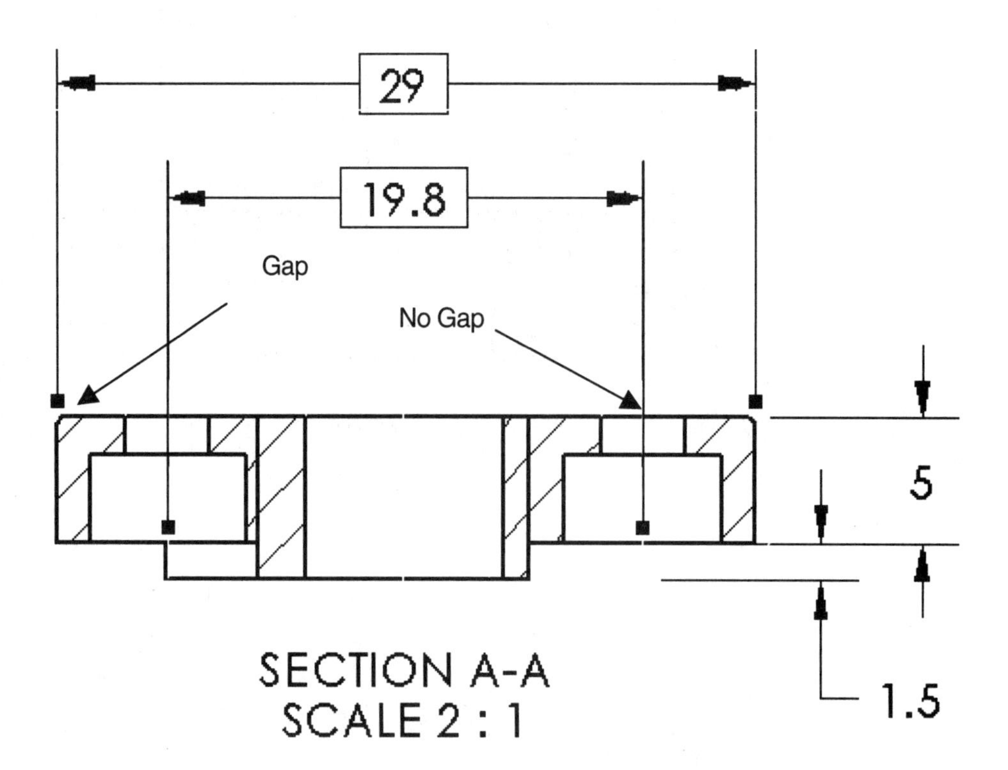

Open the Part.

312) Right-click in the **Section** view.

313) Click **Open COVERPLATE-PROJECT1**. Review the geometric relations. What additional dimensions are required in the drawing? Answer: A dimension is required to linearly locate the Counterbore hole with respect to the Center Hole. A Chamfer dimension is required.

Create Dimensions.

314) Create a horizontal dimension. Click the **Centerline** for the Hole.

315) Click the **Centerline** for the right Counter Bore.

316) Drag the **dimension** off the profile. There is no Gap between the extension line and the profile line. Space and stagger the horizontal dimension text for clarity.

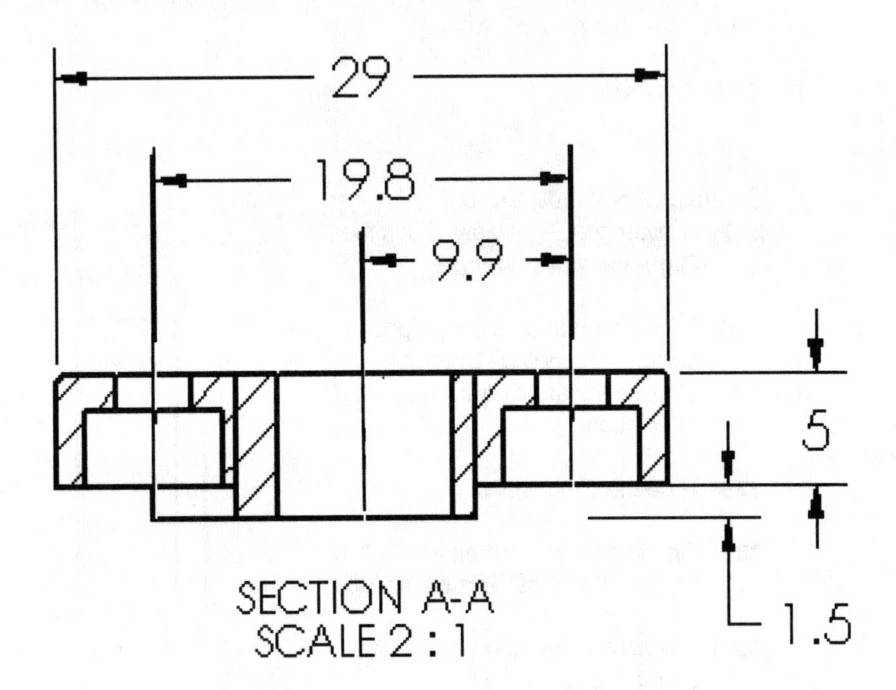

317) A Radial dimension is required in the Front view. Right-click the ∅**38** text.

318) Click Display options.

319) Click Display as Radius.

320) Drag the **R19** text aligned with the radial to the center point of the circle.

321) Enter **4X** before the R19 text in the Dimension Text box.

322) Click **OK**.

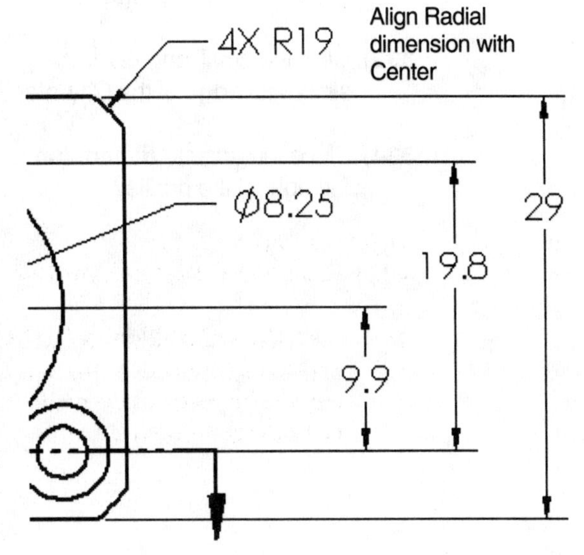

323) Activate the Front view. Click the **Front view boundary**. Create a vertical dimension.

324) Click the **circumference** of the right Counterbore.

325) Click the **center circle**.

326) Position the **dimension** to the right of the profile. There is no gap between the top extension line and the profile line. Space and stagger the other vertical dimensions.

Create a Chamfer dimension.

327) Activate the Right view. Click the **Right view boundary**.

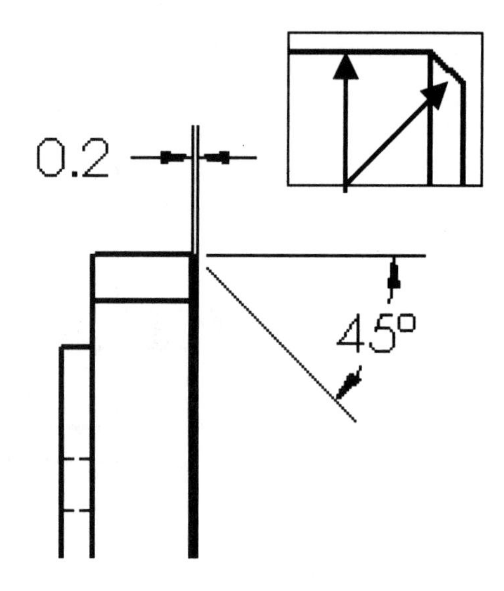

328) Click **Zoom in** on the top right corner. The default Chamfer dimensions are created in a different orientation.

329) **Hide** the dimensions.

330) Create a vertical dimension. Click the two right most **vertical lines**.

331) Place the **dimension** above the profile line.

332) Create the angular dimension. Click the top **horizontal line**.

333) Click the small angular silhouette **edge** of the Chamfer.

334) Place the angular **dimension** to the right of the profile.

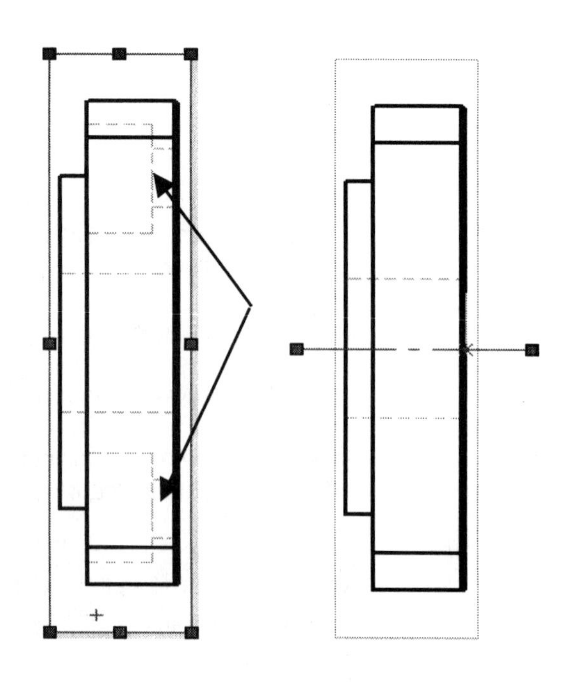

335) View the hidden lines in the Right view. Activate the Right view. Click the **Right view boundary**.

336) Click Hidden Lines Visible.

337) Hide the Counter bore hidden line. Press the **Ctrl** key.

338) Select the gray **horizontal** and **vertical lines** of the Counter bores.

339) Right-click **Hide Edge**. Release the **Ctrl** key.

340) Repeat until all **edges** of the Counter bores are remove.

341) Click **Centerline** from the Annotations toolbar.

342) Click the **top horizontal** line.

343) Click the bottom horizontal line.

344) Click **OK**.

Space horizontal and vertical dimensions with the Grid.

345) Use the Grid to evenly space multiple horizontal and vertical dimensions and views. Click **Grid** in the Sketch toolbar.

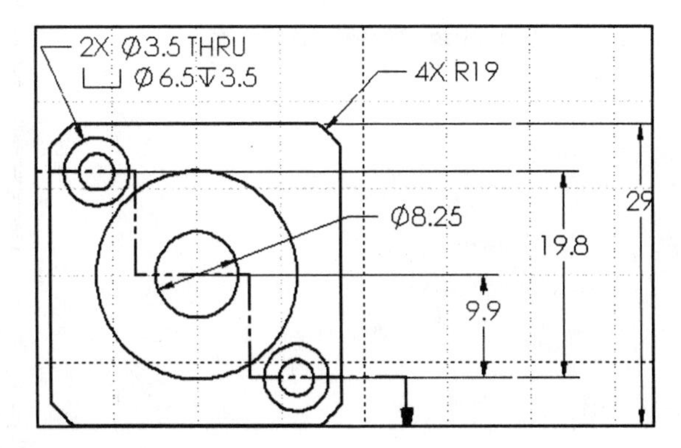

346) Check Display Grid.

347) Reposition any dimensions. Click **Grid** in the Sketch toolbar.

348) Turn off the Grid. Uncheck **Display Grid**.

349) Save the COVERPLATE drawing. Click **Save**.

Add a New Sheet.
350) Click COVERPLATE-Sheet1.

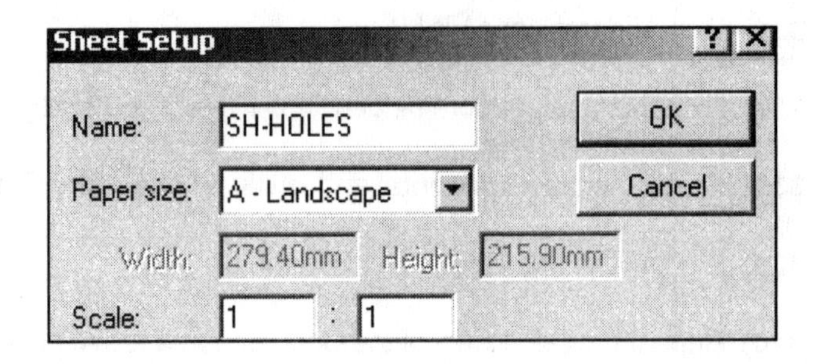

351) Right-click on the **Sheet1 tab**.

352) Click **Add Sheet**. Sheet3 is the default Sheet Name.

353) Enter **SH-HOLES** for sheet Name. A new sheet named, SH-HOLES is added.

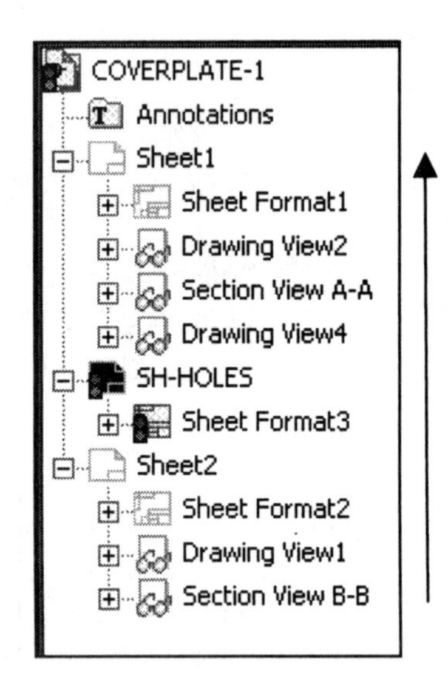

354) Change sheet order. Drag **SH-HOLES** upward in the FeatureManager.

355) Position **SH-HOLES** below Sheet1.

356) Open the COVERPLATE part. Click **Open**, **COVERPLATE**.

357) Select the **With Nose Holes** configuration.

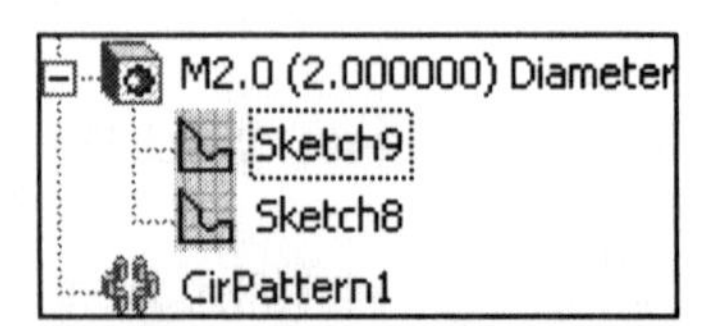

358) Expand the **M2.0 Hole** feature.

359) Display the Sketch used to position the holes. Click **Show**.

Insert a Named Front view into the Drawing.

360) Click Windows, Tile Horizontal.

361) Click inside COVERPLATE-SH-HOLES.

362) Click Named View.

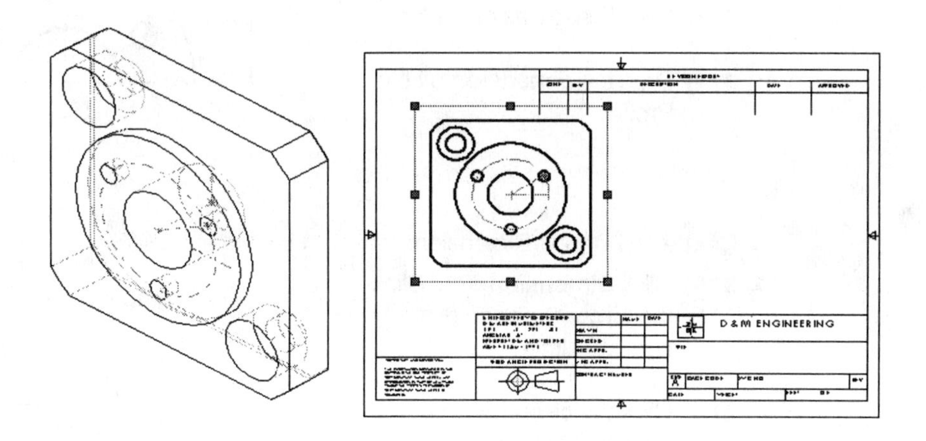

363) Click inside the **COVERPLATE** Graphics window.

364) Select **Front** from the View Orientation dialog box.

365) Create a Custom Scale of **2:1**.

366) Click **OK**.

Insert Dimensions for M2.0.

367) Click the **top right hole** (Seed feature).

368) Click Insert Model Items.

369) Click the **Hole Wizard Location** Check Box.

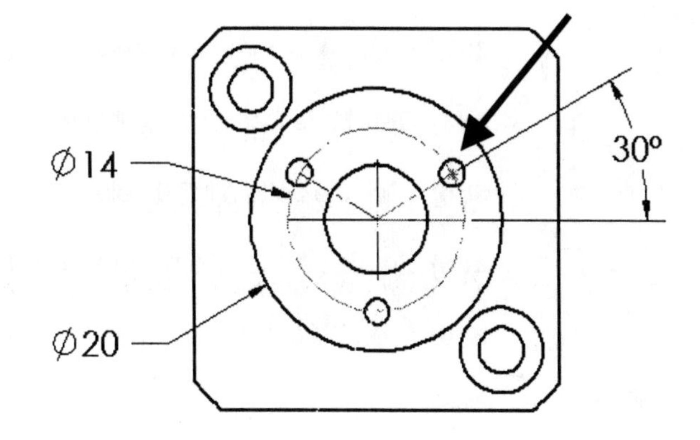

370) Click Insert Model Items.

371) Drag the **30°** dimension text off the profile.

372) Right-click the ∅**14** dimension text. Click **Display Options**.

373) Click Display as diameter.

374) Drag the **dimension** off the Profile.

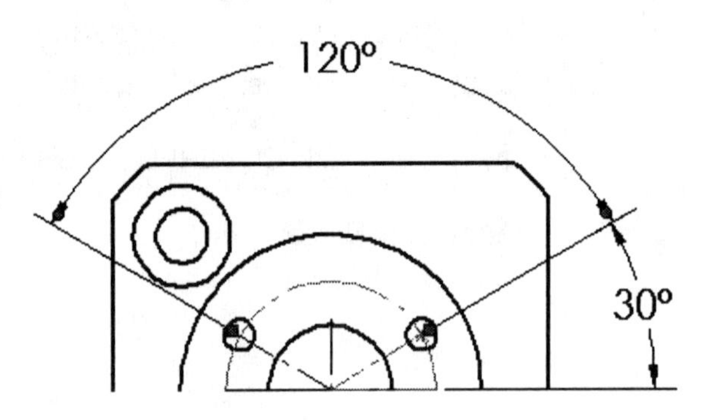

Create an Angular Dimension.

375) Click **Dimension** . Click the **circumference** of the right M2 circle.

376) Click the **circumference** of the left M2 circle.

377) Click the **Origin**.

378) Drag the **mouse pointer** downward to display the 120° dimension text.

379) Position the **120°** dimension text above the profile line.

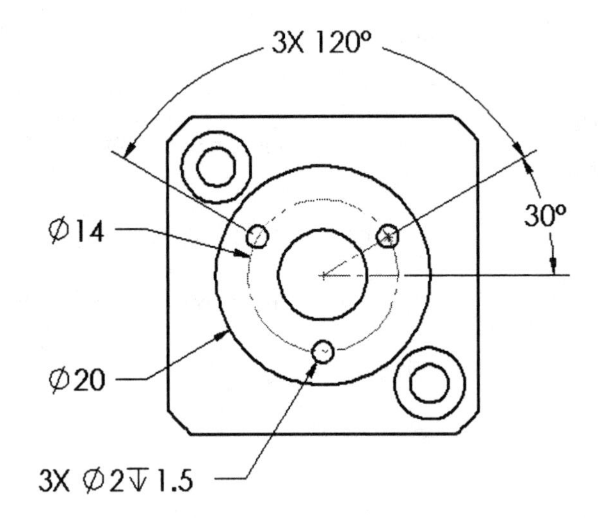

380) Sketch a **centerline** from the Origin to the center of the left circle.

381) Right-click the **120°** text.

382) Enter **3X** before the dimension text.

383) Click **Hole Callout** to activate.

384) Click the bottom **M2.0** hole.

385) Position the text off the profile. Click **Hole Callout** to deactivate.

386) Enter **3X** before the dimension text.

A Cosmetic Thread requires the COVERPLATE part configuration to be active.

387) Open the part from the drawing view. Right-click inside the **Front view boundary**.

388) Click Open COVERPLATE.

389) Click Configuration.

390) Click With Holes Configuration.

391) Return to the drawing. Click **Window, COVERPLATE-Sheet1** from the Main toolbar.

392) Add a Cosmetic Thread. Right-click the **left M2 hole** (Edge<1>).

393) Click Dimension Annotations, Cosmetic Thread.

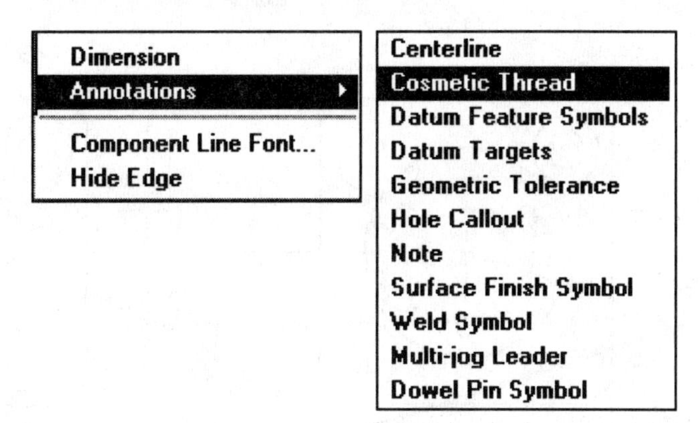

394) Select the **right M2 hole** (Edge<2>) and the **bottom M2 hole** (Edge<3>).

395) Enter **2.40** for Major Diameter.

396) Click **OK**.

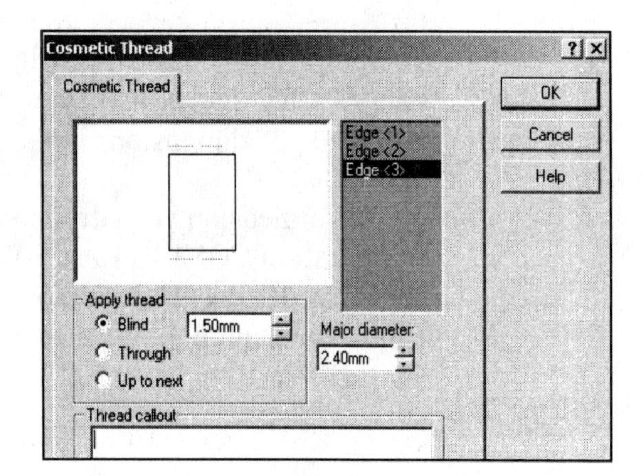

397) Open the **COVERPLATE** part. The Cosmetic Thread has been added to the M2 Hole feature.

Note: The part configuration must be active before the Cosmetic Thread is inserted into the drawing view.

Modifying Features

The design process is dynamic. We do not live in a static world. Explore the following changes to the COVERPLATE, With Nose Holes configuration in the exercises at the end of this Project.

Decrease the Boss diameter to ∅17. The drawing dimension decreases. The Dimension text position is unchanged. Feature changes that modify size require simple or no changes in the drawing.

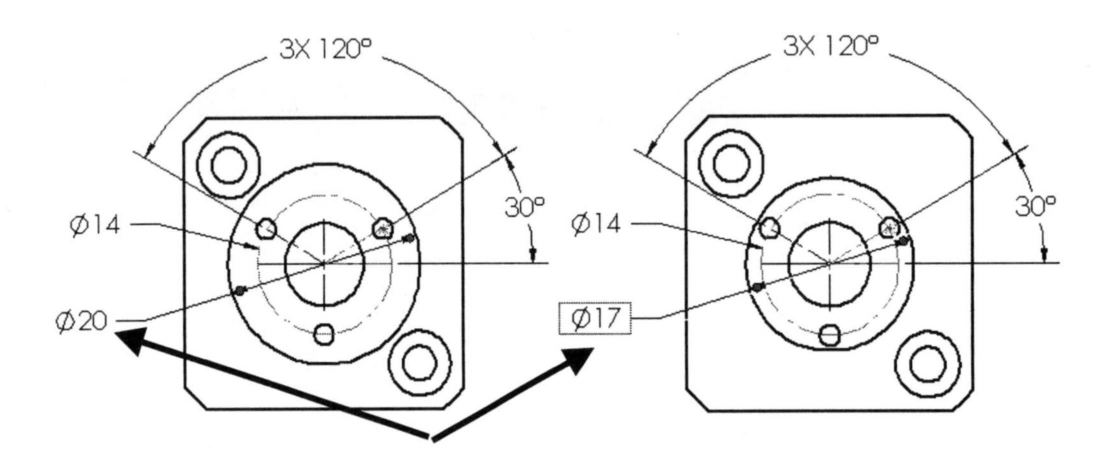

When a part has multiple configurations, dimension changes affect one or more configurations. Modify the 30° to 45° dimensions in the drawing.

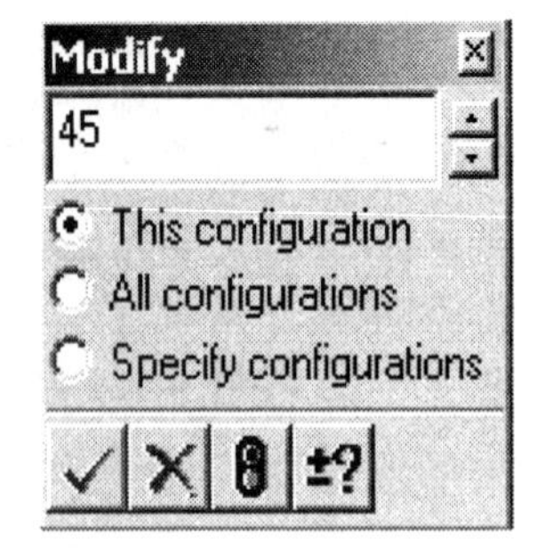

The 30° dimension is controlled by the part configuration COVERPLATE-With Nose Holes. The COVERPLATE-With Nose Holes is the current configuration of the part.

Note: If you receive a warning message, the part was saved in a different configuration. Open the part and set the configuration to With Nose Holes before you change a dimension.

Dimensions require no repositioning when you modify the M2 hole location from 30° to 45°. Change the number of holes from 3 to 4 in the Circular Pattern.

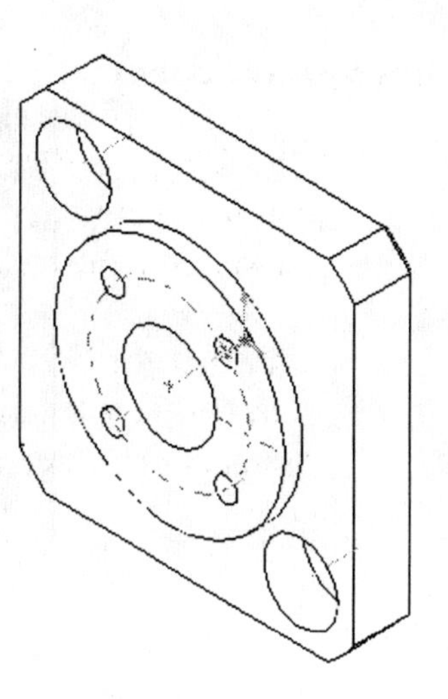

The change from 3 to 4 instances requires drawing modification. The new instance is inserted after the 1st hole.

The number of instances changes form 3 to 4 in two dimensions. The dimensions are repositioned. Add a new centerlines and center marks. Redefine quantity of holes.

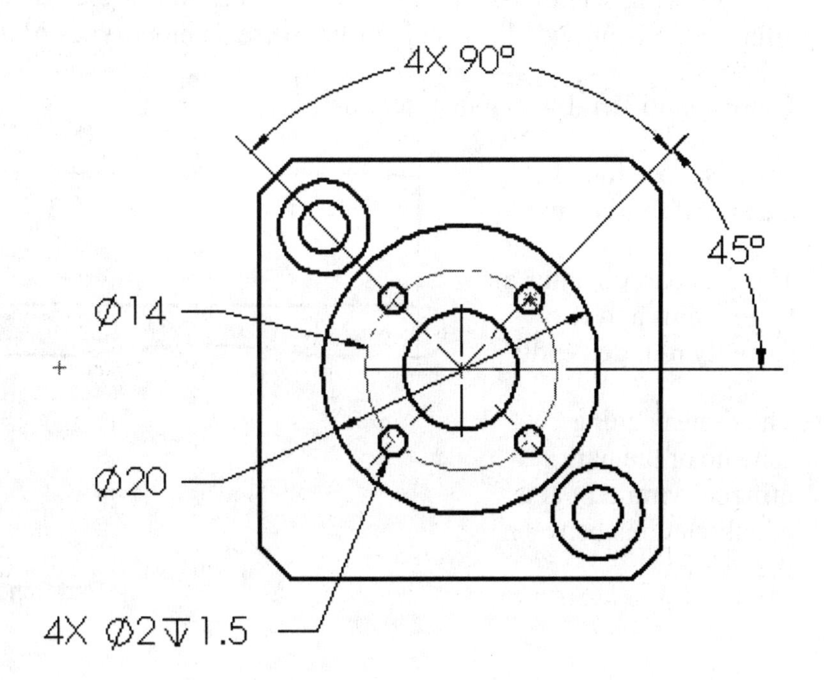

When additional dimensions must be added to the drawing, utilize the model item, Reference geometry, Axes and Planes.

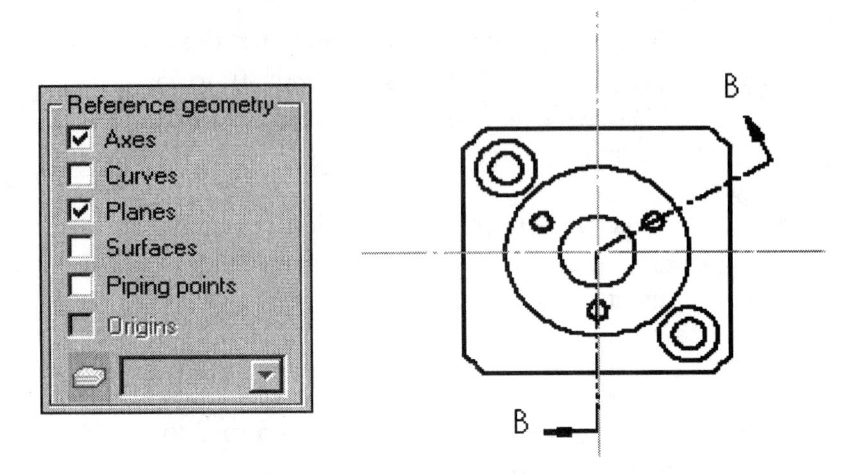

Dimensioning Features

The following section is not related to the CYLINDER assembly. The information is provided to you in order to create other types of dimensions.

There are no step-by-step instructions.

Exercises are found at the end of this section.

The first part contains a large radius and two partially rounded ends.

The dimensioning scheme of the part utilized symmetry and equal relations between the two arcs.

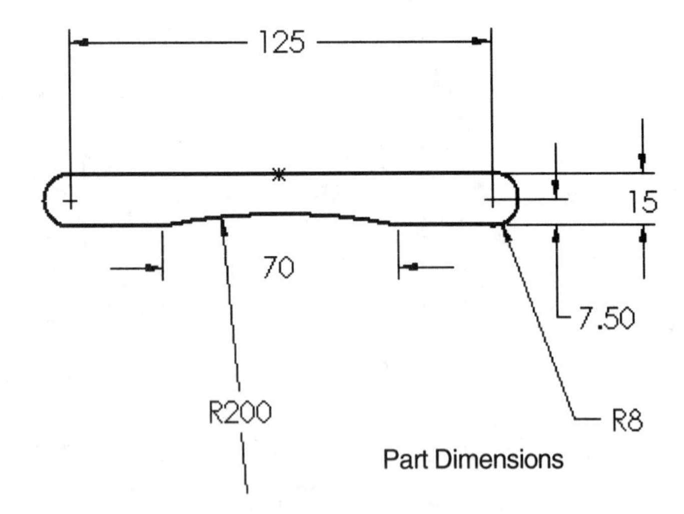

Part Dimensions

Foreshorten Radii: Large radii are drawn outside the sheet boundary or overlap a second view. A Foreshorten radius inserts three line segments on the leader line. Right-click Properties.

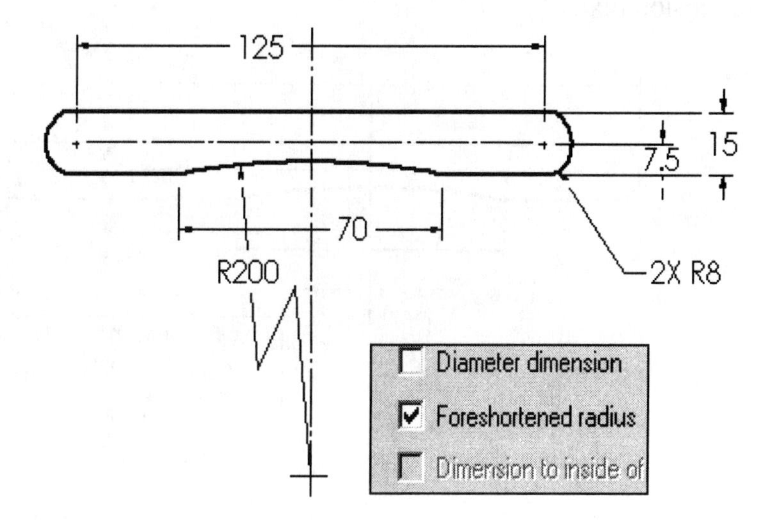

Check the Foreshortened radius check box.

Drag the endpoints of the foreshortened radius if required to fit within the view.

Partially Rounded Ends:
The overall length of the part is measured between the two center points of the arc. Return to the part sketch to redefine the dimension scheme.

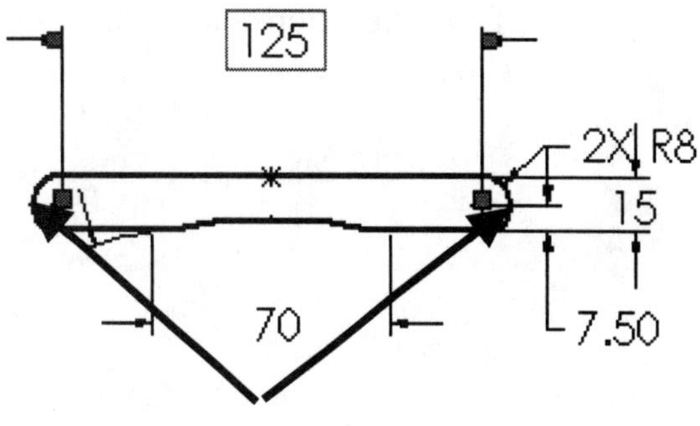

Delete the overall dimension. Select each arc to redefine the overall length. Do not select the center point. The dimension is displayed from the center points.

Exit the sketch. Save the part and return to the drawing.

Update the drawing. Rebuild the drawing. Insert model items again to reflect the new dimension scheme.

Modify the dimension in the drawing. Right-click Properties on the overall dimension text.

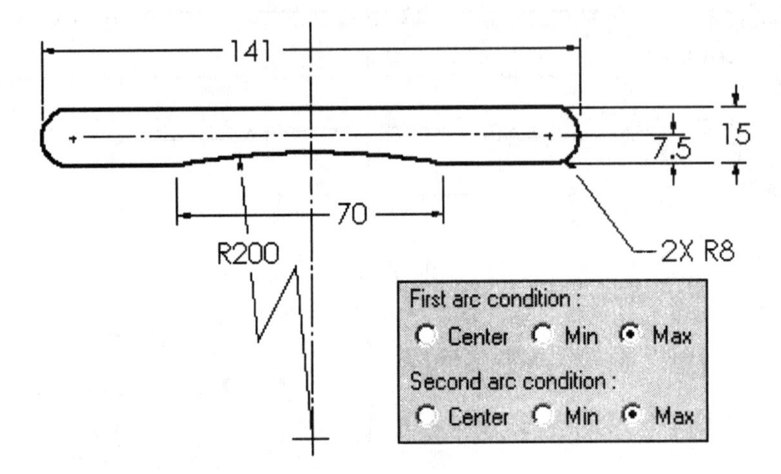

Click Max for the First arc condition. Click Max for the Second arc condition.

Enter 2X for the radius text. Add center marks to indication the center of the radii.

Create a gap.

Drag the extension lines. Add centerlines to complete the drawing.

Slotted Holes:
Slotted Holes utilized symmetry in the part. Redefine the dimensions for the Slot Cut according to the ASME 14.5M standard. The ASME 14.5M standard requires the dimension to the outside of the arc.

The Radius value is not dimensioned. Select each arc to create a center/min/max arc condition in the part.

Insert the part into the drawing. The end radii are indicated.

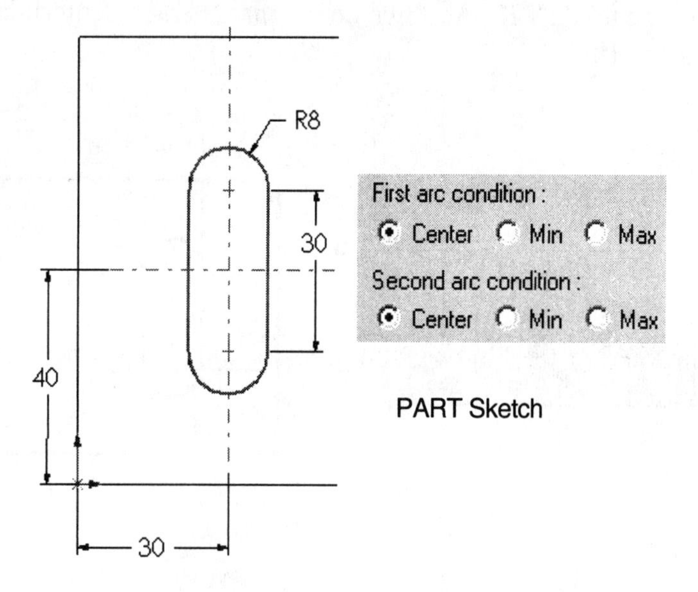

PART Sketch

No dimension is labeled. There are three methods to dimension a slot in the drawing.

Modify the dimension properties to create one of the following:

Method 1: Select Center for the First Arc condition from Properties. Select Center for the Second Arc condition from Properties.

Create a linear dimension between the two vertical lines of the slot. Create a radial dimension. Delete the radius value.

Enter the text 2X R. Add two center marks and a centerline between the two center marks.

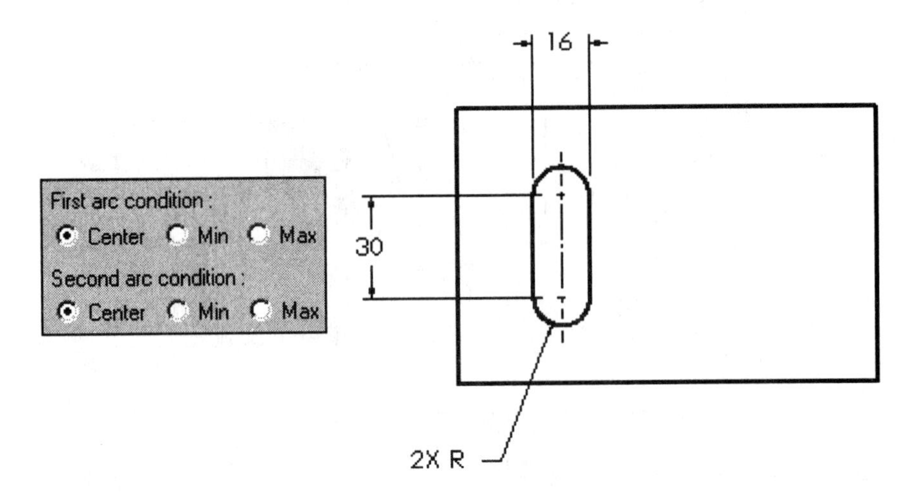

Method 2: Use a Note with leader. Enter the text of the overall width and height of the Slot. Use a radial dimension.

Enter the 2XR text. Add two intersecting centerlines.

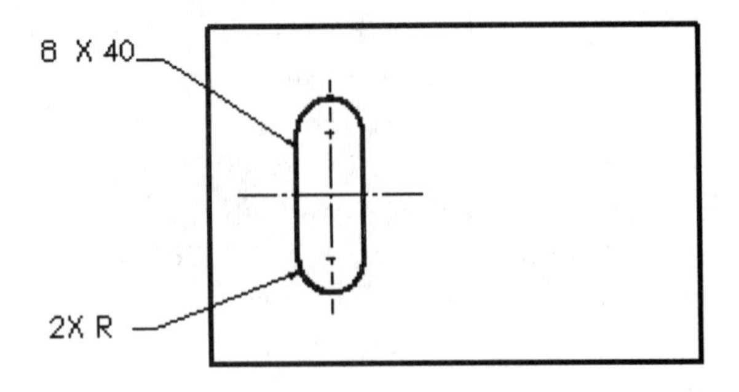

Method 3: Select Max for the First Arc condition from Properties. Select Max for the Second Arc condition from Properties.

Create a linear dimension between the two vertical lines of the slot.

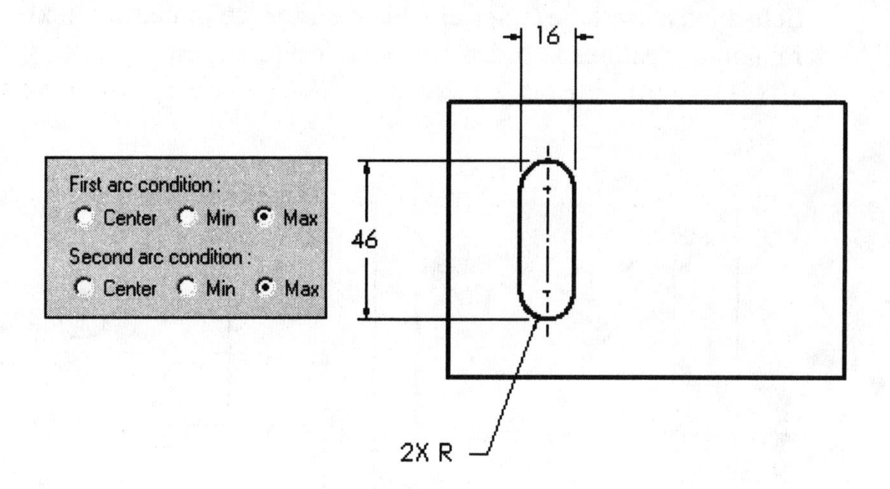

Create a radial dimension. Delete the radius value. Enter 2X R text.

Location of Features

The following section is not related to the CYLINDER assembly. The information is provided to you in order to explore other methods of locating features.

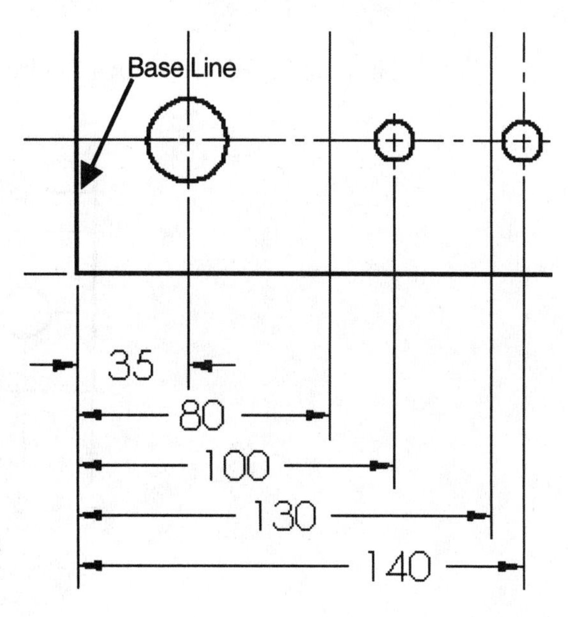

Rectangular coordinate dimensioning locates features with respect to one another, from a datum or an origin.

The ASME Y14.5M Rectangular Coordinate Dimensioning standard specifies linear distances in a coordinate direction from two perpendicular planes.

The method corresponds to Base Line Dimensioning in SolidWorks.

Base Line Dimensioning: Create Base Line Dimensions. Select a Base Line. Select a feature(hole). Select a location for the dimension text. Select the remaining features in order from smallest to largest.

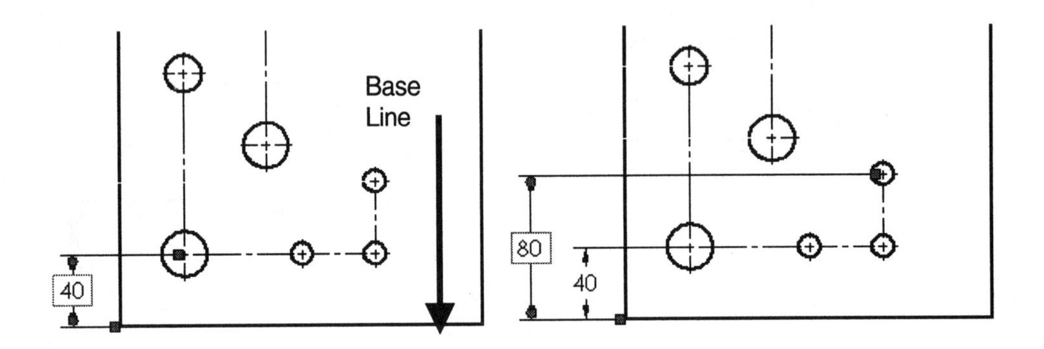

The ASME Y14.5M Rectangular Coordinate Dimensioning standard Without Dimension

Lines displays an extension line without the use of dimension lines or arrows.

This method corresponds to Ordinate Dimensioning in SolidWorks.

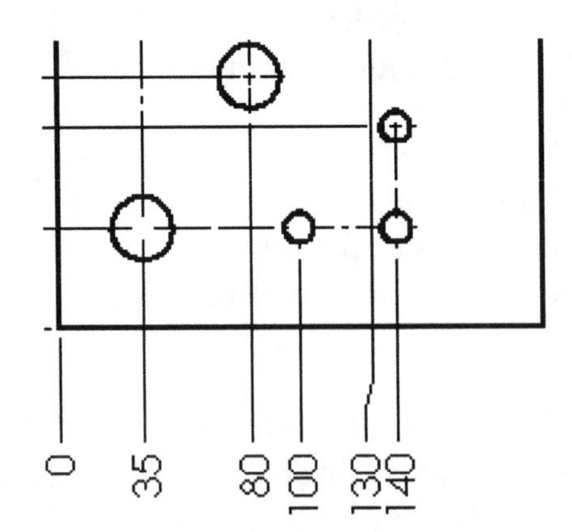

Ordinate Dimension: Create Horizontal Ordinate Dimensions. Select Tools, Dimension, Horizontal Ordinate.

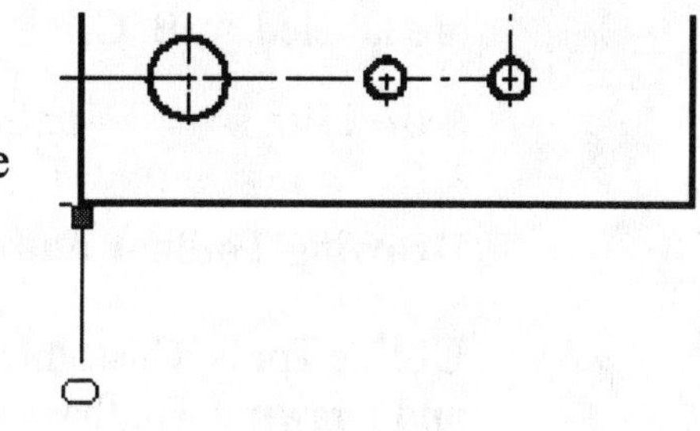

Select the Origin or vertex for a zero location. All other dimensions are measure from this location.

Select a location for the dimension text off the profile. Select a feature (hole). Select the remaining features in order from smallest to largest.

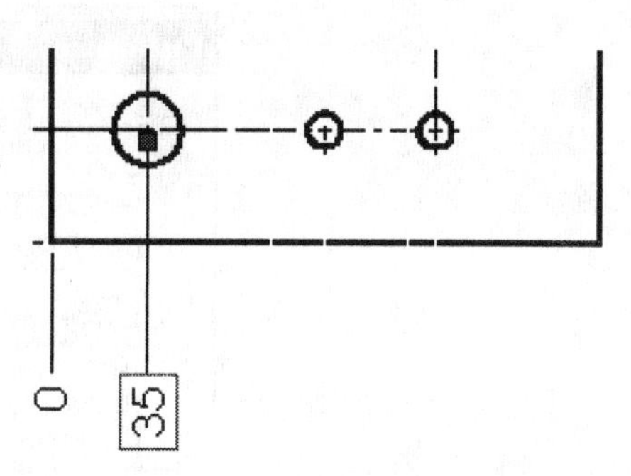

Create Vertical Ordinate Dimensions. Select Vertical Ordinate. Select the Origin of vertex for zero location.

Select a location for the dimension text off the profile. Select a feature (hole).

Select the remaining features in order from smallest to largest.

Extension lines will jog to fit dimension text.

View the .AVI file. Select Help, Ordinate Dimensions.

Baseline Dimensioning and Ordinate Dimensioning produce reference dimensions. They cannot be modified. Part features drive Baseline and Ordinate dimensions.

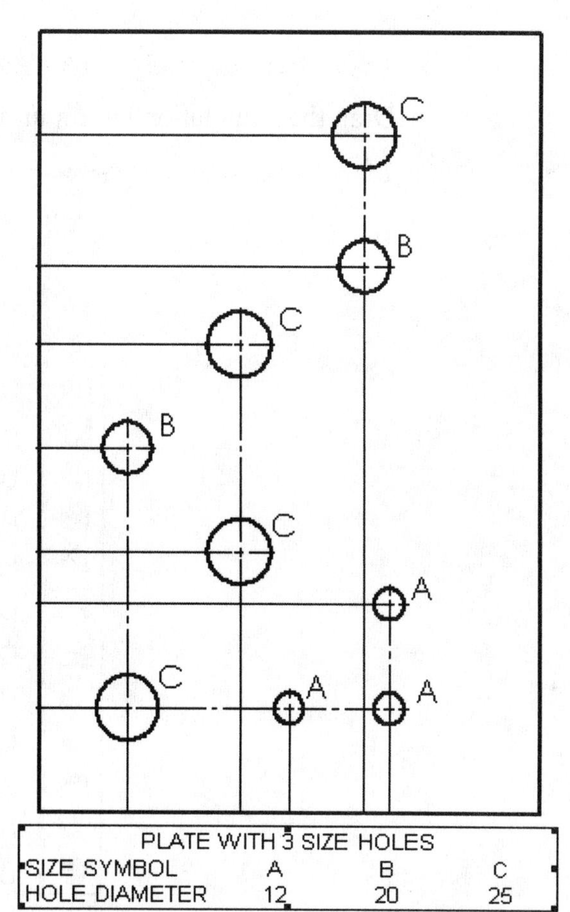

PLATE WITH 3 SIZE HOLES			
SIZE SYMBOL	A	B	C
HOLE DIAMETER	12	20	25

Design tables created in the part/assembly can be inserted into the drawing. Combine design tables with notes in a drawing to label similar features. Holes are labeled A, B, C.

Drawing Toolbar and Annotation Toolbar:

Utilize Tools, Customize, Commands to add icons to the Annotation Toolbar and Drawing Toolbar

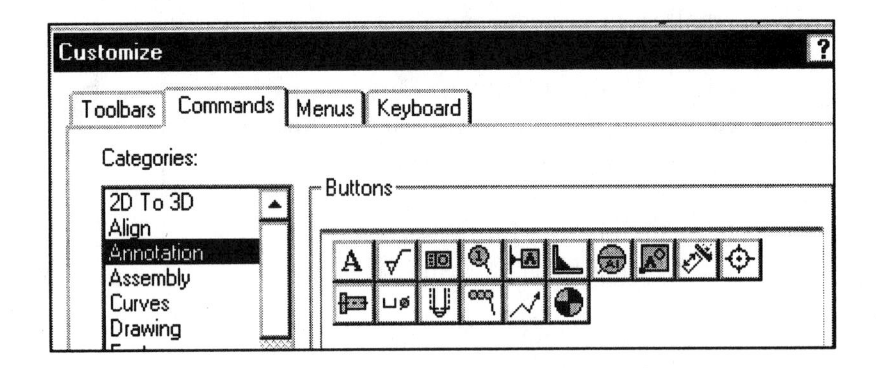

Drag the Annotation button into the Annotation Toolbar.

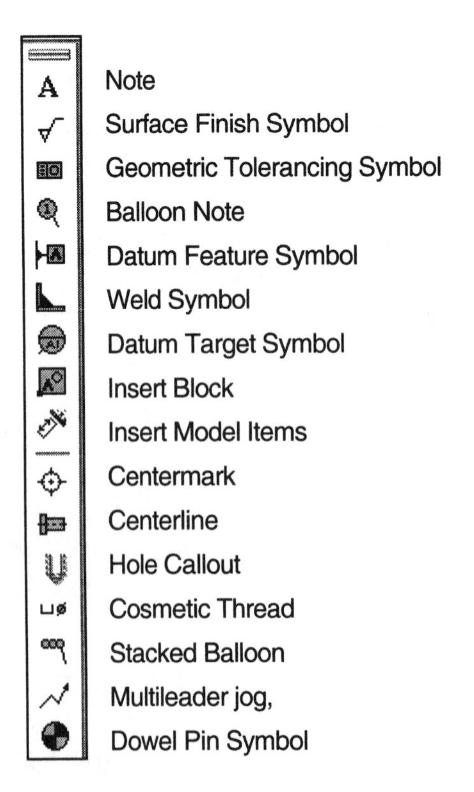

Drag the Drawing button into the Drawing Toolbar.

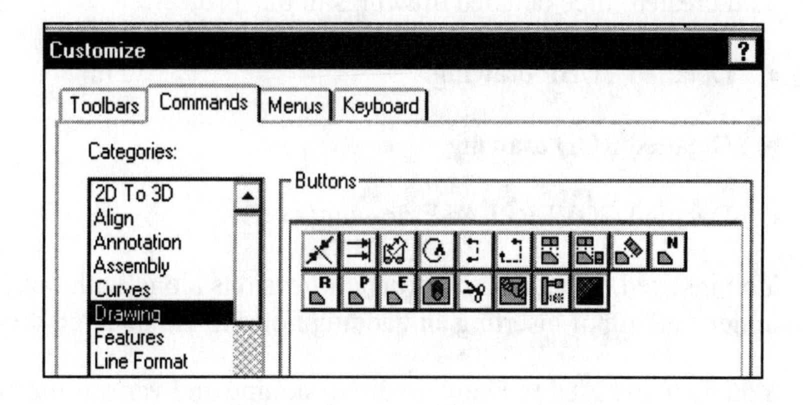

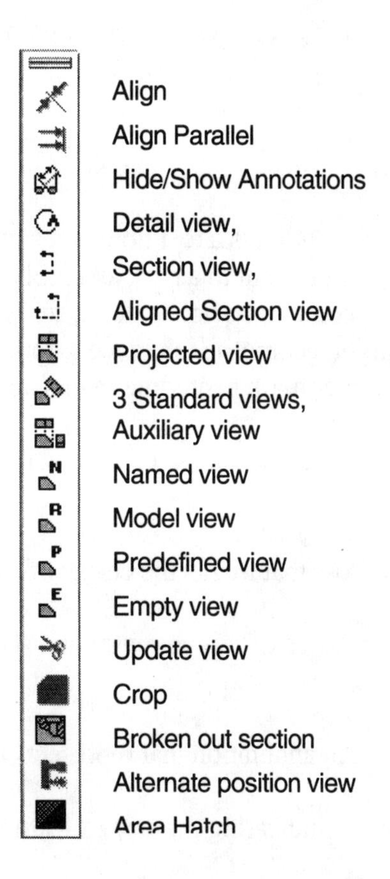

Additional information on Detailing is found in the SolidWorks Help with the On-line Tutorials, Getting Started Manual, SolidWorks On-line Users Guide, Index and Glossary.

Project Summary

You created three detailed drawings in this project:

- Detailed TUBE drawing.
- Detailed ROD drawing.
- Detailed COVERPLATE drawing.

You inserted, added and modified dimensions along with obtaining an understanding of inserting and adding notes in the detailed drawings.

You were exposed to Drawing dimensioning and various methods to move, hide and recreate dimensions that adhere to a drawing standard.

Perform a few project exercises on your own before moving on to Project 4.

Project Terminology

Annotation: An annotation is a text note or a symbol that adds specific information and design intent to a part, assembly, or drawing. Annotations in a drawing include Specific note, hole callout, surface finish symbol, datum feature symbol, datum target, geometric tolerance symbol, weld symbol, balloon, and stacked balloon, center mark centerline marks, area hatch, and block.

Baseline dimensions: Dimensions referenced from the same edge or vertex in a drawing.

Center mark: A cross that marks the center of a circle or arc.

Centerline marks: An axis of symmetry in a sketch or drawing displayed in a phantom font.

Cosmetic thread: An annotation that represents threads.

Dimension: A value indicating the size of feature geometry.

Dimension Line: A linear dimension line references the dimension text to extension lines indicating the feature being measured.

Extension line: The line extending from the profile line indicating the point from which a dimension is measured.

Fit tolerance: The tolerance between a hole and a shaft.

Leader: A solid line from an annotation to the referenced feature.

Linked Note: An note that references a SolidWorks property or custom property from another document.

Model item: A dimension or annotation of feature geometry that can be used in detailing drawings.

Ordinate dimensions: Chain of dimensions referenced from a zero ordinate in a drawing or sketch.

Reference dimension: Displays a measurement of an item. A referenced dimension cannot drive the model and its value cannot be modified. Reference dimension update when model dimensions change.

Precision: Controls the number of decimal places displayed in a dimension.

Silhouette edge: A curve representing the extent of a cylindrical or curved face when viewed from the side.

Copy and Paste: Utilize copy/paste for annotations, parametric notes, linked notes and design table text.

Questions:

1. Dimensions in a drawing are _______________ from a part or _________ in the drawing.
2. Drawing notes, Hole Callout, and Center Mark tools are found in the

   SolidWorks _____________ toolbar.
3. Identify the order in which dimensions are inserted into a drawing with the various views: Standard views, Section views and Detail views.
4. True or False. Feature dimensions are always inserted into the correct view.
5. Identify the command to select when dimensions are no longer required.
6. Describe a Reference dimension in SolidWorks.
7. Describe the procedure to move a part dimension from one view to a second view.
8. List and describe the three methods to dimension simple holes and circular geometry.
9. True or False. Cosmetic Threads created in the Drawing are automatically added to the referenced part.
10. A drawing references multiple part configurations. A dimension is changed on a drawing. Identify the three Modify options that appear in the Modify dialog box.
11. Name three different types of Reference geometry created in a part or assembly that are inserted into a drawing.
12. Describe how a Foreshorten radius is created.
13. Identify the arc conditions that are required when dimensioning the overall length of a slot.
14. Provide a definition for Baseline dimensioning.

Exercises

Exercise 3.1:

Open the COVERPLATE part. Decrease the Boss diameter to ∅17.

Open the COVERPLATE drawing. Modify the M2 hole location from 30° to 45°. Change the number of M2 holes from 3 to 4 in the Circular Pattern.

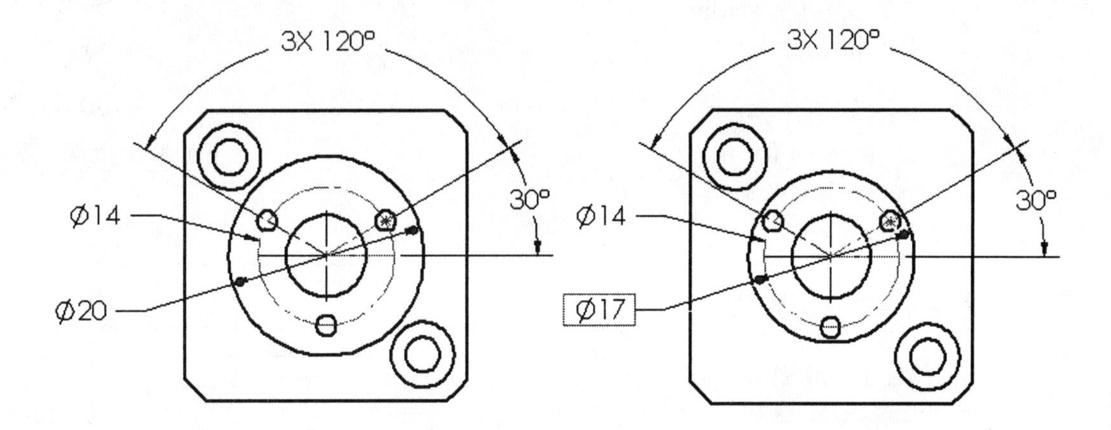

Add a new centerline and redefine the dimension. Reposition the dimensions and edit the note quantity of M2 holes from 3 to 4.

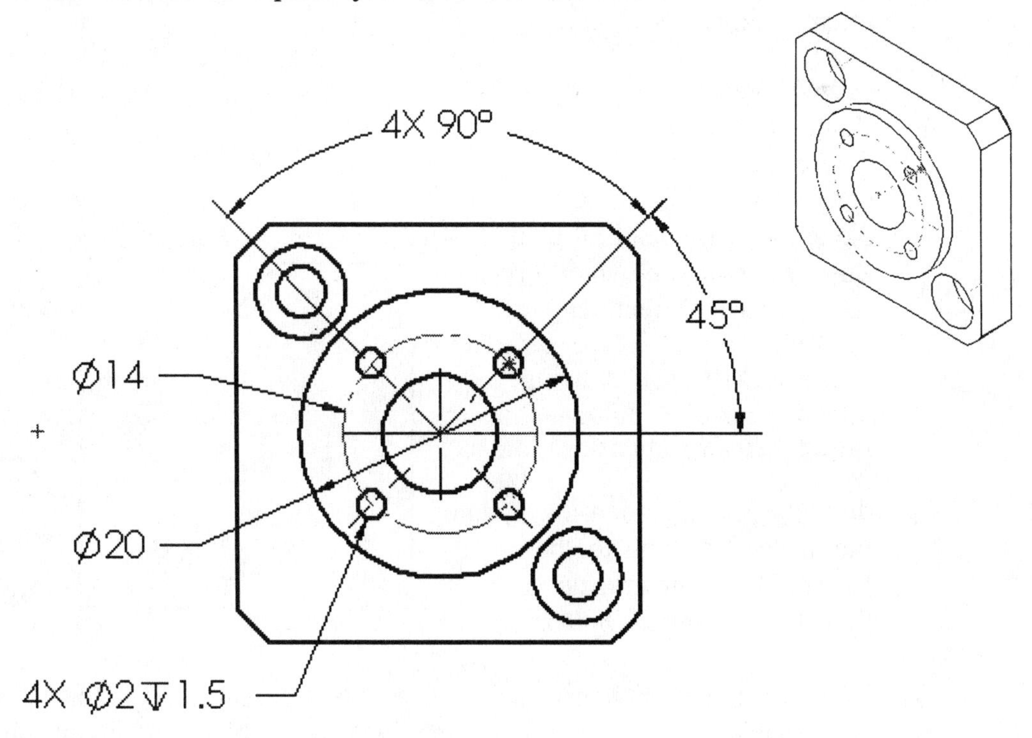

Exercise 3.2:

Create a new drawing for the RADIUS-ROUNDED END part located in the 2003drwparts file folder. Add a Foreshorten Radius for the R200.

Create an overall dimension for the partially rounded ends. Modify the Radius text according to the ASME standard. Add centerlines and center marks.

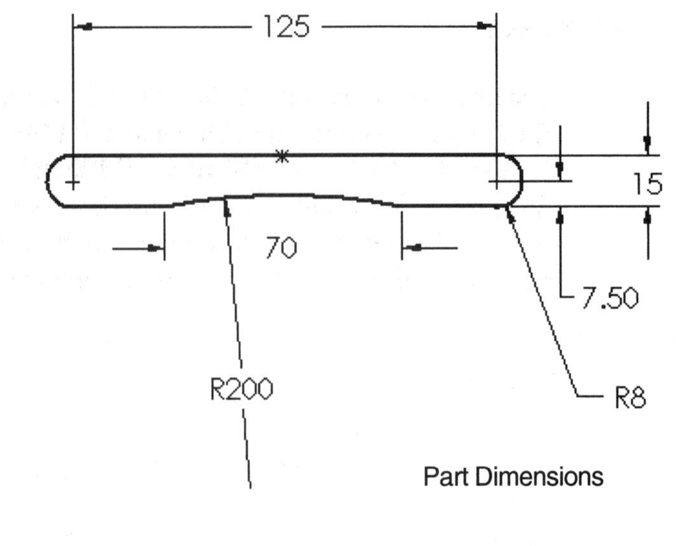

Part Dimensions

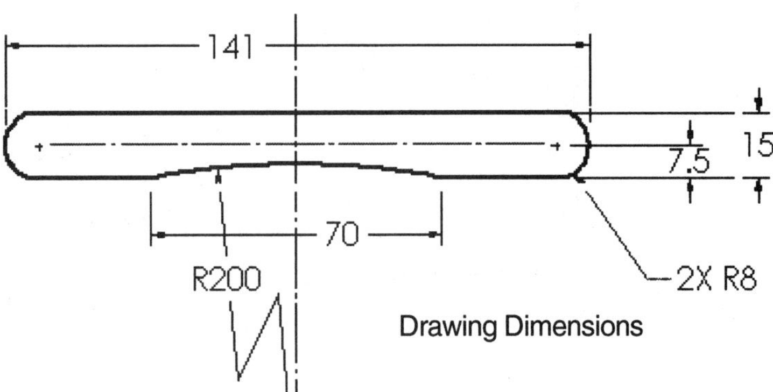

Drawing Dimensions

Exercise 3.3:

Dimension chamfers in drawings. Create a new part called C-BLOCK. Sketch an L-shaped Extruded Base feature. Add two different size CHAMFER features.

10 X 45.0°

5 X 30.0°

Create a C-BLOCK drawing.

Insert chamfer dimensions into a drawing. Click Dimension on the Sketch Relations toolbar. Right-click in the graphics area. Select the Chamfer Dimension.

The mouse pointer changes to . Select the chamfered edge. Select one of the lead-in edges. Click in the graphics area to place the dimension and to display the CHAMFER.

Exercise 3.4:

Create a new drawing for the SLOT-PLATE part located in the 2003drwparts file folder. Redefine the dimensions for the Slot Cut according to the ASME 14.5M Standard. The ASME 14.5M Standard requires an outside dimension for a slot. The Radius value is not dimensioned. Select each arc to create a center/min/max arc condition in the part. Insert the part into the drawing. The end radii are indicated but no dimension is labeled.

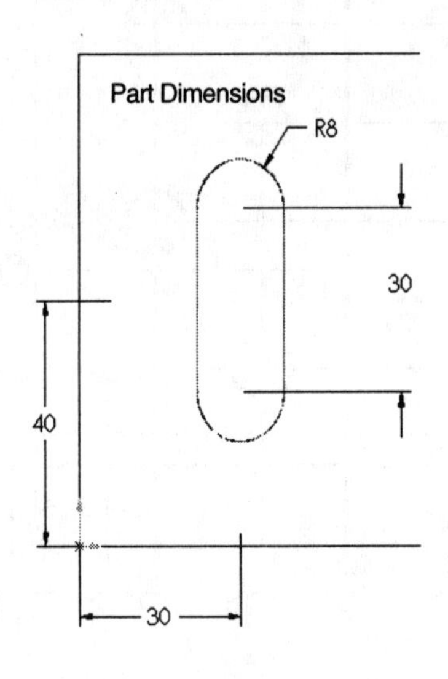

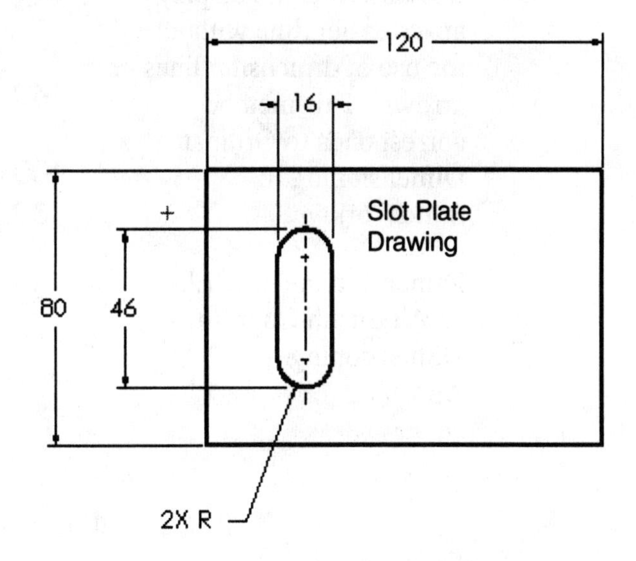

Exercise 3.5:

Create a new drawing for the TABLE-PLATE part located in the 2003drwparts file folder. Rectangular coordinate dimensioning locates features with respect to one another, from a datum or an origin. ASME Y14.5M Rectangular Coordinate Dimensioning specifies linear distances in a coordinate direction from two perpendicular planes. This method corresponds to Base Line Dimensioning in SolidWorks. Dimension the TABLE-PLATE with Base Line Dimensioning.

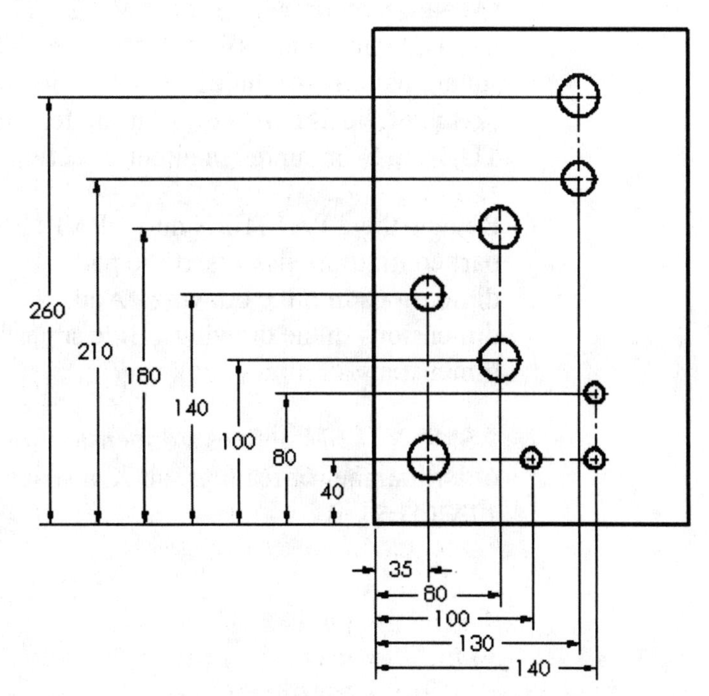

Exercise 3.6:

Create a new drawing for the TABLE-PLATE part located in the 2003drwparts file folder. ASME Y14.5M Rectangular Coordinate Dimensioning Without Dimension Lines displays an extension line without the use of dimension lines or arrows. This method corresponds to Ordinate Dimensioning in SolidWorks.

Dimension the TABLE-PLATE with Coordinate Dimensioning.

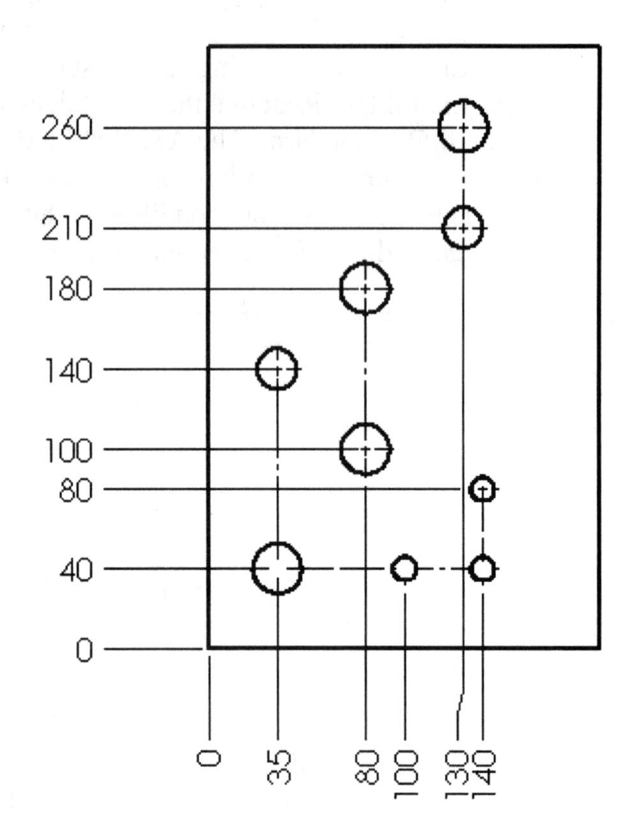

Exercise 3.7:

Create a new drawing for the MOUNTINGPLATE-CYLINDER part located in the 2003drwparts file folder. The MOUNTING-CLYLINDER contains a square pattern of 4 holes. The location of the holes corresponds to the initial sketch of the TUBE Base Extrude feature. Four holes are used in order for the TUBE to be mounted in either direction.

Review the 2 PATTERN and 3 PATTERN part configurations. Insert the part dimensions into the drawing. Add dimensions in the drawing. Hide superfulous dimensions.

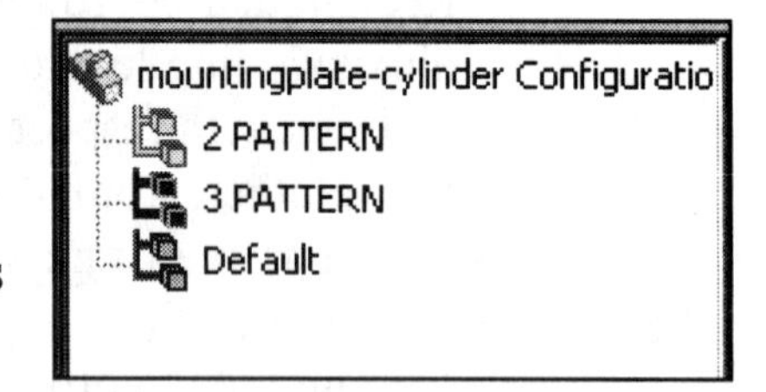

ASME Y14.5M defines a dimension for a repeating feature in the following order: number of features, an X, a space and the size of the feature. Example: 12X ∅3.5.

Create a drawing called MOUNTINGPLATE-CYLINDER with two sheets. Use an empty A size Drawing Template.

a) SHEET1 contains the 3 PATTERN configuration.

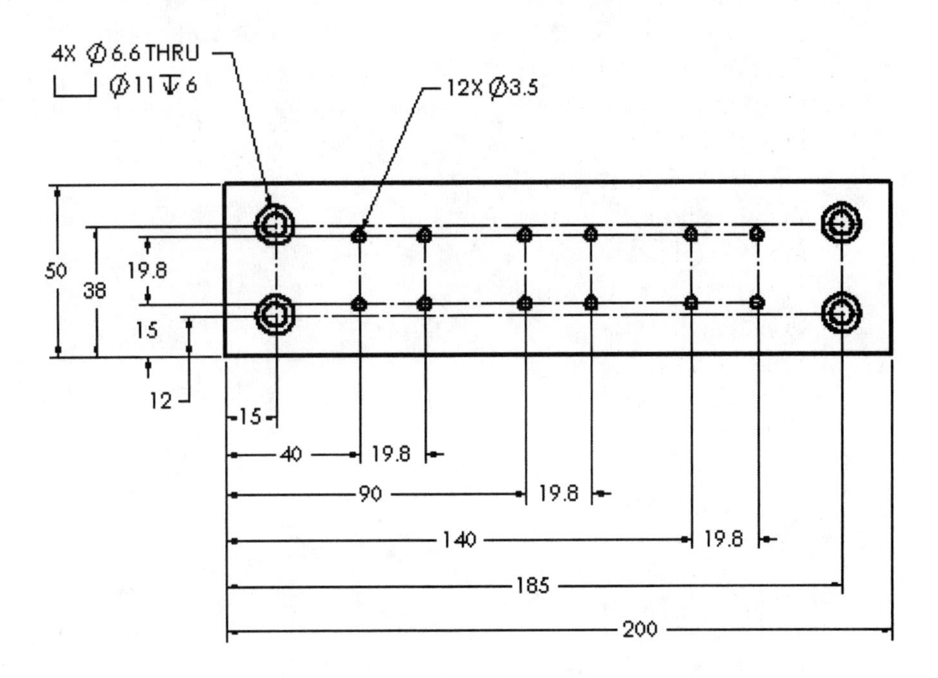

b) SHEET2 contains the 2 PATTERN configuration.

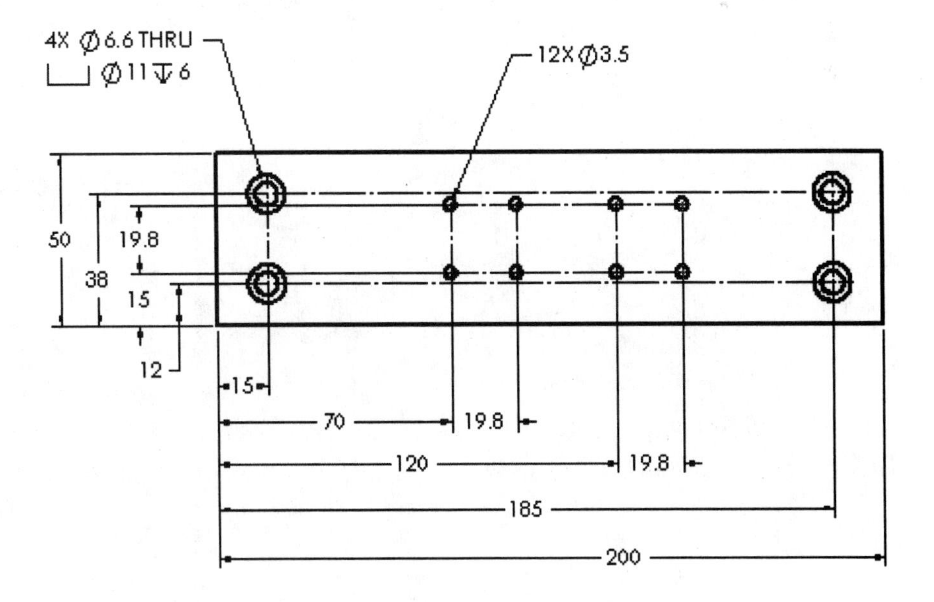

Notes:

Project 4

Assembly Drawing

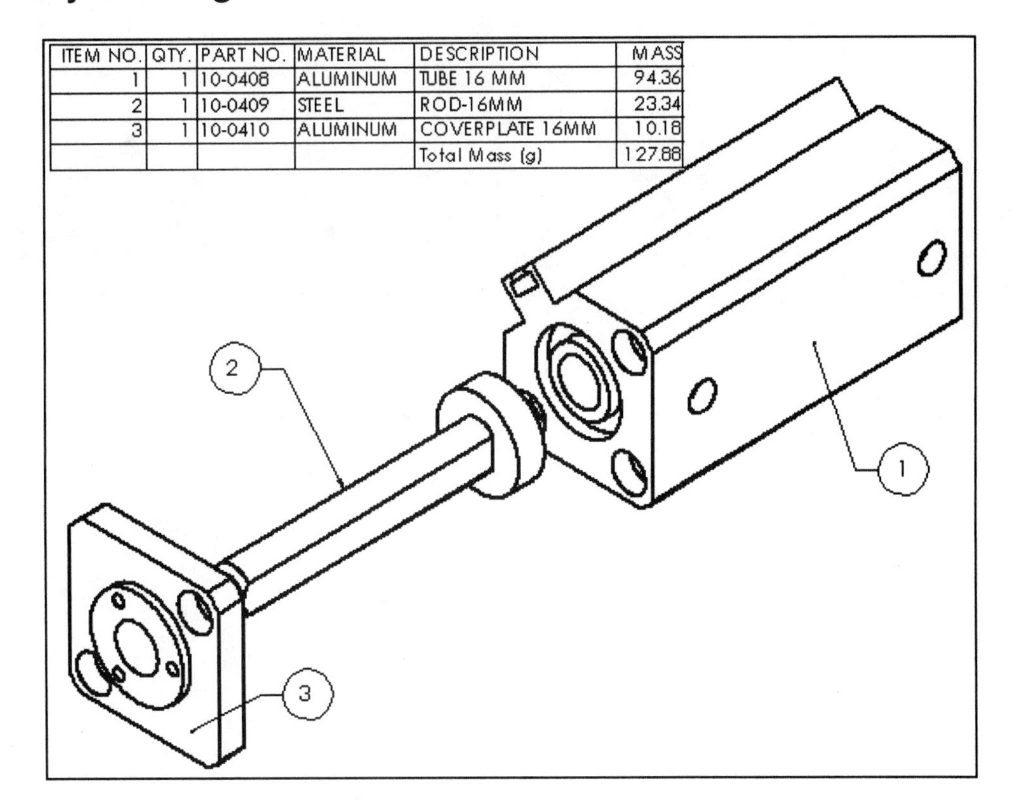

ITEM NO.	QTY.	PART NO.	MATERIAL	DESCRIPTION	MASS
1	1	10-0408	ALUMINUM	TUBE 16 MM	94.36
2	1	10-0409	STEEL	ROD-16MM	23.34
3	1	10-0410	ALUMINUM	COVERPLATE 16MM	10.18
				Total Mass (g)	127.88

Below are the desired outcomes and usage competencies based on the completion of this Project. An Assembly drawing refers to a SolidWorks drawing that contains a SolidWorks Assembly.

Project Desired Outcomes:	**Usage Competencies:**
CYLINDER Assembly. Custom Properties. Design Table. CYLINDER Drawing.	Ability to create an assembly with multiple configurations.
	An understanding of Custom Properties and SolidWorks Properties.
	Ability to create an assembly drawing using multiple part configurations.
Bill of Matcrials.	Knowledge to develop and incorporate a Bill of Materials into a drawing.

Notes

Project 4 – Assembly Drawing

Project Objective

In this project create the following:

- CYLINDER assembly.
- Custom Properties.
- Design Table.
- CYLINDER drawing.
- Bill of Materials.

The CYLINDER assembly consists of the TUBE, ROD and COVERPLATE parts.

Add Custom Properties to each part to describe Material, Mass and Description of the part configurations.

Create six different configurations of the CYLINDER assembly with an Excel Design Table.

Add a Design Table. Add a Bill of Materials with Custom Properties. Insert six different configurations of the CYLINDER assembly.

Insert two different configurations of the TUBE, ROD and COVERPLATE parts into the drawings.

On the completion of this project, you will be able to:

- Add Balloons.
- Create Assemblies.
- Insert Assemblies.
- Insert a Bill of Materials.
- Add Custom Properties.
- Insert and Edit Design Tables.

Project Situation

You need to create a CYLINDER drawing for the manufacturing and marketing department. Manufacturing requires a Bill of Materials for the assembly configuration.

Marketing requires part and assembly views for their on-line catalog. The CYLINDER drawing consists of multiple configurations of the CYLINDER assembly and the Bill of Materials.

Create the CYLINDER assembly. The CYLINDER assembly consists of the TUBE, ROD and COVERPLATE parts.

Add Custom Properties to each part to describe Material, Mass and Description of the part configurations. Create multiple configurations of the CYLINDER assembly.

Add a Design Table. Add a Bill of Materials with Custom Properties. Insert six different configurations of the CYLINDER assembly.

Insert two different configurations of the TUBE, ROD and COVERPLATE parts.

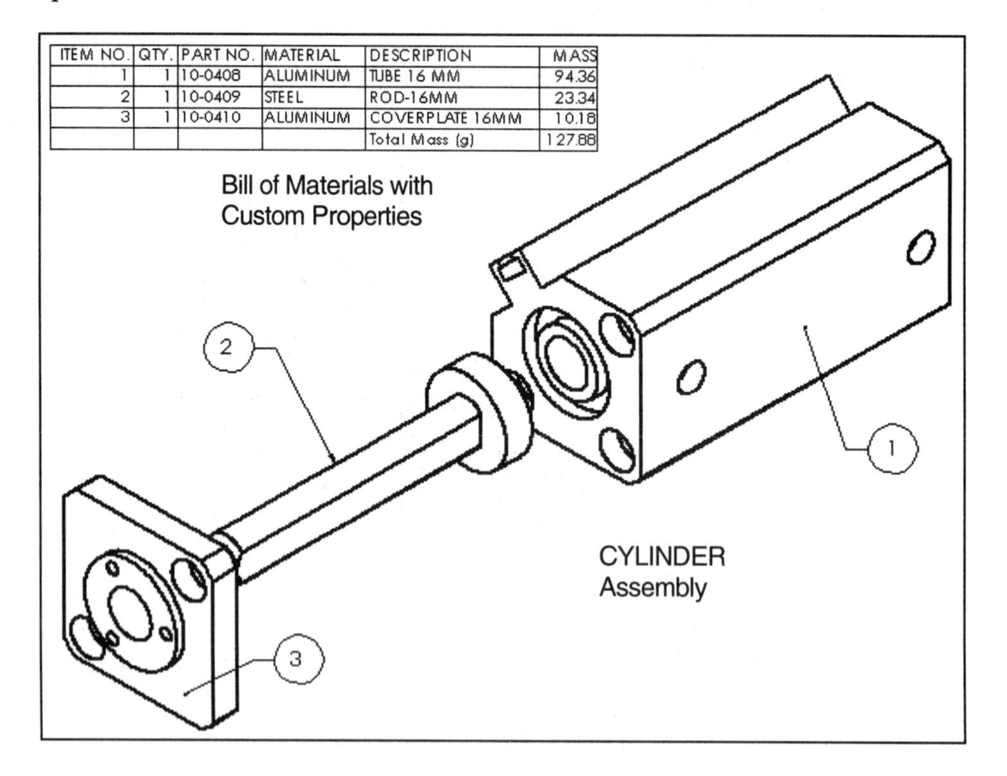

ITEM NO.	QTY.	PART NO.	MATERIAL	DESCRIPTION	MASS
1	1	10-0408	ALUMINUM	TUBE 16 MM	94.36
2	1	10-0409	STEEL	ROD-16MM	23.34
3	1	10-0410	ALUMINUM	COVERPLATE 16MM	10.18
				Total Mass (g)	127.88

Isometric View

Project Overview

In this project you will create the following:

- CYLINDER Assembly.
- Custom Properties.
- Design Table.
- CYLINDER Drawing.
- Bill of Materials.

Create the CYLINDER assembly. Create an Exploded view.

Add Balloons to the TUBE, ROD and COVERPLATE parts.

Insert the assembly into sheet 1 of 2 of the CYLINDER drawing.

Insert a Bill of Materials.

Add Custom Properties to the TUBE, ROD and COVERPLATE parts.

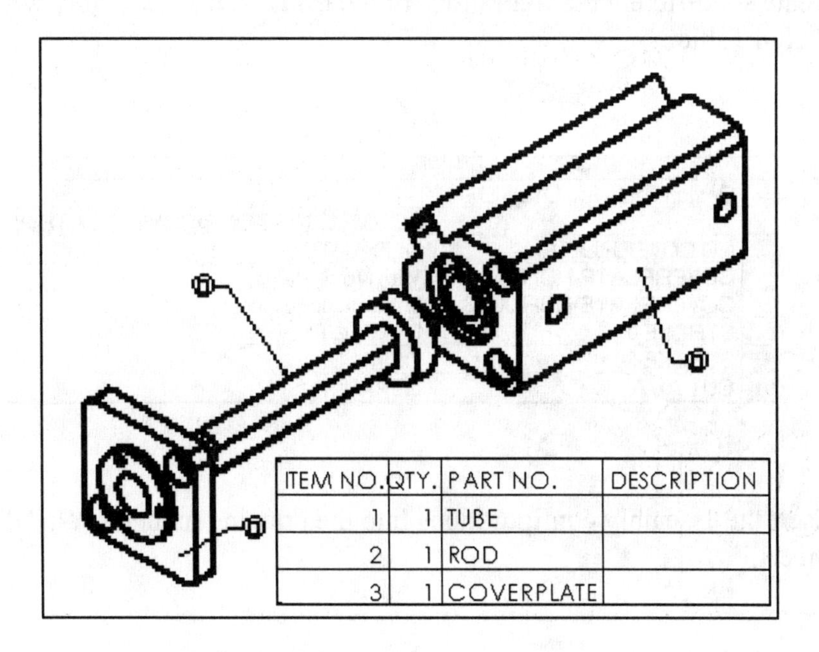

ITEM NO.	QTY.	PART NO.	DESCRIPTION
1	1	TUBE	
2	1	ROD	
3	1	COVERPLATE	

		CUSTOM PROPERTIES:		
PART:	Config. PART NO.	DESCRIPTION:	MATERIAL:	MASS:
TUBE. Default	10-0408	TUBE-16MM	ALUMINUM	"SW-Mass@@Default@TUBE.SLDPRT"
ROD				
Default	10-0409	ROD-16MM	STEEL	"SW-Mass@@Default@ROD.SLDPRT"
Short Rod	10-0409S	ROD-16MM-SHORT	STEEL	"SW-Mass@@Short Rod@ROD.SLDPRT"
Long Rod*	10-0409L	ROD-16MM-LONG	STEEL	"SW-Mass@@Long Rod@ROD.SLDPRT"
COVER PLATE				
Default	10-0410	COVERPLATE-16MM	ALUMINUM	"SW-Mass@@Default@ COVERPLATE.SLDPRT"
With Nose Holes	10-0410A	COVERPLATE-16MM W/HOLES	ALUMINUM	"SW-Mass@@With Nose Holes@COVERPLATE.SLDPRT"
Without Nose Holes	10-0410B	COVERPLATE-16MM-NOHOLES	ALUMINUM	"SW-Mass@@Without Nose Holes@COVERPLATE.SLDPRT"

Create six different configurations of the CYLINDER assembly with an Excel Design Table.

	A	B
1	Design Table for: CYLINDER	
2		$CONFIGURATION@COVERPLATE<1>
3	NO COVERPLATE	DEFAULT
4	COVERPLATE-HOLES	With Nose Holes
5	COVERPLATE-NOHOLES	Without Nose Holes
6	STROKE 0	DEFAULT
7	STROKE 30	DEFAULT
8	CUT AWAY	DEFAULT

Insert the assembly configurations into the drawing named CYLINDER-Sheet1.

The configuration names are: NO COVERPLATE, COVERPLATE-HOLES and COVERPLATE-NO HOLES. Add an Excel Custom Bill of Materials that corresponds to each assembly configuration.

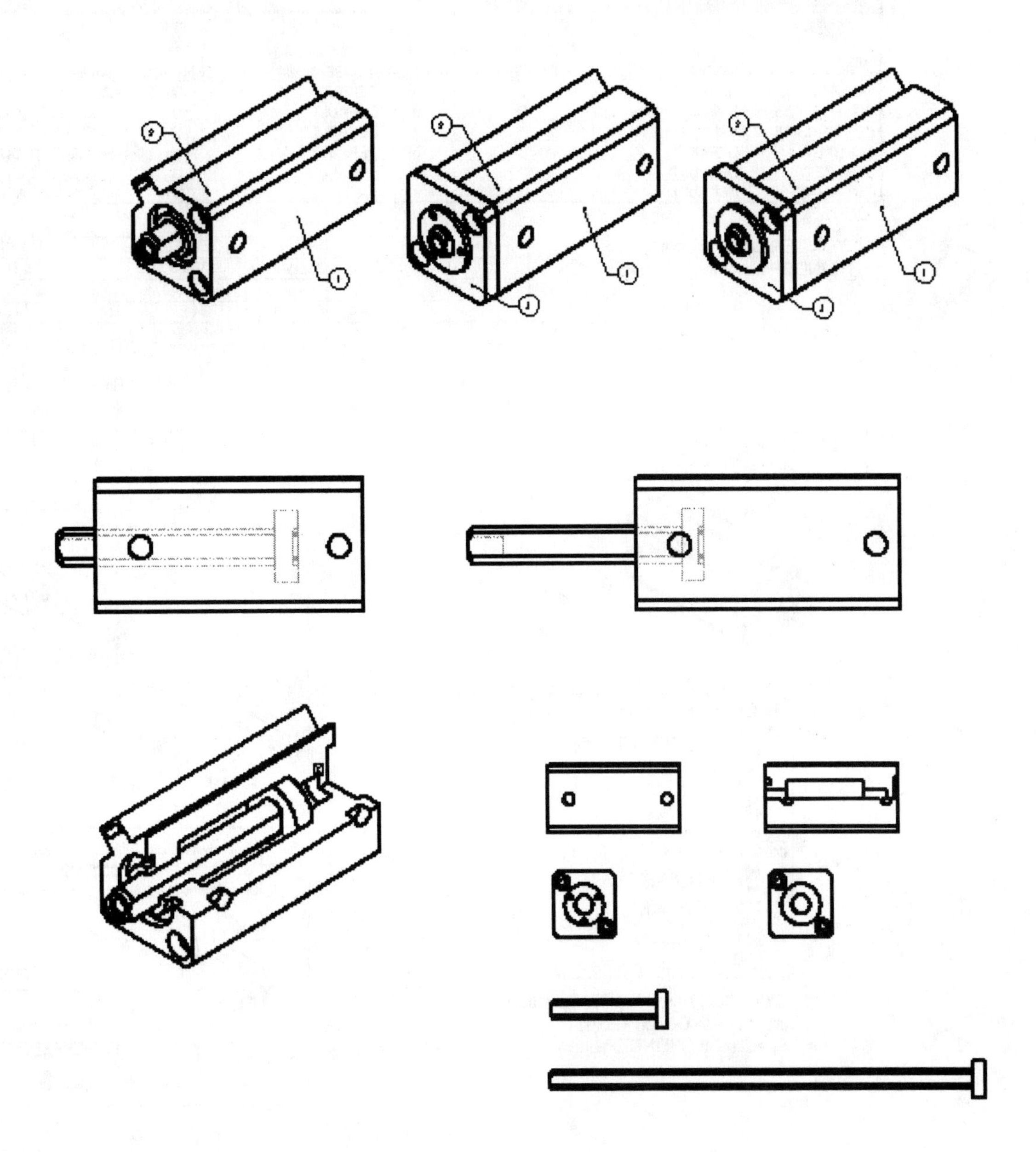

Insert the assembly configurations into the drawing named CYLINDER-Sheet2. The configuration names are: STROKE 0, STROKE 30 and CUT AWAY.

Insert two different part configurations into the drawing named CYLINDER-Sheet2, for the TUBE, ROD and COVERPLATE parts.

Parts, assemblies, drawings, Custom Properties, Excel Design Tables and Excel Bill of Materials are incorporated in this Project. Review the following document structure before starting.

A	B	C	D	E	F	G
Design Table for: CYLINDER						
	$CONFIGURATION @COVERPLATE<1>	$STATE@COVER PLATE<1>	stroke@ distance1	$CONFIGURATION @TUBE<1>	$STATE@CON CENTRIC2	$USER_NOTES
NO COVERPLATE	DEFAULT	S	0	DEFAULT	U	BASIC
COVERPLATE-HOI	With Nose Holes	R	0	DEFAULT	U	ANODIZED COVER
COVERPLATE-NOI	Without Nose Holes	R	0	DEFAULT	U	ANODIZED COVER
STROKE 0	DEFAULT	S	0	DEFAULT	U	NO COVER ROD AT HOME
STROKE 30	DEFAULT	S	30	DEFAULT	U	NO COVER ROD FULLY EX
CUT AWAY	DEFAULT	R	0	Section Cut	S	SEE INTERNAL FEATURES

CYLINDER Design Table
CYLINDER assembly

ITEM NO.	QTY.	PART NO.	MATERIAL	DESCRIPTION	MASS
1	1	10-0408	ALUMINUM	TUBE 16 MM	94.36
2	1	10-0409	STEEL	ROD-16MM	23.34
3	1	10-0410	ALUMINUM	COVERPLATE 16MM	10.18
				Total Mass (g)	127.88

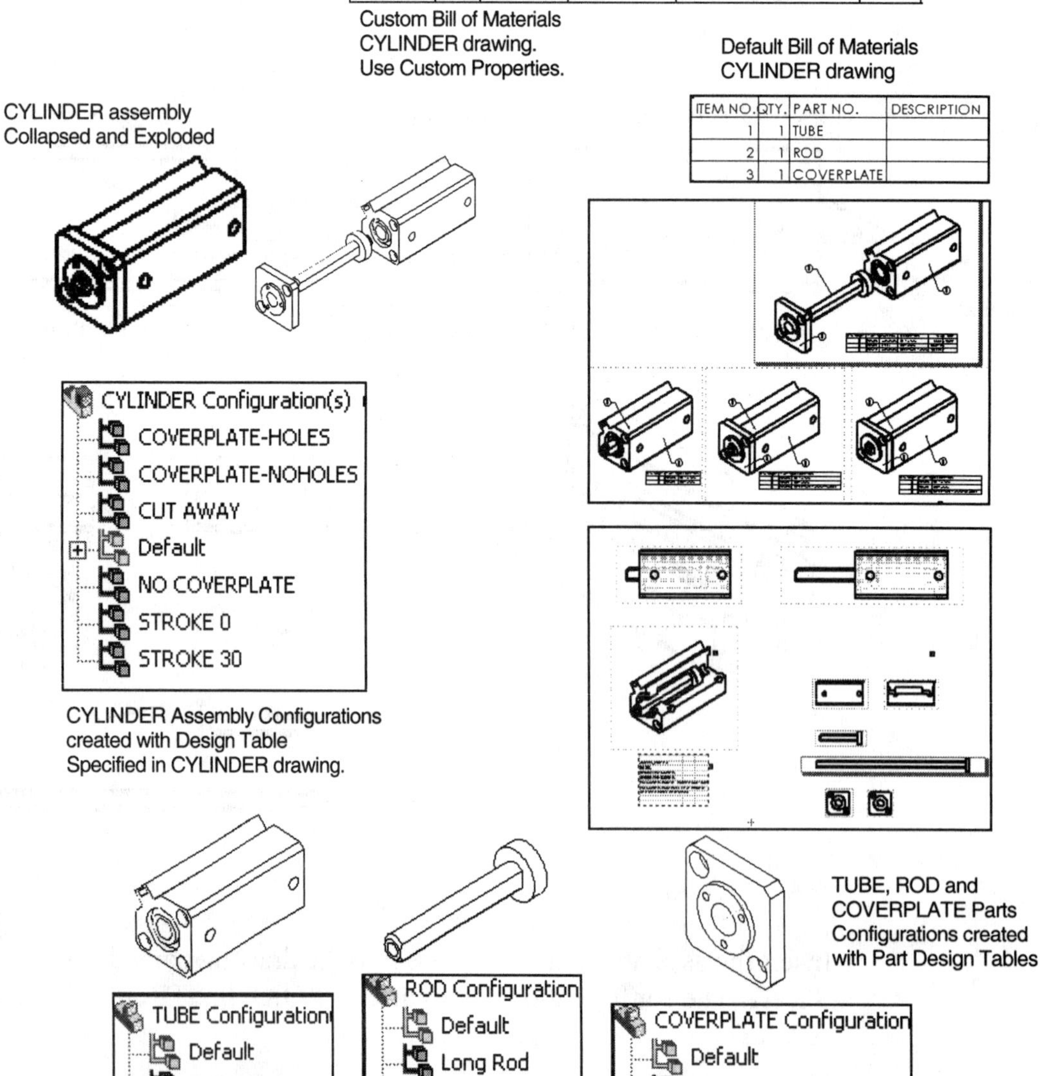

Custom Bill of Materials
CYLINDER drawing.
Use Custom Properties.

Default Bill of Materials
CYLINDER drawing

ITEM NO.	QTY.	PART NO.	DESCRIPTION
1	1	TUBE	
2	1	ROD	
3	1	COVERPLATE	

CYLINDER assembly
Collapsed and Exploded

CYLINDER Assembly Configurations created with Design Table Specified in CYLINDER drawing.

TUBE, ROD and COVERPLATE Parts Configurations created with Part Design Tables

SolidWorks Tools and Commands

The following SolidWorks Assembly tools are utilized in this Project:

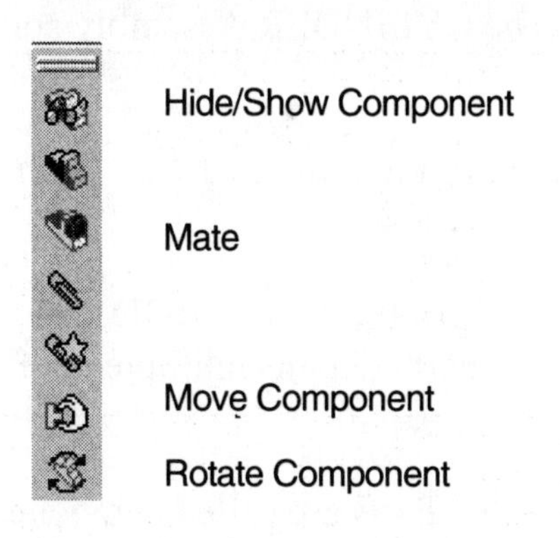

Other SolidWorks tools and commands are utilized in this Project:

SolidWorks Tools and Commands:		
Configurations	Properties	View Orientation
Explode View	Balloon and Stacked Balloon	Bill of Materials
Mass Properties	Design Table	Custom Property

Create the CYLINDER Assembly and CYLINDER Drawing

The CYLINDER Assembly consists of three parts. The three parts are: TUBE, ROD and COVERPLATE. The CYLINDER Assembly contains multiple configurations.

The CYLINDER-Sheet1 drawing contains the Exploded view of the CYLINDER assembly and a Bill of Materials.

Create a new CYLINDER Assembly. Open the TUBE, ROD, and COVERPLATE parts. Recall that each part contains multiple configurations that were created with a Design Table.

Set each part to its Default configuration. Parts are called components in an assembly.

The TUBE part is the base component in the CYLINDER assembly. The TUBE part is the first part inserted into the CYLINDER assembly and is fixed to the Origin. Insert the ROD part.

Mate the ROD part to the TUBE part. A Mate is a geometric relationship between components in an assembly. Insert the COVERPLATE part. Mate the COVERPLATE part to the TUBE part.

Assembly modeling requires practice and time. Below are a few helpful assembly modeling techniques:

- Create an assembly layout structure. This will organize the sub-assemblies and components.
- Insert sub-assemblies and components as lightweight components. Lightweight components conserve on file size, rebuild time and complexity. Set Lightweight components in the Tools, Options command.

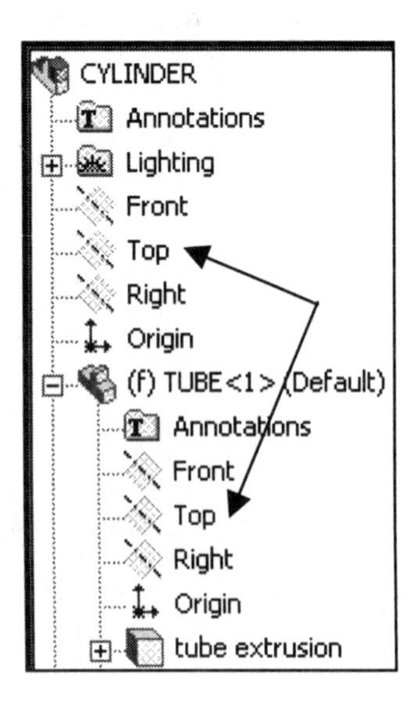

- Use the Zoom and Rotate commands to select the geometry in the mate process. Zoom to select the correct face. Filters are useful to select geometry.
- Improve display; apply different colors to features and components.
- Mate with reference planes when addressing complex geometry.

- Activate Temporary axis and Planes from the View menu.
- Select reference planes from the FeatureManager. Expand the component to view the plane locations. Identify the location of a component in 3D space relative to the assembly planes.
- Remove display complexity. Hide components when not required. Suppress features when not required.
- Position the component in the proper orientation.
- Conserve time. Use Preview during the mate operation. Suppress unwanted features. Create additional flexibility into a mate. Use a distance mate with a zero value.
- Remove unwanted entries. Use the Delete key from the Assembly Mating Items Selected text box.
- Verify the position of the mated components. Use Top, Front, Right and Section views.
- Caution should be used when viewing the color red in an assembly. Red indicates that you are editing a part in the context of the assembly.
- Avoid unwanted references. Verify your selections with the PropertyManager.

In dynamics, motion of an object is described in linear and rotational terms. Components possess linear motion along the x, y and z-axes and rotational motion around the x, y, and z-axes. In an assembly, each component has 6 degrees of freedom: 3 translational (linear) and 3 rotational. Mates remove degrees of freedom. All components are rigid bodies. The components do not flex or deform.

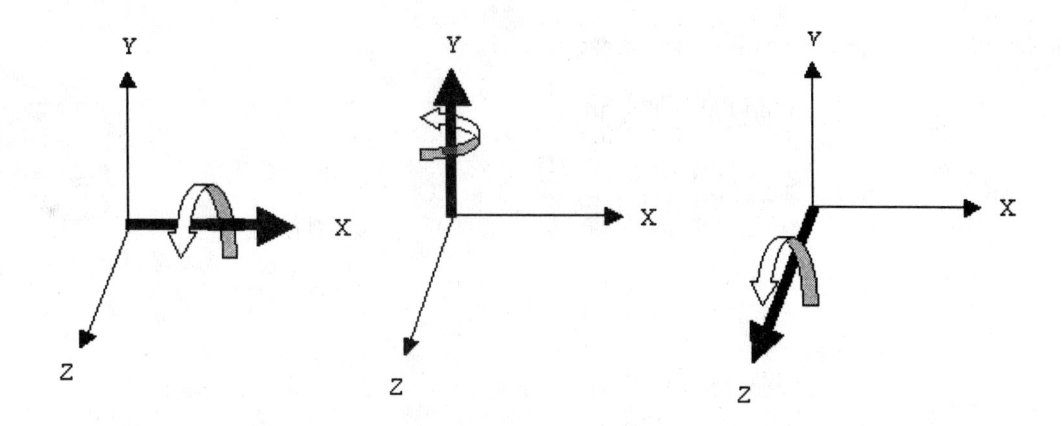

Create the CYLINDER Assembly.

1) Click **File**, **New**.

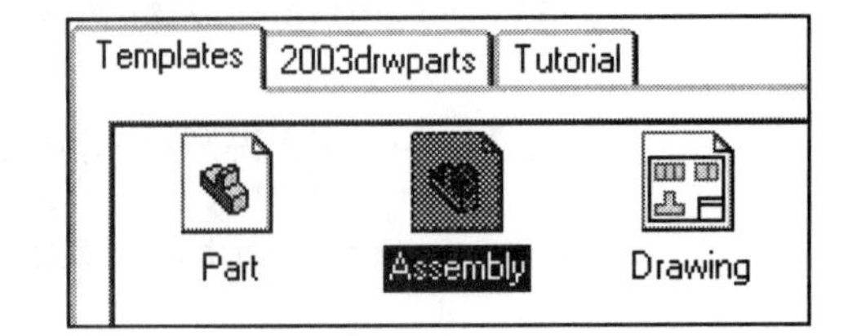

2) Click the **Templates** tab.

3) Click **Assembly**. Click **OK**.

4) Set the Assembly Document Template options. Click **Tools**, **Options**, **Document Properties**.

5) Select **Millimeters** for Units.

6) Select **ANSI** for the Detailing Dimensioning Standard.

7) Save the empty assembly. Click **File**, **Save**.

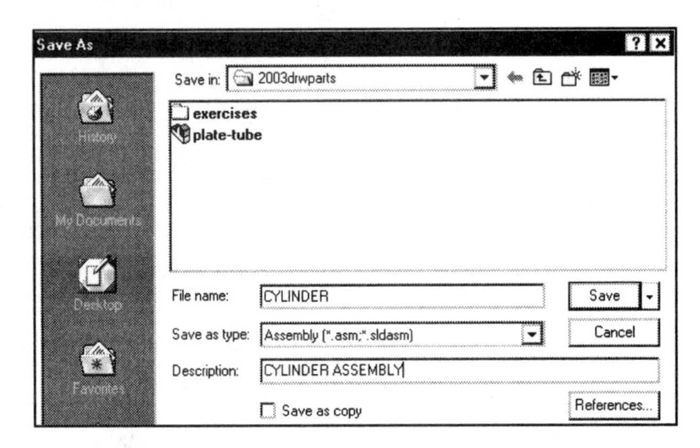

8) **Browse** to the **2003drwparts** file folder.

9) Enter **CYLINDER** for file name.

10) Enter **CYLINDER ASSEMBLY** for Description.

Open the TUBE, ROD and COVERPLATE parts. Set the default configuration.

11) **Open** the TUBE part.

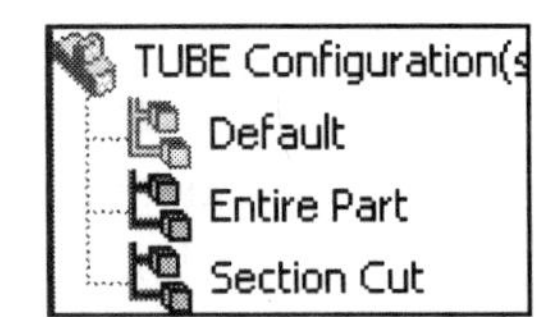

12) Click **Configuration**. The TUBE has three configurations: Default, Entire Part and Section Cut.

13) Set the configuration. Click **Default**.

14) Click **FeatureManager**.

15) **Open** the ROD part.

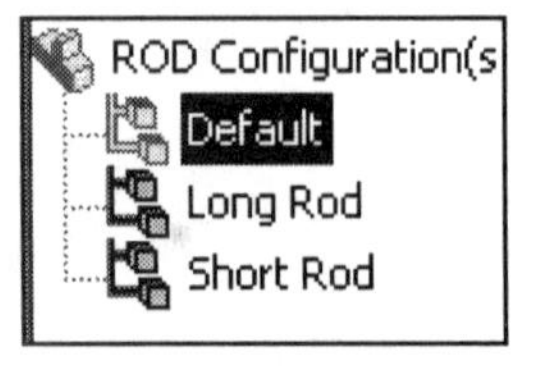

16) Click **Configuration**. The ROD has three configurations: Default, Long Rod and Short Rod.

17) Set the configuration. Click **Default**.

18) Click **FeatureManager**.

19) **Open** the COVERPLATE part.

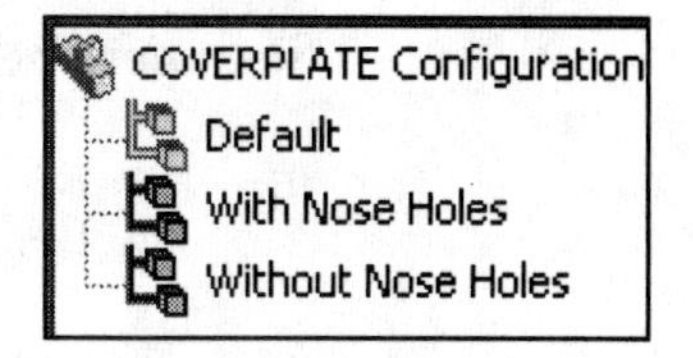

20) Click **Configuration**. The COVERPLATE has three configurations: Default, With Nose Holes and Without Nose Holes.

21) Set the configuration. Click **Default**.

22) Click **FeatureManager**.

The current configuration is saved with the part. Example: Set the configuration of the TUBE part to the Section Cut configuration.

Save the part and exit SolidWorks. Open the TUBE part. The Section Cut configuration is displayed.

Save parts in their Default configuration for the assembly. Set the configuration for each part in the assembly.

Select Component Properties, Referenced configuration, Use named configuration option. Select the configuration name from the drop down list.

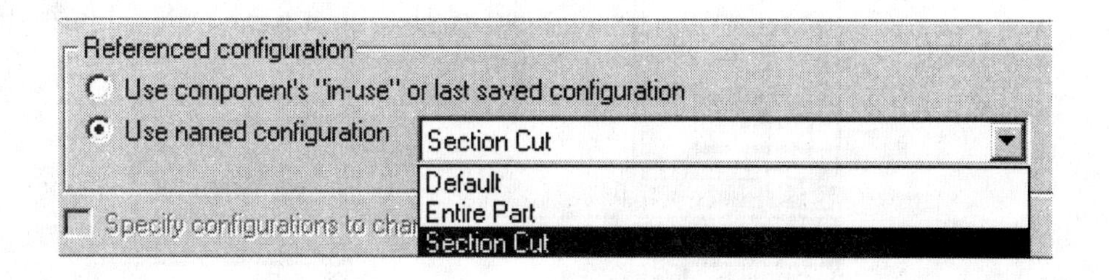

Save parts in their Default configuration for the drawing. Set the configuration for each part in the drawing. Select Configuration information, Used named configuration option. Select the configuration name from the drop down list.

Save parts and assemblies in their Default configuration.

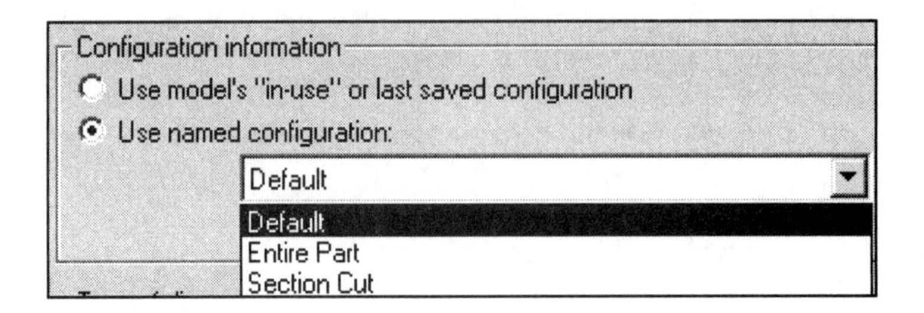

Insert the TUBE component.

23) Display all four documents. Click **Windows**, **Tile Horizontal**.

24) Click the **TUBE** TUBE (Default) icon from the top of the FeatureManager.

25) Drag the **TUBE** to the Origin of the CYLINDER assembly window. The mouse pointer displays when position on the Origin. Display the TUBE component in the assembly Graphics window. Release the **mouse button**.

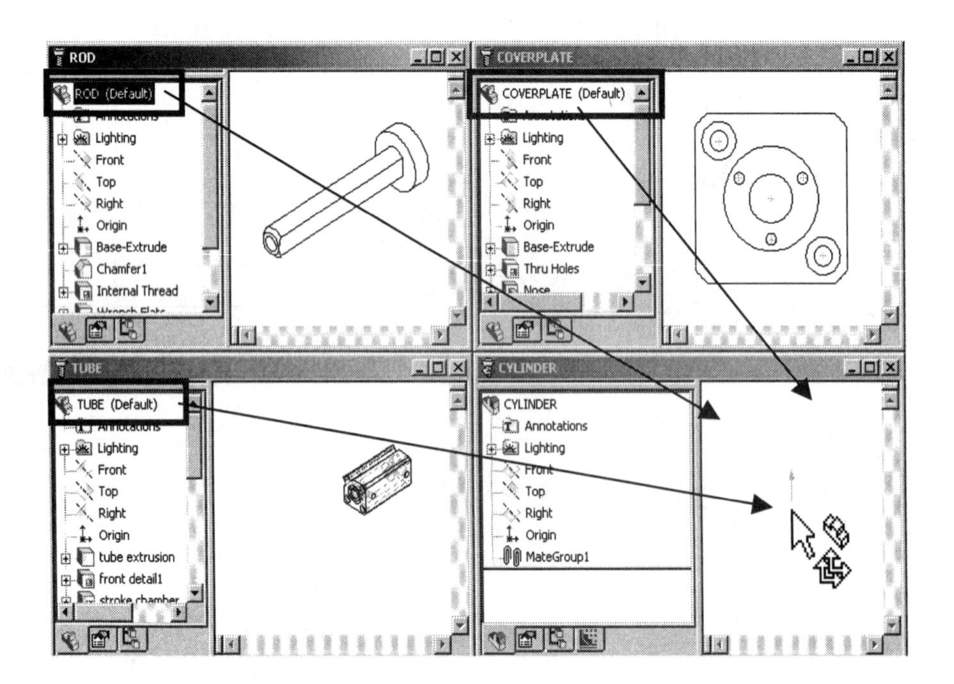

Note: The Front view of the TUBE is displayed in the CYLINDER assembly.

The TUBE name is added to the CYLINDER assembly FeatureManager with the symbol (f).

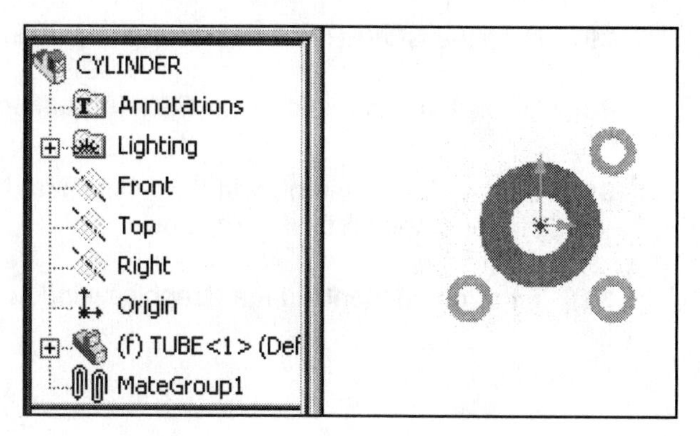

The symbol (f) represents a fixed component. A fixed component cannot move and is locked to the assembly Origin.

To remove the fixed state, Right-click a component name in the FeatureManager.

Click Float.

The symbol <#> indicates the number of copies in the assembly.

The symbol <1> indicates the original component, "TUBE" in the assembly.

The (Default) indicates the configuration used in the assembly.

Insert the ROD component.

26) Click the **ROD** ROD (Default) icon from the top of the FeatureManager.

27) Drag the **ROD** to the right of the TUBE. Do not select the Origin.

Insert the COVERPLATE component.

28) Click the **COVERPLATE** COVERPLATE icon from the top of the FeatureManager.

29) Drag the **COVERPLATE** above the TUBE. Do not select the Origin.

30) Display an Isometric view. Click **Isometric**.

31) Enlarge the assembly window. Click **Maximize** in the upper right hand corner of the CYLINDER window.

32) Fit all components in the Graphics window. Press the **f** key.

Note: Click Window from the Main menu to switch between documents. The CYLINDER Assembly, CYLINDER-Sheet 1 drawing, TUBE, ROD and COVERPLATE parts all remain open during this Project.

✔ 1 CYLINDER - Sheet1
2 COVERPLATE
3 ROD
4 TUBE
5 CYLINDER

33) Save the CYLINDER assembly. Click **Save**.

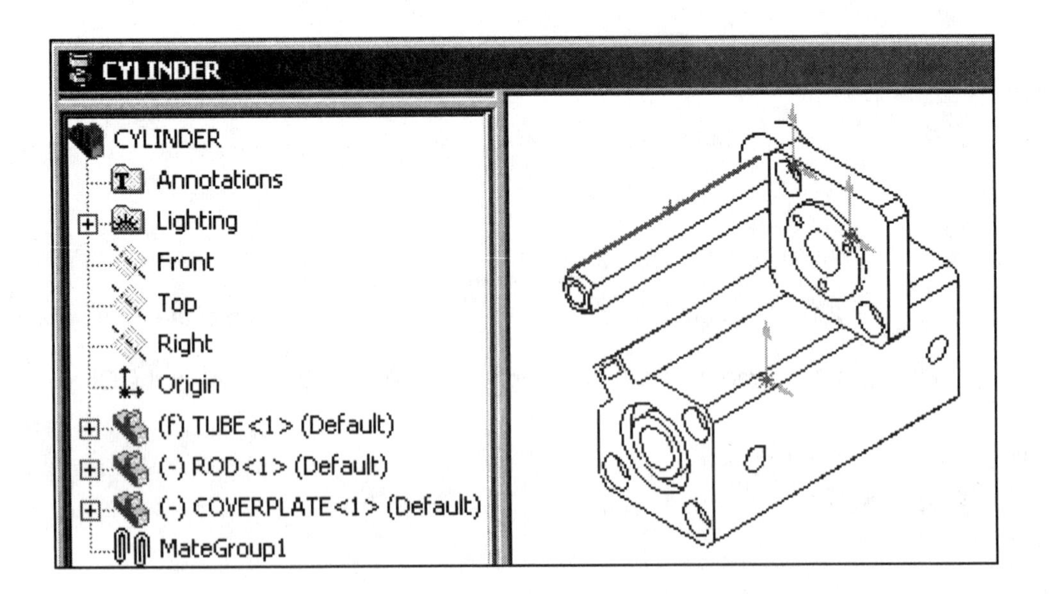

Move the ROD Component.

34) The ROD Component is located in an assembly and is free to move. Click the **ROD** Component.

35) Click the **Move Component** from the Assembly toolbar. The PropertyManager is displayed on the left side of the Graphics window. The mouse pointer displays .

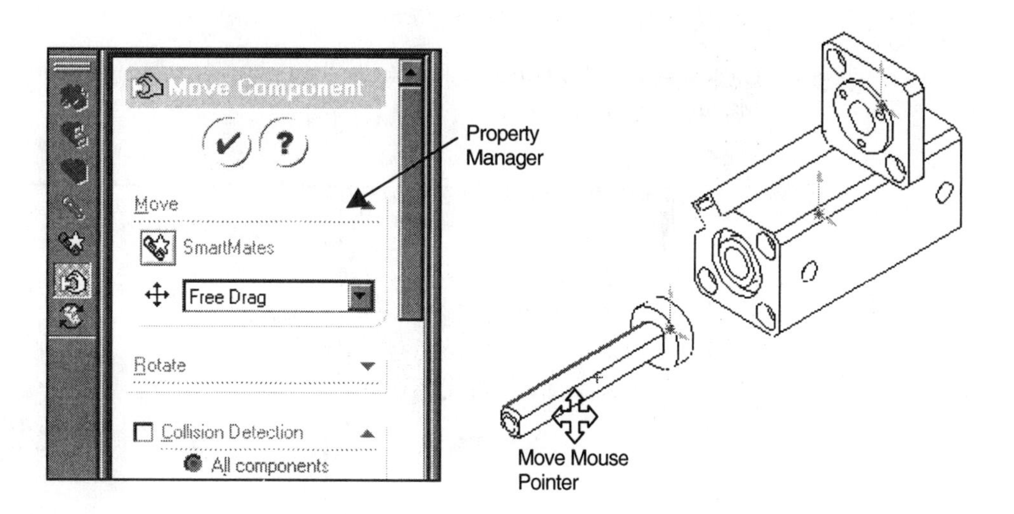

36) Position the **ROD** Component in front of the TUBE Component.

37) Click **OK**.

Mates are relationships that align and fit components in an assembly. Create three Mates in this section.

Components are inserted into an assembly with various intuitive options: Coincident, Parallel, Tangent, Concentric, Distance, Angle and Perpendicular.

Establishing the correct component relationship in an assembly requires forethought on component interaction.

Hide the COVERPLATE component to view all surfaces during the mate process.

Hide the COVERPLATE Component.

38) Right-click the **COVERPLATE** Component in the CYLINDER FeatureManager.

39) Click **Hide Components**.

40) **Zoom in** on the back of the ROD Component and the front of the TUBE Component.

Create the first Mate.

41) Click **Mate** from the Assembly toolbar. The Assembly Mating Property Manager is displayed.

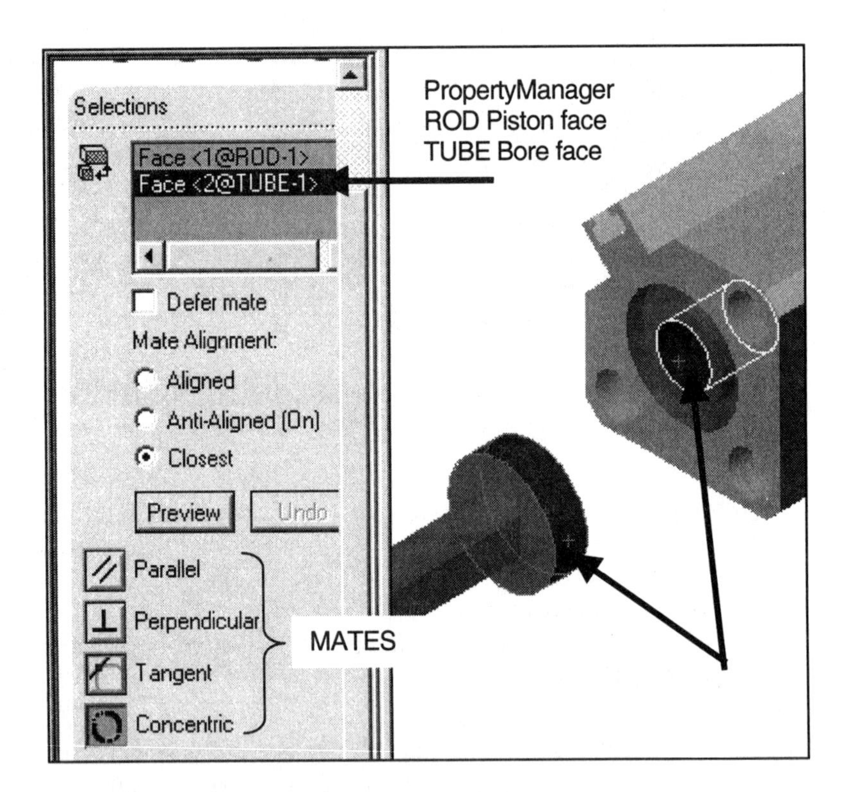

42) Create a Concentric Mate. Click the **Piston cylindrical face** of the ROD.

43) Click the **BORE cylindrical face** of the TUBE. The Concentric Mate is the default mate type.

44) Click **Preview**.

45) Click **OK**.

Move and Rotate the ROD Component.

46) Click **Move Component** from the Assembly toolbar.

47) Drag the **ROD** in a horizontal direction. The ROD rotates around an imaginary axis concentric with the TUBE. The ROD linearly translates in and out of the TUBE.

48) Click **OK**.

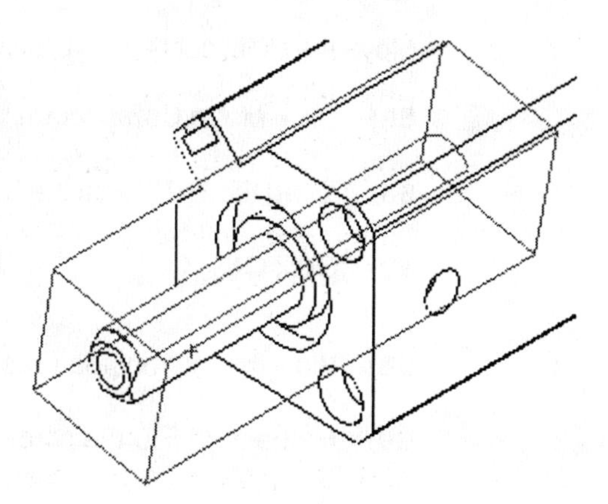

Create the second **Parallel** Mate.

49) Click **Mate** from the Assembly toolbar.

50) Click the **right** face of the TUBE.

51) Click the flat **Wrench Cut face** of the ROD.

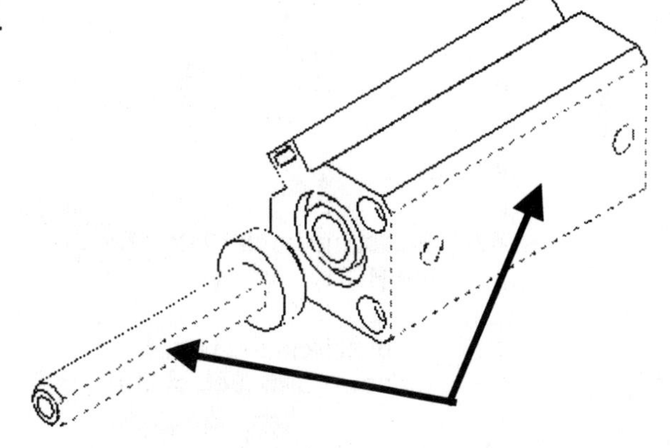

52) Click **Parallel** from the Assembly the Mate PropertyManager.

53) Click **Preview**.

54) Click **OK**.

When selecting faces, position the mouse pointer in the middle of the face. Do not position the pointer near the edge of the face. If the wrong face or edge is selected, click the face or edge again to remove it from the Items Selected text box.

Right-click in the Graphics window. Click Clear Selections to remove all geometry from the Items Selected text box. The next step is an example of the Select Other option.

Move the ROD out of the TUBE.

55) Click **Move Component** from the Assembly toolbar.

56) Drag the **ROD** until the back Piston face is outside the Bore of the TUBE.

57) Click **OK**.

58) View the internal features of the TUBE. Click **Hidden Lines Visible** .

59) Fit the model to the screen. Click the **f** key.

Create the Third Mate.

60) **Zoom in** on the TUBE.

61) Click **Mate** .

62) Select the **back Piston face** of the ROD.

63) Right-click in the **Graphics** window. Click **Select Other**.

The Select icon is displayed from the Pop-up menu.

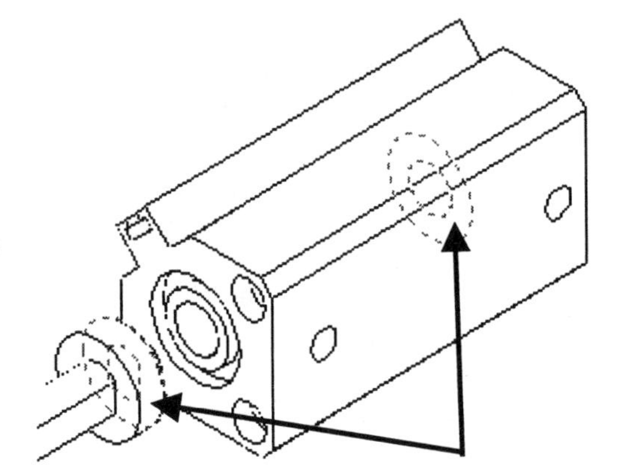

64) Right-click **N** until the correct geometry is selected. The system highlights the back Piston face of the ROD.

65) Accept the counter bore face. Click **Y**.

66) Select the back Bore face of the TUBE. Right-click in the **Graphics** window.

67) Click **Select Other**. The Select icon is displayed from the Pop-up menu.

68) Right-click **N** until the correct geometry is selected. The system highlights the back Bore face of the TUBE.

69) Create a Distance Mate. Click **Distance**.

70) Enter **0.0** for Distance.

71) Click **Preview**.

72) Click **OK**. The ROD is fully defined.

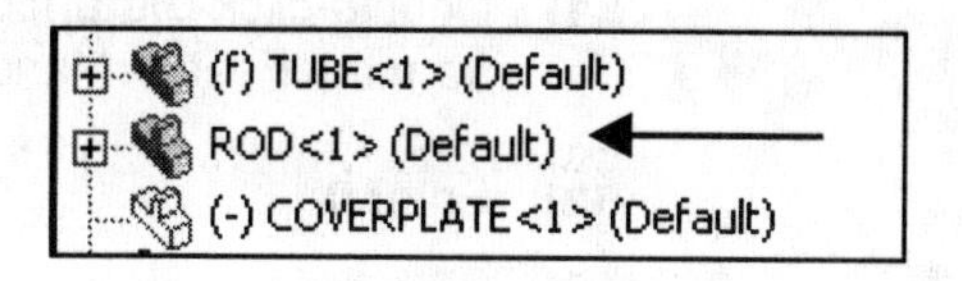

73) Display the mate types. Double-click on **Mates** in the FeatureManager. Display the full mate names.

74) Drag the **vertical FeatureManager border** to the right to display the full name.

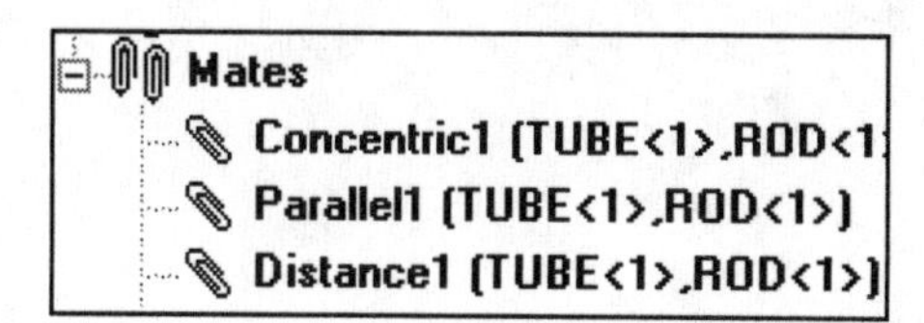

The Stroke is the distances that the ROD travels within the TUBE. View the ROD Stroke by modifying the Distance Mate.

75) Modify the Distance mate. Double-click **Distance1** Mate from the FeatureManager.

76) Double-click **0**, the blue dimension text inside the TUBE.

77) Enter **30**.

78) Click **Rebuild**.

79) Drag the **dimension text** outside the TUBE.

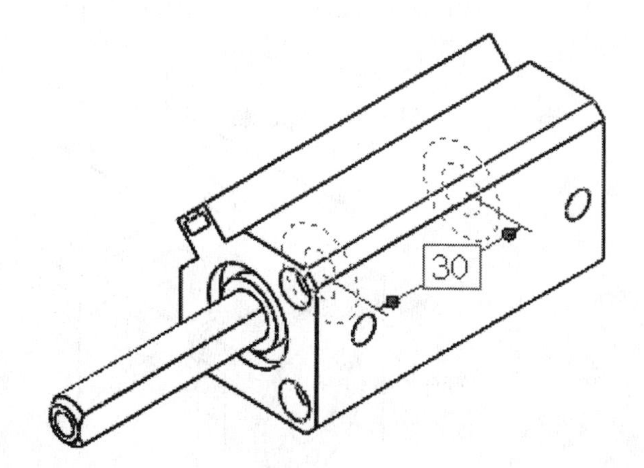

80) Return to the original position.

81) Double-click **Distance1** Mate from the FeatureManager.

82) Enter **0**.

83) Click **Rebuild**.

84) Save the CYLINDER assembly. Click **Save**.

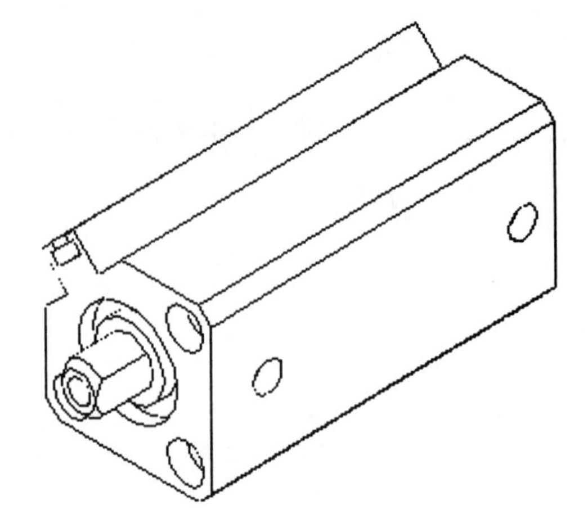

Mate the COVERPLATE Component to the TUBE Component.

85) Show the COVERPLATE. Right-click on the **COVERPLATE** in the FeatureManager.

86) Click **Show Components**. Fit the model to the screen. Press the **f** key.

87) Move and Rotate the COVERPLATE. Click **Move Component**.

88) Drag the **COVERPLATE** in front of the TUBE.

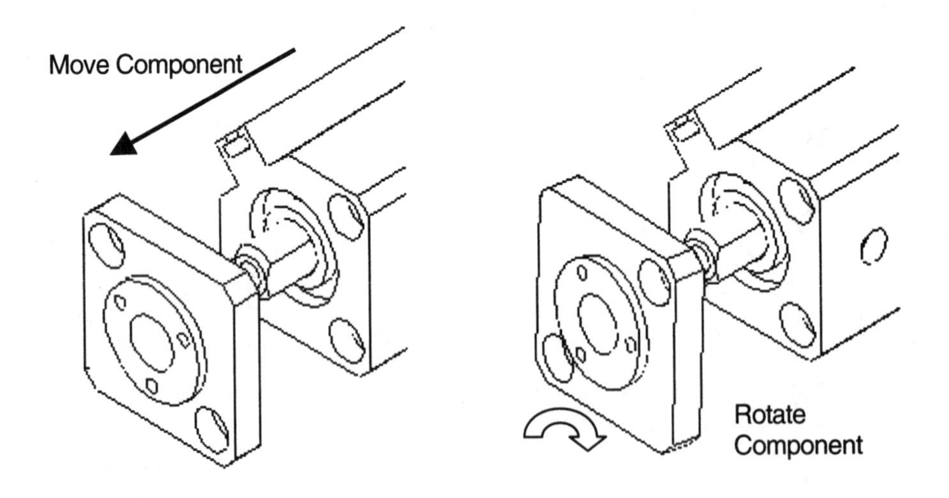

89) Click **OK**.

90) Rotate the component. Click **Rotate Component** .

91) Drag the **COVERPLATE** component until the Cbore holes are approximately aligned with the TUBE Cbore holes.

92) Click **OK**.

93) **Zoom in** on the COVERPLATE Component and the front face of the TUBE Component.

94) Create the first Mate. Click **Mate** . Click the **COVERPLATE cylindrical face** of the Cbore. Click the **TUBE cylindrical face** of the Cbore hole. Click **Concentric** from the Mate PropertyManager.

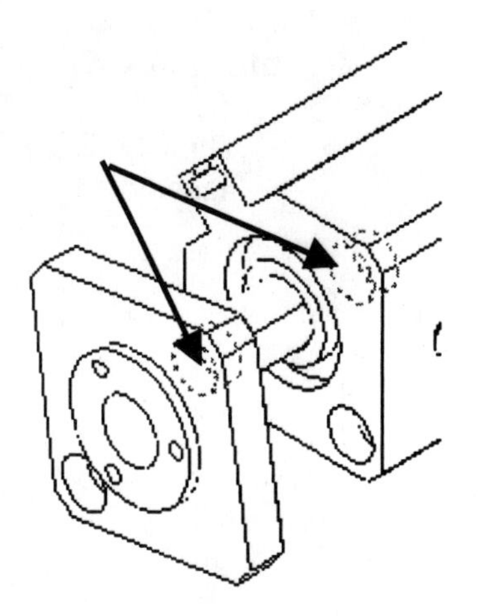

95) Click **Preview**.

96) Click **OK**.

97) Create the second Mate. Click **Mate** . Right-click the **COVERPLATE back face**. Click **Select Other**.

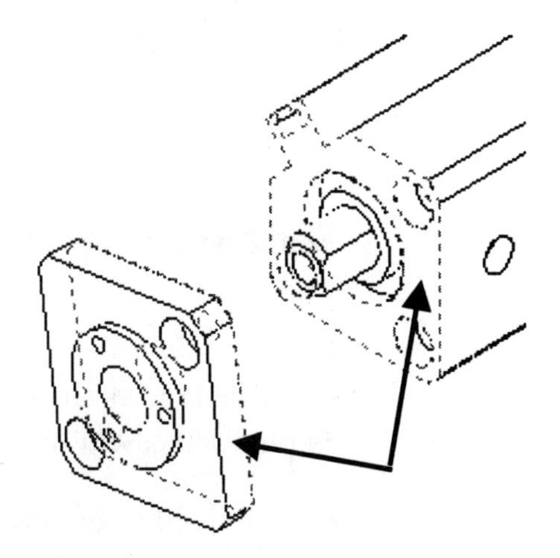

98) Click **Y** when the back face is displayed. Click the **TUBE front face**.

99) Click **Distance** from the Mate PropertyManager.

100) Enter **0**.

101) Click **Preview**.

102) Click **OK**.

Note: Use a Distance Mate with a 0 value over a Coincident Mate. The Distance Mate provides additional flexibility. The entered value can be modified.

103) Create the third Mate. Click **Mate** .

104) Click the **COVERPLATE right face**.

105) Click the **TUBE right face**.

106) Click **Parallel** from the Mate PropertyManager.

107) Click **Preview**.

108) Click **OK**.

The COVERPLATE is fully defined. The CYLINDER components are fully assembled.

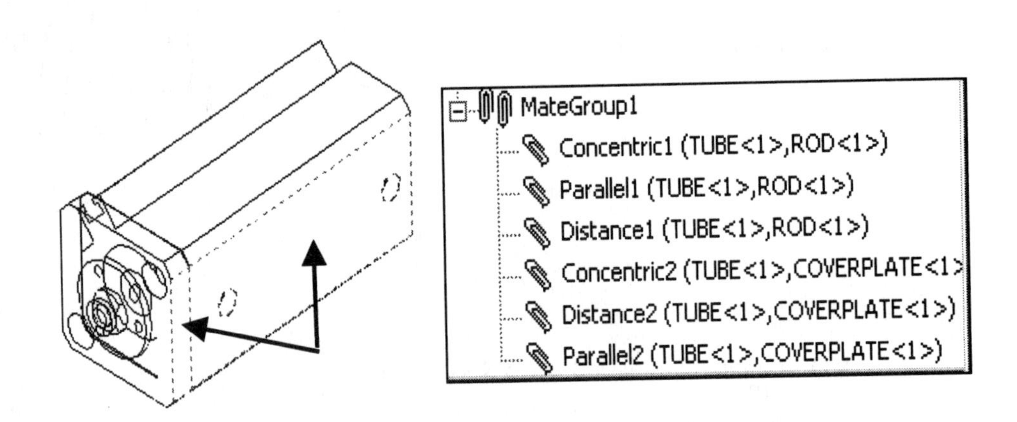

109) Save the CYLINDER assembly. Click **Save**.

Exploded views assist the designer in the viewing of the design creation. You explode an assembly in a single or multi step procedure.

AutoExplode explodes an assembly in a single step procedure. In a multi step procedure, you create individual explode steps.

Explode the CYLINDER assembly with the AutoExplode approach.

Edit steps in the Exploded view with the following Step Editing tools. From left to right the following icons represents: .

- New Step.
- Edit Previous Step.
- Edit Next Step.
- Undo Changes to Current Step.
- Delete Current Step.

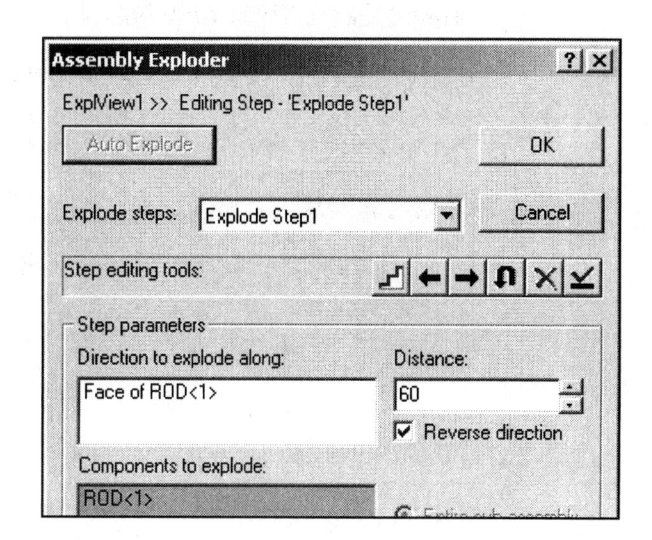

Create an Exploded view.

110) Click **Insert**, **Exploded View** from the Main toolbar.

111) Click **AutoExplode**. Two steps are automatically created.

112) Explode Step1 is displayed. The ROD is currently selected. Enter **60** for Distance.

113) Click **Apply** .

114) Modify the second step. Click **Next Step** .

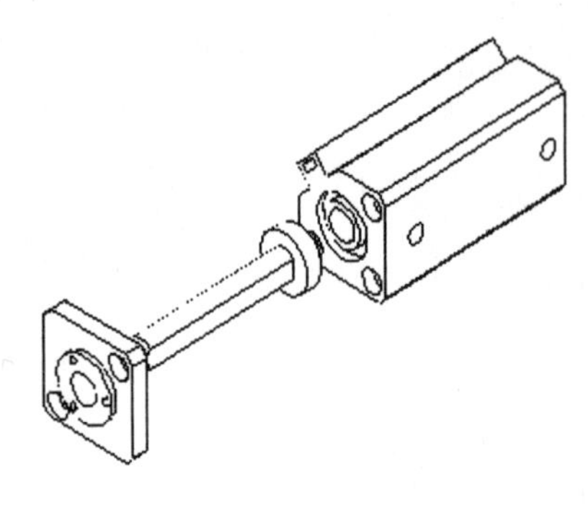

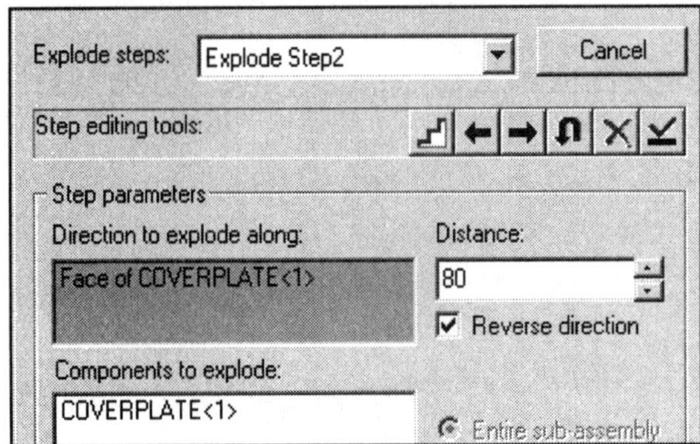

115) Enter **80**.

116) Click **Apply** .

117) Click **OK**.

118) Fit the model to the screen. Press the **f** key.

119) Display the Exploded view. Click **Configuration** .

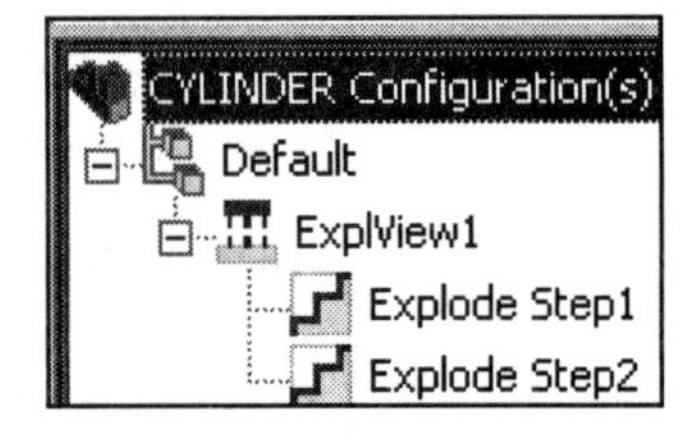

120) Expand **Default**.

121) Expand **ExplView1** to display Explode Step1 and Explode Step2.

122) Right-click on **ExplView1**. View the unexploded state.

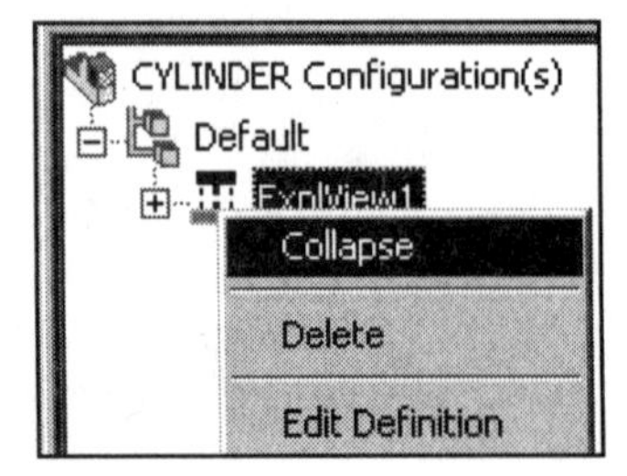

123) Click **Collapse**.

124) Return to the CYLINDER assembly. Click the **Assembly FeatureManager** .

125) Save the CYLINDER assembly. Click **Save**.

Create a New CYLINDER drawing.

126) Click **File**, **New**.

127) Click the **2003drwparts** file folder.

128) Click the **C-SIZE-ANSI-MM-EMPTY** drawing template.

129) Click **No Sheet Format**.

130) Click **C-Landscape** for Paper Size.

131) Click **OK**.

132) Save the empty drawing. Click **File**, **Save**.

133) Enter **CYLINDER** for Filename.

134) Enter **CYLINDER ASSEMBLY DRAWING** for Description.

Add an Isometric view of the CYLINDER assembly to the CYLINDER drawing.

135) Click **Named** view from the Drawing toolbar. Select the CYLINDER assembly. Click **Window**, **CYLINDER**.

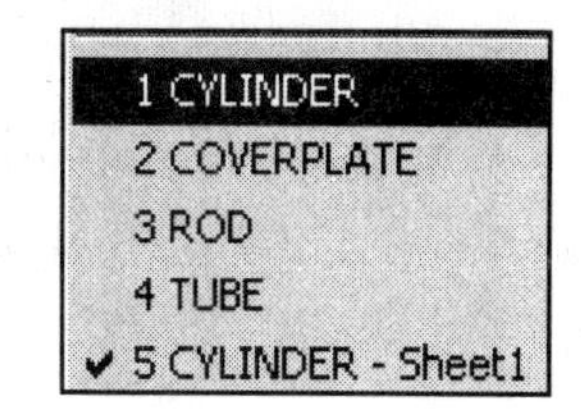

136) Click the **CYLINDER** CYLINDER icon in the CYLINDER Assembly FeatureManager.

137) Click the **position** in the center of the CYLINDER drawing to locate the Isometric view.

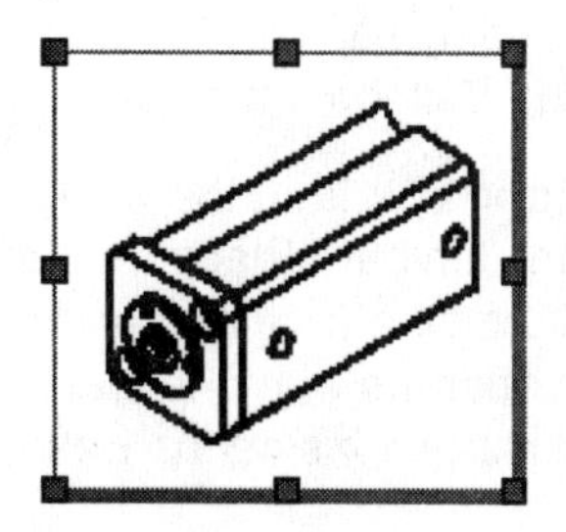

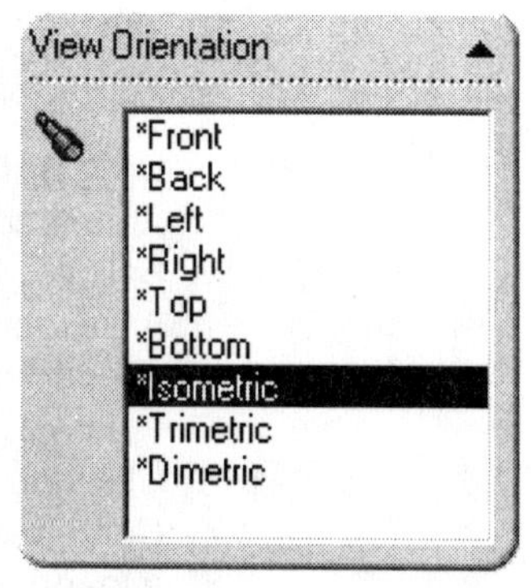

138) Click **Isometric** for the **View Orientation** dialog box. Enter **2:1** for the Custom Sheet Scale.

139) Drag the **view boundary** to the upper right corner of the sheet.

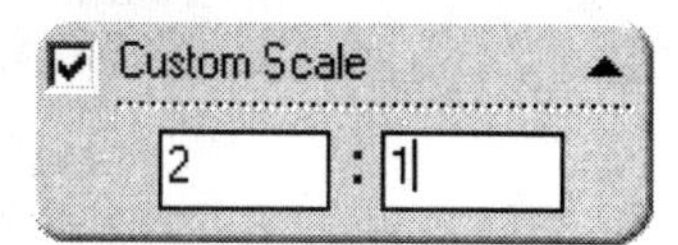

140) Click **Zoom to Selection**.

Display the Exploded view state.

141) Right-click **Properties** inside the Isometric view.

142) Check **Show in explode state** in the Configuration information text box.

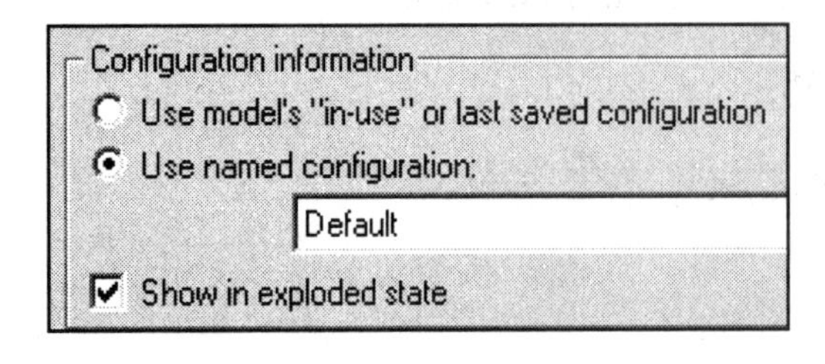

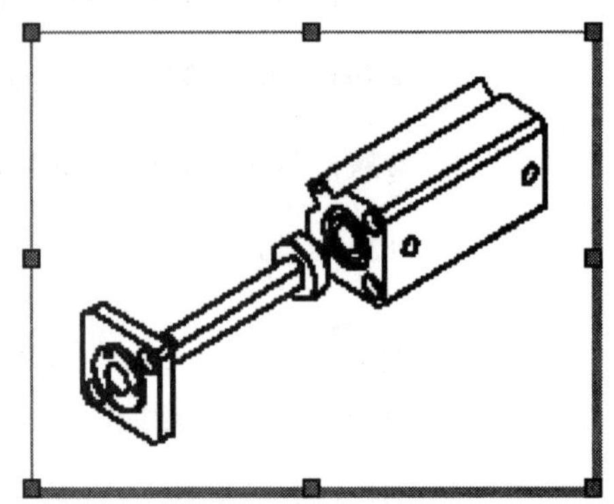

143) Click **OK**.

144) Fit the model to the screen. Press the **f** key. The Isometric view displays the Exploded view state that was created in the CYLINDER assembly.

Use Balloon annotations to label components in an assembly. A unique item number is placed inside the balloon. Item numbers are listed in the Bill of Materials.

Add a Balloon to Label each component.

145) Activate the Balloon annotation.

Click **Balloon** from the Annotations toolbar. Place the first Balloon. Click the **face** of the TUBE. A Balloon appears with a leader line and the number 1 inside the Balloon.

146) Click the **face** of the COVERPLATE Component. A Balloon appears with a leader line and the number 3 inside the Balloon.

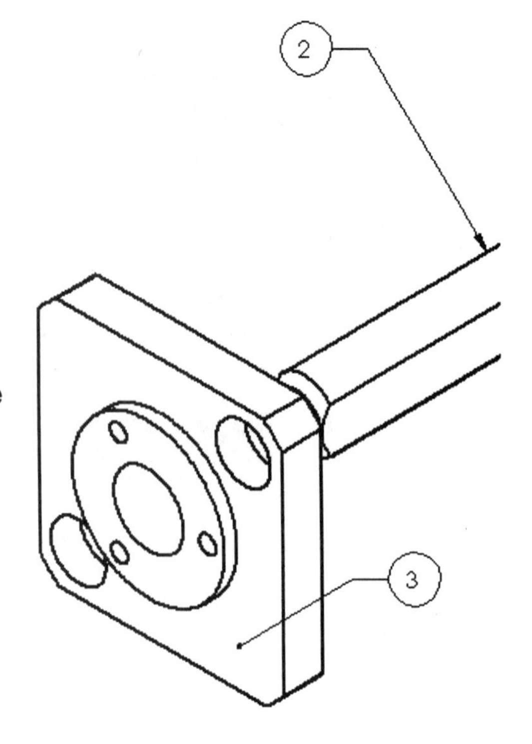

147) Click the **edge** of the ROD. A Balloon appears with a leader line and the number 2 inside the Balloon.

148) Click **OK**.

149) Turn the Balloon annotation off. Click **Balloon** .

150) Click and drag each **Balloon** off the CYLINDER components.

151) Save the CYLINDER drawing. Click **Save**.

By default, automatic Balloon numbering is determined by the order in which components are placed into the assembly.

Balloon numbers correspond to the Item numbers in the Bill of Materials.

The Bill of Materials is a CUSTOM PROPERTY spreadsheet. The Bill of Materials records the parts used in an assembly.

Modify the Balloon display. Create a Custom Property and Custom Bill of Materials.

The default Balloon arrow style displays a dot when a face is select and a filled arrow when an edge is selected.

The arrow style is defined in the document template.

Stacked Balloons contain multiple item numbers with a single leader.

Stacked Balloons are created in an exercise at the end of this Project.

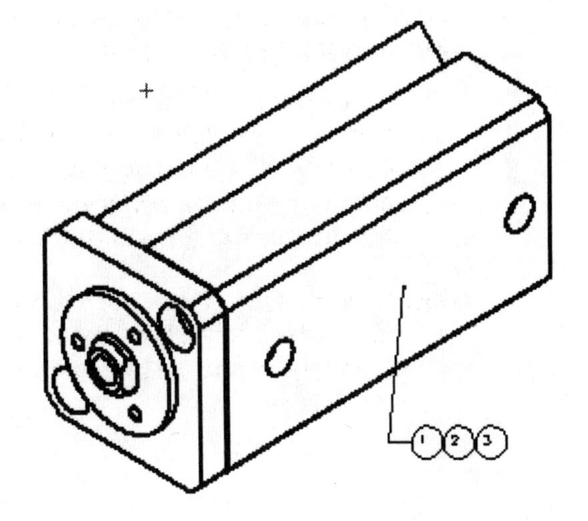

Utilize the default Bill of Materials (BOMTEMP.XLS) and Custom Bill of Materials (BOM-MASS-DM.XLS).

Explore Custom Properties in the TUBE, ROD and COVERPLATE parts, CYLINDER assembly and CYLINDER drawing.

Insert the default Bill of Materials.

152) Click inside the **Isometric view boundary**. The view boundary is displayed in green.

153) Click **Insert**, **Bill of Materials**.

154) Select **BOMTEMP.xls**. The Excel document BOMTEMP contains the Property column headings: ITEM NO., QTY, PART NO. and DESCRIPTION.

155) Click **Open**.

Note: By default, ITEM NO. is defined by the component order in the assembly. QTY is defined by the occurrences of a component in an assembly. PART NO. is the SolidWorks filename. DESCRIPTION is not defined.

bomtemp

	A	B	C	D	E
1	ITEM NO.	QTY.	PART NO.	DESCRIPTION	$$END
2					
3					
4					

156) Display the Bill of Material inside the current view. Uncheck **Use table anchor point** from the Anchor point text box.

157) Click **OK**.

158) Drag the **Bill of Material** above the horizontal view boundary.

159) Click **Zoom to Selection** on the Bill of Materials.

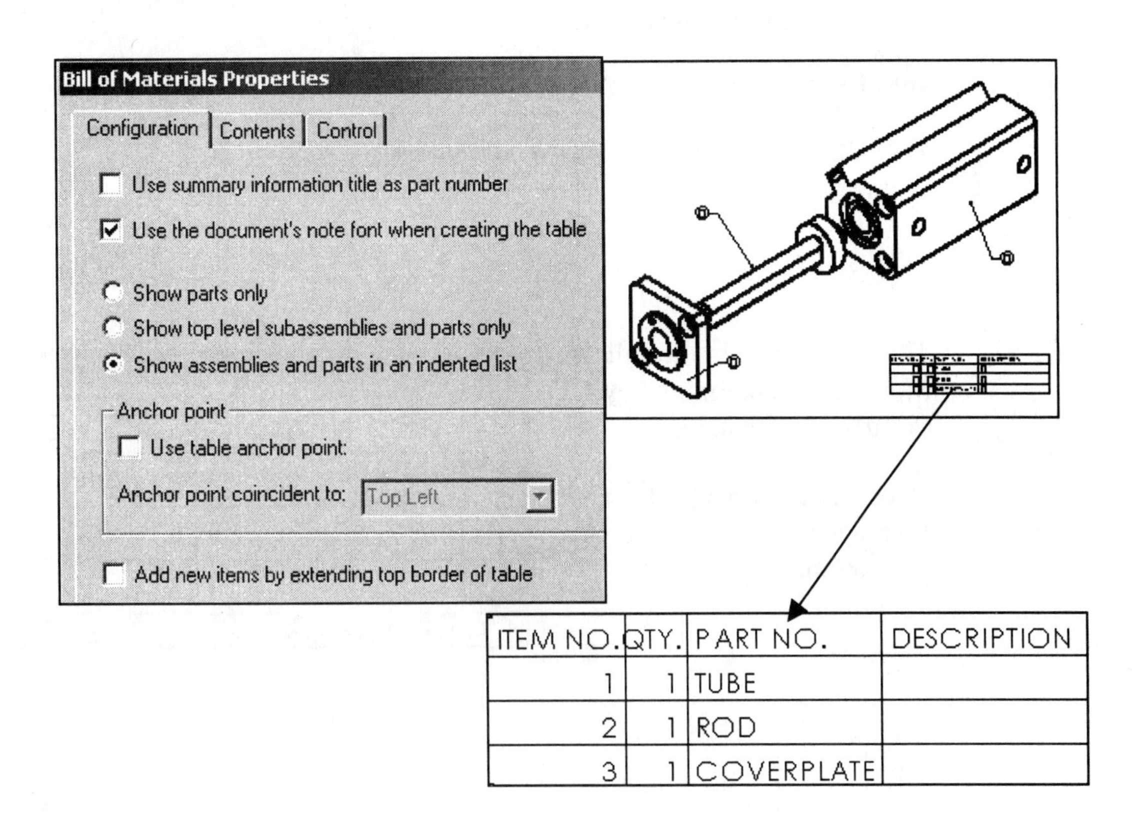

ITEM NO.	QTY.	PART NO.	DESCRIPTION
1	1	TUBE	
2	1	ROD	
3	1	COVERPLATE	

The Contents tab specifies which items appear in the Bill of Materials. The Control tab specifies how row numbers are assigned and what happens when a component is deleted. The Control tab also specifies how to split long Bill of Material tables for larger assemblies.

160) Enter a DESCRIPTION for each component in the Bill of Material. Double-click the **Bill of Material**. You are in Excel.

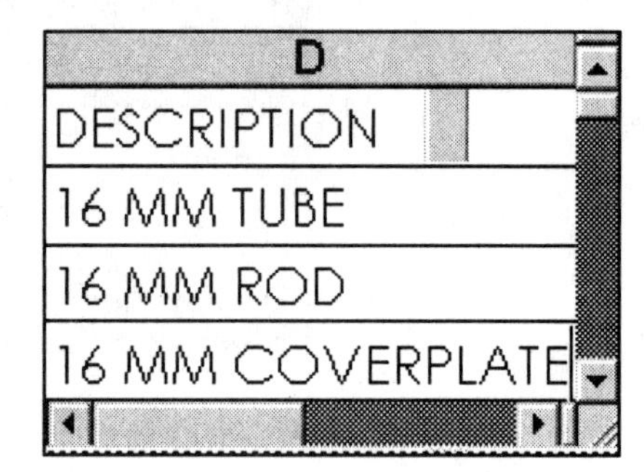

D
DESCRIPTION
16 MM TUBE
16 MM ROD
16 MM COVERPLATE

161) Enter **16 MM TUBE** in cell D2.

162) Enter **16 MM ROD** in cell D3.

163) Enter **16 MM COVERPLATE** in cell D4.

164) Format Column D. Click **Format** from the Excel Main toolbar.

165) Click **Auto-Format**.

166) Click the **Option** button.

167) Check the **Width/Height** check box.

168) **Uncheck** all other options.

169) Exit and return to SolidWorks. Click **outside** the Bill of Material.

170) Save the CYLINDER drawing. Click **Save**.

The PART NO. in the Bill of Materials is determined by the SolidWorks filename.

Replace the PART NO. filename text with your own part number.

	A	B	C
1	ITEM NO.	QTY.	PART NO.
2	1	1	10-0408
3	2	1	10-0409
4	3	1	10-0410

Sheet1

Enter the new PART NO.: 10-0408 in cell C2, 10-0409 in cell C3, 10-0410 in cell C4. Everything looks great!

You are done for the day. You exit SolidWorks.

Open the CYLINDER drawing. The Bill of Materials reverts back to the SolidWorks filename in the PART NO. column.

ITEM NO.	QTY.	PART NO.	DESCRIPTION
4	1	TUBE	TUBE-16MM
5	1	ROD	ROD-16MM
6	1	COVERPLATE	COVERPLATE-16MM

Why did the entries in the Bill of Materials change? How do you control the Bill of Materials entries? Answer: Use Configuration Properties.

How do you add additional Properties such as Material and Mass? Answer: Use Custom Properties.

Properties

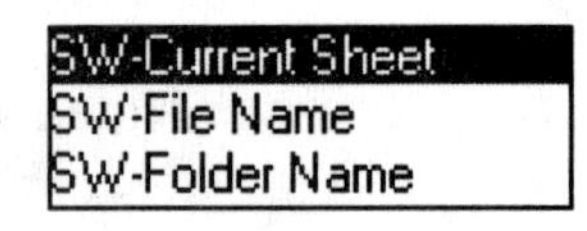

Properties were added to the Sheet Format in Project1. You created a Custom Property to define a CONTRACT NUMBER on the Drawing Template. Properties are shared between multiple documents.

Create three Custom Properties: DESCRIPTION, MATERIAL and MASS. Modify the Bill of Materials to include these Properties.

Link the MASS Property to the mass of each component. Enter the density for each component. Calculate the mass properties for each component.

The TUBE and COVERPLATE is manufactured from Aluminum. The density of Aluminum is 2.640 g/cm^3. The ROD is manufactured from Steel. The density of Steel is 7.842 g/cm^3.

A numeric PART NO. is required in the Bill of Materials. Modify the PART NO. to use a specified name in the Bill of Material.

Open a Bill of Material Excel File.

171) Double-click **BOMTEMP-MASS-DM.xls** in the 2003drwparts file folder. The SolidWorks default BOMTEMP CUSTOM PROPERTY file has been modified to contains three additional Column Headings: MATERIAL, MASS and COST. Define MATERIAL, DESCRIPTION and MASS as Customer Properties in each Part. COST will be entered at a later date. The COST column has been formatted to display the $ symbol.

BOMTEMP-MASS-DM

	A	B	C	D	E	F	G	H
1	ITEM NO.	QTY.	PART NO.	MATERIAL	DESCRIPTION	MASS	COST	$$END
2								
3								

172) Minimize the BOMTEMP-MASS-DM.xls.

Calculate the Mass Properties for the TUBE Default configuration.

173) Open the TUBE part.

174) Click **Tools**, **Mass Properties**. Click the **Options** button.

175) Enter **5** for Decimal Places.

176) Enter **0.00264** g/mm^3 for Density.

177) Click **Close**.

Mass properties of TUBE

Output coordinate System : -- default --

Density = 0.00264 grams per cubic millimeter

Mass = 94.35945 grams

Volume = 35742.21510 cubic millimeters

Surface area = 14507.56968 square millimeters

Calculate the Mass Properties for the ROD Default, Short Rod and Long Rod configurations.

178) Open the ROD part.

179) Click **Configuration**.

180) Click **Tools**, **Mass Properties**.

181) Click the **Options** button.

182) Enter **5** for Decimal Places.

183) Enter **0.00784**g/mm^3 for Density.

184) Click **Close**.

185) Click the **Short Rod** Configuration.

186) Click **Tools**, **Mass Properties**.

187) Click **Close**.

188) Click the **Long Rod** Configuration.

189) Click **Tools**, **Mass Properties**.

190) Click **Close**.

Mass properties of ROD

Output coordinate System : -- default --

Density = 0.007840 grams per cubic millimeter

Mass = 23.340125 grams

Volume = 2977.056743 cubic millimeters

Surface area = 1963.979302 square millimeters

Calculate the Mass Properties for the COVERPLATE Default, With Nose Holes and Without Nose Holes configurations.

191) Open the COVERPLATE part.

192) Click **Tools**, **Mass Properties**.

193) Click the **Options** button.

194) Enter **5** for Decimal Places.

195) Enter **0.00264**g/mm^3 for Density. Click **Close**.

196) Click **With Nose Holes** Configuration.

197) Click **Tools, Mass Properties**. The Mass is 10.57225g.

198) Click **Close**.

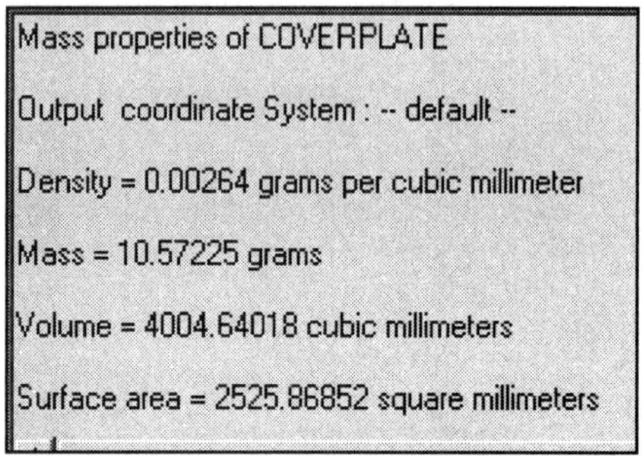

Mass properties of COVERPLATE

Output coordinate System : -- default --

Density = 0.00264 grams per cubic millimeter

Mass = 10.57225 grams

Volume = 4004.64018 cubic millimeters

Surface area = 2525.86852 square millimeters

199) Click **Without Nose Holes** Configuration.

200) Click **Tools**, **Mass Properties**. The Mass is 10.60957g.

201) Click **Close**.

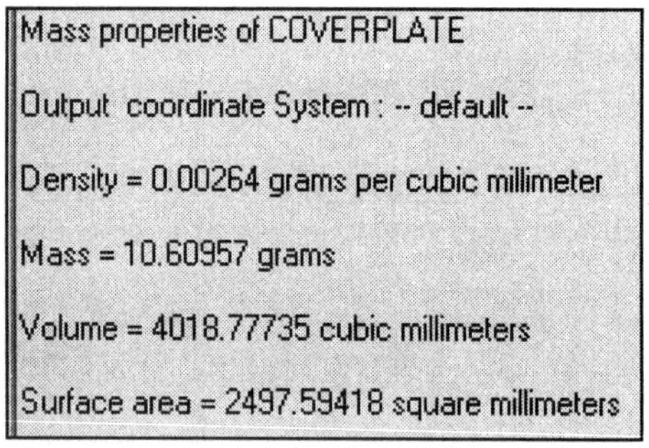

Mass properties of COVERPLATE

Output coordinate System : -- default --

Density = 0.00264 grams per cubic millimeter

Mass = 10.60957 grams

Volume = 4018.77735 cubic millimeters

Surface area = 2497.59418 square millimeters

Create the following Properties:

PART:	Config. PART NO.	DESCRIPTION:	MATERIAL:	MASS:
TUBE Default	10-0408	TUBE-16MM	ALUMINUM	"SW-Mass@@Default@TUBE.SLDPRT"
ROD				
Default	10-0409	ROD-16MM	STEEL	"SW-Mass@@Default@ROD.SLDPRT"
Short Rod	10-0409S	ROD-16MM-SHORT	STEEL	"SW-Mass@@Short Rod@ROD.SLDPRT"
Long Rod*	10-0409L	ROD-16MM-LONG	STEEL	"SW-Mass@@Long Rod@ROD.SLDPRT"
COVER PLATE				
Default	10-0410	COVERPLATE-16MM	ALUMINUM	"SW-Mass@@Default@COVERPLATE.SLDPRT"
With Nose Holes	10-0410A	COVERPLATE-16MM W/HOLES	ALUMINUM	"SW-Mass@@With Nose Holes@COVERPLATE.SLDPRT"
Without Nose Holes	10-0410B	COVERPLATE-16MM-NOHOLES	ALUMINUM	"SW-Mass@@Without Nose Holes@COVERPLATE.SLDPRT"

*Long Rod Properties are set in the Exercises at the end of this Project.

Add Custom Properties to the TUBE.

202) Open the TUBE part.

203) Click **Configuration** 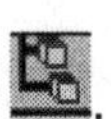.

204) Right-click **Properties** on the Default configuration.

205) Click the **Custom** button from the Configuration Properties dialog box.

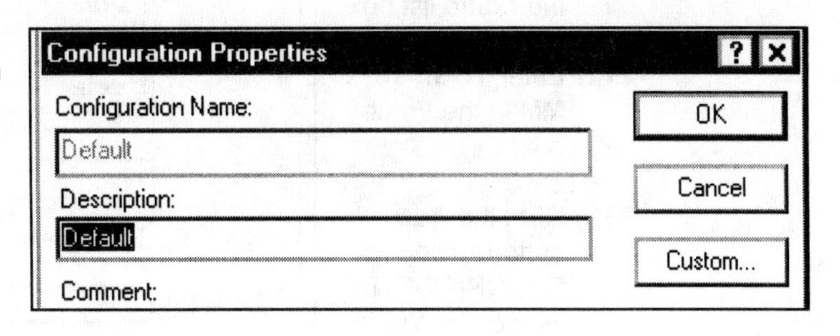

206) Click the **Configuration Specific** tab.

207) Add the Property, named MASS for the Default Configuration. Enter **MASS** in the Name text box.

208) Click the **MASS Properties** button.

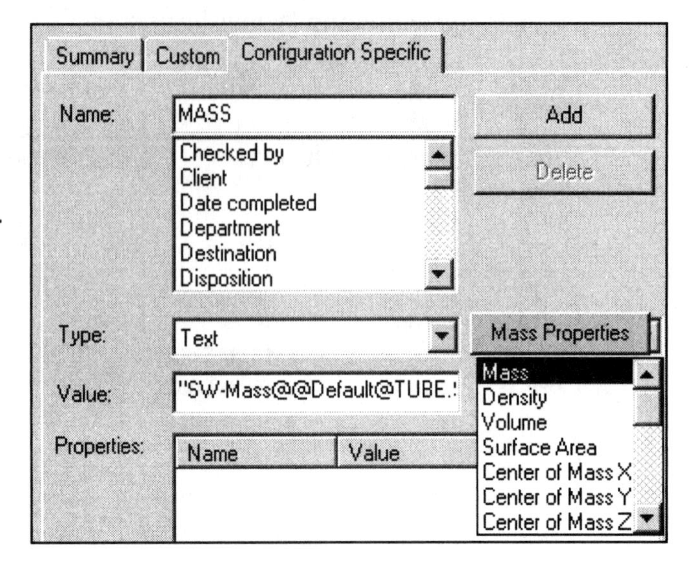

209) Select **Mass** from the drop down list. The SW-Mass@ @Default@ TUBE.SLDPRT value is added to the Value text box.

SW-Mass is the Mass Property name. Default is the Configuration name. TUBE is the part name.

210) Click **Add**. The Mass Name and Value are added to the Properties box.

Properties:

Name	Value	Type
MASS	94.3594	Text

211) Add the Property, named Description for the Default Configuration. Select **Description** from the Name list box.

212) Enter **TUBE 16 MM** in the Value box.

213) Click the **Add** button. The DESCRIPTION Name and Value are added to the Properties box.

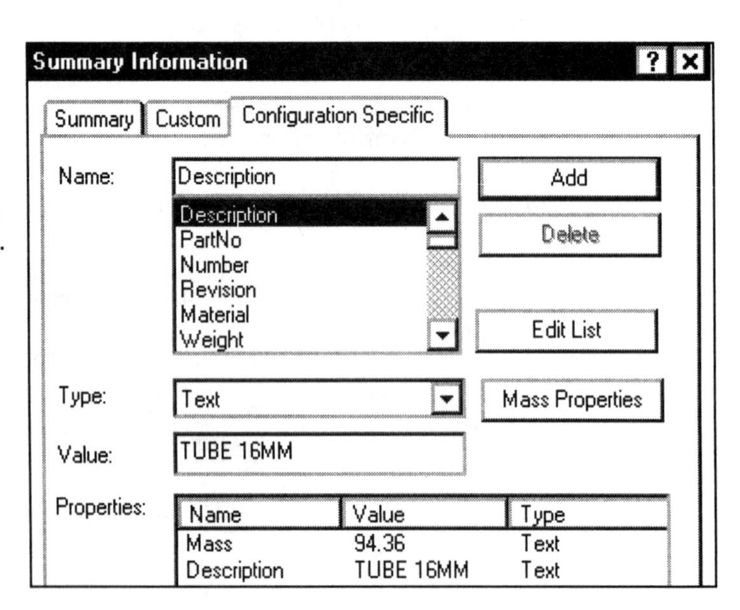

214) Add the Property, named Material for the Default Configuration. Select **Material** from the Name list box.

215) Enter **ALUMINUM** in the Value box.

216) Click the **Add** button. The MATERIAL Name and Value are added to the Properties box.

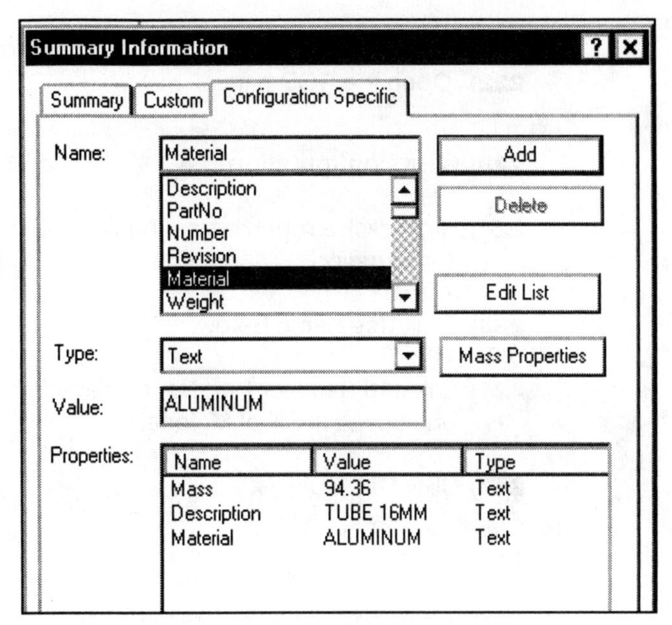

217) The three Configuration Specific Custom Properties for the Default TUBE are listed in the Properties box. Click **OK**.

218) Add the PART NO. for the TUBE Default configuration. Click the **User Specified Name** from the drop down list.

219) Enter **10-0408** for PART NO. displayed when used in Bill of Materials.

220) Click **OK** from the Configuration Properties box.

The User Specified Name is displayed next to the Default configuration.

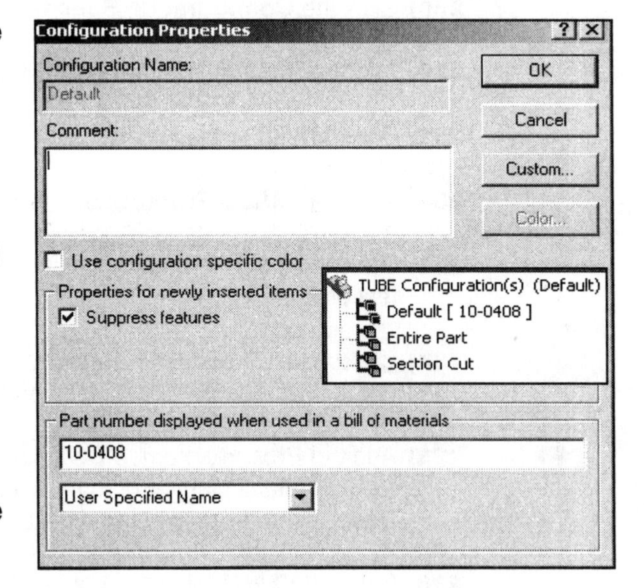

221) Save the TUBE part. Click **Save**.

Add the PART NO. for the Default Rod configuration.

222) **Open** the ROD part.

223) Click **Configuration**.

224) Right-click **Properties** on the Default configuration.

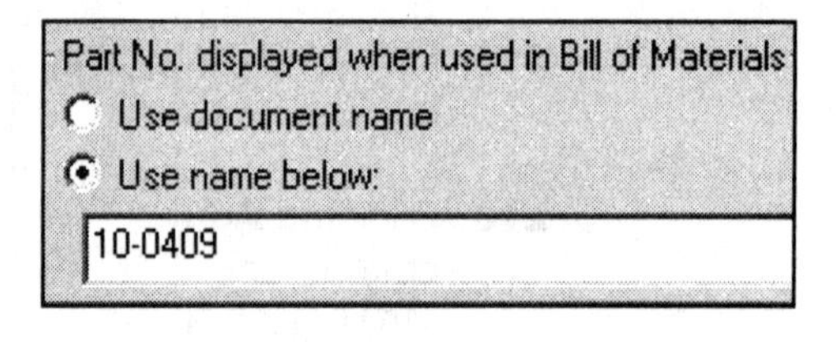

225) Click **Use name below**.

226) Enter **10-0409** for Part No. displayed when used in Bill of Materials.

227) Click **OK**.

Add the Custom Properties for the Default configuration.

228) Click the **Custom** button from the Configuration Properties dialog box.

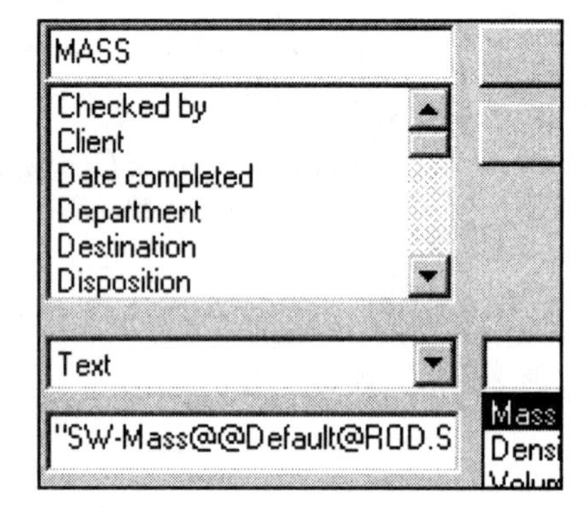

229) Click the **Configuration Specific** tab.

230) Add the Property, named MASS for the Default Configuration.

231) Enter **MASS** in the Name text box.

232) Click the **MASS Properties** button.

233) Select **Mass** from drop down list. The SW-Mass value is added to the Value text box.

234) Click **Add**. The Mass Name and Value are added to the Properties box.

Name	Value	Type
DESCRIPTION	ROD-16MM	Text
MATERIAL	STEEL	Text
MASS	23.340125	Text

235) Add the Property, named Description for the Default Configuration. Select **Description** from the Name list box.

236) Enter **ROD 16 MM** in the Value box.

237) Click the **Add** button. The Description Name and Value are added to the Properties box.

238) Add the Property, named MATERIAL for the Default Configuration. Select **Material** from the Name list box.

239) Enter **STEEL** in the Value box.

240) Click the **Add** button. The Material Name and Value are added to the Properties box. The three Configuration Specific Custom Properties are listed in the Properties box.

241) Click **OK** from the Summary dialog box.

Add the PART NO. for the Short Rod configuration.

242) Double-click the **Short Rod Configuration**.

243) Right-click **Properties** on the **Short Rod** configuration.

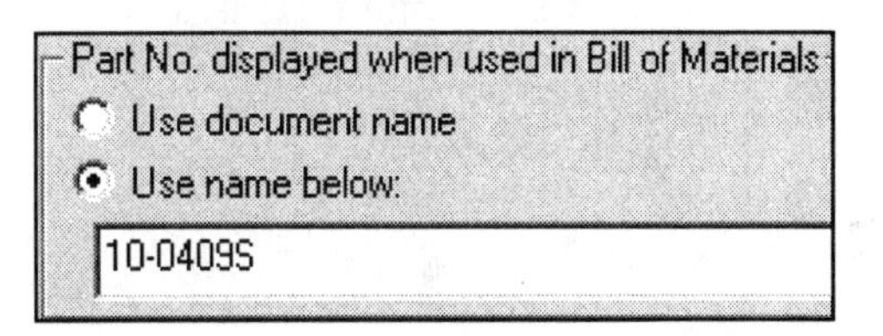

244) Click **Use name below**.

245) Enter **10-0409S** for the Part No. displayed when used in the Bill of Materials.

246) Click **OK**.

Add the Custom Properties for the Short Rod configuration.

247) Click the **Custom** button from the Configuration Properties dialog box.

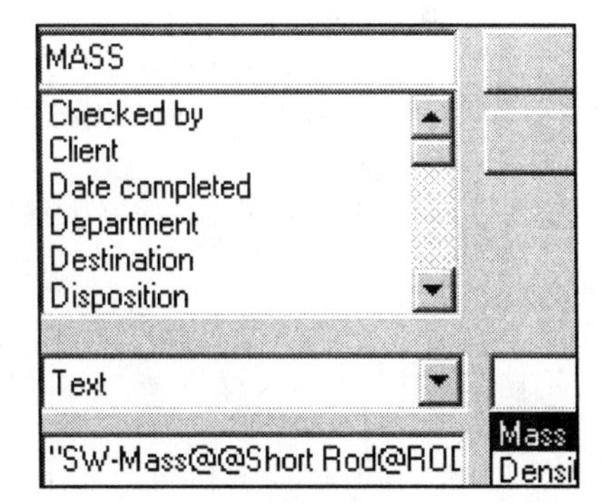

248) Click the **Configuration Specific** tab.

249) Add the Property, named MASS for the Default Configuration. Enter **MASS** in the Name text box.

250) Click the **MASS Properties** button.

251) Select **Mass** from drop down list. The SW-Mass value is added to the Value text box.

252) Click **Add**. The Mass Name and Value are added to the Properties box.

253) Add the Property, named DESCRIPTION for the Default Configuration. Select **Description** from the Name list box.

254) Enter **ROD 16 MM SHORT** in the Value box.

255) Click the **Add** button. The Description Name and Value are added to the Properties box.

256) Add the Property, named MATERIAL for the Default Configuration. Enter **MATERIAL** in the Name text box.

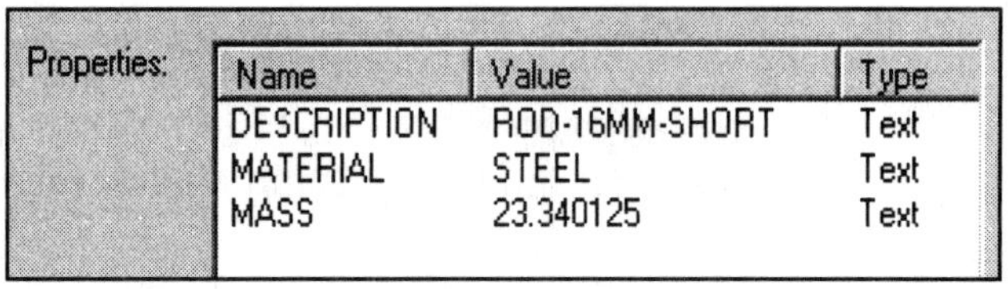

Properties:

Name	Value	Type
DESCRIPTION	ROD-16MM-SHORT	Text
MATERIAL	STEEL	Text
MASS	23.340125	Text

257) Enter **STEEL** in the Value box.

258) Click the **Add** button. The Material Name and Value are added to the Properties box. The three Configuration Specific Custom Properties are listed in the Properties box.

259) Click **OK** from the Summary dialog box.

260) Click **OK** from the Configuration Properties dialog box.

261) Save the ROD part. Click **Save**.

Add the PART NO. for the Default COVERPLATE configuration.

262) **Open** the COVERPLATE part.

263) Click **Configuration** .

264) Right-click **Properties** on the Default configuration.

265) Click **Use name below**.

266) Enter **10-0410** for Part No. displayed when used in Bill of Materials.

267) Click **OK**.

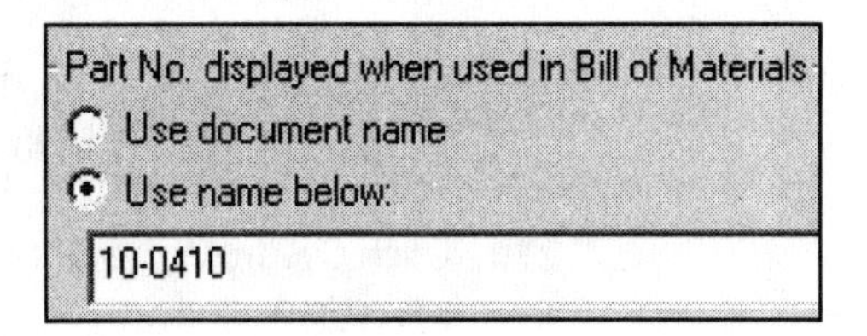

Add the Custom Properties for the COVERPLATE Default configuration.

268) Click the **Custom** button from the Configuration Properties dialog box.

269) Click the **Configuration Specific** tab. Add the Property, named MASS for the Default Configuration. Enter **MASS** in the Name text box.

270) Click the **MASS Properties** button.

271) Select **Mass** from the drop down list. The SW-Mass value is added to the Value text box. Click **Add**. The Mass Name and Value are added to the Properties box.

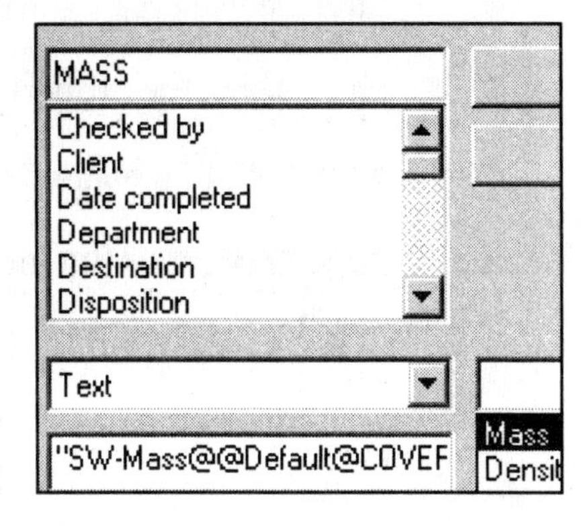

272) Add the Property, named DESCRIPTION for the Default Configuration. Select **Description** in the Name text box.

Name	Value	Type
MASS	10.57225	Text
DESCRIPTION	COVERPLATE 16 MM	Text
MATERIAL	ALUMINUM	Text

273) Enter **COVERPLATE 16MM** in the Value box.

274) Click the **Add** button. The Description Name and Value are added to the Properties box.

275) Add the Property, named MATERIAL for the Default Configuration. Select **Material** in the Name text box.

276) Enter **ALUMINUM** in the Value box.

277) Click the **Add** button. The Material Name and Value are added to the Properties box. The three Custom Properties are listed in the Properties box.

278) Click **OK** from the Summary dialog box.

279) Click **OK** from the Configuration Properties dialog box.

Add the PART NO. for the COVERPLATE With Nose Holes configuration.

280) Double-click the **With Nose Holes** configuration.

281) Right-click **Properties** on the **With Nose Holes** configuration.

282) Click **Use name below**.

283) Enter **10-0410A** for Part No. displayed when used in Bill of Materials.

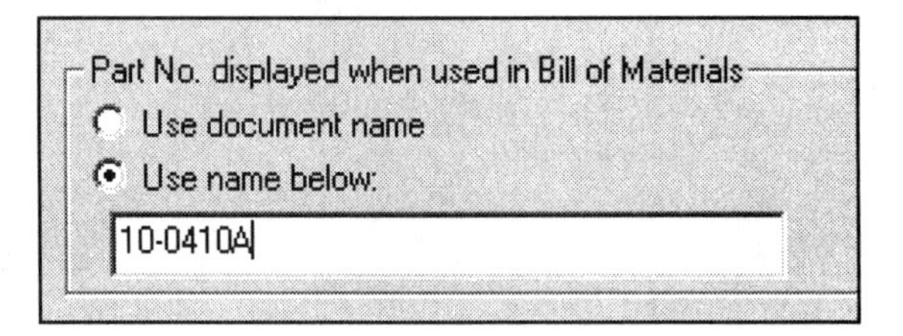

Add the Custom Properties for the With Nose Holes configuration.

284) Click the **Custom** button from the Configuration Properties dialog box.

285) Click the **Configuration Specific** tab. Add the Property, named MASS for the Default Configuration.

286) Enter **MASS** in the Name text box.

287) Click the **MASS Properties** button.

Properties:

Name	Value
MASS	10.57225
DESCRIPTION	COVERPLATE-16MM W/HOLES
MATERIAL	ALUMINUM

288) Select **Mass** from the drop down list. The SW-Mass value is added to the Value text box.

289) Click **Add**. The Mass Name and Value are added to the Properties box.

290) Add the Property, named DESCRIPTION for the Default Configuration. Select **Description** from the Name text box.

291) Enter **COVERPLATE 16MM W/HOLES** in the Value box.

292) Click the **Add** button. The Description Name and Value are added to the Properties box.

293) Add the Property, named MATERIAL for the Default Configuration. Select **MATERIAL** from the Name text box.

294) Enter **ALUMINUM** in the Value box.

295) Click the **Add** button. The Material Name and Value are added to the Properties box. The three Custom Properties are listed in the Properties box.

296) Click **OK**.

Add the PART NO. for the COVERPLATE Without Nose Holes configuration.

297) Double-click the **Without Nose Holes** Configuration.

298) Right-click **Properties** on the **Without Nose Holes** configuration.

299) Click **Use name below**.

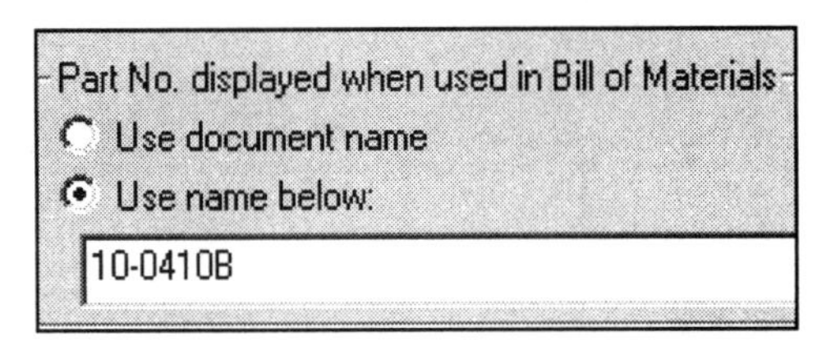

300) Enter **10-0410B** for Part No. displayed when used in Bill of Materials.

301) Click **OK**.

Add the Custom Properties for the Without Nose Holes configuration.

302) Click **Custom** button from the Configuration Properties dialog box.

303) Click the **Configuration Specific** tab. Add the Property, named MASS for the Default Configuration.

304) Enter **MASS** in the Name text box.

305) Click the **MASS Properties** button.

Properties:	Name	Value
	MASS	10.60957
	DESCRIPTION	COVERPLATE 16MM NO HOLES
	MATERIAL	ALUMINUM

306) Select **Mass** from drop down list. The SW-Mass value is added to the Value text box.

307) Click **Add**. The Mass Name and Value are added to the Properties box.

308) Add the Property, named DESCRIPTION for the Default Configuration. Select **Description** from the Name text box.

309) Enter **COVERPLATE-16MM-NO HOLES** in the Value box.

310) Click the **Add** button. The Description Name and Value are added to the Properties box.

311) Add the Property, named MATERIAL for the Default Configuration. Select **Material** from the Name text box.

312) Enter **ALUMINUM** in the Value box.

313) Click the **Add** button. The Material Name and Value are added to the Properties box. The three Configuration Specific Custom Properties are listed in the Properties box.

314) Click **OK** from the Summary dialog box.

315) Click **OK** from the Configuration Properties box.

316) Save the COVERPLATE part. Click **Save**.

Properties and PART NO's. have been assigned. Execute the Bill of Materials BOMTEMP-MASS-DM and view the results.

Add the New Bill of Materials with Custom Properties.

317) **Open** the CYLINDER-Sheet1 drawing.

318) Click the old **Bill of Materials**.

319) Press the **Delete** key.

320) Click the Isometric **view boundary**.

321) Insert the new Bill of Materials. Click **Insert**, **Bill of Materials**.

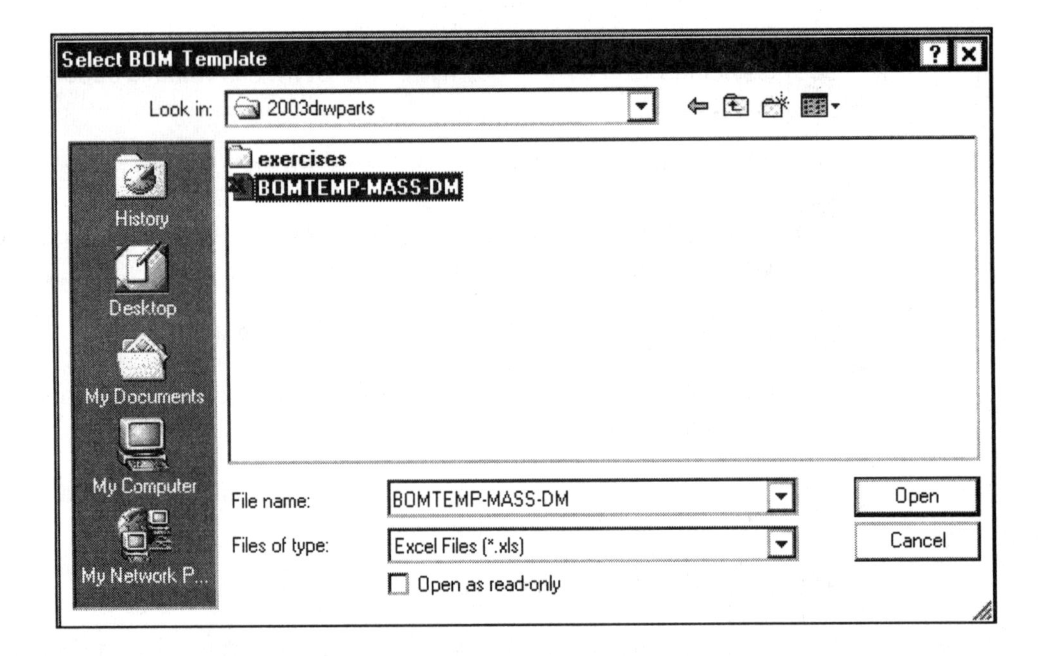

322) Click **BOMTEMP-MASS-DM.XLS** from the 2003drwparts file folder.

323) Click **Open**.

324) Display the Bill of Material inside the view. Uncheck **Use table anchor point** from the Anchor point text box.

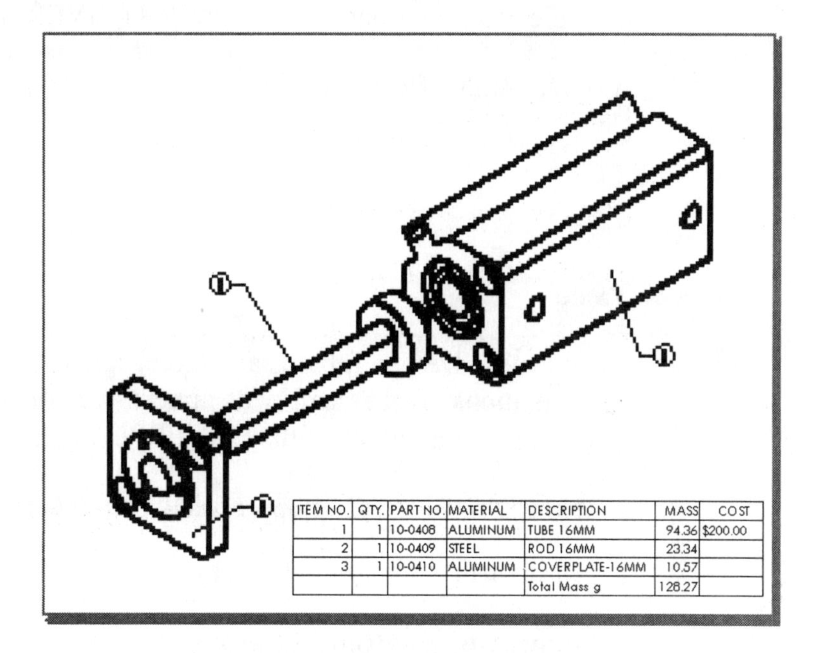

325) Click **OK**.

326) Drag the **Bill of Materials** to the left of the COVERPLATE.

327) Double-click the **Bill of Materials**.

328) Enter **200** in the COST column. The value $200.00 is displayed. The Column was formatted in Excel to $ currency.

ITEM NO.	QTY.	PART NO.	MATERIAL	DESCRIPTION	MASS	COST
1	1	10-0408	ALUMINUM	TUBE 16MM	94.36	$200.00
2	1	10-0409	STEEL	ROD 16MM	23.34	
3	1	10-0410	ALUMINUM	COVERPLATE-16MM	10.57	
				Total Mass g	128.27	

329) Enter **Total Mass(g)** in cell E5.

330) Click cell **F5**.

331) Calculate the total mass. Click **Sum** Σ. Exit Excel.

332) Click inside the SolidWorks **Graphics** window.

The CYLINDER assembly contains the default configuration for the TUBE, ROD and COVERPLATE parts. How do you modify the assembly to support multiple configurations of the ROD and COVERPLATE in a drawing?

Answer: With a Design Table.

Design Table

An assembly Design Table is and Excel spread sheet that controls multiple configurations of an assembly document. Create the Design Table with three different configurations of the COVERPLATE:

- No COVERPLATE (Suppressed Component).
- COVERPLATE With Nose Holes.
- COVERPLATE Without Nose Holes.

Insert six CYLINDER assembly configurations into a drawing.

Edit the ROD Distance Mate variable name.

333) Collapse the Exploded view. Right-click Collapse in the Graphics window.

334) Click the **CYLINDER assembly** window.

335) Expand the **Mates** entry in the FeatureManager.

336) Double-click the **Distance1 (TUBE<1>, ROD<1>)**.

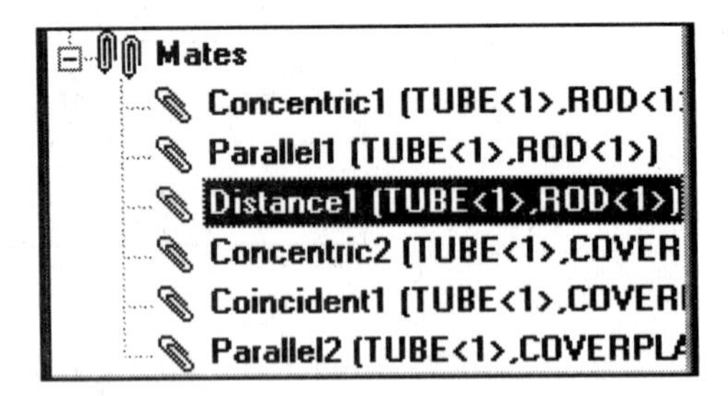

337) Right-click **Properties** on the **0** text inside the CLYINDER assembly.

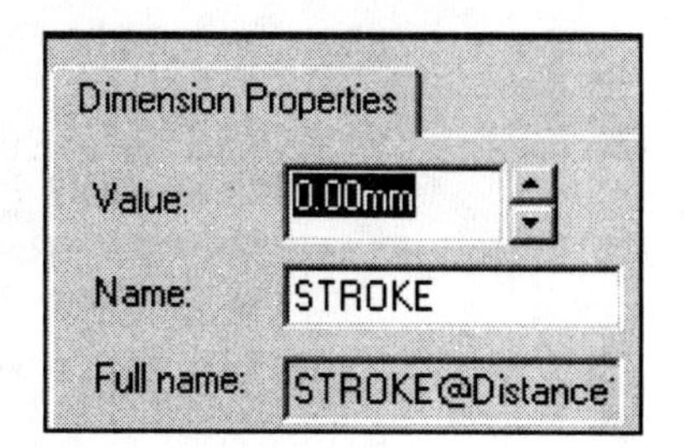

338) Enter **STROKE** for Name. The full dimension name is STROKE@Distance1.

339) Click **OK**.

Create a Design Table in the CYLINDER assembly.

340) The TUBE, ROD and COVERPLATE parts are set to their Default configurations. Click **Insert**, **New Design Table** from the Main toolbar.

341) Click **Blank**.

342) Click **OK**.

343) **Delete** the First Instance.

344) Enter **NO COVERPLATE** in Cell A3.

345) Enter **COVERPLATE-HOLES** in Cell A4.

346) Enter **COVERPLATE-NOHOLES** in Cell A5.

347) Drag the **column bar** between Column A and Column B to the right until the full Configuration Names are displayed.

	A	B	C
1	Design Table for: CYLINDER		
2		$CONFIGURATION@COVERPLATE<1>	$STATE@COVERPLATE<1>
3	NO COVERPLATE	DEFAULT	S
4	COVERPLATE-HOLES	With Nose Holes	R
5	COVERPLATE-NOHOLES	Without Nose Holes	R

348) Enter **$CONFIGURATION@COVERPLATE<1>** in Cell B2.

349) Enter **DEFAULT** in Cell B3.

350) Enter **With Nose Holes** in Cell B4.

351) Enter **Without Nose Holes in** Cell B5.

352) Enter **$STATE@COVERPLATE<1>** in Cell C2.

353) Enter **S** for Suppress in Cell C3.

354) Enter **R** for Resolved in Cell C4 and C5.

Note: Configuration names in the Design Table correspond to the exact configuration name in the FeatureManager.

355) Exit the Design Table. Click a **position** outside the Design Table. Three assembly CYLINDER configurations are created.

356) Click **OK**. Verify the configurations.

357) Click the **Configuration** icon.

358) Double-click on **NO COVERPLATE**.

359) Double-click on **COVERPLATE-HOLES**.

360) Double-click on **COVERPLATE-NOHOLES**.

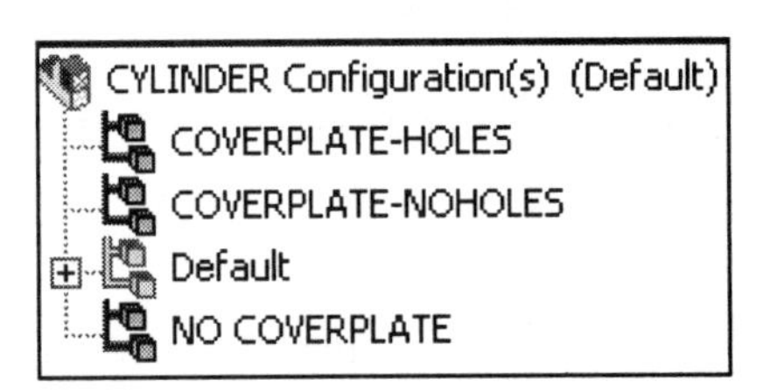

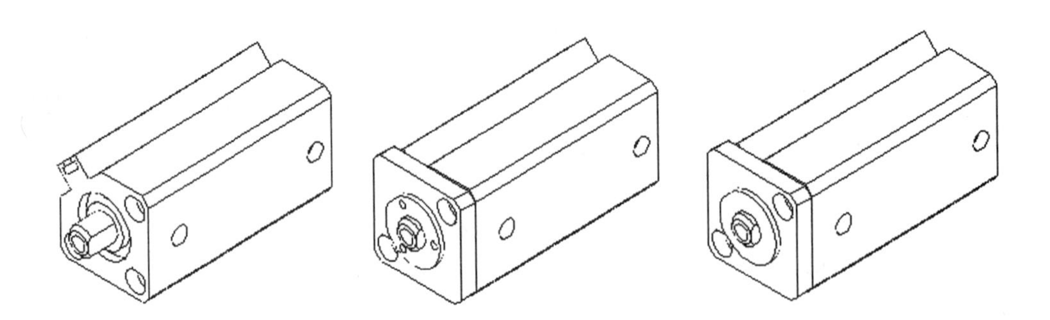

361) Return to the Default. Double click the **Default** Configuration.

362) Edit the Design Table. Right-click **Design Table** in the FeatureManager.

363) Click **Edit Table**.

364) Click **Cancel** to the Add Rows and Columns dialog box.

	A	B	C
1	Design Table for: CYLINDER		
2		$CONFIGURATION@COVERPLATE<1>	$STATE@COVERPLATE<1>
3	NO COVERPLATE	DEFAULT	S
4	COVERPLATE-HOLES	With Nose Holes	R
5	COVERPLATE-NOHOLES	Without Nose Holes	R
6	STROKE 0	DEFAULT	S
7	STROKE 30	DEFAULT	S
8	CUT AWAY	DEFAULT	R

365) Add three more configurations. Enter **STROKE 0** in cell A6.

366) Enter **STROKE 30** in cell A7.

367) Enter **CUT AWAY** in cell A8.

368) Enter **Default** in cell B6, B7 and B8.

369) Enter **S** in Cell C6 and C7.

370) Enter **R** in cell C8.

Add the STROKE@DISTANCE1 variable.

371) Enter **STROKE@DISTANCE1** in cell D2.

372) Enter **0** in cells D3 through D6 and D8.

373) Enter **30** in Cell D7.

374) Verify the STROKE 0 and STROKE 30 configurations. Click a **position** outside the Design Table.

375) Click **OK**.

	A	D
1	Design Table for: CYLINDER	
2		STROKE@DISTANCE1
3	NO COVERPLATE	0
4	COVERPLATE-HOLES	0
5	COVERPLATE-NOHOLES	0
6	STROKE 0	0
7	STROKE 30	30
8	CUT AWAY	0

376) Double-click **STROKE 0** from the Configuration Manager.

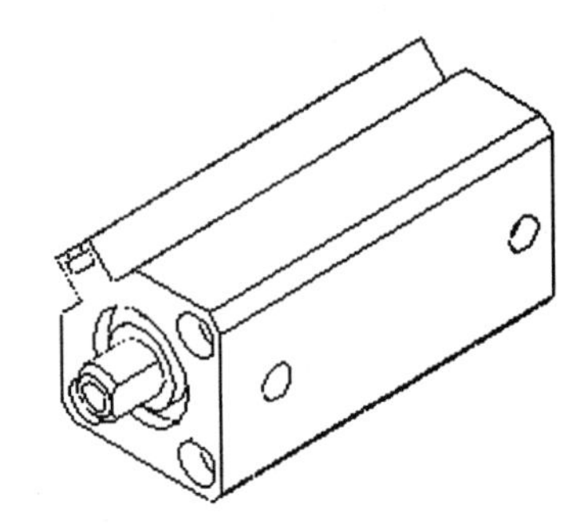

377) Double-click **STROKE 30** from the Configuration Manager.

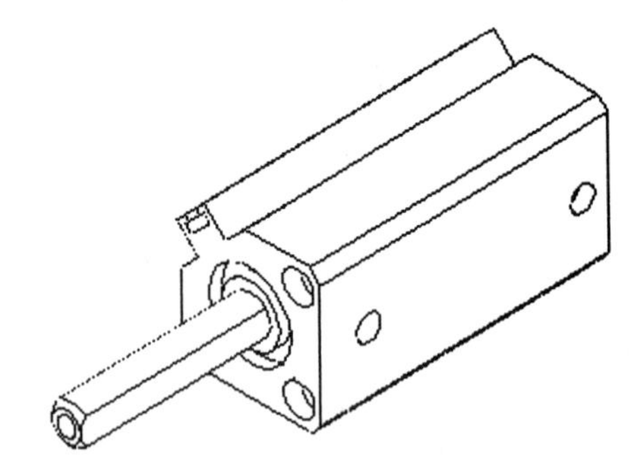

378) Return to the **Default** configuration.

379) Edit the Design Table.

380) Right-click **Design Table** in the FeatureManager.

381) Click **Edit Table**.

Add $CONFIGURATION@TUBE<1> and $STATE@CONCENTRIC2.

382) Enter **$CONFIGURATION@TUBE<1>** in cell E2.

383) Enter **Section Cut** in Cell E8.

384) Enter **DEFAULT** in cells E3 through E7.

385) Enter **$STATE@CONCENTRIC2** in cell F2.

386) Enter **U** for Unsuppressed cells F3 through F7.

387) Enter **S** for Suppressed in cell F8.

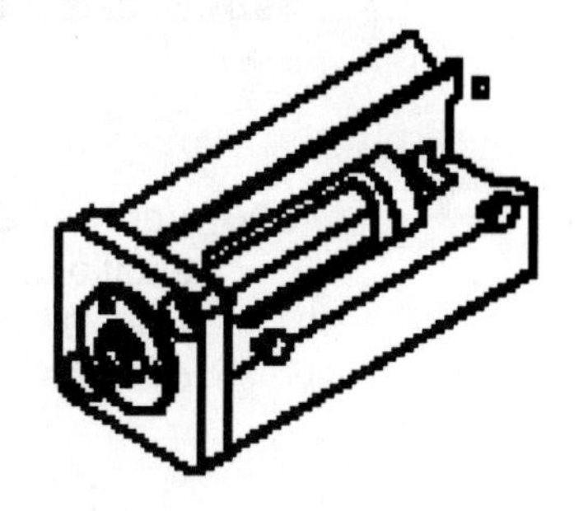

	A	E	F
1	Design Table for: CYLINDER		
2		$CONFIGURATION@TUBE<1>	$STATE@CONCENTRIC2
3	NO COVERPLATE	DEFAULT	U
4	COVERPLATE-HOLES	DEFAULT	U
5	COVERPLATE-NOHOLES	DEFAULT	U
6	STROKE 0	DEFAULT	U
7	STROKE 30	DEFAULT	U
8	CUT AWAY	Section Cut	S

Add User_Notes.

388) Enter **$USER_ NOTES in** Cell G2. SolidWorks does not calculate $USER_NOTES.

389) Enter the following notes displayed in Cells G3 through G8. Enter **BASIC** in Cell G3.

390) Enter **ANODIZED COVER** in Cell G4.

391) Enter **ANODIZED COVER** in Cell G5.

392) Enter **NO COVER ROD AT HOME POSITION** in Cell G6.

393) Enter **NO COVER ROD FULLY EXTENDED** in Cell G7.

G
$USER_NOTES
BASIC
ANODIZED COVER
ANODIZED COVER
NO COVER ROD AT HOME POSITION
NO COVER ROD FULLY EXTENDED
SEE INTERNAL FEATURES

394) Enter **SEE INTERNAL FEATURES** in Cell G8.

395) Update the configurations. Click a **position** outside the Design Table.

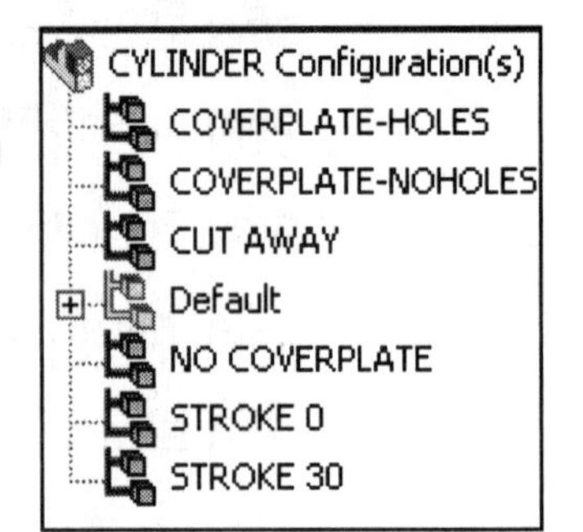

396) Verify all CYLINDER configurations before creating additional views in the CYLINDER drawing. Double-click each configuration name. Double-click the **Default** configuration.

397) Save the CYLINDER assembly. Click **Save**.

Multiple Configurations in the Drawing

Multiple configurations created in the assembly Design Table allow you to insert various configurations into the drawing.

Specify the configuration in the Properties of the current view. The CYLINDER drawing contains six configurations of the CYLINDER assembly and two configurations of the TUBE, ROD and COVERPLATE parts.

Add the NO COVERPLATE configuration to the CYLINDER drawing.

398) **Open** the CYLINDER-Sheet1 drawing.

399) Click the Exploded Isometric **view boundary**.

400) Copy the view. Press **Ctrl C**.

401) Click a **position** in the left corner of Sheet1.

402) Paste the view. Press **Ctrl V**.

403) Right-click **Properties** on the view boundary.

404) Select **NO COVERPLATE** from the Configuration information text box.

405) Click **OK**.

Add the COVERPLATE-HOLES Configuration.

406) Click a **position** to the right of the NO COVERPLATE configuration.

407) Paste the view. Press **Ctrl V**.

408) Right-click **Properties** on the view boundary.

409) Select **COVERPLATE-HOLES** from the Configuration information text box.

410) Click **OK**.

Add the COVERPLATE-NOHOLES Configuration.

411) Click a **position** to the right of the COVERPLATE-HOLES configuration.

412) Paste the view. Press Ctrl **V**.

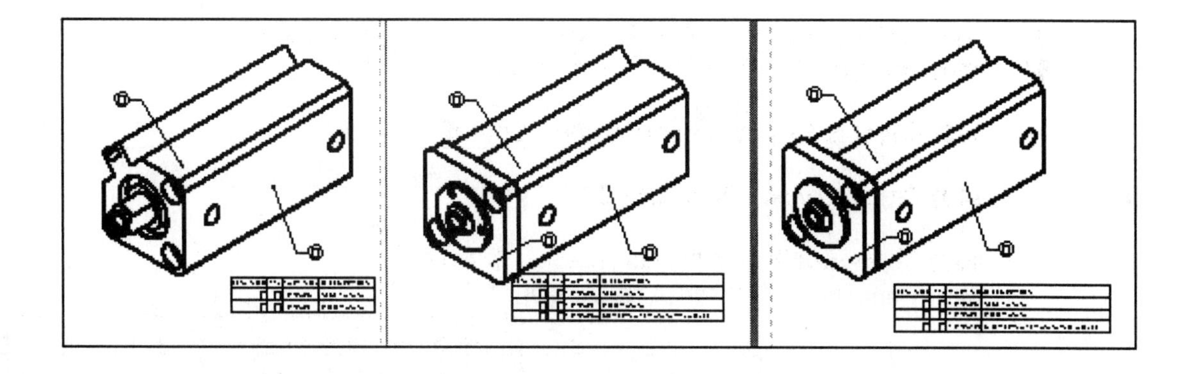

413) Right-click **Properties** on the view boundary.

414) Select **COVERPLATE-NOHOLES** from the Configuration information text box.

415) Click **OK**.

416) Click the **NO COVERPLATE view boundary**.

417) Click **Insert**, **Bill of Materials**.

418) Select the default **BOMTEMP.XLS**.

419) Uncheck **Use table anchor point** from the Anchor point text box.

420) Click **OK**.

421) Drag the **Bill of Materials** to the right corner of the view. The Bill of Materials reflects the Design Table configurations and the Customer Properties created in each part.

422) Click the **COVERPLATE-HOLES view boundary**.

423) Click **Insert**, **Bill of Materials**.

ITEM NO.	QTY.	PART NO.	DESCRIPTION
1	1	10-0408	TUBE 16 MM
2	1	10-0409	ROD-16MM

424) Select the default **BOMTEMP.XLS**.

425) Uncheck **Use table anchor point** from the Anchor point text box.

426) Click **OK**.

427) Drag the **Bill of Materials** to the right corner of the view.

ITEM NO.	QTY.	PART NO.	DESCRIPTION
1	1	10-0408	TUBE 16 MM
2	1	10-0409	ROD-16MM
3	1	10-0408A	COVERPLATE-16MM-W/HOLES

428) Repeat for **COVERPLATE-NOHOLES**.

ITEM NO.	QTY.	PART NO.	DESCRIPTION
1	1	10-0408	TUBE 16 MM
2	1	10-0409	ROD-16MM
3	1	10-0410B	COVERPLATE-16MM-NO HOLES

429) The Balloons reflect only two components in the NO COVERPLATE configuration. Click the **2 Balloon**.

430) Drag the **attachment point** to the front of the ROD.

431) Save the CYLINDER drawing. Click **Save**.

432) Add Sheet2. Right-click on the **Sheet 1** tab.

433) Click **Add Sheet**.

434) Click **OK**.

435) Click a **position** in the top left corner of Sheet2. Past the view. Press **Ctrl V**. Right-click **Properties** on the view boundary.

436) Select **STROKE 0** from the Configuration information text box. Click **OK**.

437) **Delete** the **Balloons**.

438) Click a **position** to the right of the STROKE 0 configuration.

439) Paste the view. Press **Ctrl V**.

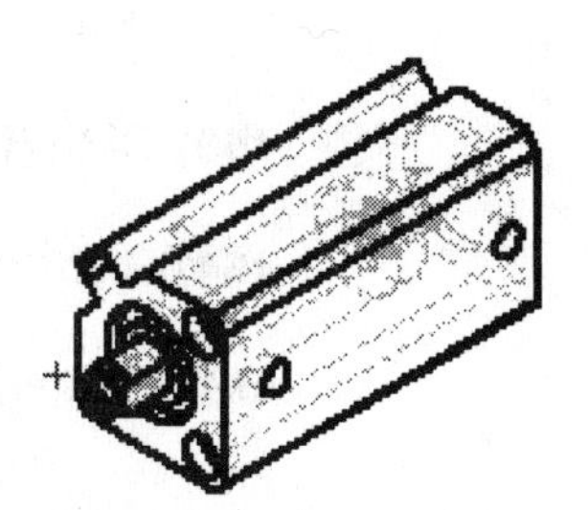

440) Right-click **Properties** on the view boundary.

441) Select **STROKE 30** from the Configuration information text box.

442) Click **OK**.

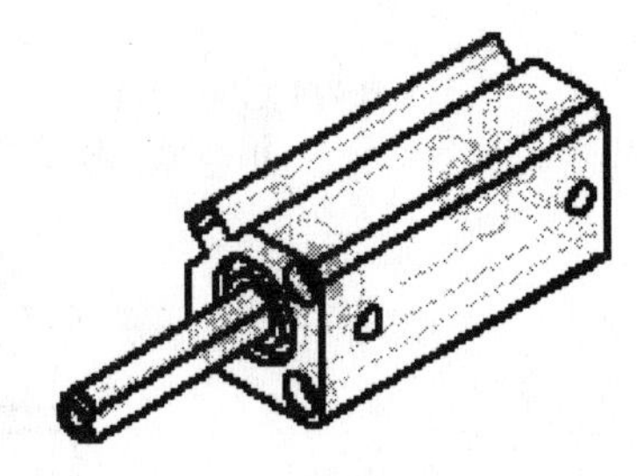

443) **Delete** the Balloons.

444) Click inside the STROKE 0 configuration **view boundary**.

445) Double-click **Right** for the View Orientation dialog box.

446) Click **Hidden Lines Removed**.

447) Right-click **Properties**.

448) Click the **Show Hidden Edges** tab.

449) Click the **ROD** component.

450) Click **OK** to display the hidden edges of the ROD inside the CYLINDER.

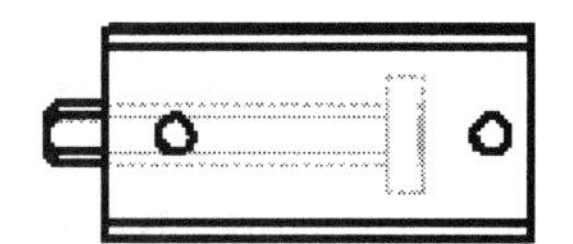

451) Click inside the STROKE 30 configuration **view boundary**.

452) Double-click **Right** for the View Orientation dialog box.

453) Click **Hidden Lines Removed**.

454) Right-click **Properties**.

455) Click the **Show Hidden Edges** tab.

456) Click the **ROD** component.

457) Click **OK** to display the hidden edges of the ROD inside the CYLINDER.

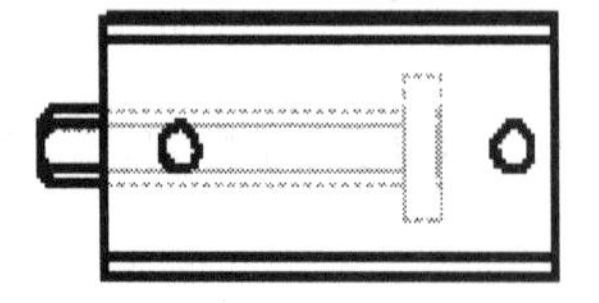

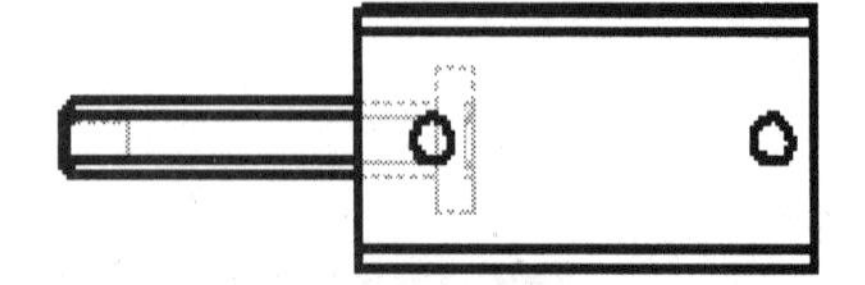

Add a CUT AWAY Configuration.

458) Click a **position** below the STROKE 0 configuration. Paste the view. Press **Ctrl V**.

459) Right-click **Properties** on the view boundary.

460) Select **CUT AWAY** from the Configuration information text box.

461) Click **OK**.

462) Hide the COVERPLATE in the drawing. Right-click **Properties** on the view boundary.

463) Click the **Hide/Show Component** tab.

464) Select the **COVERPLATE** inside the CUT AWAY view.

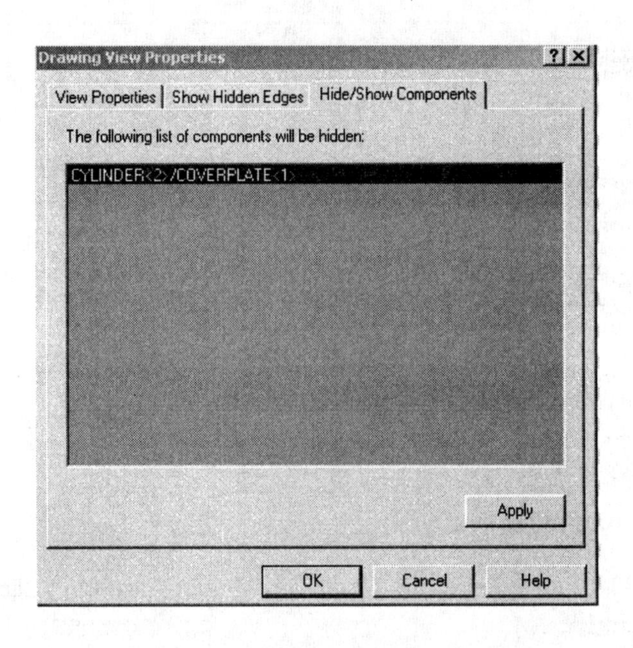

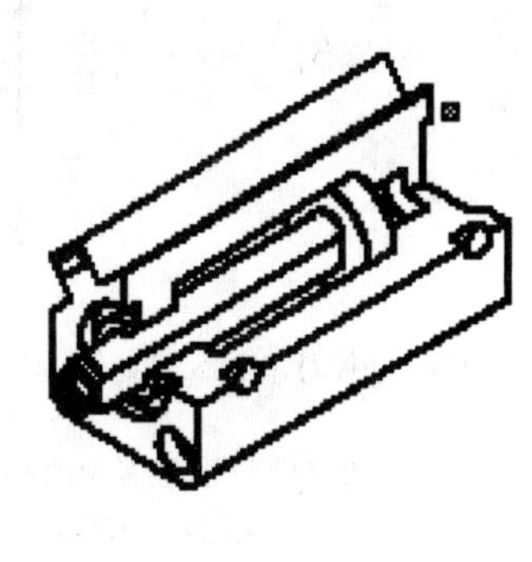

465) Hide the COVERPLATE and expose the ROD. Click **OK**.

466) **Delete** the Balloons.

Add Part views.

467) **Close** all files except for the CYLINDER assembly and the CYLINDER-Sheet2 drawing. Click **Tile Horizontal**.

468) Click **Default** from the Configuration Manager. Display the FeatureManager.

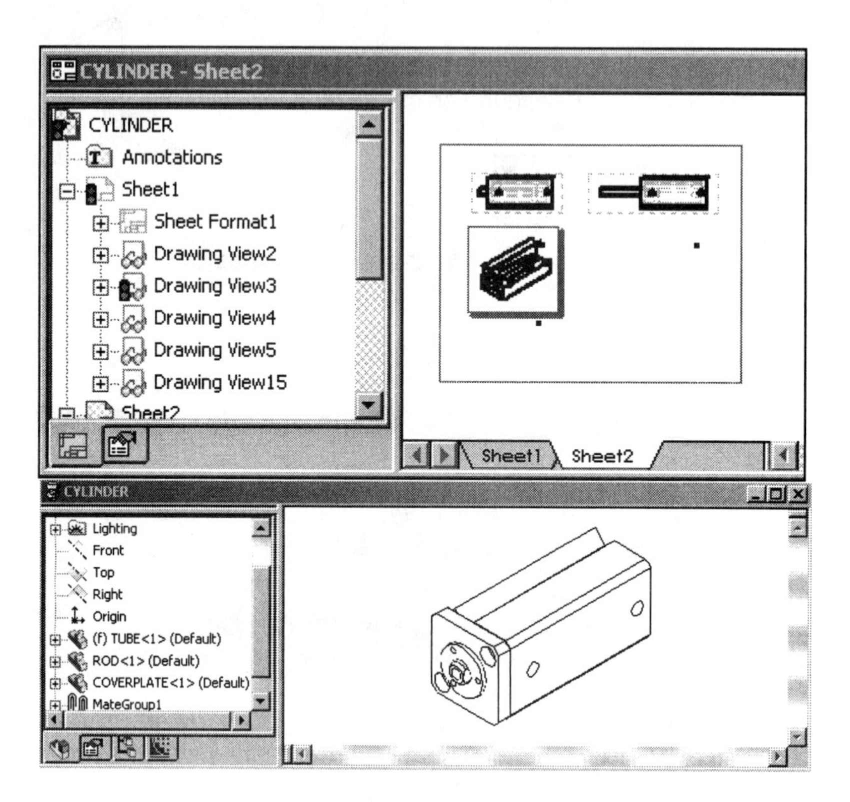

469) Click **Assembly FeatureManager** .

470) Add the COVERPLATE Front view to the CYLINDER-Sheet2 drawing. Click the **Sheet2** tab.

471) Click **Named** view from the Drawing toolbar.

472) Click the **COVERPLATE<1>** from the CYLINDER FeatureManager.

473) Click a **position** to the right of the CUT AWAY view.

474) Click **Front** from the View Orientation box.

475) Add the TUBE Right view to the CYLINDER-Sheet2 drawing. Click **Named** view from the Drawing toolbar.

476) Click **TUBE<1>** from the CYLINDER FeatureManager.

477) Click a **position** to the right of the CUT AWAY view.

478) Click **Right** from the View Orientation box.

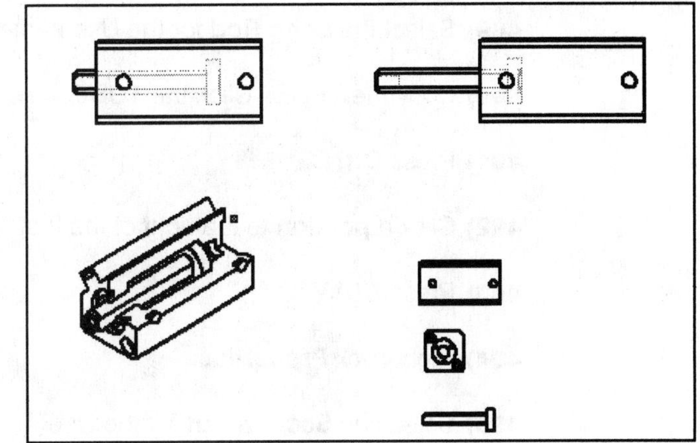

479) Add the ROD Right view to the CYLINDER-Sheet2 drawing. Click **Named** view from the Drawing toolbar.

480) Click **ROD<1>** from the CYLINDER FeatureManager.

481) Click a **position** to the right of the CUT AWAY view.

482) Click **Right** from the View Orientation box.

483) **Zoom in** on the part views.

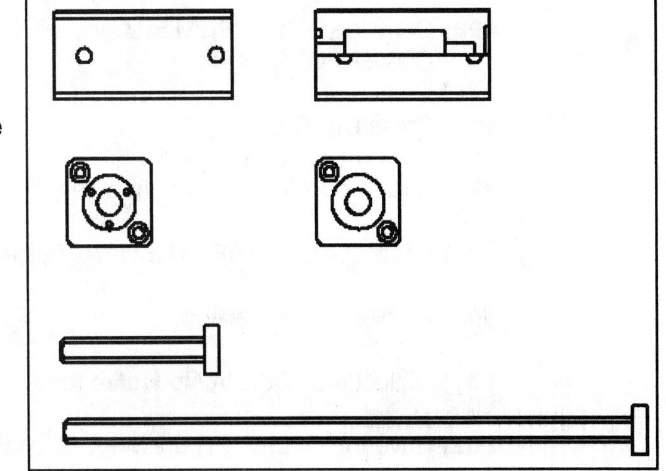

Copy the part views.

484) Copy the ROD. Click the **ROD view boundary**.

485) Copy the ROD view. Press **Ctrl C**.

486) Click a **postion** below the first ROD.

487) Paste the ROD view. Press **Ctrl V**.

488) Right click **Properties**.

489) Select the **Long Rod** for the Use named configuration.

490) Copy the TUBE. Click the **TUBE view boundary**. Copy the TUBE view.

491) Press **Ctrl C**.

492) Click a **postion** to the right of the first TUBE. Paste the TUBE view.

493) Press **Ctrl V**.

494) Right-click **Properties**.

495) Select the **Section Cut** for the Use named configuration.

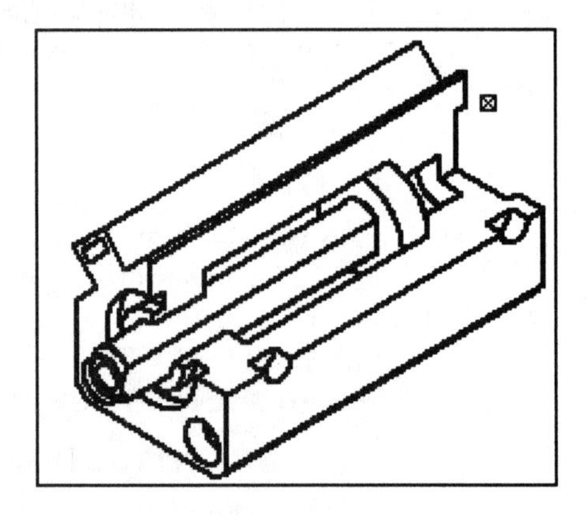

496) Copy the COVERPLATE. Click the **COVERPLATE view boundary**. Copy the COVERPLATE view.

497) Press **Ctrl C**.

498) Click a **postion** to the right of the first COVERPLATE.

499) Paste the COVERPLATE view. Press **Ctrl V**.

500) Right-click **Properties**.

501) Select the **Without No Holes** for the Use named configuration.

502) Save the CYLINDER drawing. Click **Save**.

503) Insert the Design Table into the Drawing. Right-click the **Cut Away view bounday**.

504) Click **Insert**, **Design Table** from the Main toolbar.

Design Table for: CYLINDER		
	$CONFIGURATION@COVERPLATE<1>	$STATE@COVERPLATE<1>
NO COVERPLATE	DEFAULT	S
COVERPLATE-HOLES	With Nose Holes	R
COVERPLATE-NOHOLE	Without Nose Holes	R
STROKE0	DEFAULT	S
STROKE30	DEFAULT	S
CUT AWAY	DEFAULT	R

505) View the Design Table. Right-click **Properties**.

506) Enter **2** for Design Table Scale in the OLE Object Property dialog box.

507) Click **OK**.

508) Save the CYLINDER drawing. Click **Save**.

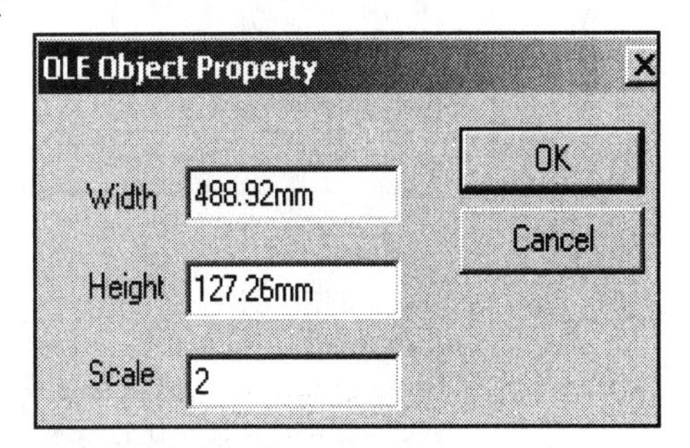

The Design Table can be modified in Excel to display only the specific rows and columns of information. Modifying Design Tables will be explored in Project 5.

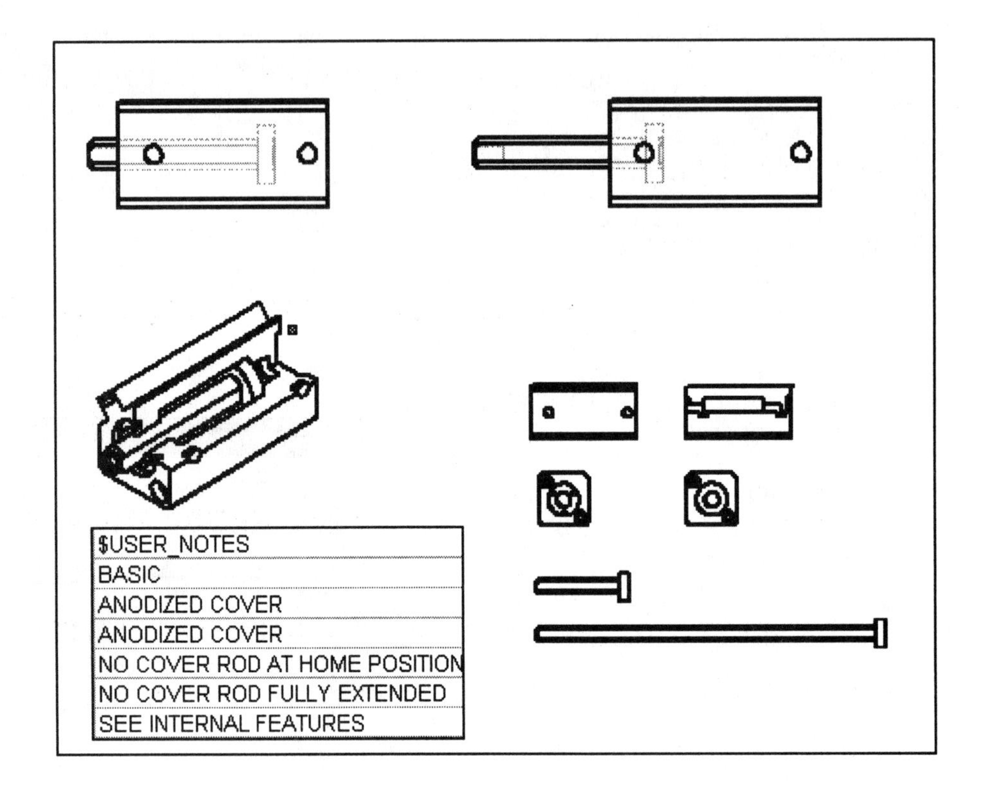

Project Summary

You developed the following in this project:

- CYLINDER assembly.
- Custom Properties.
- Design Table.
- CYLINDER drawing.
- Bill of Materials.

The CYLINDER assembly consisted of the TUBE, ROD and COVERPLATE parts.

You added Custom Properties to each part to describe Material, Mass and Description of the part configurations.

You created six different configurations of the CYLINDER assembly with an Excel Design Table.

You added a Design Table and a Bill of Materials with Custom Properties.

Try the exercises at the end of this project before going on to Project 5.

Project Terminology

Anchor point: The origin of the Bill of Material in a sheet format.

Attachment Point: An attachment point is the end of a leader that attaches to an edge, vertex, or face in a drawing sheet.

Assembly: An assembly is a document in which parts, features, and other assemblies (sub-assemblies). When a part is inserted into an assembly it is called a component. Components are mated together. The extension for a SolidWorks assembly file name is .SLDASM.

Balloon: A balloon labels the parts in the assembly and relates them to item numbers on the bill of materials (BOM) added in the drawing. The balloon item number corresponds to the order a component was inserted into an assembly.

Bill of Materials: A table inserted into a drawing to keep a record of the parts used in an assembly.

Component: A part or sub-assembly within an assembly.

Configurations: Variations of a part or assembly that control dimensions, display and state of a model.

Design Table: An excel spreadsheet utilized to control multiple configurations of a part or assembly.

Exploded view: A configuration in an assembly that displays its components separated from one another.

Mass Properties: The physical properties of a model based upon geometry and material.

Copy and Paste: Utilize copy/paste for annotation note, parametric notes or linked notes or notes in a design table.

Questions:

1. Create an assembly from a ______________.
2. Each component in an assembly has ____________ degrees of freedom.
3. Identify the method to remove degrees of freedom in an assembly.
4. True or False. A Design Table is used to create multiple configurations in an assembly.
5. True or False. A Design Table is used to create multiple configurations in a part.
6. True or False. A Design Table is used to create multiple configurations in a drawing.
7. Describe five assembly modeling techniques.
8. Describe the (f) symbol functionality in the FeatureManager.
9. Identify the three Mate options used to mate the ROD to the TUBE.
10. True or False. A Distance Mate provides additional flexibility over a Coincident Mate. Why?
11. Identify the location to create an Exploded view.
12. Describe the function of a Balloon annotation.
13. The PART NO. column in the default Bill of Materials is determined by ______________.
14. Describe the procedure to add additional Properties such as Material and Mass to the Bill of Materials.
15. Describe the function of the $STATE variable in the part Design Table.
16. A drawing can contain multiple configurations of a _____________ and an ___________.

Exercises

Exercise 4.1:

Calculate the Mass Properties for the Long Rod Configuration. Add Properties to the Long Rod Configuration: DESCRIPTION, MATERIAL and MASS.

Enter 10-0409L for Use name below.

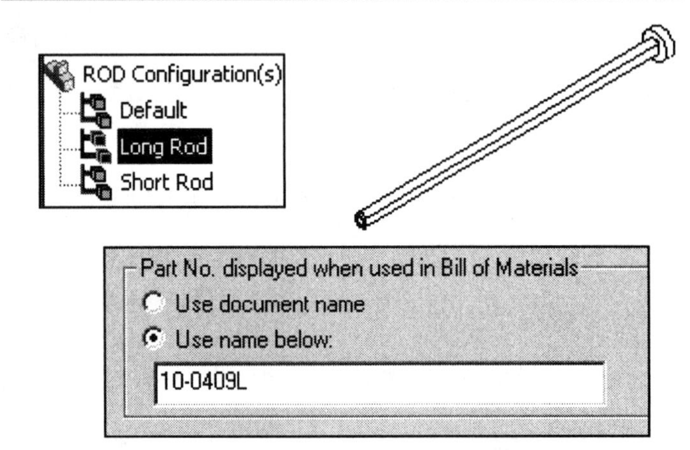

ROD Long Rod*	10-0409L	ROD-16MM-LONG	STEEL	"SW-Mass@@Long Rod@ROD.SLDPRT"

Note: Exercise 4.2 through 4.6 must be completed in order.

Exercise 4.2:

Create a RACK assembly.

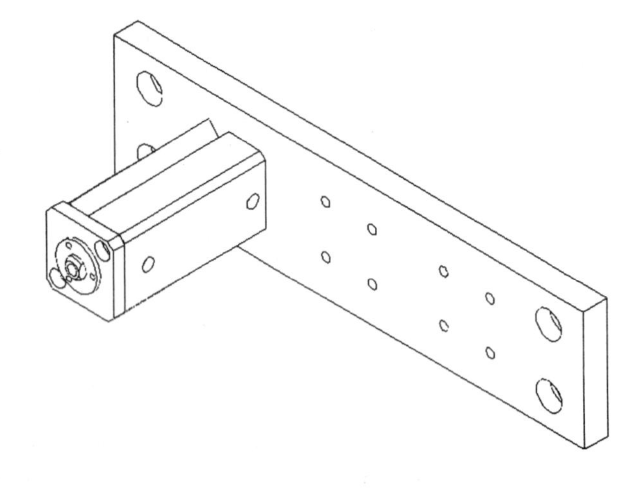

a) Insert the MOUNTINGPLATE-CYLINDER as the base component in the assembly, fixed to the Origin. The default configuration contains three sets of holes. Insert the CYLINDER component. The CYLINDER is a sub-assembly comprised of the TUBE, ROD and COVERPLATE components.

b) Mate the Cbore hole of the TUBE with the bottom left hole of the MOUNTINGPLATE-CYLINDER. Mate the back face of the TUBE and the front face of the MOUNTINGPLATE-CYLINDER. Add a parallel Mate.

c) Create a Component Pattern in the RACK assembly. Click Insert, Component Pattern, Use and Existing Feature Pattern(Derived). Select CYLINDER as the Seed Component from the assembly FeatureManager. Select LPattern1 as the Pattern feature in the MOUNTINGPLATE-CYLINDER.

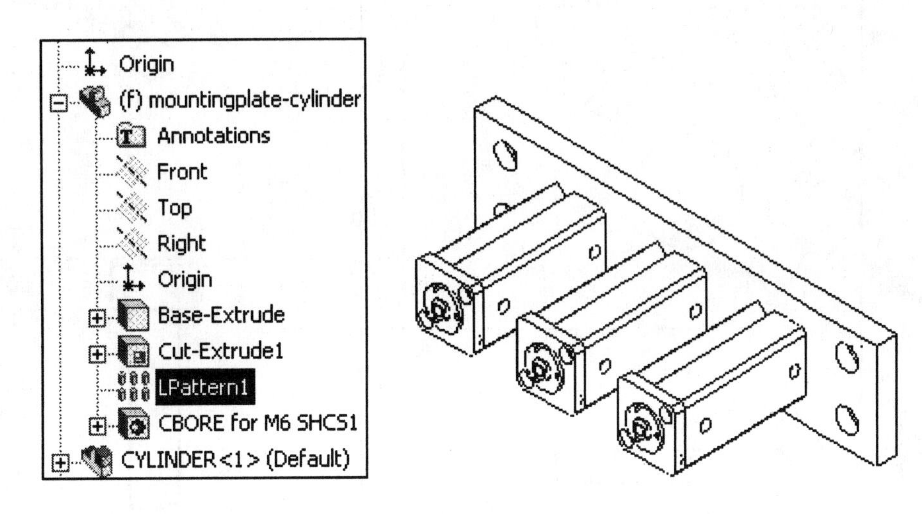

Exercise 4.3:

Add hardware to the RACK assembly. Utilize the SolidWorks Toolbox to obtain hardware part files. The SolidWorks Toolbox is a set of industry standard bolts, screws, nuts, pins, washers, structural shapes, bearings, PEM inserts and retaining rings.

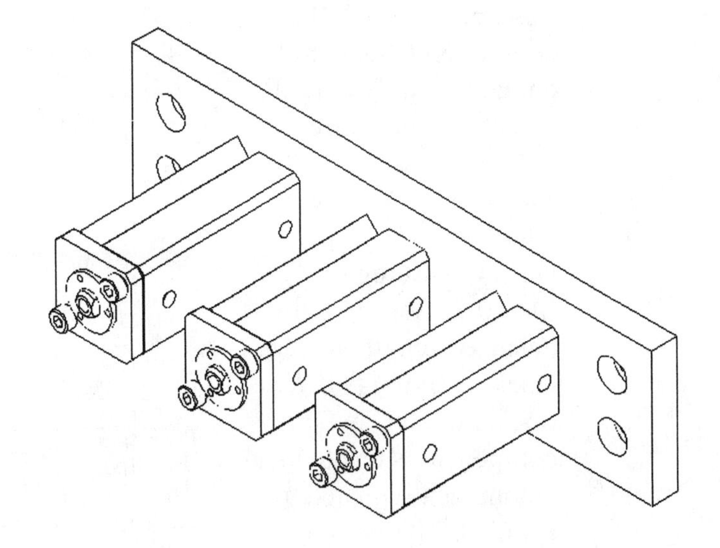

These SolidWorks parts are inserted into an assembly. An M4x0.7x16 and M4x0.7x20 socket head cap screws and M4 washers complete the RACK assembly. Download the SolidWorks Toolbox from the SolidWorks web site (www.SolidWorks.com). Note: For a 30 day trial.

Note: Below are the part dimensions for the actual hex head cap screw and washer.

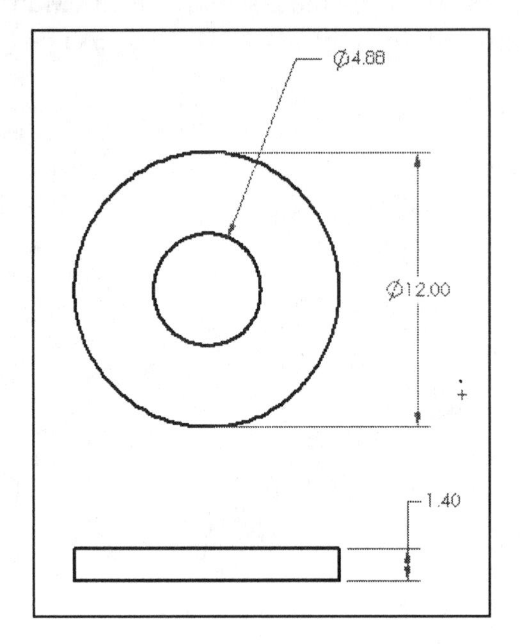

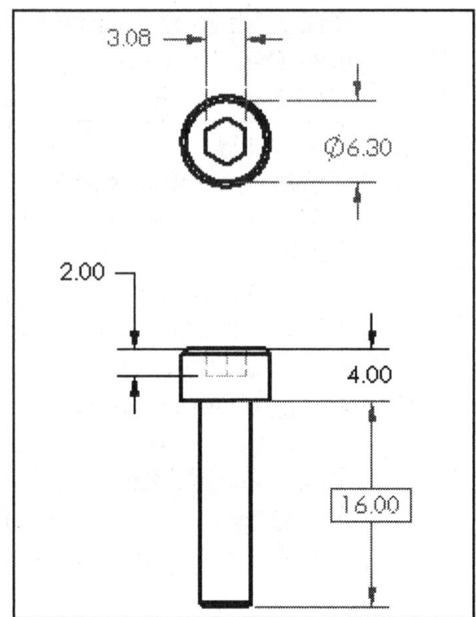

a) Click Toolbox from the Main toolbar. Insert an M4X0.7X16 Socket Head Cap Screw. Click the bottom left Cbore circle on the first CYLINDER. Select the Hardware option. Click the Bolts and Screws Tab. Select ANSI Metric, Socket Head Cap Screw, M4X0.7. Select 16 for Fastener Length. Click Create. The M4X0.7 Socket Head Cap Screw component is created and automatically mated to the CYLINDER assembly. Repeat for the second Cbore circle on the first CYLINDER.

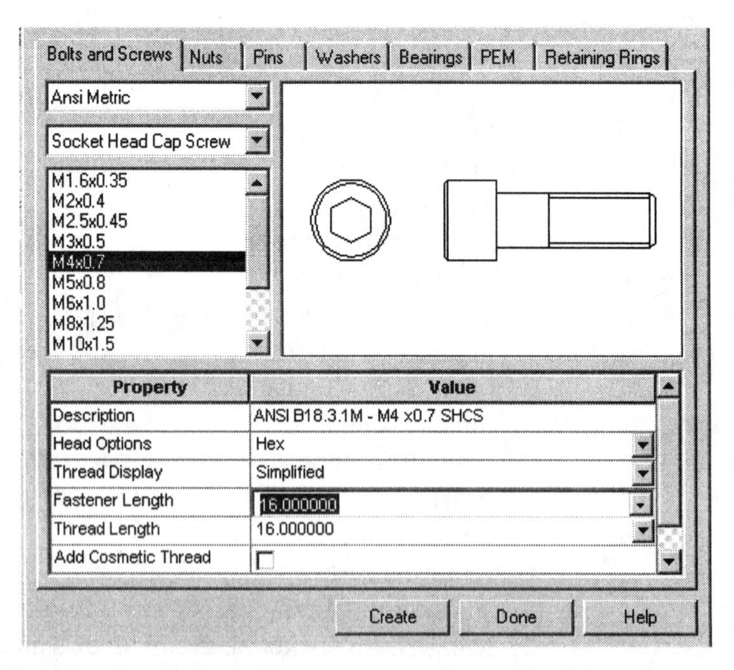

b) Edit the derived Component pattern. Select the two M4x0.7 Socket Head Cap Screws for the Seed Component from the FeatureManager. The Socket Head Caps Screws are displayed on the other components.

c) Insert a M4 Washer and a M4X0.7X20 Socket Head Cap Screw to the back face of the MOUNTINGPLATE-CLYLINDER.

d) Before submitting the MOUNTINGPLATE-CYLINDER drawing to the machine shop, check material and hardware availability with the parts department. No metric size socket head cap screws are in the stock room. There are only ¼″ and ½″ flat plate stock available in the machine shop. What do you do? Modify your design to reflect inch components.

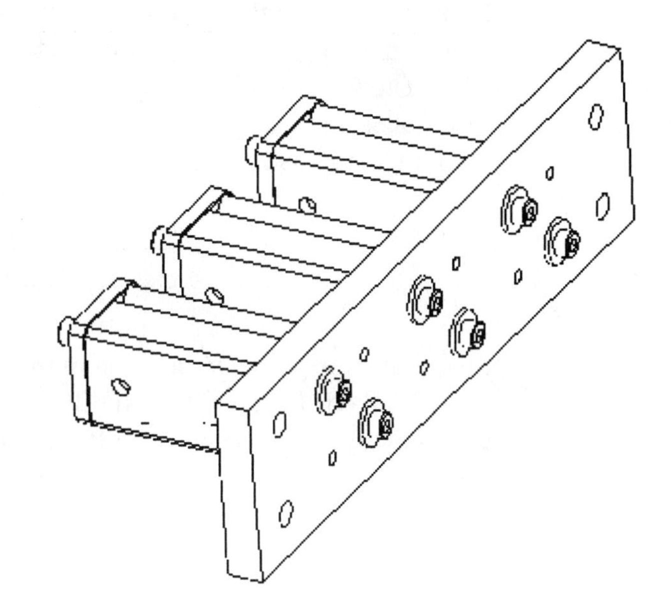

Exercise 4.4:

a) Create an Exploded view for the RACK assembly.

b) Create a new drawing, RACK. Add Balloons to the MOUNTINGPLATE-CYLINDER, CYLINDER and M4X0.7X16 Socket Head Cap Screw. Add a Stacked Balloon to the M4X0.7X20 Socket Head Cap Screw. Select the M4Washer.

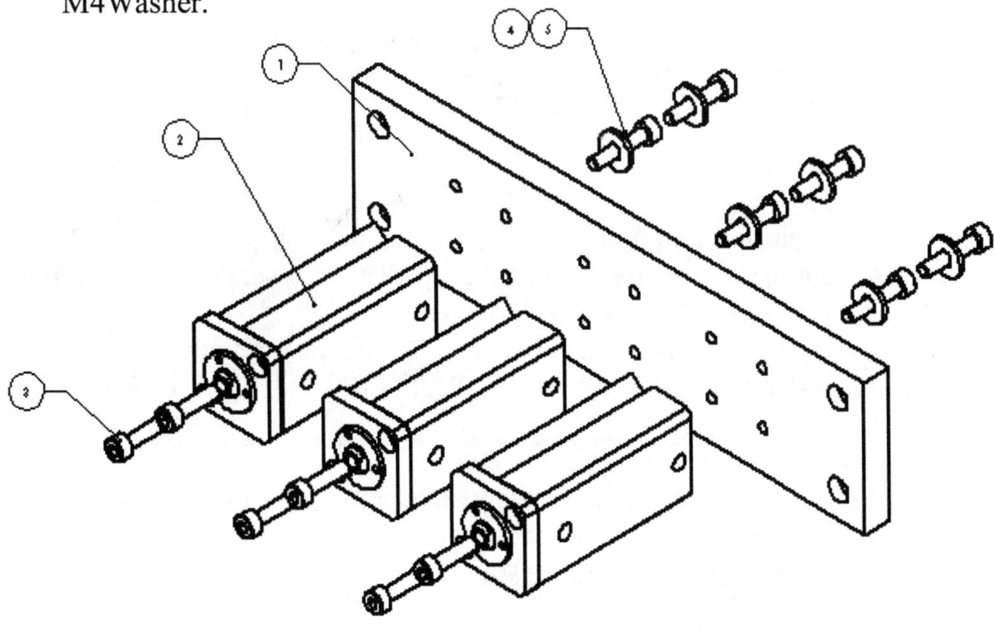

c) Insert a Bill of Materials to the Exploded view in the RACK drawing. Click Show assemblies and parts as an indented list.

ITEM NO.	QTY.	PART NO.	DESCRIPTION
1	1	50-052-1	MOUNT PLATE
2	3	99-0531	COMPACT AIR CYLINDER
	1	10-0408	TUBE 16 MM
	1	10-0409	ROD-16MM
	1	10-0410	COVERPLATE 16MM
3	6	5126-16	SHCS-M4X0.7-16
4	6	5126-20	SHCS-M4X0.7-20
5	6	5226-1	WASHER 4 MM

d) Add the Property Description in the Configuration Manager for the MOUNTINGPLATE-CYLINDER and hardware. Add the PART NO. to the components necessary to complete the Bill of Materials

Exercise 4.5:

a) The MOUNTINGPLATE-CYLINDER part has two configurations named: 2 PATTERN and 3 PATTERN. Review the configurations for the MOUNTING PLATE-CYLINDER part.

Design Table for: mountingplate-cylinder		
	D1@LPattern1	
2 PATTERN	2	
3 PATTERN	3	

b) Control the configurations in the RACK assembly. Create a Design Table for the RACK assembly. Add two RACK assembly configurations: 2 CYLINDER and 3 CYLINDER.
Enter $CONFIGURATION@mountingplate-cylinder<1>.
Enter configuration names: 2 PATTERN and 3 PATTERN.

Design Table for: RACK	
	$CONFIGURATION@mountingplate-cylinder<1>
2 CYLINDER	2 PATTERN
3 CYLINDER	3 PATTERN

c) Sheet2 contains the MOUNTINGPLATE-CYLINDER, 2 PATTERN configuration.

Only one Exploded view can exist per assembly. Delete the Stacked Balloons. The hardware quantities are displayed in the Bill of Materials.

ITEM NO.	QTY.	PART NO.	DESCRIPTION
1	1	2PATTERN	
2	2	CYLINDER	
	1	10-0408	TUBE 16 MM
	1	10-0409	ROD-16MM
	1	10-0410	COVERPLATE 16MM
3	4	AM_shcs_M4x0.7_16.0_16.0_He_Sim_1	
4	4	AM_shcs_M4x0.7_20.0_20.0_He_Sim_1	
5	4	AM_fw8_M4	

d) Add new Explode Steps to display the hardware in both a vertical and horizontal direction. Exploded view lines are 3D curves that are added to the Exploded view in the assembly.

Hint. Click Insert, Explode Line Sketch. The Route Line PropertyManager is displayed. Click the circular edges of the corresponding cap screw and holes in order from left to right. Click OK. Repeat for the remaining hardware.

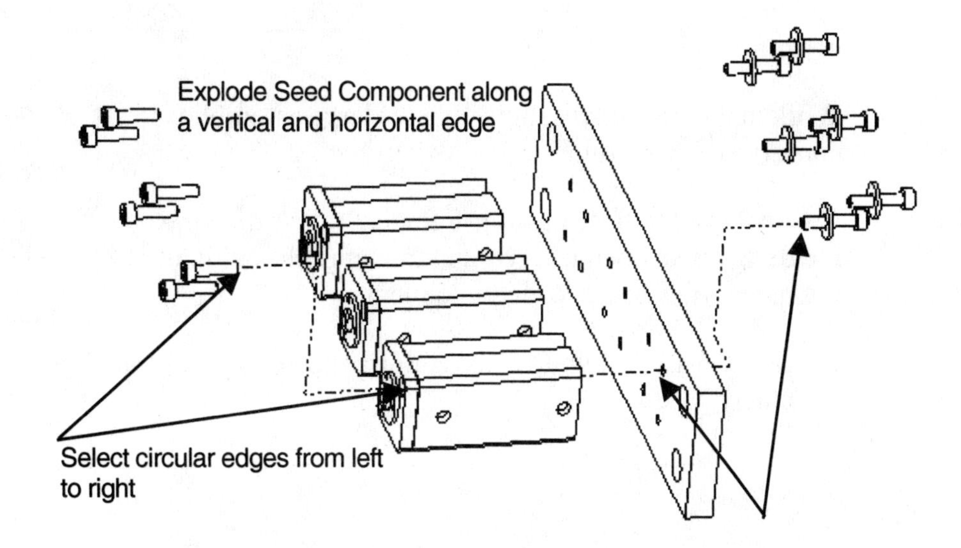

Exercise 4.6:

Add the Property Description in the Configuration Manager for the MOUNTINGPLATE-CYLINDER and hardware. Add the PART NO. to the components necessary to complete the Bill of Materials for the 2PATTERN configuration.

ITEM NO.	QTY.	PART NO.	DESCRIPTION
1	1	50-052-2	2 PATTERN MOUNT PLATE
2	2	99-0531	COMPACT AIR CYLINDER
	1	10-0408	TUBE 16 MM
	1	10-0409	ROD-16MM
	1	10-0410	COVERPLATE 16MM
3	4	5126-16	SHCS-M4X0.7-16
4	4	5126-20	SHCS-M4X0.7-20
5	4	5226-1	WASHER M4

Assembly Drawings contain hundreds of components. Utilize Large Assembly Mode, Drawing settings to optimize performance.

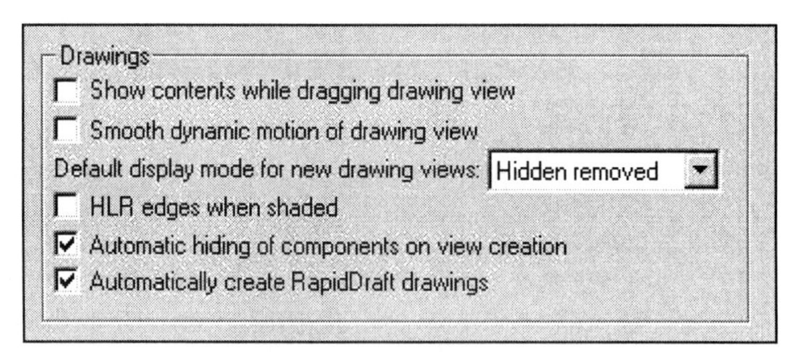

Work in the Shaded display mode for large assemblies to conserve drawing time.

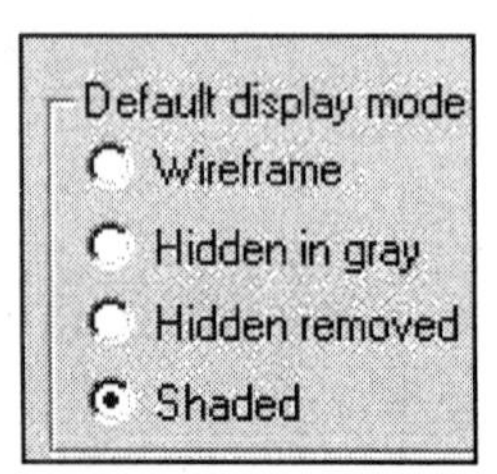

Create RapidDraft™ drawings. RapidDraft™ drawings are designed to open and work in drawing files without the model files being loaded into memory.

Refer to On-Line Help for more information on RapidDraft™ and Large Assembly Mode.

Project 5

Applied Geometric Tolerancing and Other Drawing Symbols

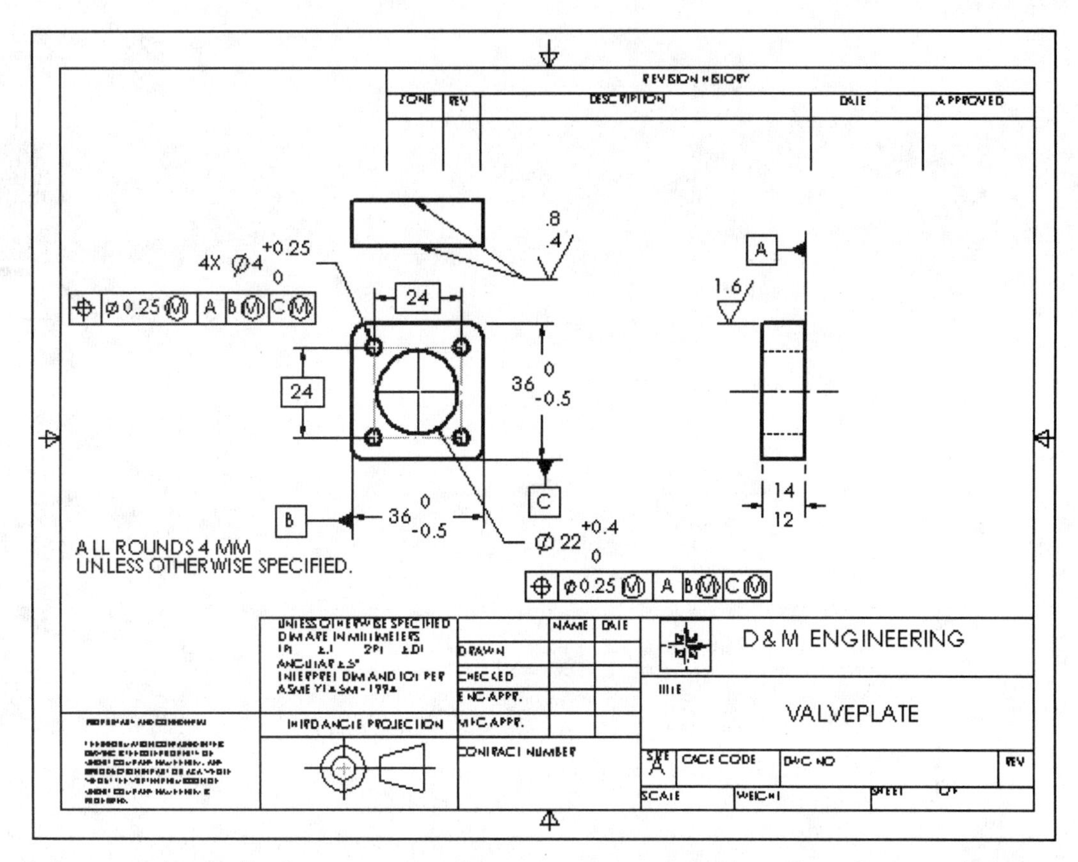

Below are the desired outcomes and usage competencies based on the completion of this Project. A SolidWorks eDrawing is a compressed document that does not require the corresponding part or assembly. SolidWorks eDrawings are animated to display multiple views and dimensions.

Project Desired Outcomes:	Usage Competencies:
VALVEPLATE drawing. VALVEPLATE eDrawing.	Ability to apply Datum Feature Symbols, Geometric Tolerance Symbols, Surface Finish Symbols and Weld Symbols.
PLATE-TUBE drawing. PLATE-CATALOG drawing.	Ability to modify the Design Table in EXCEL.
	Knowledge to Create, Save and Insert Blocks in a drawing.

Notes

Project 5 – Applied Geometric Tolerancing

Project Objective

Create three drawings and a SolidWorks eDrawing:

- VALVEPLATE drawing.
- PLATE-TUBE drawing.
- PLATE-CATALOG drawing.
- VALVEPLATE eDrawing.

The VALVEPLATE drawing consists of three standard views. Insert the dimensions. Modify the dimensions to contain Basic, Bilateral and Limit Tolerance. Create a Parametric Note.

Create a Weld Bead Assembly feature. Create a PLATE-TUBE drawing. Create Weld Symbols.

The PLATE-CATALOG drawing is used for the on-line catalog. The PLATE-CATALOG drawing utilizes a Design Table. Format the Design Table using in EXCEL.

Insert dimensions from the PLATE-CATALOG Default part. Modify the PLATE-CATALOG drawing to contain symbolic representation of the dimensions.

The VALVEPLATE eDrawing is a compressed document that does not require the corresponding part or assembly.

On the completion of this project, you will be able to:

- Modify dimensions to contain Basic, Bilateral, Limit Tolerance.
- Insert Datum Features, Geometric Tolerance and Surface Finish Symbols.
- Insert a Weld Bead assembly feature. Insert Weld Symbols.
- Format a Design Table in EXCEL.
- Make, Save and Insert Blocks.

Project Situation

As a designer you are required to work on multiple projects. Create the following four drawings.

1.) VALVEPLATE drawing:

The VALVEPLATE drawing requires Datum Features and Geometric Tolerances.

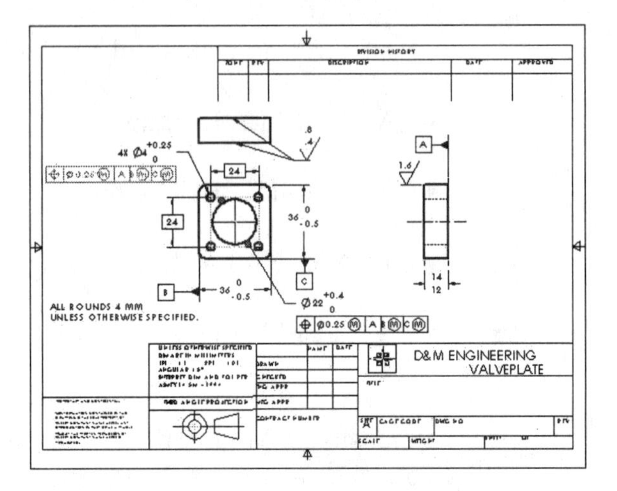

2.) SolidWorks VALVEPLATE eDrawing:

The SolidWorks VALVEPLATE eDrawing is a compressed stand alone drawing.

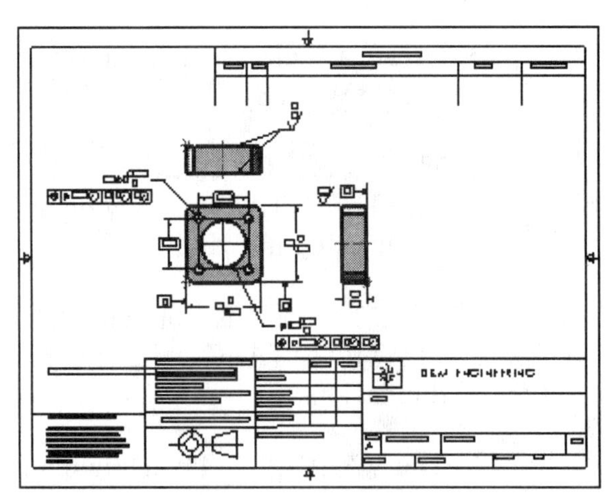

3.) PLATE-TUBE drawing:

The PLATE-TUBE drawing is a conceptual customer drawing. The customer is concerned about the cosmetic appearance of a weld. Create a Weld Bead Assembly feature. Create a PLATE-TUBE drawing. Create Weld Symbols.

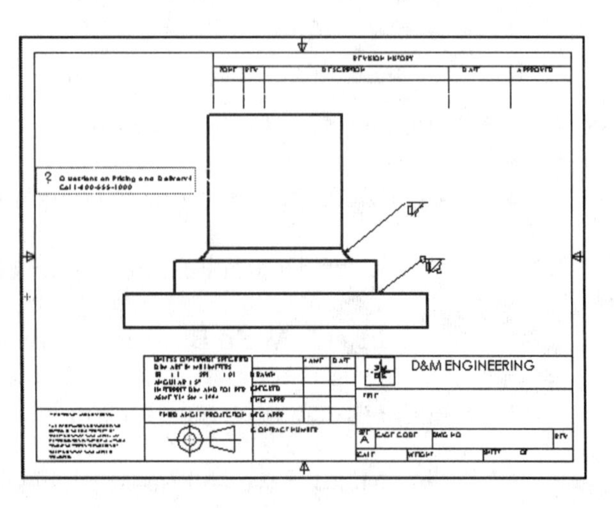

4.) PLATE-CATALOG drawing:

The PLATE-CATALOG drawing will be used for the on-line catalog. The PLATE-CATALOG drawing utilizes a Design Table. The Design Table is formatted in EXCEL.

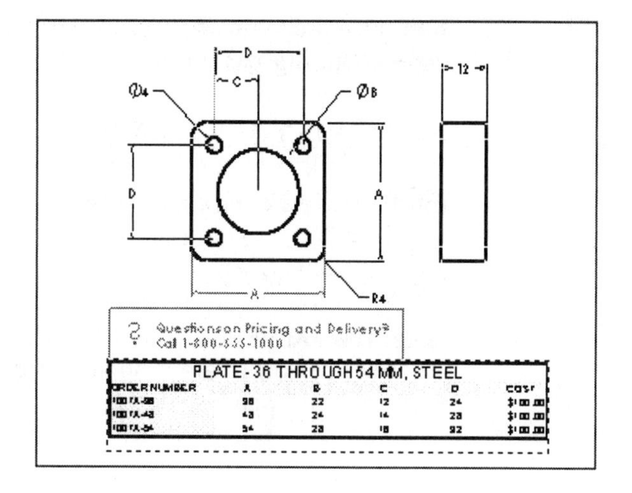

Project Overview

Review the VALVEPLATE part. Create a VALVEPLATE drawing with three standard views. Insert the dimensions. Modify the dimensions to contain Basic, Bilateral and Limit Tolerance. Create a Parametric Note to represent the 4 MM Radius.

Insert Datum Feature Symbols; A, B and C which correspond to the Primary, Secondary and Tertiary Datum Planes respectively.

Insert Geometric Tolerance Symbols. Build a Feature Control Frame. Add Surface Finish Symbols to complete the drawing.

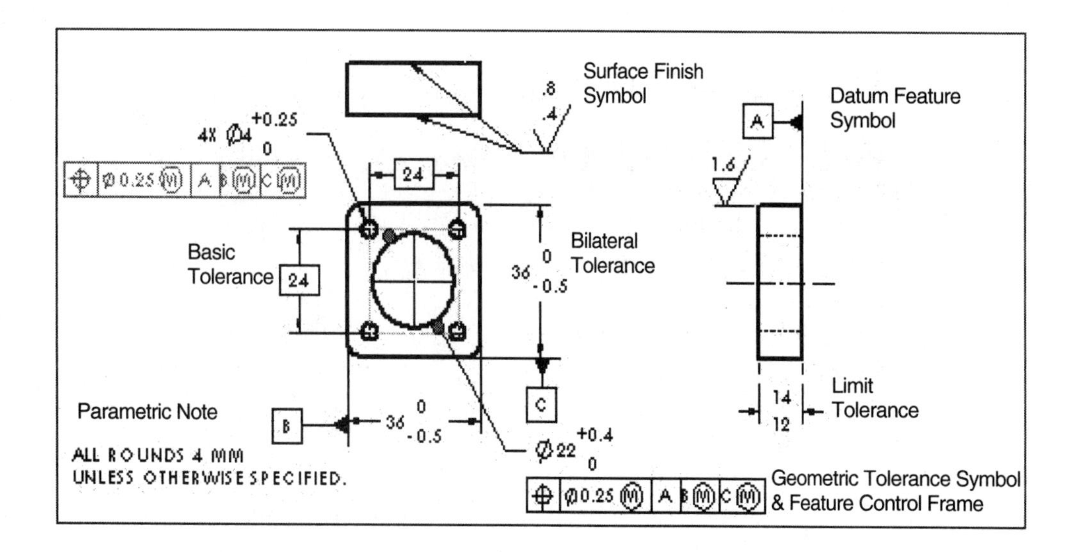

A SolidWorks VALVEPLATE eDrawing is a compressed document that does not require the corresponding part or assembly.

SolidWorks VALVEPLATE eDrawing is animated to display multiple views and dimensions.

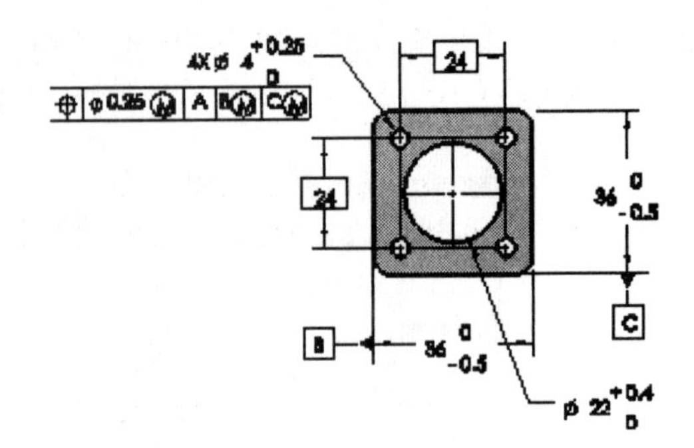

Note: The VALVEPLATE and PLATE-CATALOG drawing are based upon the ASME Y14.5M Position Tolerancing standard at MMC Relative to Datum Feature Center Planes. Used with Permission.

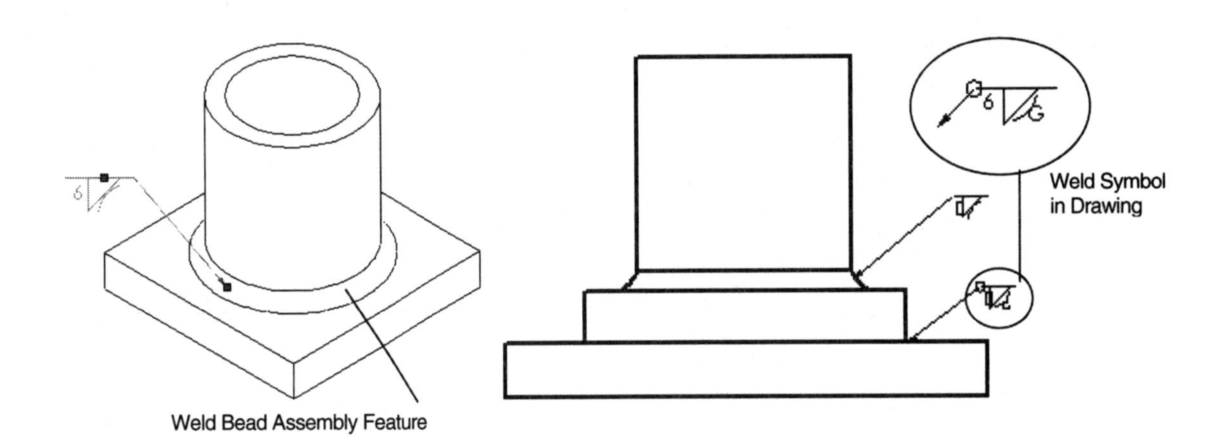

Review the PLATE-TUBE assembly. Insert a Weld Bead Assembly Feature.

Create a new PLATE-TUBE drawing. Define a Fillet Weld. Insert a Weld Symbol.

Review the PLATE-CATALOG part. There are three part configurations. Create a PLATE-CATALOG drawing. Insert Dimensions.

Represent the dimensions as symbols: A, B, C and D. Insert the Design Table. Modify the Design Table in EXCEL.

PLATE - 36 THROUGH 54 MM, STEEL					
ORDER NUMBER	A	B	C	D	COST
1007A-36	36	22	12	24	$100.00
1007A-48	48	24	14	28	$100.00
1007A-54	54	28	16	32	$100.00

Create a new Block in the PLATE-CATALOG drawing.

Insert the Block into the PLATE-TUBE drawing.

Note: The three drawings contained in this section are sample drawings; they are not complete.

Each drawing displays examples of applying various types of symbols in SolidWorks.

Dimension fonts and Note fonts are increased in size for improved visualization.

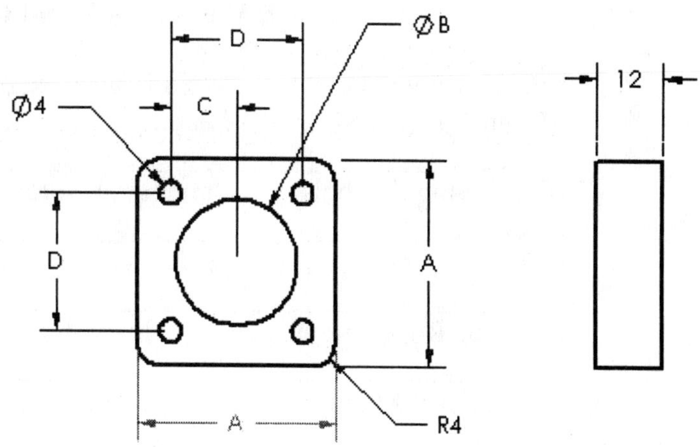

SolidWorks Tools and Commands

The following SolidWorks Annotation tools are utilized in this Project:

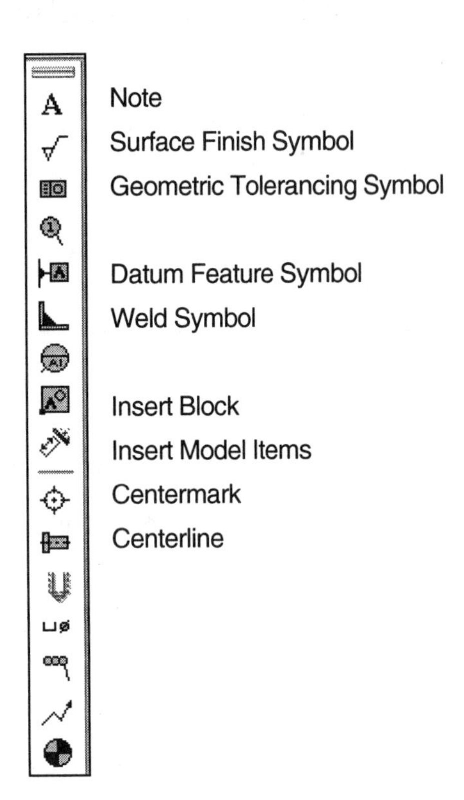

Other SolidWorks tools and commands that are utilized in this Project:

SolidWorks Tools and Commands:		
Basic, Bilateral & Limit Tolerance	Weld Bead Assembly Feature	Feature Control Frame
Modifying Symbols	Parametric Note	eDrawings
EXCEL Commands:		
Hide Row/Column	Insert Row	Merge Cells
Number/Alignment	Font	Border/Patterns

Review the VALVEPLATE Part

A drawing contains part views, geometric dimensioning and tolerances, notes and other related design information. In SolidWorks, when a part is modified, the drawing updates automatically.

When a dimension in the drawing is modified, the part is updated automatically. Perform the following tasks before starting the VALVEPLATE drawing:

- Verify the part.
- View the dimensions in each part.

Verify the part.

1) Open the part. Click **Open** from the Standard toolbar. The Part is located in the 2003drwparts folder.

2) **Browse** for the VALVEPLATE part.

3) View the part. Check the **Preview** check box.

4) Click the **Open** button. The VALVEPLATE part is displayed in the Graphics window.

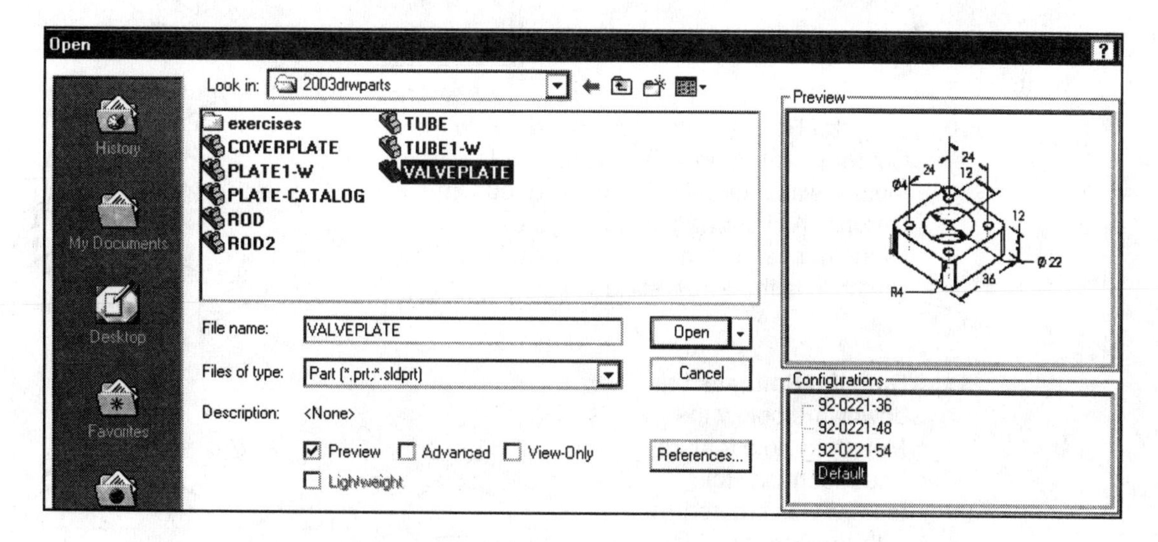

5) Expand the SolidWorks window to full screen. Click the **Maximize** icon in the top right hand corner of the Graphics window.

Display all feature dimensions.

6) Click the **Annotations** folder in the VALVEPLATE FeatureManager.

7) Click **Show Feature Dimensions**.

Review the part features. View the dimensions in each feature.

8) Position the Rollback bar. Place the **mouse pointer** over the yellow Rollback bar at the bottom of the FeatureManager design tree. The mouse pointer displays a symbol of a hand.

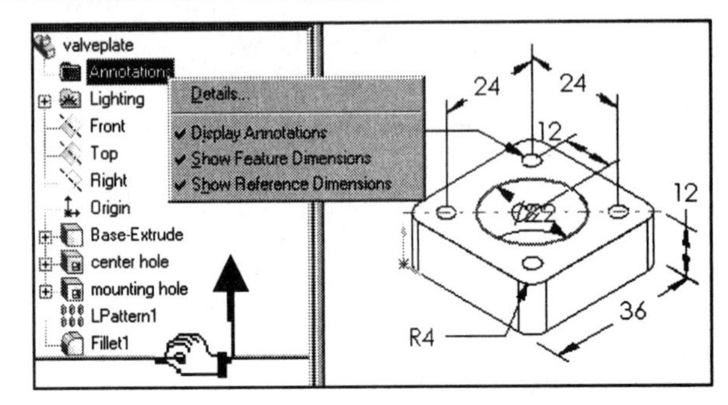

9) Drag the **Rollback** bar upward to below the Base-Extrude feature. The Base-Extrude feature is displayed. The Base-Extrude sketch contains an equal geometric relationship between the front and right side.

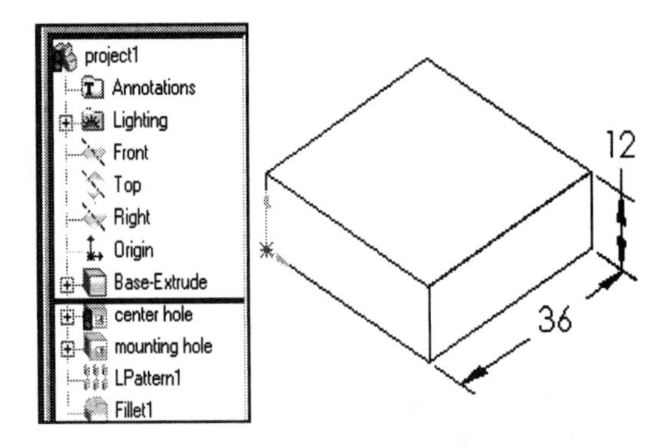

10) Drag the **Rollback** bar downward below the Center Hole feature. The Center Hole feature was created with an Extruded Cut feature. A circle was sketched at the midpoint of a diagonal construction line. The sketch is in the Show state.

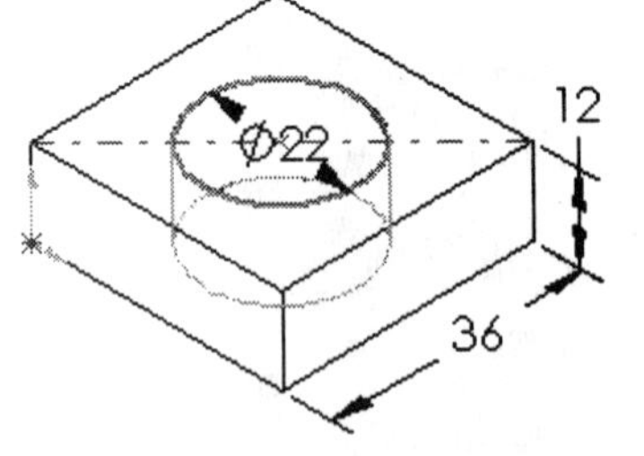

11) Drag the **Rollback** bar downward below the Mounting Hole feature. The Mounting Hole feature was created with an Extruded Cut feature. The center point of the Mounting Hole is coincident with a diagonal construction line. Two dimensions locate the hole. The sketch is fully defined with a geometric relation.

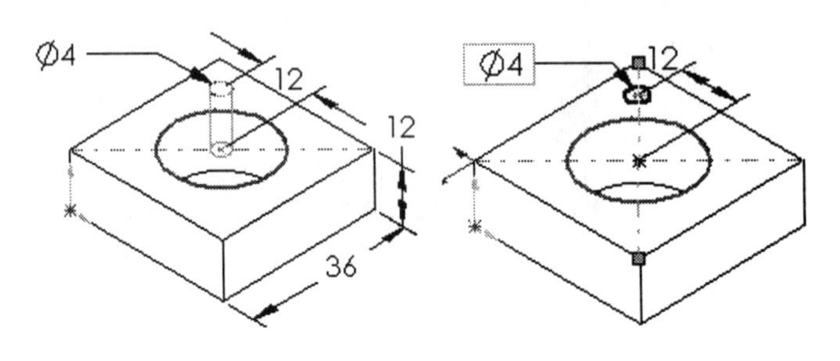

12) Right-click on the **Mounting Hole**.

13) Click **Edit Sketch**.

14) Click **Display/Delete Relations** from the Sketch Relations toolbar. View the Coincident relation on a diagonal centerline.

15) Close the Sketch. Click **Sketch**.

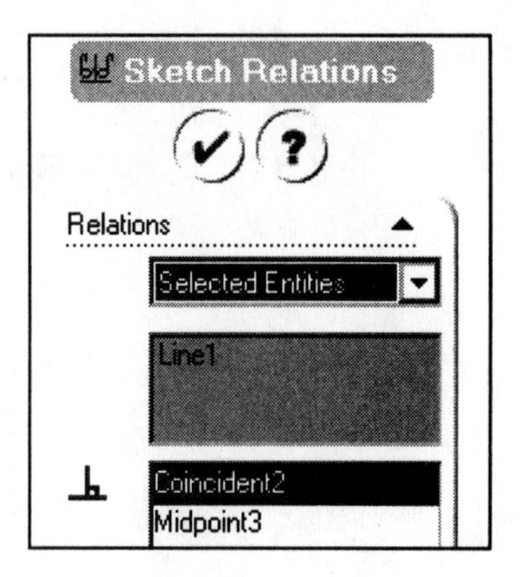

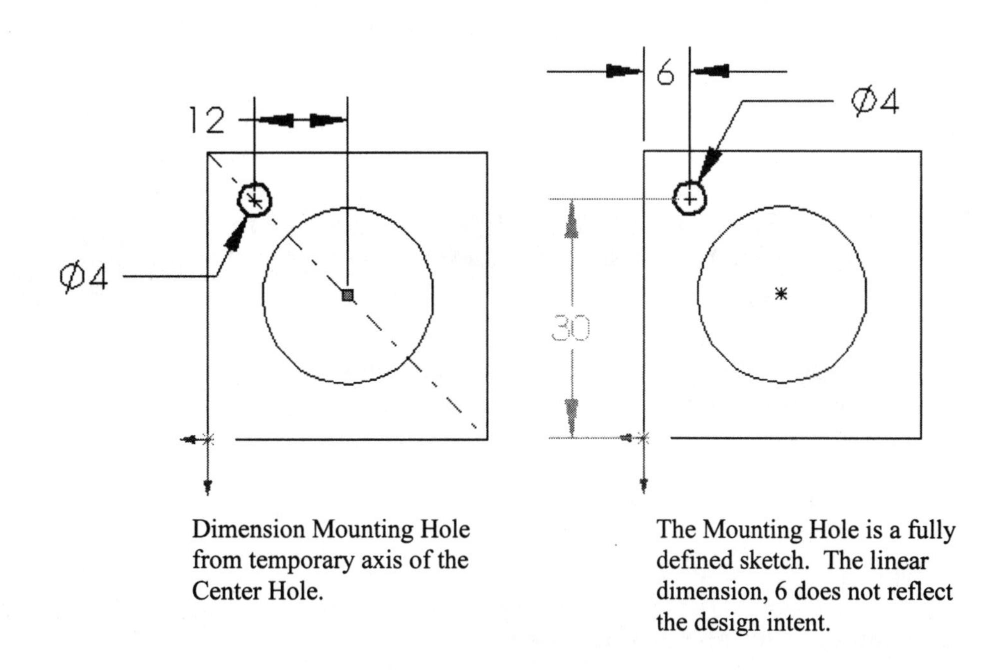

Dimension Mounting Hole from temporary axis of the Center Hole.

The Mounting Hole is a fully defined sketch. The linear dimension, 6 does not reflect the design intent.

The location of the Mounting Holes with respect to the Center Hole is critical. Do not dimension the Mounting Holes from the part edges.

Select View, Temporary Axis to display the axis for the Center Hole feature. Dimension the Mounting Hole from the Center Hole axis.

16) View the Linear Pattern. Drag the **Rollback** bar downward below the Lpattern feature. View the four edges.

17) Drag the **Rollback** bar downward to below the Fillet feature.

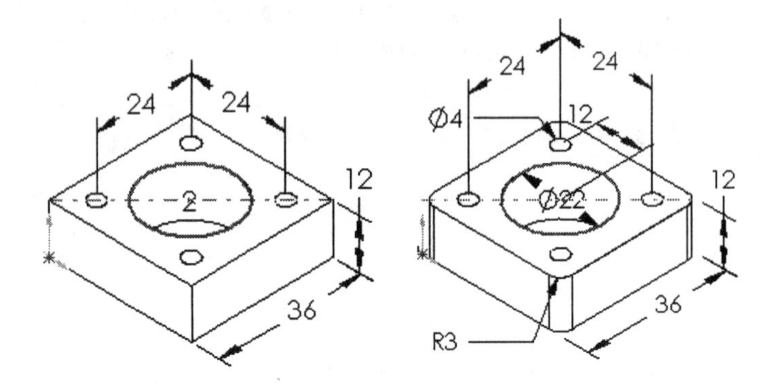

The Precision of the VALVEPLATE part is set to 0. Set the Precision to 0 in the VALVEPLATE drawing.

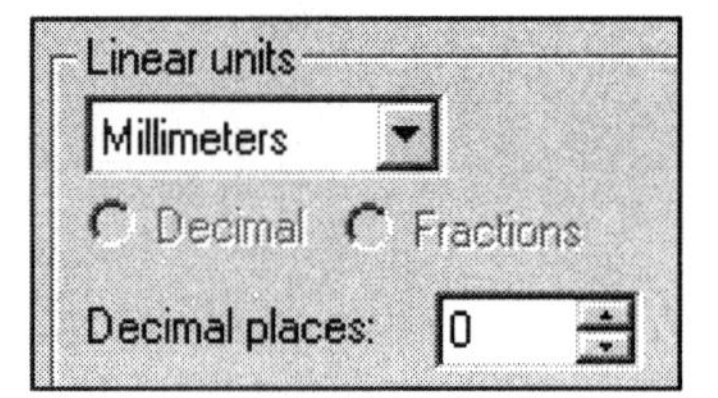

Control the Tolerance for individual dimensions in the drawing.

Create the VALVEPLATE Drawing

The VALVEPLATE drawing requires Geometric Tolerancing.

Create a New VALVEPLATE drawing.

18) Click **File**, **New**.

19) Click the **2003drwparts** file folder.

20) Click the **A-ANSI-MM** drawing template.

21) Save the empty drawing. Click **File**, **Save**.

22) Enter **VALVEPLATE** for Filename.

23) Enter **VALVEPLATE** Drawing for Description.

Set the Precision Display for the VALVEPLATE drawing.

24) Click **Tools**, **Options**, **Document Properties**, **Dimensions**.

25) Click the **Precision** button.

26) Enter **0** for Value.

27) Enter **1** for Tolerance in the Primary Units text box.

28) Click **OK**.

29) Click **OK**.

Dimensions are displayed with no decimal places. The default Tolerance is set to the None option.

Example:
The dimension 36 contains no decimal places.

Select Tolerances.
The default display is 1decimal place.
Example +0.5/-0.1.

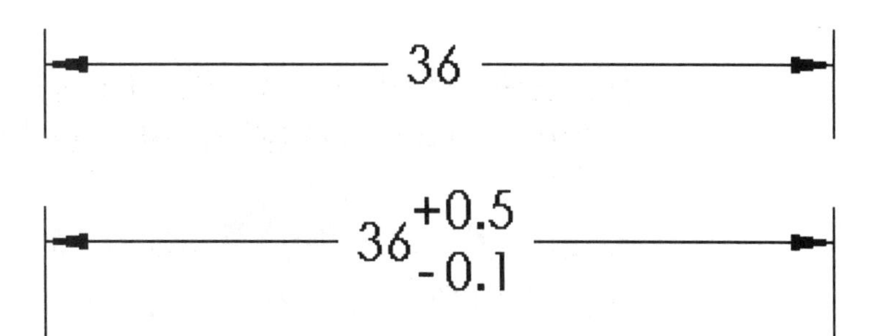

Add a Named view.

30) Click **Named** view from the Drawing toolbar.

31) Click **Window**, **VALVEPLATE**.

32) Click the center of the **VALVEPLATE** part.

33) Click a **position** in the VALVEPLATE-Sheet1 Graphics window.

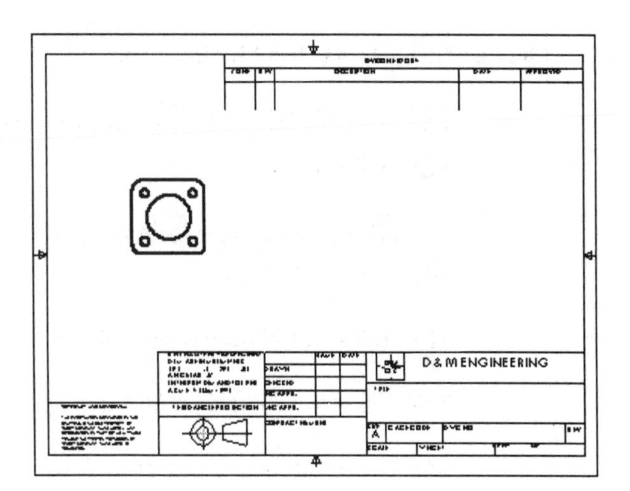

34) Double-click **Top** from the View Orientation text box. The Default Scale is 1:1. The Top Named view is the Front view in the drawing.

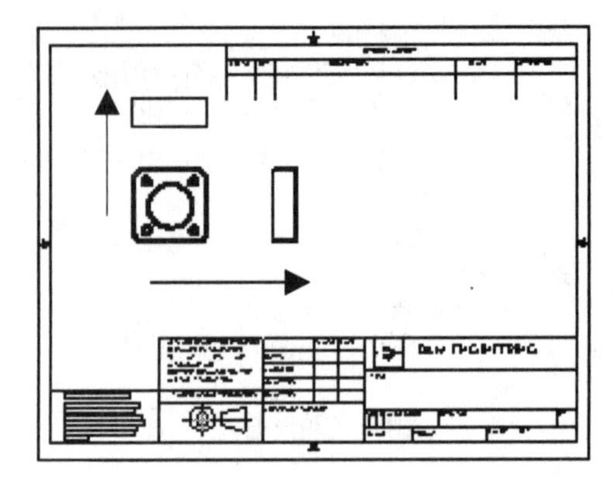

Create the Projected view.

35) Create the Right view as a projected view. Click the **border** of the Named view.

36) Click **Projected View** from the Drawing toolbar.

37) Click the **location** to the right of the Front view. The Right view is displayed.

38) Create the Top view as a projected view. Click the **border** of the Named view.

39) Click **Projected View** from the Drawing toolbar.

40) Click the **location** to the top of the Named view. The Front view, Right view and Top view are displayed.

Insert Model Dimensions into the Top view and Right view.

41) Click a **position** inside the Sheet1 boundary.

42) Click **Insert Model Items** .

43) Click **OK**.

44) Drag the **dimensions** approximately 10mm ways from the profile.

Hide the Dimensions that are not required.

45) Click the **12** dimension text in the Front view.

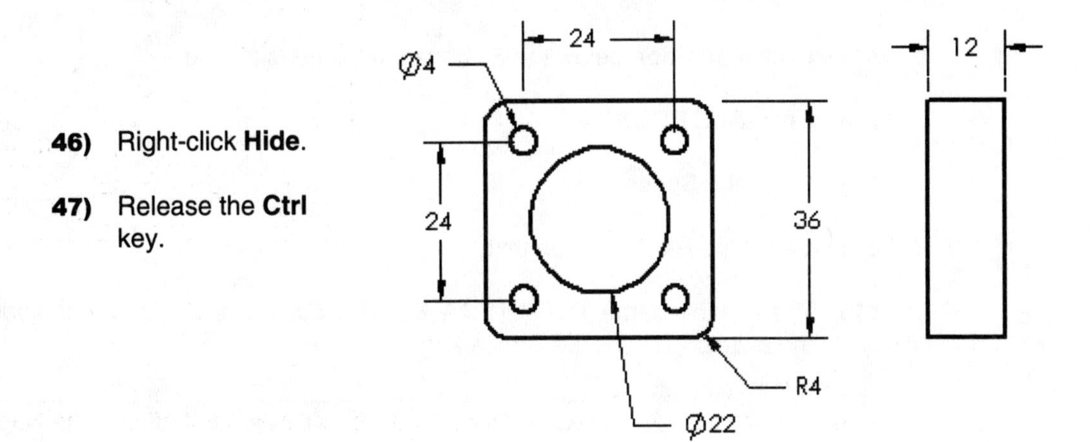

46) Right-click **Hide**.

47) Release the **Ctrl** key.

Add an overall horizontal dimension in the Top view.

48) Click **Dimension** .

49) Click the **left** and **right vertical lines**.

50) Click a **position** below the profile.

51) Drag the **extension lines** off the profile.

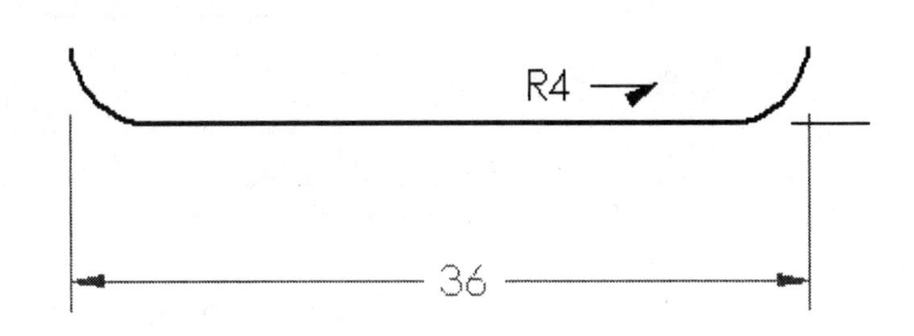

Create Parametric Notes by selecting dimensions in the drawing. Example: Specify the Fillet Radius of the VALVEPLATE part as a note in the drawing. When the Fillet Radius is modified, the corresponding note is modified.

Hide superfluous feature dimensions. Select and Hide the R4 dimension.

Do not delete. Recall hidden dimension with the View, Show Hidden command.

Create a Parametric Note.

52) Click **Note** A from the Annotations toolbar. The Note Properties dialog box is displayed.

53) Click **a position** below the 36 horizontal dimension text.

54) Enter **ALL ROUNDS.**

55) Press the **Space** key.

56) Select the **R4** dimension text.

57) The variable name "D1Fillet1@ VALVEPLATE@ DrawingView1" is added to the text box. Press the **Space** key.

ALL ROUNDS "D1@Fillet1@VALVEPLATE-1@Drawing View1"

58) Enter **MM**.

59) Press the **Space** key.

60) Enter **UNLESS OTHERWISE SPECIFIED**.

61) Click **OK**.

ALL ROUNDS 1 MM UNLESS OTHERWISE SPECIFIED.

Note: Right-click Properties on the Note in the Graphics window. Edit the text in the Note text dialog box.

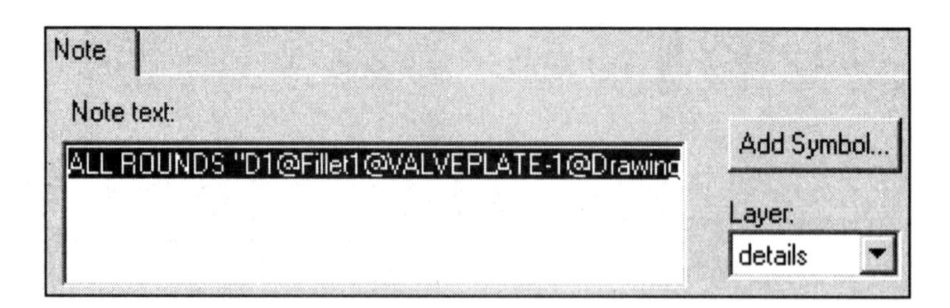

The Parametric Note specifies that the radius of all Rounds is 4MM. Do not double dimension a drawing with a note and a corresponding dimension. Hide the radial dimension.

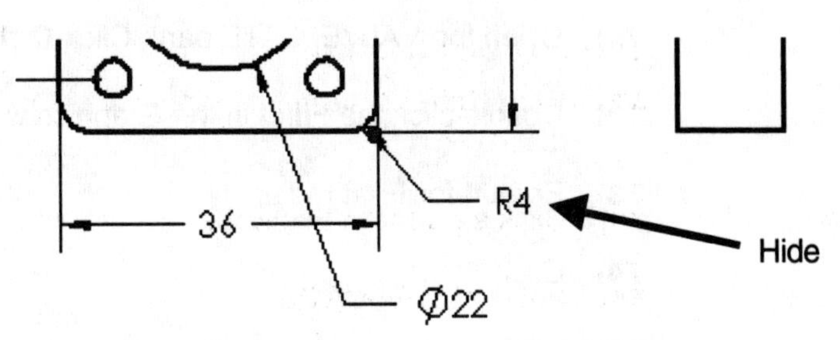

62) Remove the R4 dimension. Right-click **R4**.

63) Click **Hide**.

64) Verify the Parametric Note. Open the VALVEPLATE part. Click the **Front view boundary**.

65) Click **Open VALVEPLATE.sldprt.**

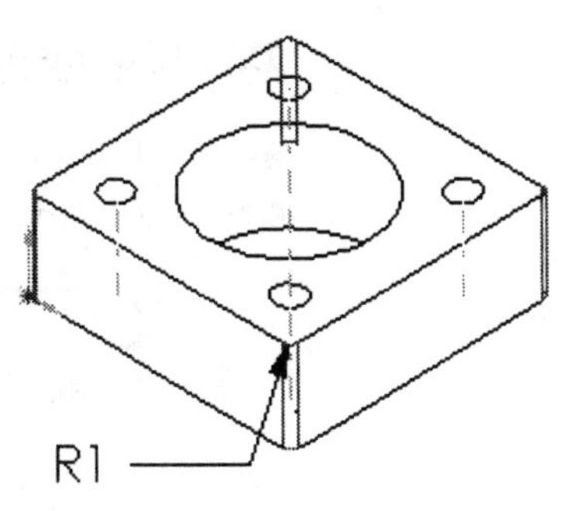

66) Double-click the **Fillet** in the Graphics window.

67) Enter **1** for Fillet radius.

68) Click **Rebuild**. Click **Save**.

69) Open the VALVEPLATE drawing. Right-click the **VALVEPLATE** part icon.

70) Click **Open Drawing**. The Parametric Note displays 1 MM. The updated views reflect the new Fillet radius.

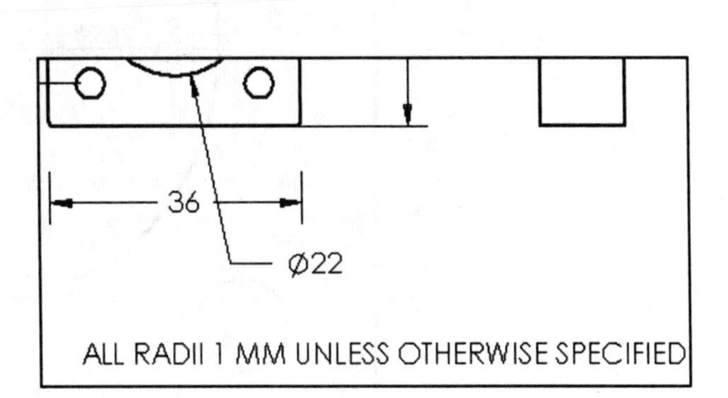

71) **Open** the VALVEPLATE part. Click **Ctrl Tab**.

72) Double-click the **Fillet** in the Graphics window.

73) Enter **4** for Fillet radius.

74) Click **Rebuild**.

75) Click **Save**.

76) Return to the VALVEPLATE drawing. Click **Ctrl Tab**. The Parametric Note displays 4 MM. The updated views reflect the new Fillet radius.

Add Center Marks.

77) Activate the Center Mark. Click **Center Mark** .

78) Click the **large circle**.

79) Click the **top left circle**.

80) Click the blue **Propagate** icon.

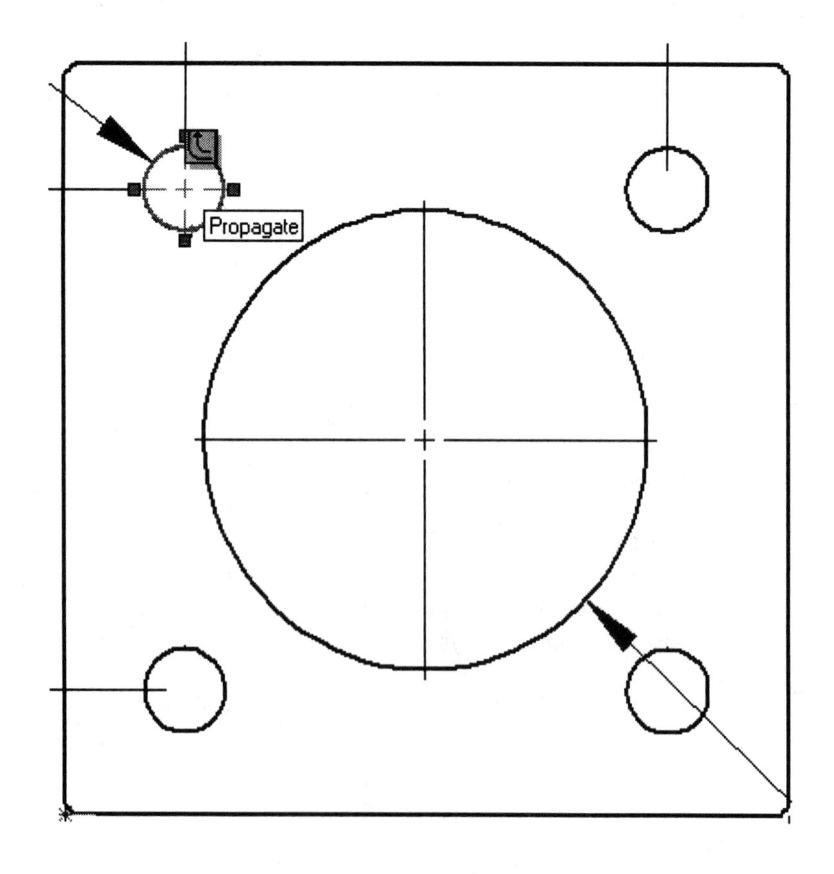

81) Four center marks and centerlines are created. Click **OK**.

Note: Select on the first hole in the Linear Pattern to propagate center marks and centerlines.

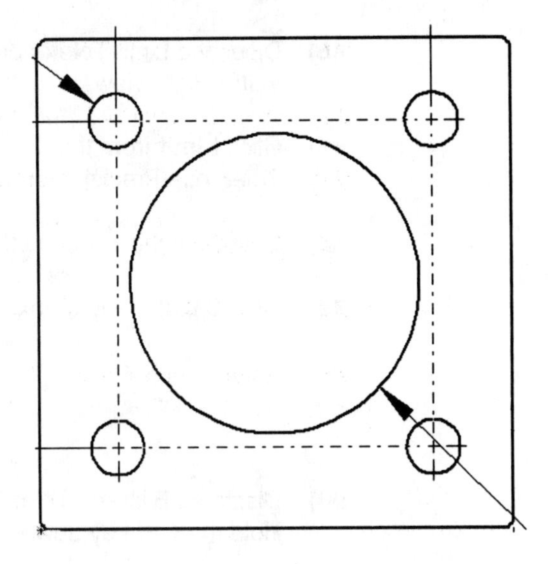

Specify Tolerance.

82) Display a Basic Tolerance. Click **Select**.

83) Hold down the **Ctrl** key. Click the horizontal and vertical **24** dimension text.

84) Release the **Ctrl** key.

85) Click **Basic** from the Tolerance/Precision drop down list.

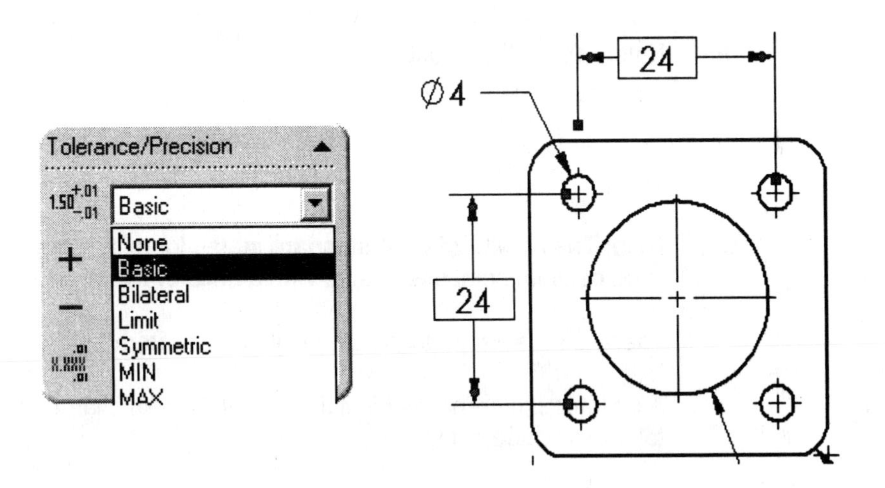

86) Display a Limit Tolerance. Click **12** in the Right view.

87) Click **Limit** from the Tolerance/Precision drop down list.

88) Enter **2** in the + text box.

89) Click **0** in the – text box.

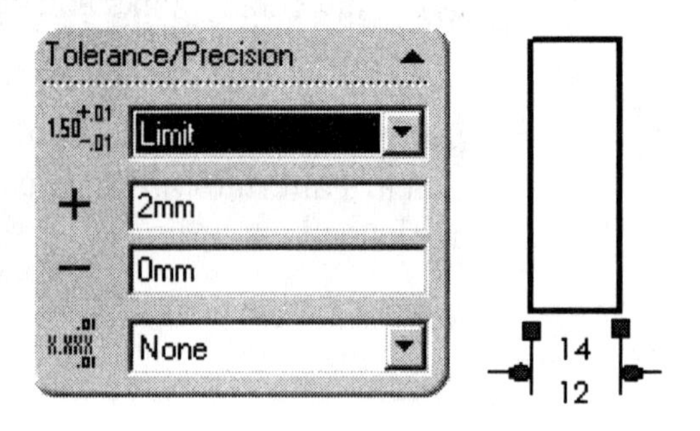

90) Display a Bilateral Tolerance. Hold the **Ctrl** key down.

91) Click the horizontal and vertical **36** dimension text.

92) Release the **Ctrl** key.

93) Click **Bilateral** from the Tolerance/Precision drop down list.

94) Enter **0.0** in the + text box.

95) Enter **–0.5** in the – text box.

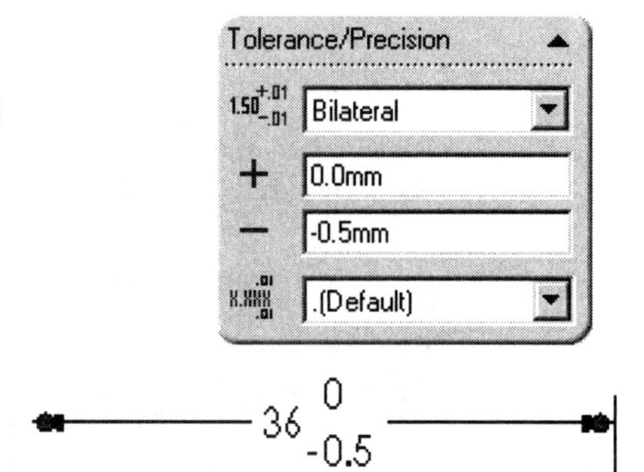

Note: The ASME Y14.5M standard states for millimeter dimensions, that there is no decimal point associated with a unilateral tolerance on a 0 value.

There is no +/- sign associated with the unilateral tolerance on a 0 value.

A unilateral tolerance is similar to a SolidWorks bilateral tolerance; with one tolerance value set to 0.

The other tolerance value contains a +/- sign. Select Bilateral Tolerance in SolidWorks when a unilateral tolerance is required.

Examples of decimal inch unilateral tolerance are provided at the end of this project.

96) Display a Bilateral Tolerance on the diameter dimension text. Click the **∅22** dimension text.

97) Click **Bilateral** in the Tolerance/Precision text box.

98) Enter **0.4** in the + text box.

99) Enter **0** in the – text box.

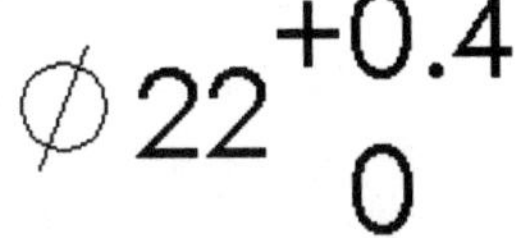

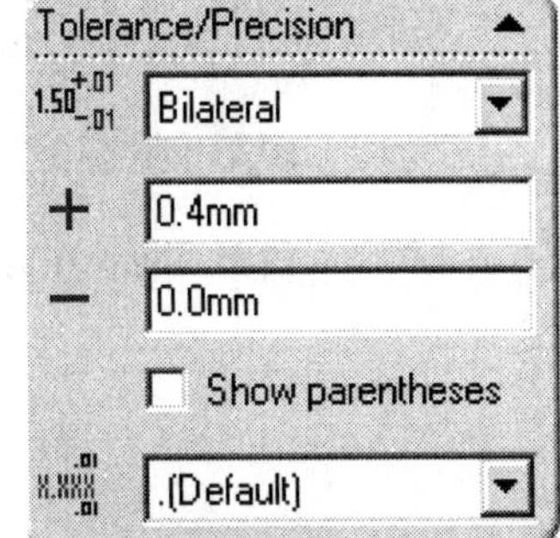

100) Display a Bilateral Tolerance on the diameter dimension text. Click the **∅4** dimension text.

101) Click **Bilateral** in the Tolerance/Precision text box.

102) Enter **0.25** in the + text box.

103) Enter **0** in the – text box.

4X ∅4 $^{+0.25}_{0}$

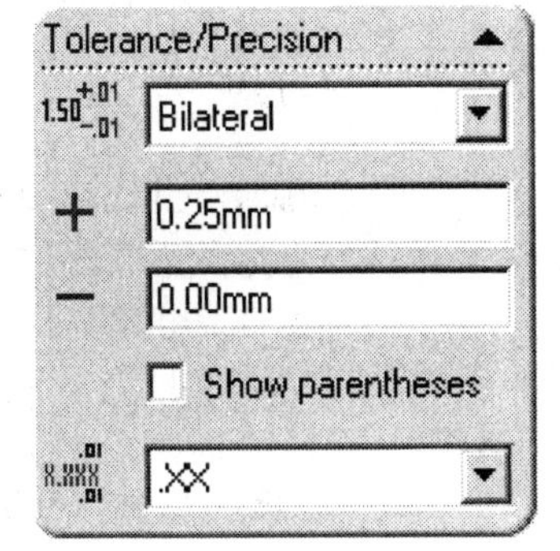

104) Right-click **Properties**. Uncheck **Use document's precision**.

105) Click the **Precision** button.

106) Enter **2** for Tolerance Primary units.

107) Click the **Modify Text** button.

108) Enter **4X** before the **∅4** dimension text.

109) Click **OK**.

110) Click **OK** from the Dimension Properties dialog box.

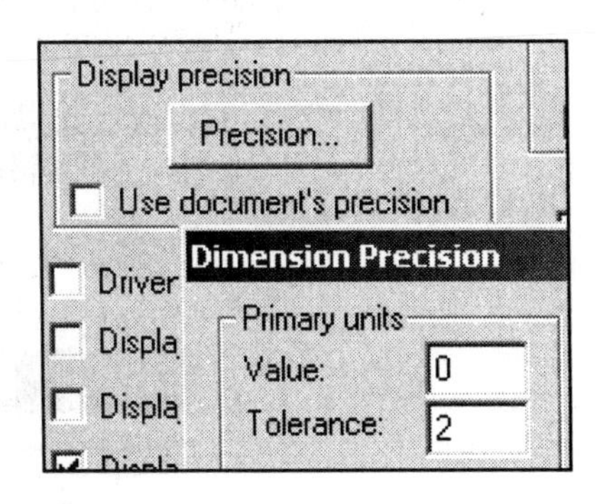

Add Center line and Hide hidden edges.

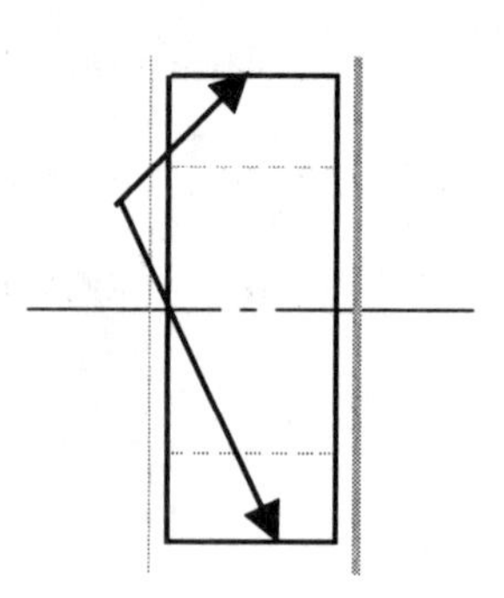

111) Click **Centerline** from the Annotations toolbar.

112) Click the **top** most horizontal edge.

113) Click the **bottom** most horizontal edge.

114) Click **Hidden Lines Visible**.

115) Hide the 4 small hole edges. Hold the **Ctrl** key down.

116) Click the 4 **silhouette edges**.

117) Right-click **Hide**.

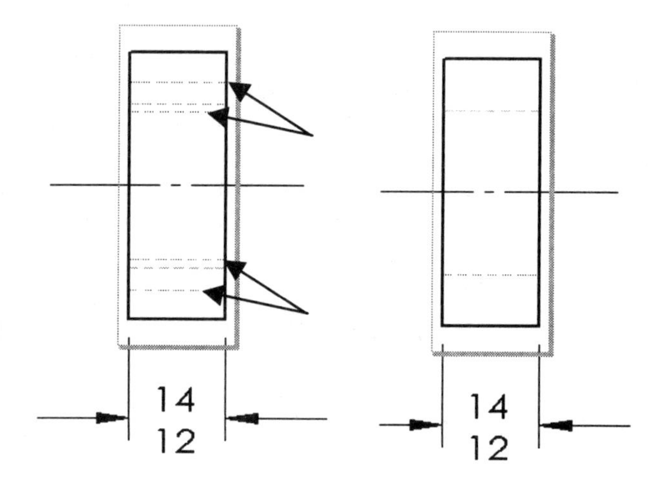

Note: How do you control the specific display of a hidden line? Answer: Display Hidden Lines Visible.

Create a Layer called Hidden. Select a thin dashed line for Style and Thickness. Select the edge and Convert Entities. Select the converted lines. Place the converted lines on a Hidden Layer. Display Hidden Lines Removed.

Create a new Layer for the Datum Features and Geometric Tolerance.

118) Click **Layers**. Click **New**.

119) Enter **GTOL** for Layer Name.

120) Enter **DATUMS& GEOTOLERANCE** for Description.

Add Datum Feature Symbols.

121) Create the Datum Feature for the Datum A Primary Reference Plane.

122) Click the **right vertical edge** in the Right view.

123) Click **Datum Feature Symbol** from the Annotations Toolbar. The Label is A.

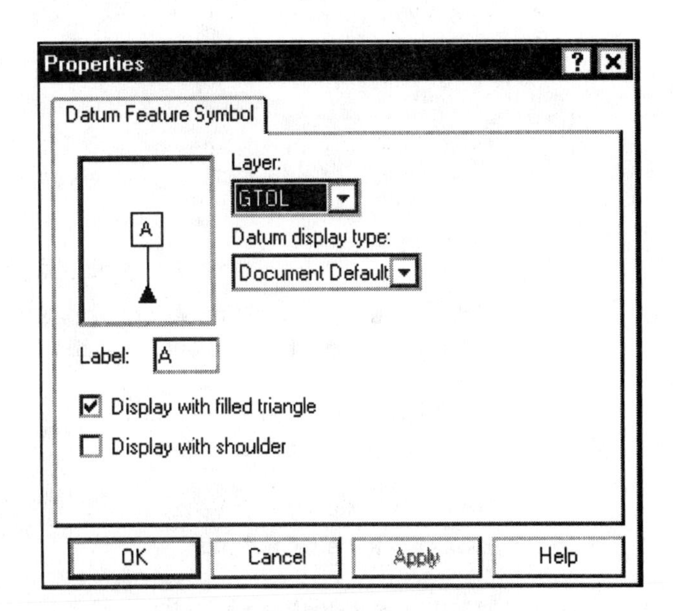

124) Drag the **Datum Feature Symbol A** above the top profile line in the Right view.

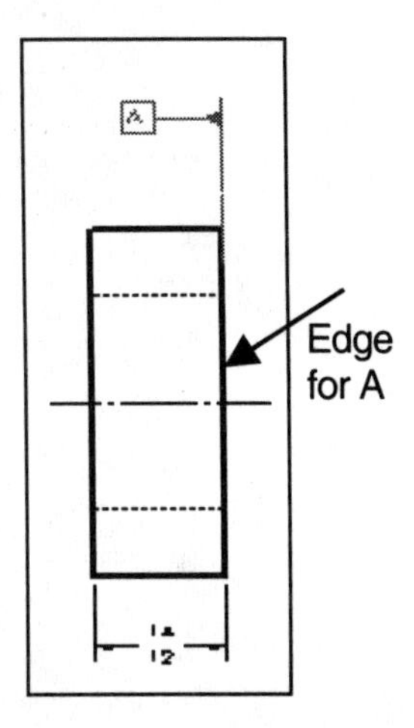

125) Create the Datum Feature for the Datum B Secondary Reference Plane. Click the **left vertical edge** in the Front view. The Label is B.

126) Drag the **Datum Feature Symbol B** below the bottom profile line in the Front view.

127) Create the Datum Feature for the Datum C Tertiary Reference Plane. Click the **bottom horizontal edge** of the Front view. The Label is C.

128) Drag the **Datum Feature Symbol C** to the right of the profile line in the Front view.

129) Click **OK**.

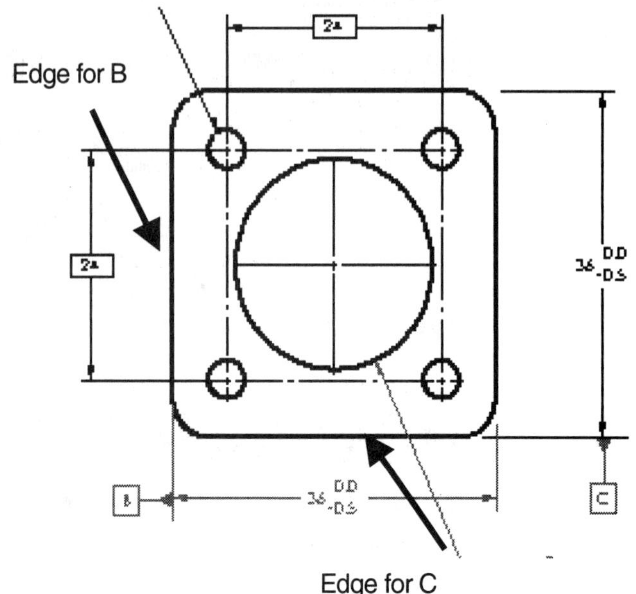

Add a Feature Control Frame for the ∅22 Hole Dimension.

130) Click a **position** below the ∅22 hole dimension text.

131) Click **Geometric Tolerance** from the Annotations toolbar.

132) Click the **GCS** button.

133) Select **Position** from the Symbol Library Geometric Tolerancing list box.

134) Click **OK**. The Feature Control Frame displays the Position symbol in the Preview box.

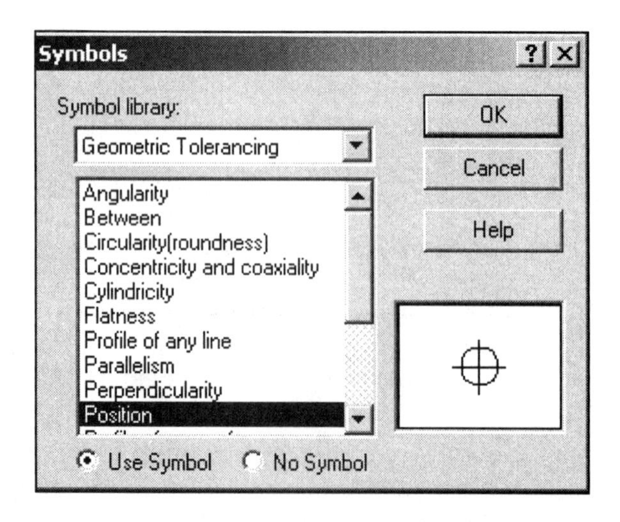

135) Click the **Diameter** ∅ button.

136) Enter **0.25** for the Tolerance1 box.

137) Click the **MC** button. The Maximum Material Condition Modifying Symbol is displayed.

138) Click **OK**. The Feature Control Frame displays ⌖ | ∅ 0.25 Ⓜ in the Preview box.

139) Enter **A** in the Primary box.

140) Enter **B** in the Secondary box.

141) Click the **MC** button. The Maximum Material Condition Modifying Symbol is displayed.

142) Click **OK**.

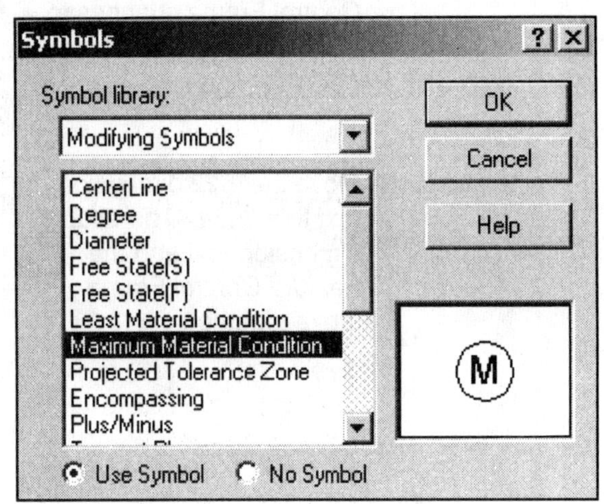

143) Enter **C** in the Tertiary box.

144) Click the **MC** button. The Maximum Material Condition Modifying Symbol is displayed.

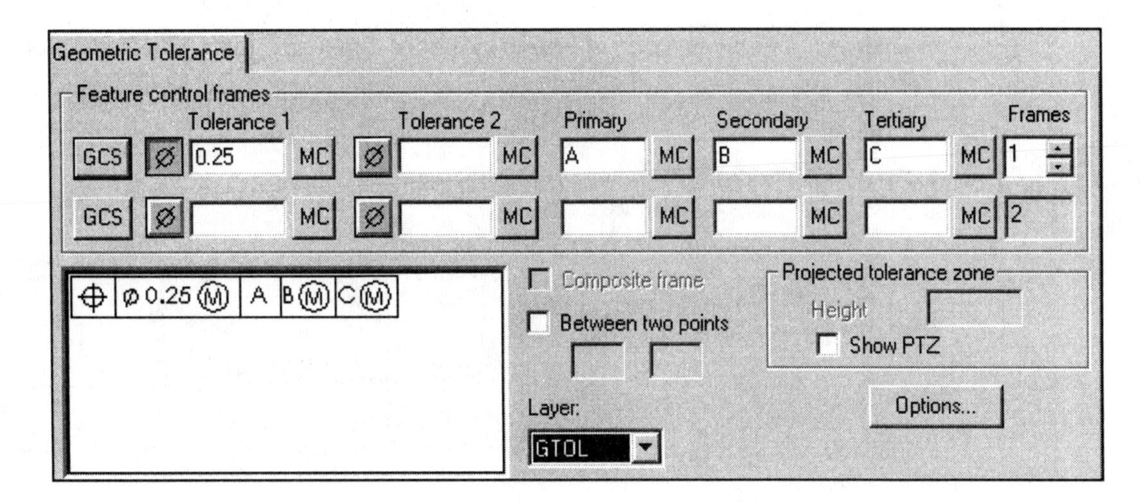

145) Click **OK**. The Feature Control Frame is complete.

146) Click **OK**.

147) Drag the **Feature Control Frame** to the Ø22 dimension text. The Dimension icon is displayed. The Feature Control Frame attaches to the Ø22 dimension text.

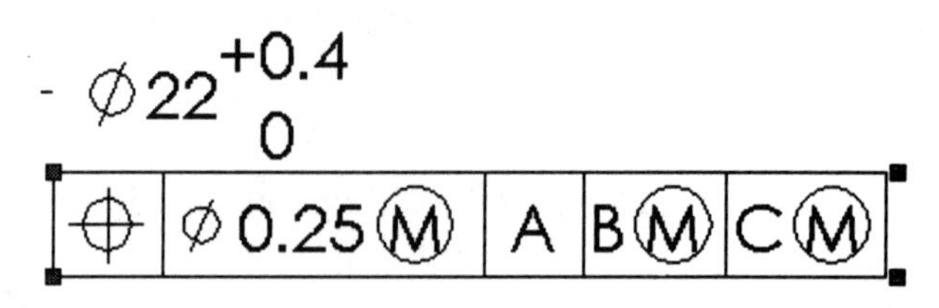

148) Move the Ø**22** dimension text to the left. The Ø22 dimension text and the Feature Control Frame move together.

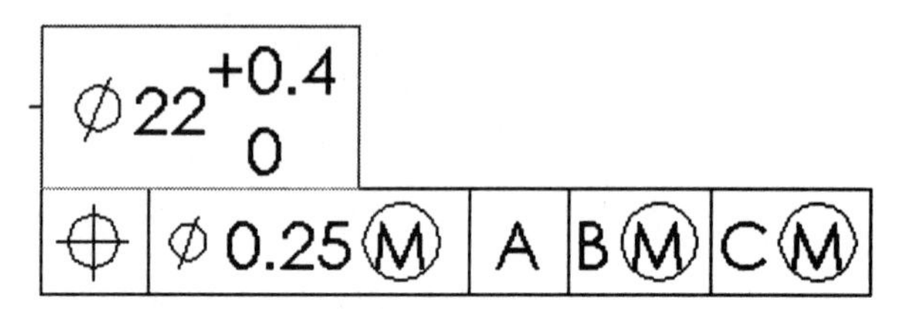

Copy the Feature Control Frame for the Ø4 Hole Dimension.

149) Click the **Feature Control Frame** for the Ø22 dimension text. The mouse pointer displays the icon.

150) Copy the Feature Control Frame. Press **Ctrl C**. Click a **position** below the Ø4 dimension text. Paste the Feature Control Frame. Press **Ctrl V**.

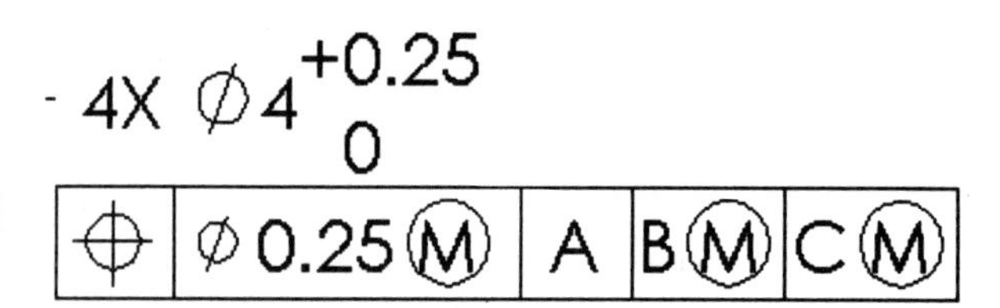

151) Drag the **Feature Control Frame** to the Ø4 dimension text. The Dimension icon is displayed. The Feature Control Frame attaches to the Ø4 dimension text.

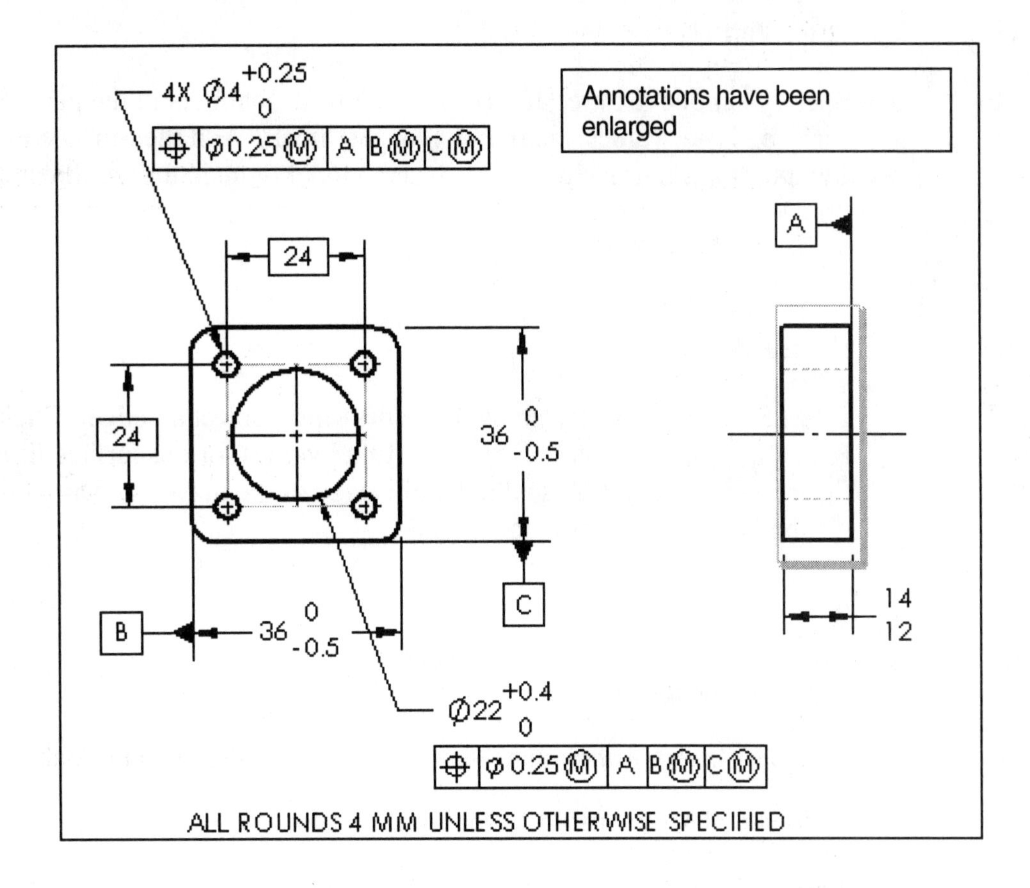

152) Save the VALVEPLATE drawing. Click **Save**.

153) Maximize the VALVEPLATE part. Click **Windows, VALVEPLATE**. The Tolerances added in the drawing are displayed in the part.

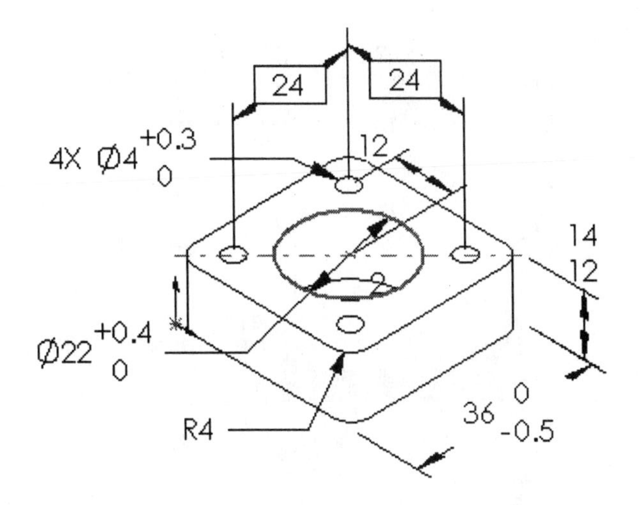

154) Hide the Annotations. Click the **Annotations** folder from the FeatureManager.

155) Uncheck **Show Feature Dimensions**.

In the Part:

Specify surface texture. Insert Surface Finish Symbols in the part. Select a surface. Enter values and options for the finish. Add the Surface Finish Symbol to the surface of a part. Insert Model Items, Surface Finish into the drawing.

In the Drawing:

Insert Surface Finish Symbols in the drawing. Select an edge. Click Surface Finish from the Annotation toolbar. Enter values for the surface finish. The Surface Finish Symbol is added to the edge of the drawing. Drag the symbol off the profile of the drawing.

Insert Surface Finish Symbol into the VALVEPLATE part.

156) Click the **top surface** of the VALVEPLATE part.

157) Click **Insert**, **Annotations**, **Surface Finish Symbol** from the Main menu.

158) Enter **0.8** micrometers for Maximum Roughness.

159) Enter **0.4** micrometers for Minimum Roughness.

160) Click **OK**. The Surface Finish Symbol is displayed.

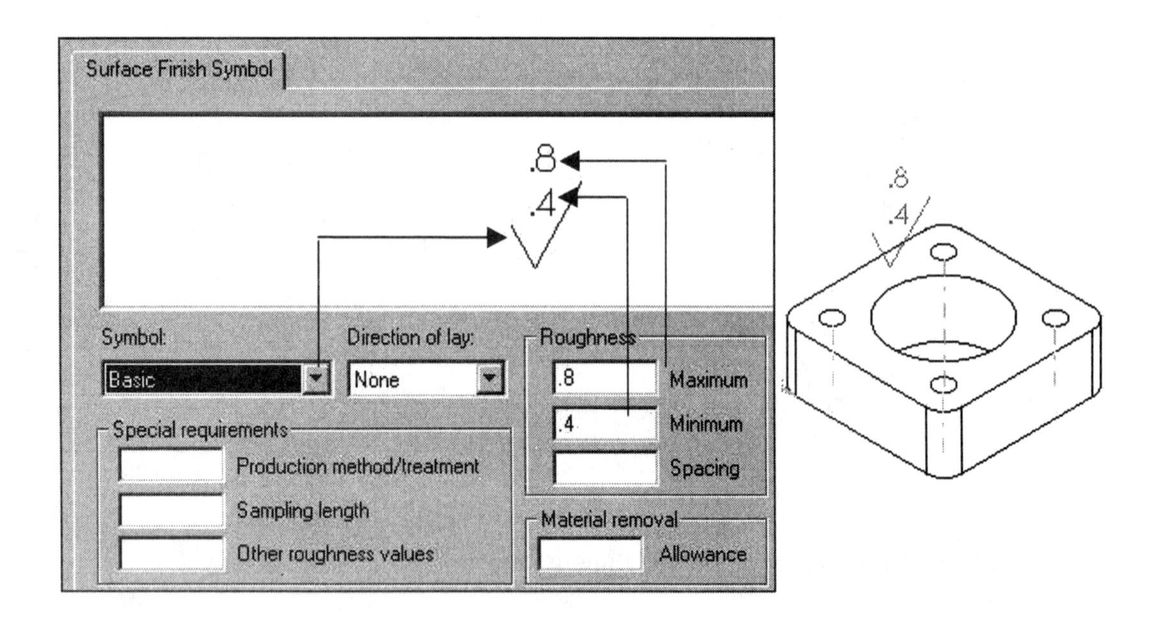

Insert the Surface Finish Symbol into the VALVEPLATE drawing.

161) Click **Windows**, **VALVEPLATE-Sheet1**.

162) Click **Insert Model Items**.

163) Check the **Surface finish** check box.

164) Uncheck the **Dimensions** check box.

165) Click **OK**. The Surface Finish Symbol is displayed in the Front view.

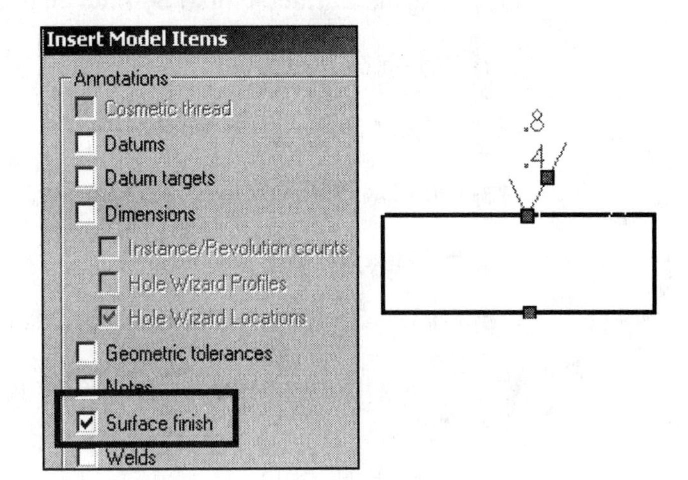

Surface Finish Symbols are displayed on an edge of a profile in the drawing.

Use multiple leaders to point to surfaces in tight spaces that use the same finish.

Move the Surface Finish Symbol.

166) Hold the **Shift** key down.

167) Drag the **Surface Finish Symbol** to the Top view.

168) Click a **position** at the midpoint of the top edge.

169) Release the **Shift** key.

Create Multiple Leaders to the Surface Finish Symbol.

170) Drag the **Surface Finish Symbol** off the profile.

171) Right-click **Properties**.

172) Select **Always show leaders**.

173) Click **OK**. The first leader is displayed.

174) Hold the **Ctrl** key down.

175) Click the tip of the **arrowhead**.

176) Display the two leaders. Drag the **arrowhead** to the bottom edge of the Top view.

177) Release the **Ctrl** key.

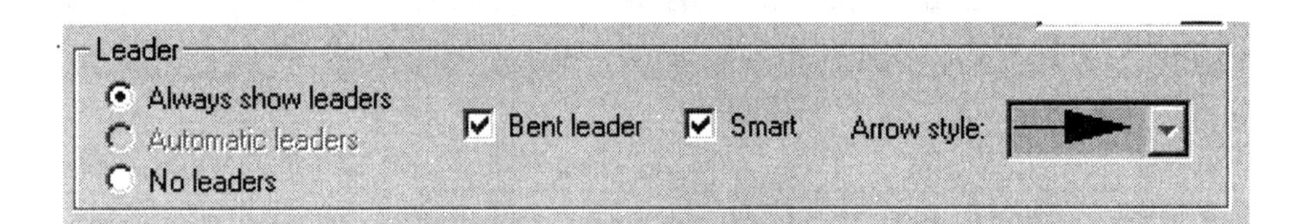

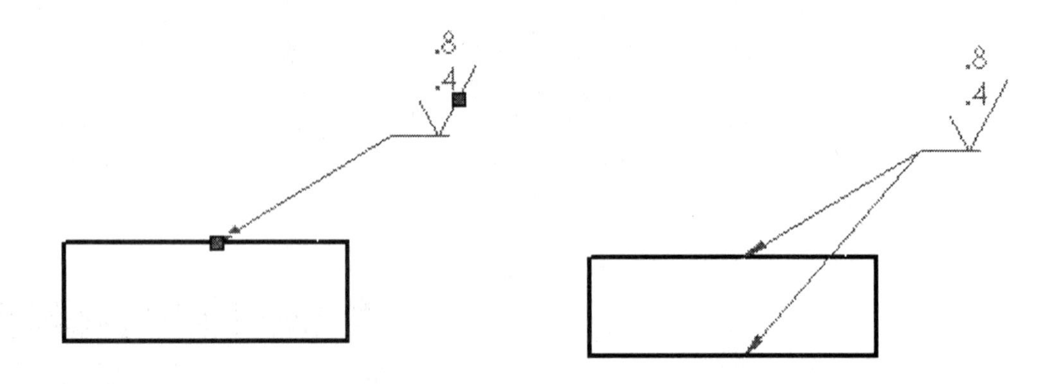

Insert Surface Finish Symbol in the VALVEPLATE drawing.
178) Click the **top edge** in the Right view.

179) Click the **Surface Finish Symbol** from the Annotation toolbar.

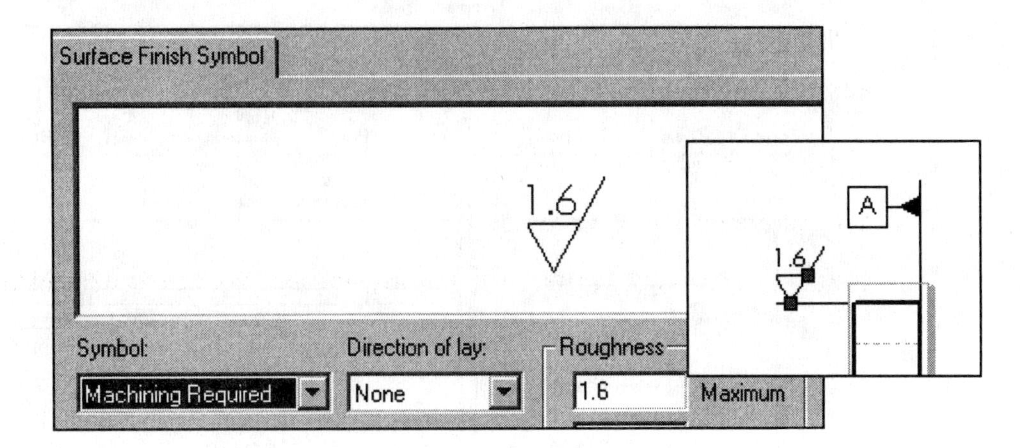

180) Select the **Machining Required** option for Symbol.

181) Enter **1.6** micrometer for Maximum Roughness.

182) Click **OK**. Produce the extension line.

183) Drag the **Surface Finish Symbol** to the left of the profile line.

184) Save the VALVEPLATE drawing. Click **Save**.

An outside machine shop will manufacture the VALVEPLATE. Send a SolidWorks eDrawing of the VALVEPLATE to the machine shop for a price quotation.

The SolidWorks eDrawing software is loaded from the installation CD. The SolidWorks eDrawing software can also be downloaded from the SolidWorks web site. Create a SolidWorks eDrawing.

Add-In a SolidWorks eDrawing.
185) Click **Tools, Add Ins.**

186) Check the **eDrawing2003** box.

187) Click **OK**.

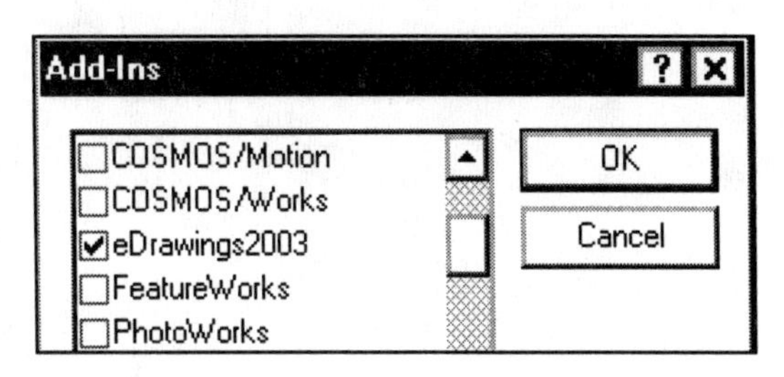

188) Create a SolidWorks eDrawing. Click the **eDrawing** icon from the eDrawing toolbar. Click **Close** in the Welcome to eDrawing dialog box.

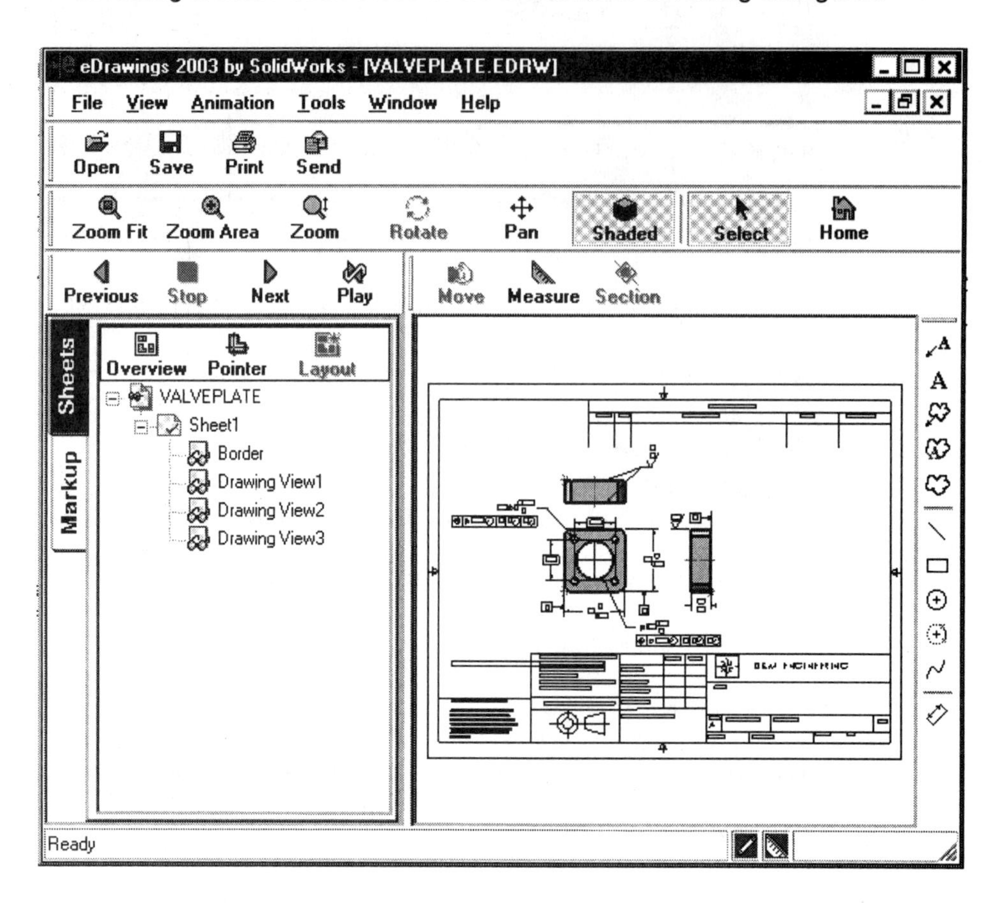

189) Click the **Next** button.

190) Display the remaining views. Click the **Play** button.

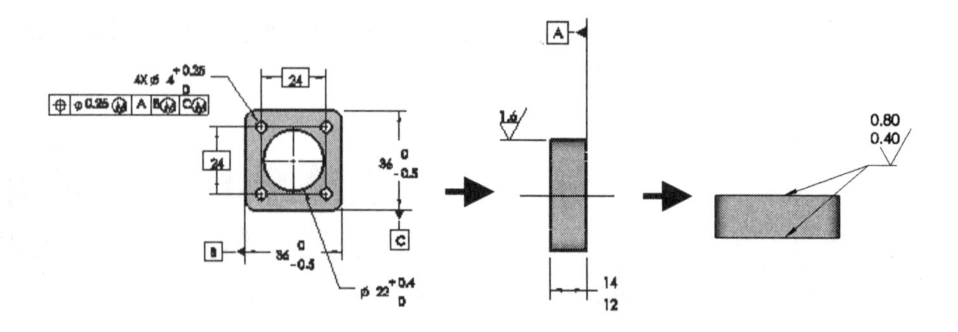

191) Save the eDrawing. Click **Save** from the eDrawing toolbar.

192) Enter **VALVEPLATE**. Click the **Save** button.

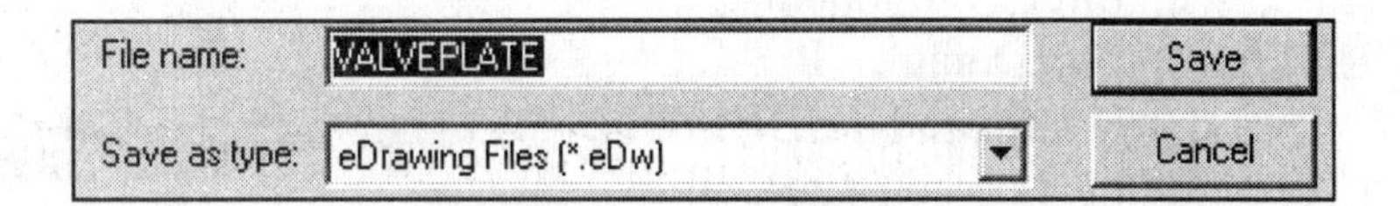

193) Exit the eDrawing module and return to SolidWorks. Click **File**, **Exit**.

194) Close all parts and drawings. Click **Windows**, **Close All**.

Review to the PLATE-TUBE assembly and create the PLATE-TUBE drawing

Review to the PLATE-TUBE assembly in the file folder 2003drwparts. Your customer requires a concept drawing of the PLATE-TUBE assembly. The TUBE part is welded to the PLATE part.

A Weld Bead is created between the parts in an assembly. The Weld Bead is an assembly feature.

A Weld Symbol automatically attaches to the Weld Bead. The Weld Symbol is inserted as a model item into the drawing.

Add a second PLATE component to the PLATE-TUBE assembly.

Create a Weld Symbol as a separate annotation in the PLATE-TUBE drawing.

Create a Weld Bead in the PLATE-TUBE Assembly.

195) Open the PLATE-TUBE Assembly. Click **Open**.

196) Select **Assembly** for Files of type.

197) Double-click **PLATE-TUBE**.

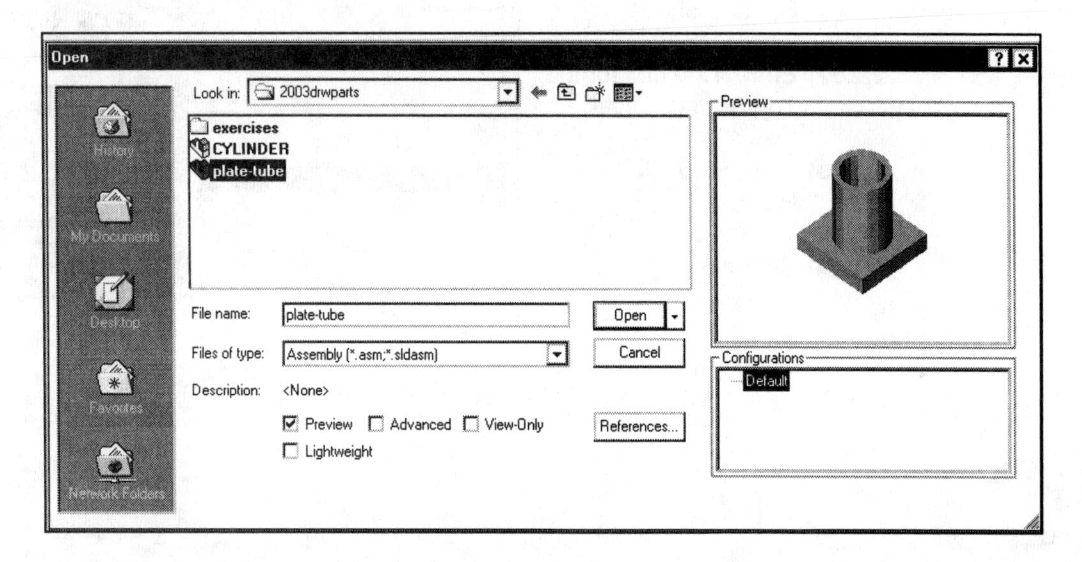

The PLATE-TUBE assembly is created with three components: PLATE1-W<1>(Default), PLATE1-W<2>(Large Plate) and TUBE1-W. PLATE1-W<2>(Large Plate). Note: PLATE1-W<2>(Large Plate) is suppressed.

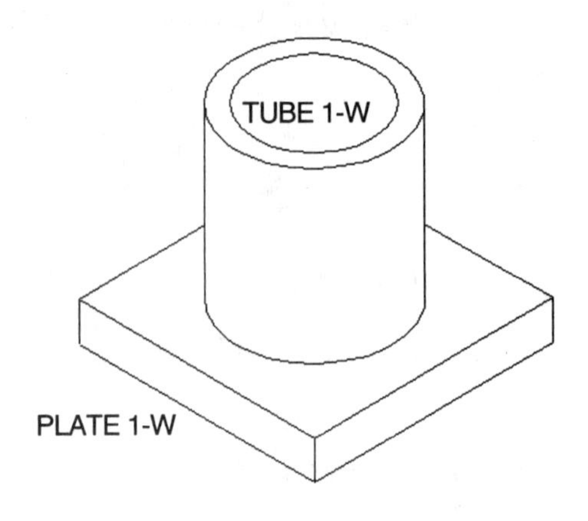

198) Create a Weld Bead between the PLATE1-W<1> and TUBE1-W components. **Insert**, **Assembly Feature**, **Weld Bead**.

199) Select **Fillet** for Weld Type.

200) Click **Next**.

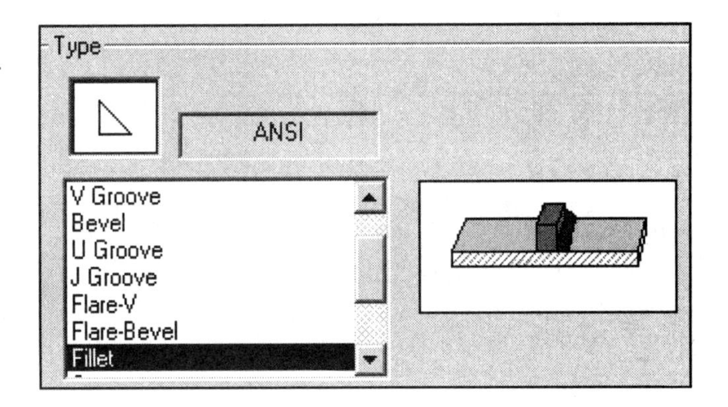

201) Select **Concave** for Surface Shape.

202) **Enter 1.00** MM for the Top Surface Delta.

203) Enter **6.00** MM for Radius.

204) Click **Next**.

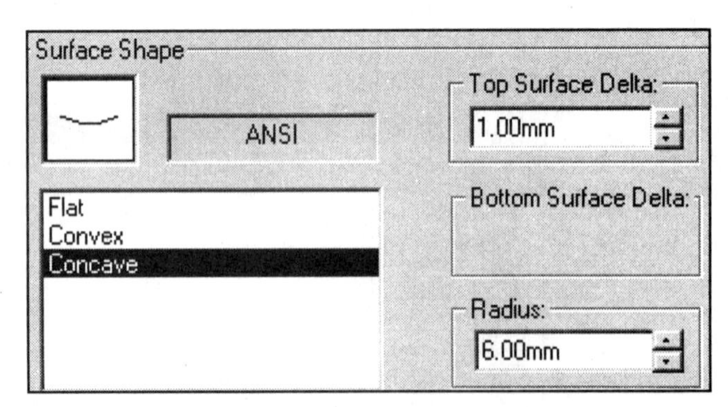

205) Click the **outside cylindrical face** of the TUBE1-W component.

206) Click the **top face** of the PLATE1-W<1> component.

207) Click **Finish**.

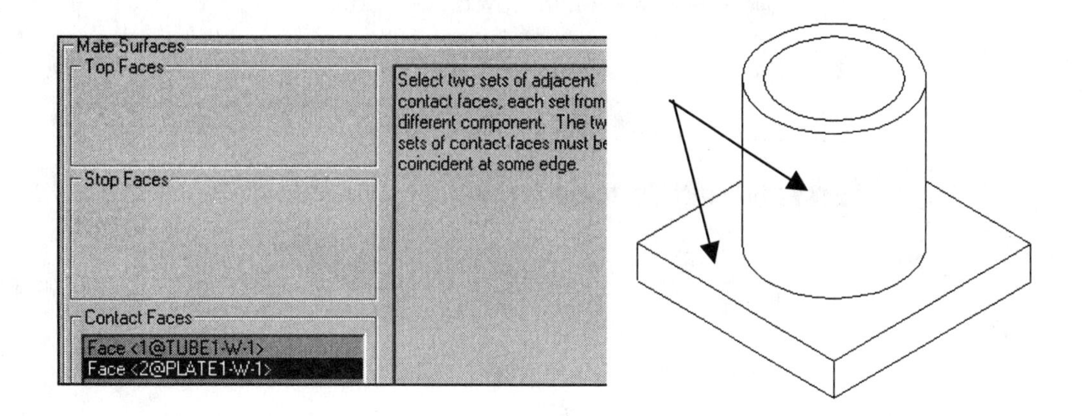

208) The default Weld Bead part name is BEAD1.SLDPRT. Accept the default part name. Click **Next**.

209) The Weld Bead is created between the PLATE1-W<1> and TUBE components. Drag the **Weld Symbol** off the profile.

Note: Right-click Set to Resolve if the Weld Bead is displayed light weight.

210) Save the PLATE-TUBE assembly. Click **Save**.

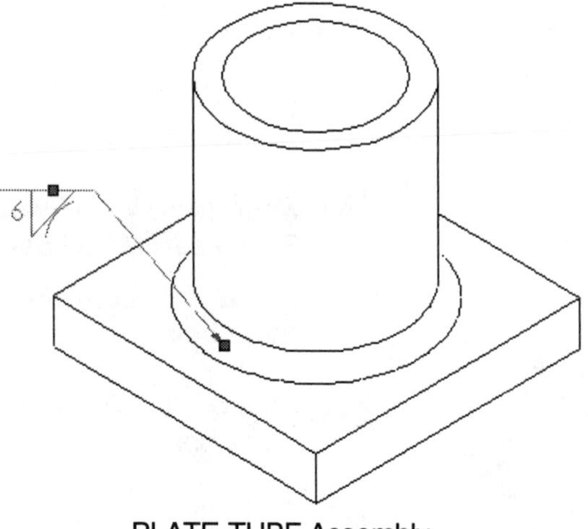

PLATE-TUBE Assembly

Create a New PLATE-TUBE drawing.

211) Click **File**, **New**.

212) Click the **2003drwparts** file folder.

213) Click the **A-ANSI-MM** drawing template.

A-ANSI-MM

214) Save the empty drawing. Click **File**, **Save**.

215) Enter **PLATE-TUBE** for Filename.

216) Enter **PLATE-TUBE DRAWING** for Description.

Add a Named view to the PLATE-TUBE drawing.

217) Click **Named** view from the Drawing toolbar. Select the PLATE-TUBE assembly.

218) Click **Window**, **PLATE-TUBE**.

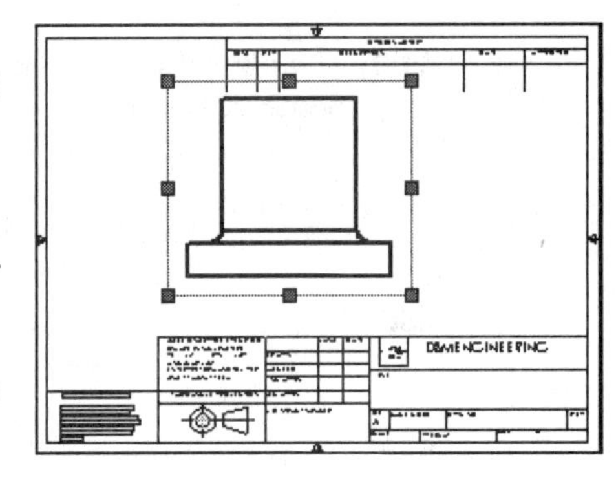

219) Click the **PLATE-TUBE** plate-tube assembly icon.

220) Click a **position** in the PLATE-TUBE-Sheet1 Graphics window.

221) Click **Front** from the View Orientation text box.

222) Click **OK**.

Insert Model Items Weld Symbols.

223) Click **Insert Model Items**. Check the **Welds** check box.

224) **Uncheck** the Dimensions check box.

225) Click **OK**.

226) Drag the **Weld Symbol** off the profile line.

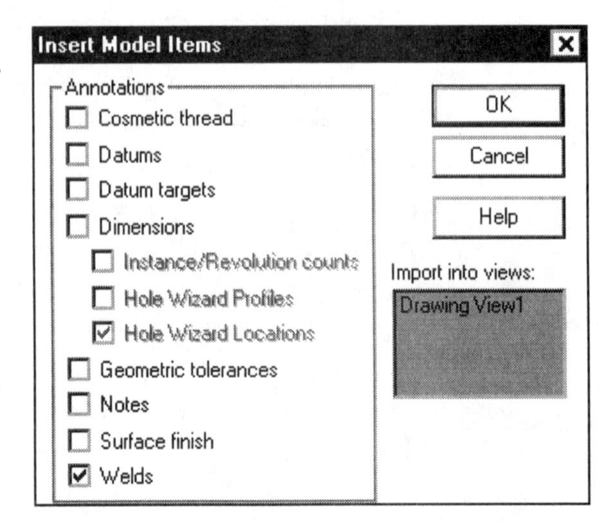

227) The arrowhead of the Weld Symbol is located in the middle of the view. This location is not correct. Select **Undo** two times to remove the symbol.

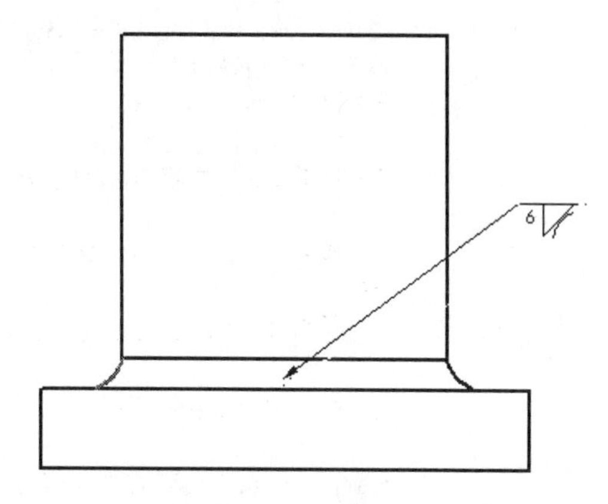

PLATE-TUBE Drawing

228) Insert the Weld Symbol in the drawing. Select the **left concave edge**.

229) Click the **Weld Symbol** from the Annotations toolbar. The ANSI Weld dialog box is displayed.

230) Select **Concave** for Contour.

231) Enter **6** for Radius.

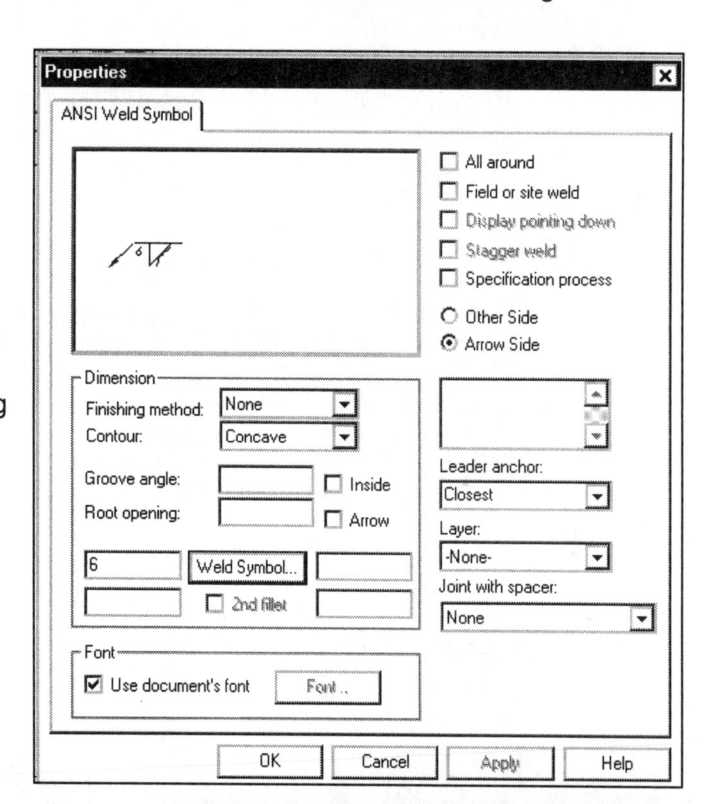

232) Click the **Weld Symbol** button.

233) Select **Fillet** for Symbols.

234) Click **OK**.

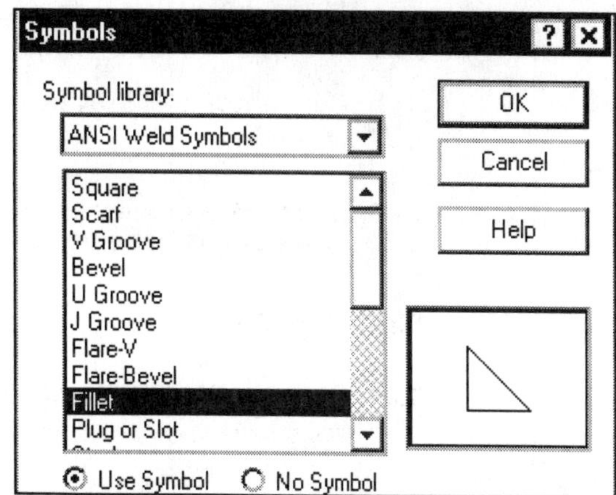

235) Display the Weld Symbol. Click **OK**.

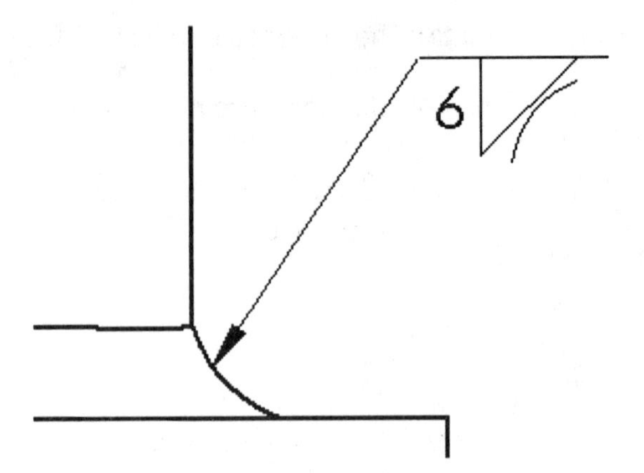

Return to the assembly. Resolve the PLATE1-W<2> component. Insert Weld Symbols directly in the drawing.

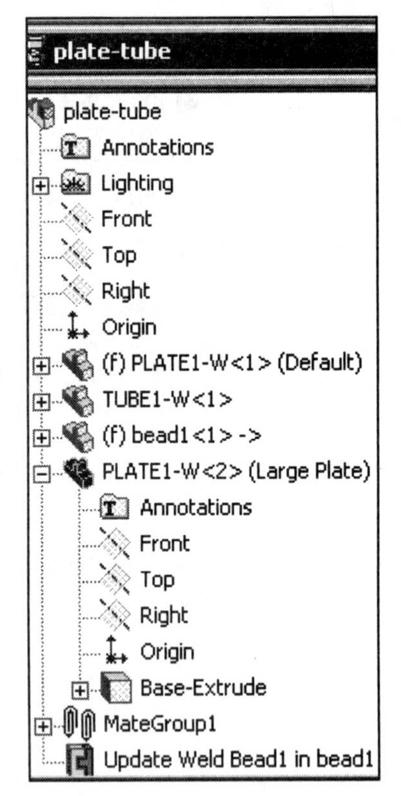

236) Return to the PLATE-TUBE assembly. Click **Window**, **PLATE-TUBE** assembly.

237) Right-click **PLATE1-W<2>(Large Plate)** in the assembly FeatureManager.

238) Click **Set to Resolve**.

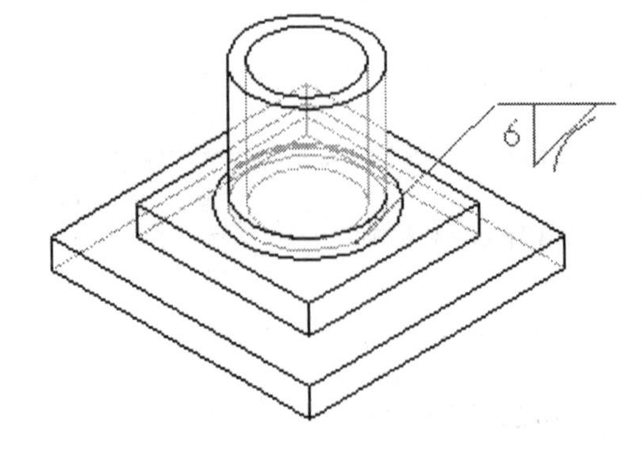

239) Return to the PLATE-TUBE drawing. Click **Ctrl Tab**.

240) Click the **right intersection**.

241) Click **Weld Symbol** . The ANSI Weld dialog box is displayed.

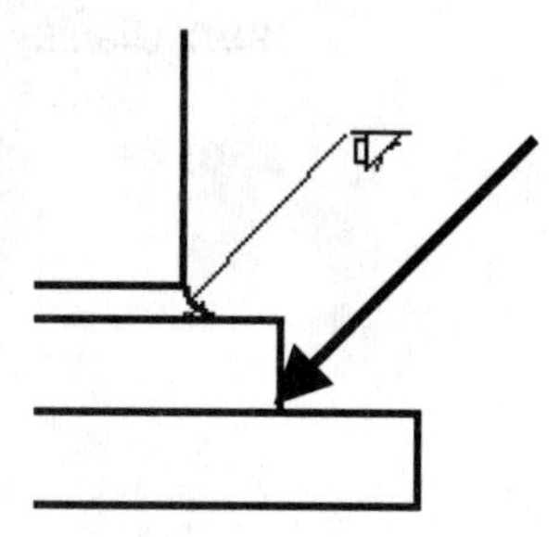

242) Click the **All around** check box to indicate that the weld extends completely around the joint.

243) Display the Weld Symbol on the same side as the arrowhead. Click **Arrow Side**.

244) Select **G-Grinding** for Finish Method.

245) Select **Convex** for Contour.

246) Click the **Weld Symbol** button.

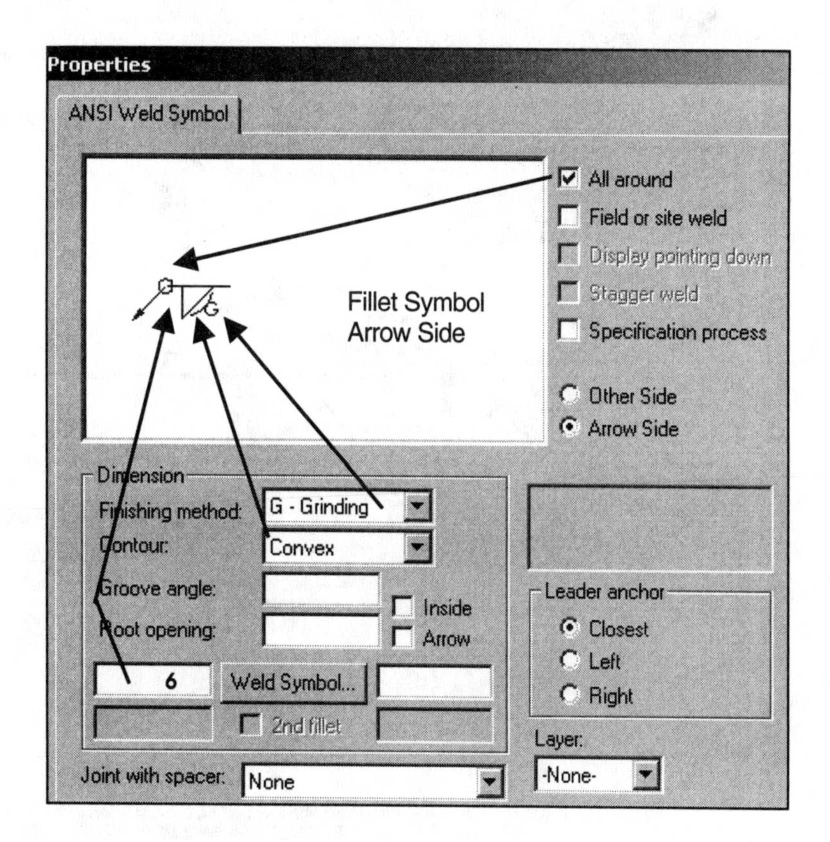

247) Select the **Fillet** option.

248) Click **OK**.

249) Display the Weld Symbol. Click **OK**.

250) Save the PLATE-TUBE drawing. Click **Save**.

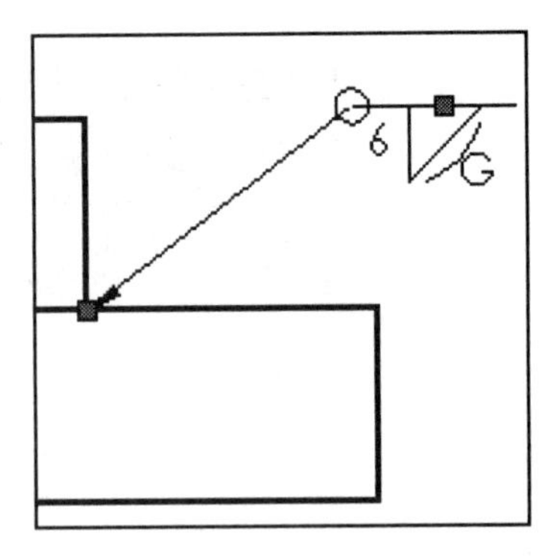

Save Rebuild time. Save Weld Symbols created in the drawing onto a weld Layer. Turn off the weld Layer when not required.

Suppress Weld Beads created in the assembly.

The Weld Symbol dialog box corresponds to the standard location of elements of a welding symbol as defined by AWS A2.4:1997.

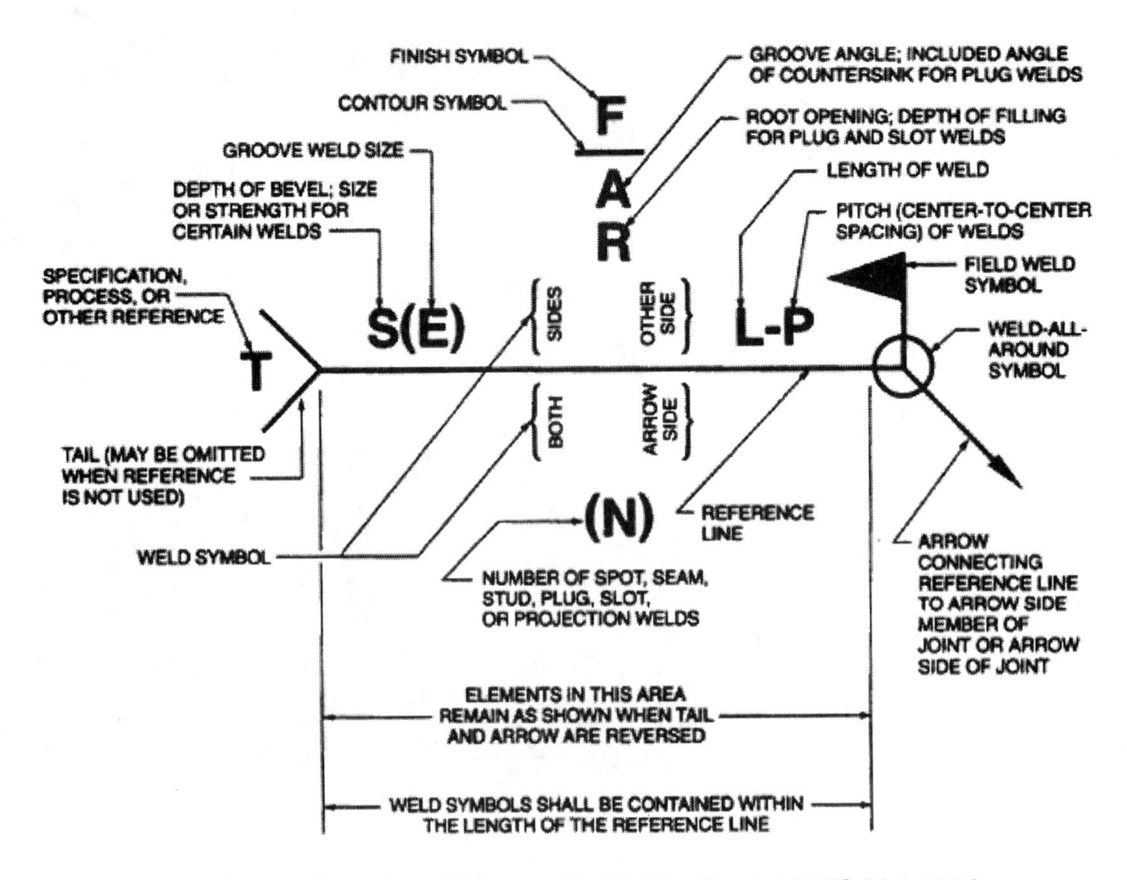

Standard Location of Elements of a Welding Symbol (AWS A2.4:1997)

When a weld is required on both sides, select Arrow Side. Enter all weld options for the Arrow Side.

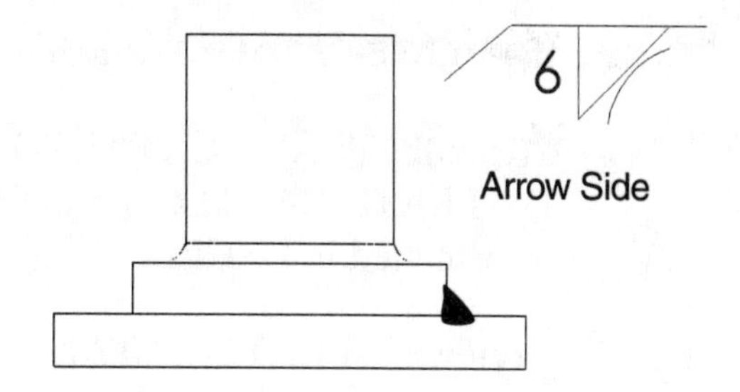

Select Other Side. Enter all options for the Other Side.

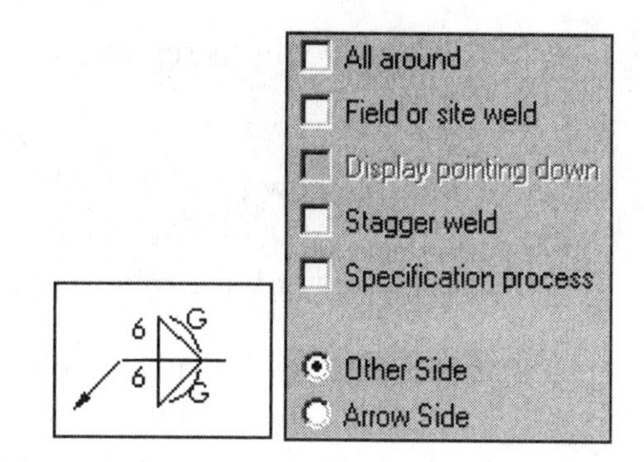

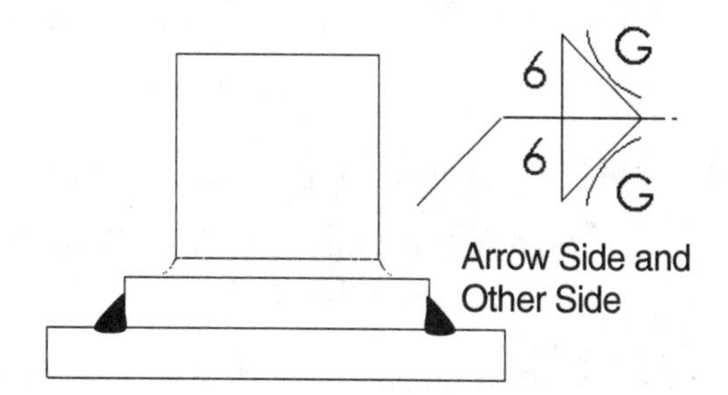

Create the PLATE-CATALOG drawing

Create the PLATE-CATALOG drawing for the company's on-line catalog. The PLATE-CATALOG drawing utilizes a Design Table. The Design Table is formatted in EXCEL.

Review the PLATE-CATALOG configurations. The PLATE-CATALOG part contains three configurations: 1007A-36, 1007A-48, 1007A-54. The configurations names represent the family part number (1007A -).

The last two digits represent the square plate size: $36mm^2$, $48mm^2$ and $54mm^2$.

Create the new PLATE-CATALOG drawing. Insert dimensions from the PLATE-CATALOG Default part. Modify the PLATE-CATALOG drawing to contain symbolic representation of the dimensions.

Example: The dimension 36 is replaced with the letter A. Insert the Design Table into the drawing. Modify the Design Table to represent the various configurations as a family of parts.

Open the PLATE-CATALOG part.

251) Open the part. Click **Open** from the Standard toolbar. The PLATE-CATALOG part is located in the 2003drwparts file folder.

252) Double-click **PLATE-CATALOG** part.

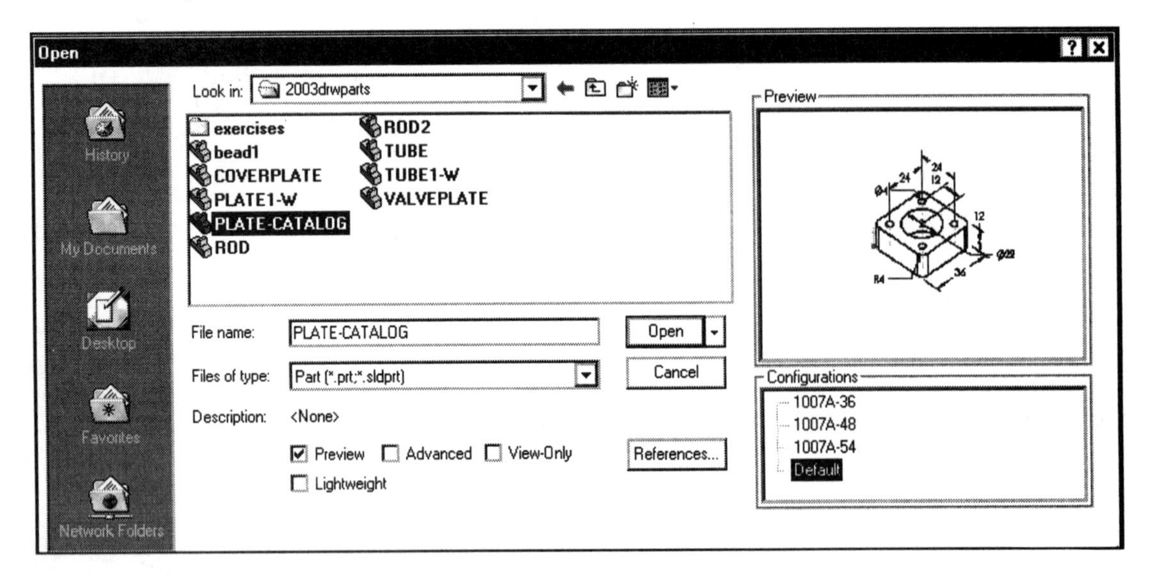

Review the PLATE-CATALOG configurations.

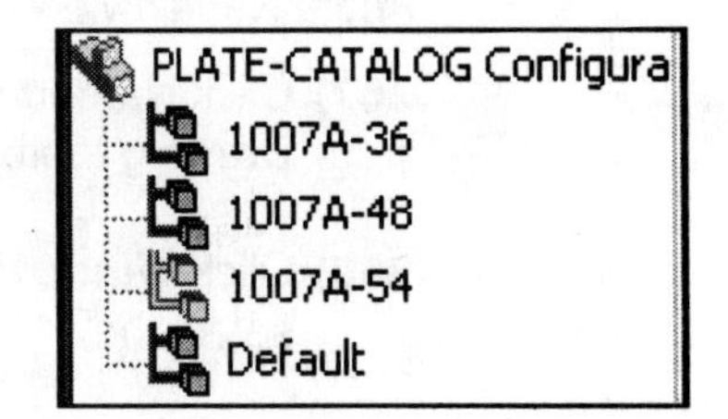

253) Click **Configuration**. The PLATE-CATALOG part has three configurations.

254) Double click **1007A-36**.

255) Repeat for **1007A-48, 1007A-54**.

256) Set the current configuration. Click **Default**. Display the FeatureManager.

257) Click the **Part** icon.

258) Click **Edit**, **Design Table** to review the variable names and configuration names for the PLATE-CATALOG.

Design Table for: PLATE-CATALOG					
	D1@Sketch1	D1@Sketch2	D2@Sketch3	D3@LPattern1	D4@LPattern1
1007A-36	36	22	12	24	24
1007A-48	48	24	16	32	32
1007A-54	54	26	18	36	36

Create a New PLATE-TUBE drawing.

259) Click **File**, **New**.

260) Click the **2003drwparts** file folder.

261) Click the **A-SIZE-ANSI-MM-EMPTY** drawing template.

262) Click **No Sheet Format**.

263) Click **A-Landscape** for Paper Size.

264) Save the empty drawing. Click **File, Save**.

265) Enter **PLATE-CATALOG** for Filename.

266) Enter **PLATE-CATALOG DRAWING** for Description.

Add a Named view.

267) Click **Named** view from the Drawing toolbar.

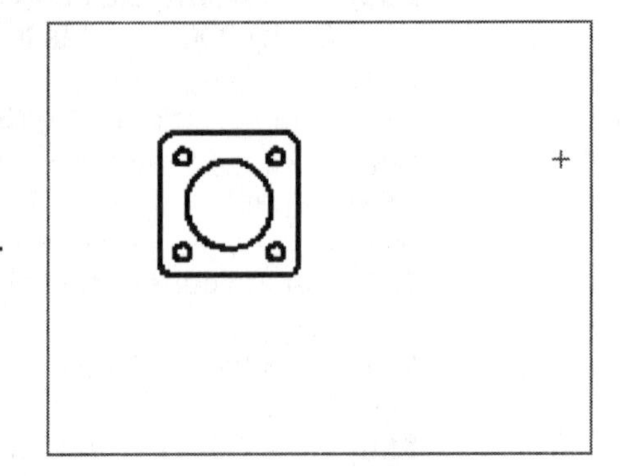

268) Click **Window**, **PLATE-CATALOG** to select the PLATE-CATALOG part.

269) Click the **PLATE-CATALOG part** icon.

270) Click a **position** inside the **PLATE-CATALOG** -Sheet1 boundary.

271) Click **Top** from the View Orientation text box.

272) Enter **2:1** for Custom Scale. The Top Named view is now the Front view in the drawing.

273) Right-click **Properties**.

274) Select **1007A-36** for the current Used Name configuration.

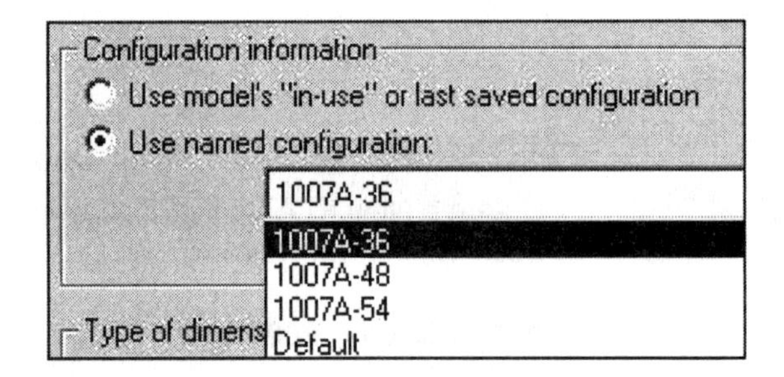

Create a Projected view.

275) Create the Right view as a projected view. Click the **boundary** of the Named view.

276) Click **Projected View** from the Drawing toolbar.

277) Click a **location** to the right of the Top view. The Top and Right views are displayed.

The font size is too small. Increase the font size for the entire drawing.

278) Click **Tools**, **Options**, **Document Properties, Annotations Font**.

279) Click **Dimensions**.

280) Click the **Font** button.

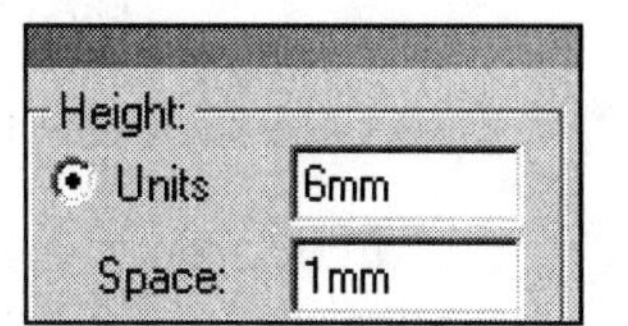

281) Enter **6mm** for Height in the Units text box.

282) Click **OK**.

283) Click **OK**.

Insert Model Dimensions into the Top and Right views.

284) Click a **position** inside the sheet boundary.

285) Click **Insert Model Items** .

286) Click **OK**.

287) Drag the **dimensions** approximately 10mm ways from the profile.

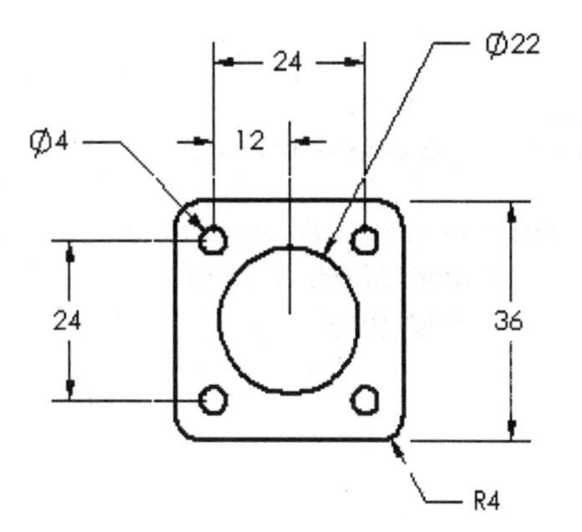

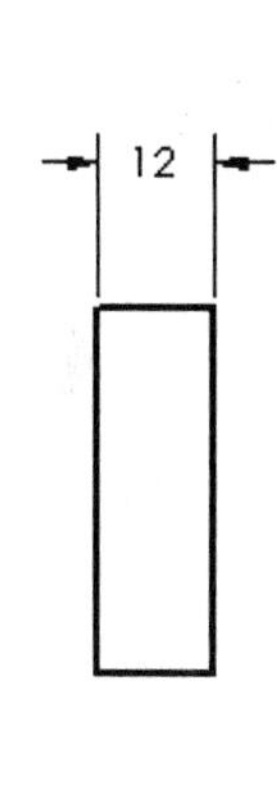

Add an overall horizontal dimension in the Top view.

288) Click **Dimension** .

289) Click the **left** and **right vertical lines**.

290) Click a **position** below the profile for the 36.00 horizontal dimension text.

291) Drag the **extension lines** off the profile.

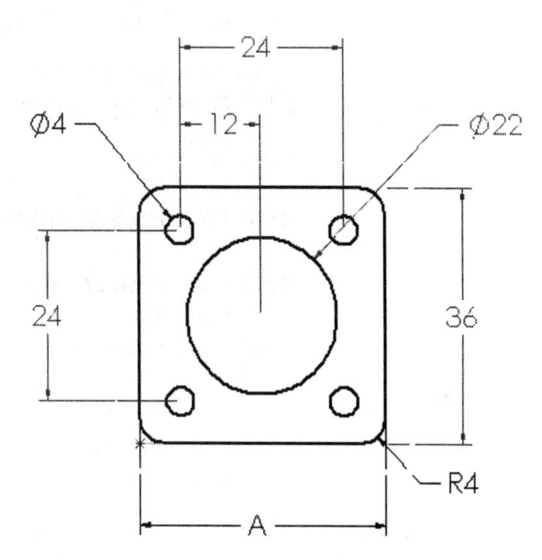

Edit dimensions to text symbols.

292) Click the **36** horizontal dimension.

293) Select the **<DIM>** text.

294) Select **Delete**.

295) Enter **A**.

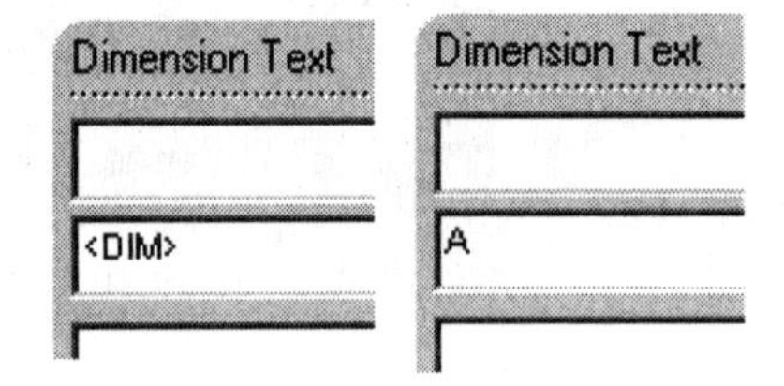

296) Click the **36** vertical dimension.

297) Select the **<DIM>** text.

298) Select **Delete**.

299) Enter **A**.

300) Click the ∅**22** dimension text.

301) Select the **<DIM>** text.

302) Select **Delete**. Do not delete the diameter symbol ∅.

303) Enter **B**.

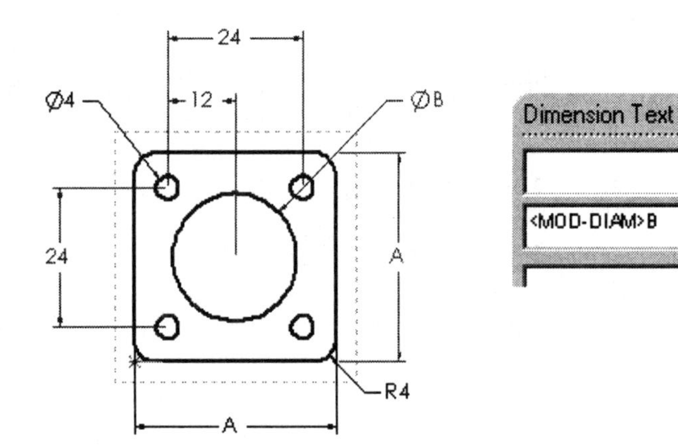

304) Click the 12 dimension text.

305) Select the **<DIM>** tex

306) Select **Delete**.

307) Enter **C**.

308) Click the **24** horizontal dimension text.

309) Select the **<DIM>** text.

310) Select **Delete**.

311) Enter **D**.

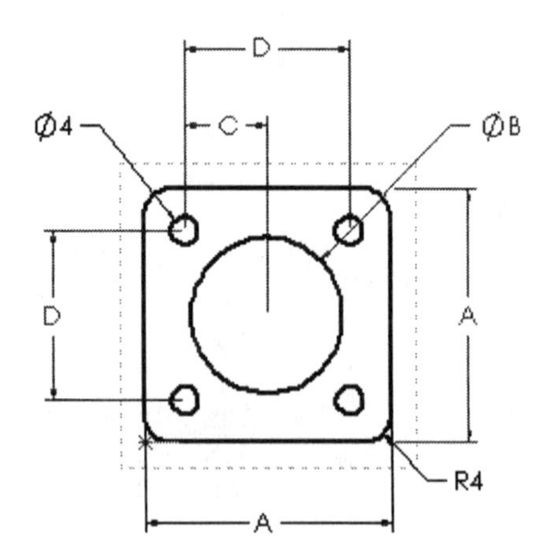

312) Repeat for the **24** vertical dimension text.

When letters replace the <DIM> placeholder, the part dimensions are updated to reflect both the letter and the dimension value. Example: 36 is replaced with A36.

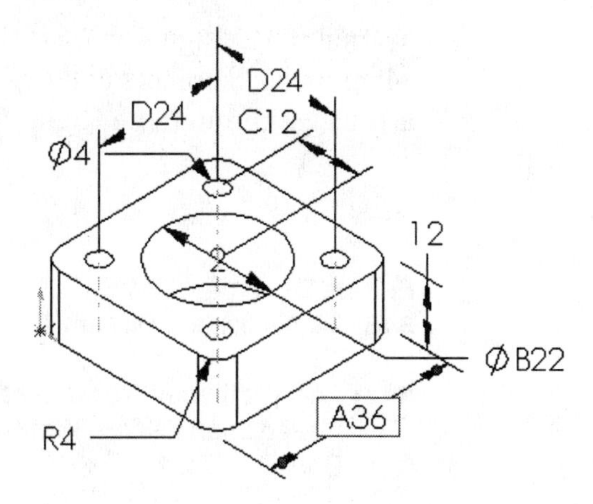

Insert a Design Table into Named view.

313) Click the **Front view boundary**.

314) Click **Insert**, **Design Table** from the Main toolbar.

315) Drag the **Design Table** below the 36 horizontal dimension text.

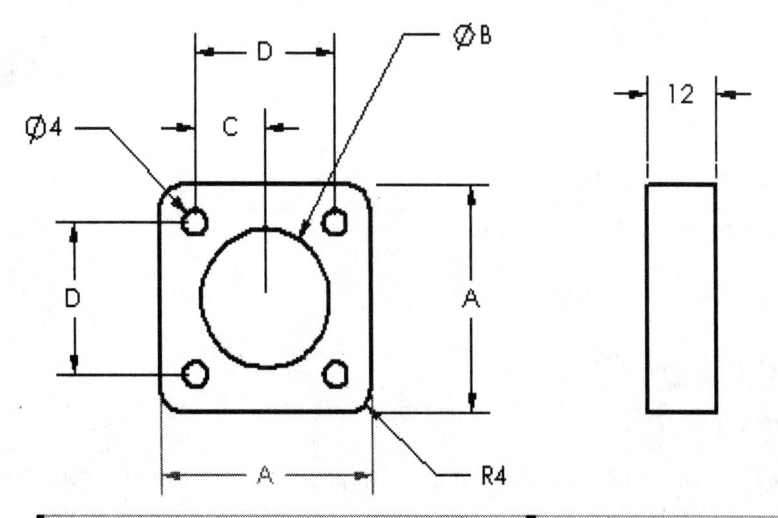

Design Table for: PLATE-CATALOG					
	D1@Sketch1	D1@Sketch2	D2@Sketch3	D3@LPattern1	D4@LPattern1
1007A-36	36	22	12	24	24
1007A-48	48	24	16	32	32
1007A-54	54	26	18	36	36

Format the Design Table in EXCEL. EXCEL functions are executed from the Main menu, Toolbars and Right-click. The commands utilized in this section are access through the Right-click.

Edit the Design Table in EXCEL.

316) Right-click the **Design Table**.

317) Click **Edit with Worksheet**. The Design Table is placed in an EXCEL Worksheet. The default name is Book 1.

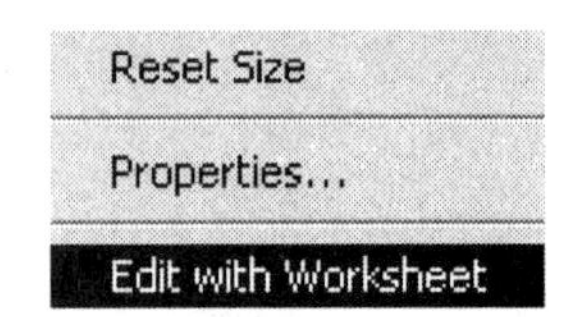

Hide Row 1 and Row 2.

318) Click **Row 1**. Drag the mouse pointer to **Row 2** in the Row frame. Both Row 1 and Row 2 are selected.

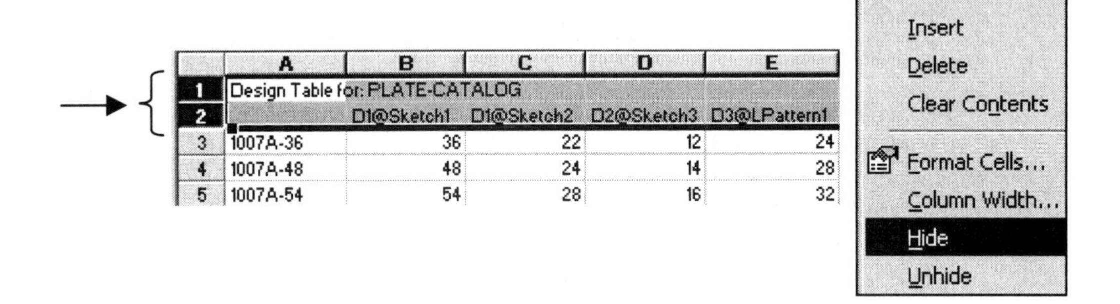

	A	B	C	D	E
1	Design Table for: PLATE-CATALOG				
2		D1@Sketch1	D1@Sketch2	D2@Sketch3	D3@LPattern1
3	1007A-36	36	22	12	24
4	1007A-48	48	24	14	28
5	1007A-54	54	28	16	32

319) Right-click **Hide**. Row 3 is now displayed as the first displayed row.

Insert 2 Rows.

320) Click **Row 3** in the Row frame.

321) Create a new row. Right-click **Insert**. Row 3 is the new row. The remaining rows move downward.

322) Create a new row. Right-click **Insert**. Row 3 is the new row. The remaining rows move downward.

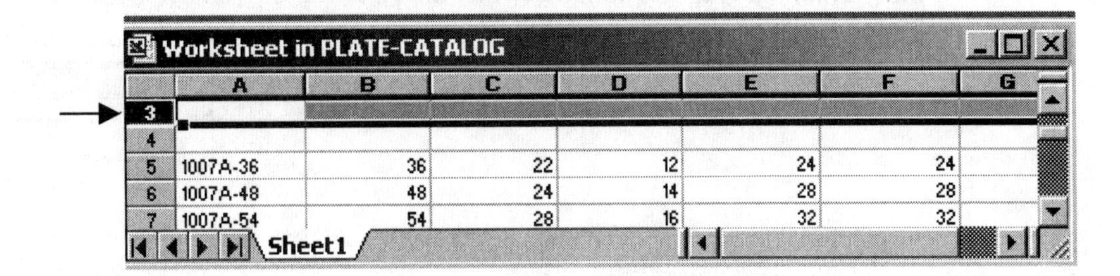

Hide Column F.

323) Click **Column F**.

324) Right-click **Hide**. Column G is displayed to the right of Column E.

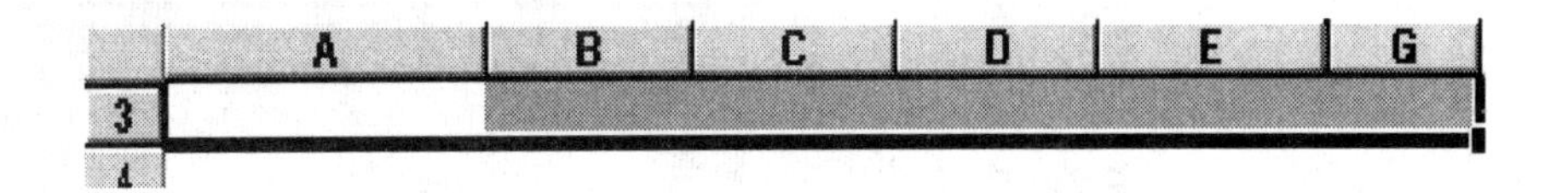

Add Title and Headers.

325) Click **Cell A3.**

326) **Hold** the left mouse button down.

327) Drag the mouse pointer to **Cell G3**

328) **Release** the left mouse button.

329) Right-click **Format Cells.**

330) Click the **Alignment** tab.

331) Click the **Merge Cell** check box.

332) Click **Center** for Horizontal.

333) Click **OK.**

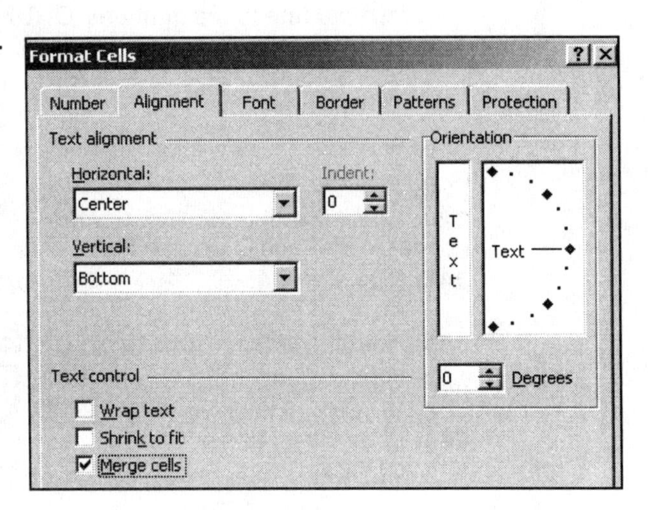

334) Enter **PLATE – 36 THROUGH 54 MM, STEEL** for the title.

	A	B	C	D	E	G
3	PLATE - 36 THROUGH 54 MM, STEEL					

335) Click **Cell A4**.

336) Enter **ORDER NUMBER**.

337) Click **Cell B4**.
Enter **A**.

338) Click **Cell C4**.
Enter **B**.

339) Click **Cell D4**.
Enter **C**.

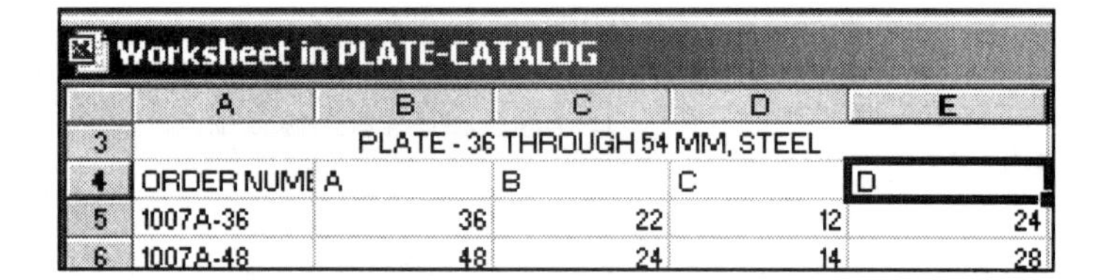

Worksheet in PLATE-CATALOG

	A	B	C	D	E
3	PLATE - 36 THROUGH 54 MM, STEEL				
4	ORDER NUME	A	B	C	D
5	1007A-36	36	22	12	24
6	1007A-48	48	24	14	28

340) Click **Cell E4**. Enter **D**. The entry ORDER NUMBER is not completely displayed.

341) Increase the Column A width. Click the **vertical line** between Column A and Column B in the Column frame. Drag the **vertical line** to the right until ORDER NUMBER is completely displayed.

	A	
3		PLATE
4	ORDER NUMBER	A
5	1007A-36	
6	1007A-48	

Format A, B, C and D Column text.

342) Click **Cell B4.**

343) **Hold** the left mouse button down.

344) Drag the mouse pointer to **Cell E7**.

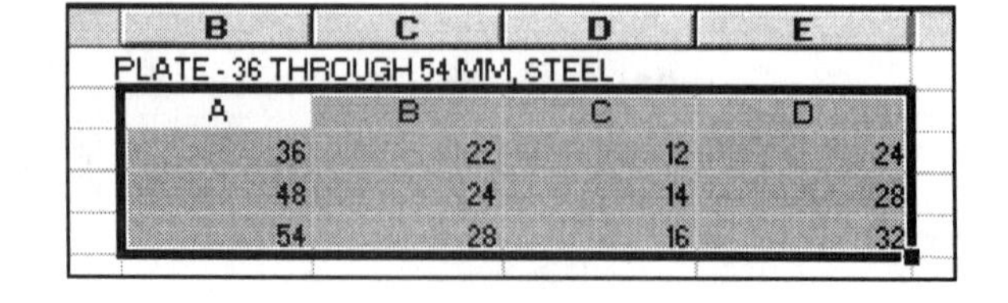

B	C	D	E
PLATE - 36 THROUGH 54 MM, STEEL			
A	B	C	D
36	22	12	24
48	24	14	28
54	28	16	32

345) **Release** the left mouse button.

346) Right-click **Format Cells**.

347) Click the **Alignment** tab.

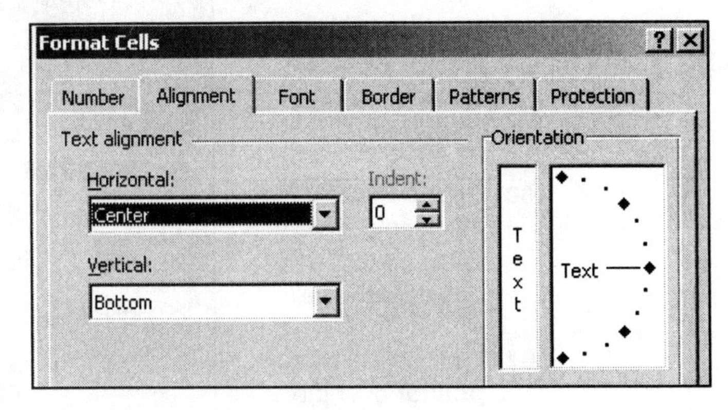

348) Click **Center** from the Horizontal drop down list.

349) Click **OK**.

Format Column G.

350) Click **Cell G4**.

351) Enter **COST**.

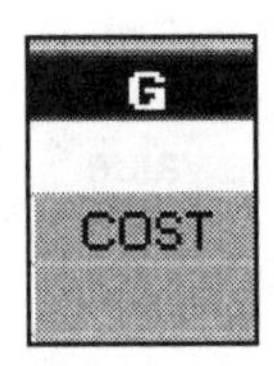

352) Drag the mouse pointer to **Cell G7**.

353) Release the **left mouse button**.

354) Right-click **Format Cells**.

355) Click the **Number** tab.

356) Click **Currency** from the Category.The default Decimal places is 2.

357) Click **OK**.

358) Click **Cell G5**.

359) Enter **100**. The currency value $100.00 is displayed.

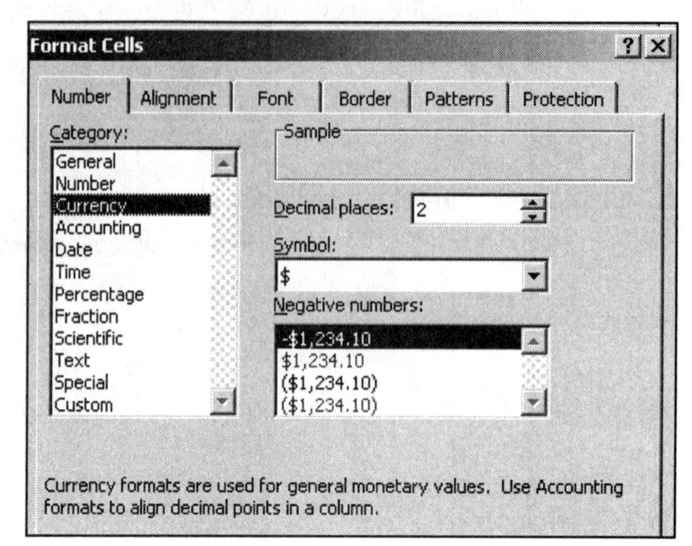

360) Click **Cell G5**. The Fill handle is located in the bottom right corner of the cell.

361) Position the **mouse pointer** over the small black square. The mouse pointer displays a black cross.

362) Click the **left mouse button**.

363) Drag the mouse pointer downward to **Cell G7.**

364) **Release** the left mouse button. The value $100.00 is displayed in Cell G6 and Cell G7.

Remove Gridlines.

365) Click **Tools**, **Options**.

366) Click the **View** tab.

367) Uncheck the **Gridlines** check box.

368) Click **OK**.

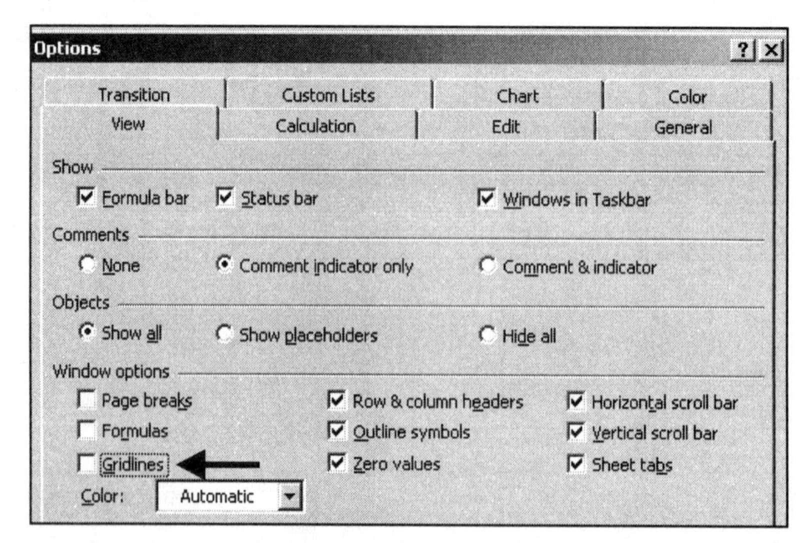

Increase Title Font.

369) Click **Cell A3**.

370) Right-click **Format Cells**.

371) Click the **Font** tab.

372) Select **14** for Size.

373) Click **OK**.

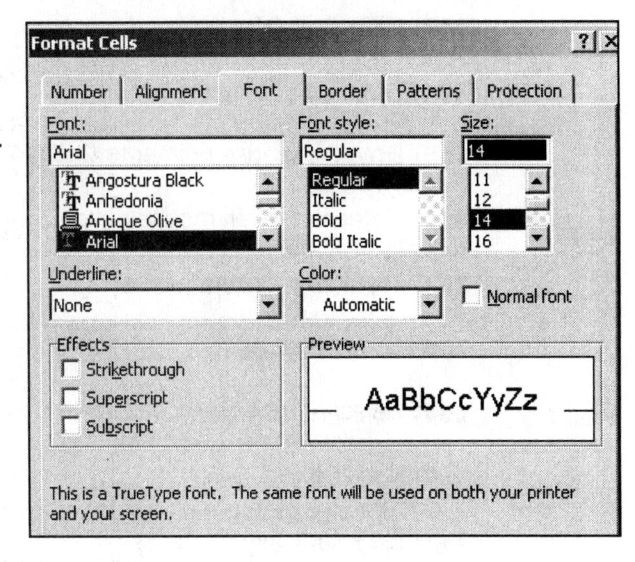

Add Borders and Color.

374) Click **Cell A3.**

375) Hold down the left mouse button down.

376) Drag the mouse pointer to **Cell G7**

377) Release the left mouse button.

378) Right-click **Format Cells**.

379) Click the **Border** tab.

380) Click the **Outside** button.

381) Select the **double line border** Style.

382) Click **OK**.

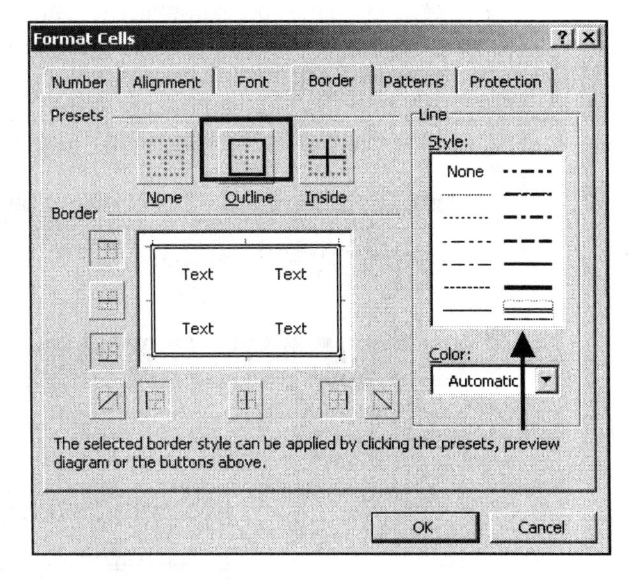

383) Click **Cell A4**.

384) **Hold** the left mouse button down.

385) Drag the mouse pointer to **Cell G4**

386) **Release** the left mouse button.

387) Right-click **Format Cells**.

388) Click **Patterns**.

389) Select a **light blue color** for shading.

390) Click **OK**.

391) Click **Cell A4**

392) **Hold** the left mouse button down.

393) Drag the mouse pointer to **Cell A7**

394) **Release** the left mouse button.

395) Right-click **Format Cells**.

396) Click **Patterns**. Select a **light blue color** for shading.

397) Click **OK**.

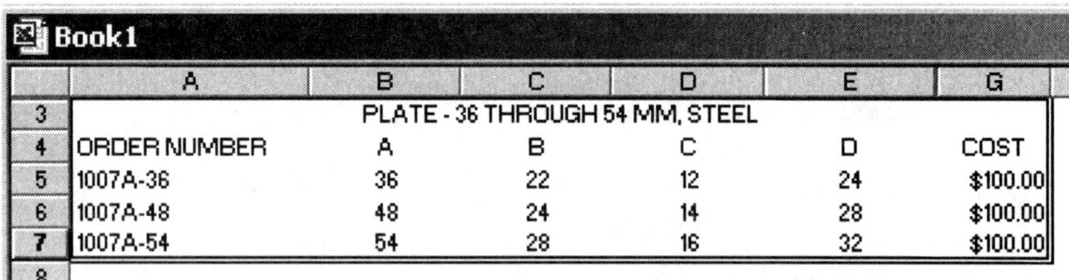

Book1

	A	B	C	D	E	G
3	PLATE - 36 THROUGH 54 MM, STEEL					
4	ORDER NUMBER	A	B	C	D	COST
5	1007A-36	36	22	12	24	$100.00
6	1007A-48	48	24	14	28	$100.00
7	1007A-54	54	28	16	32	$100.00
8						

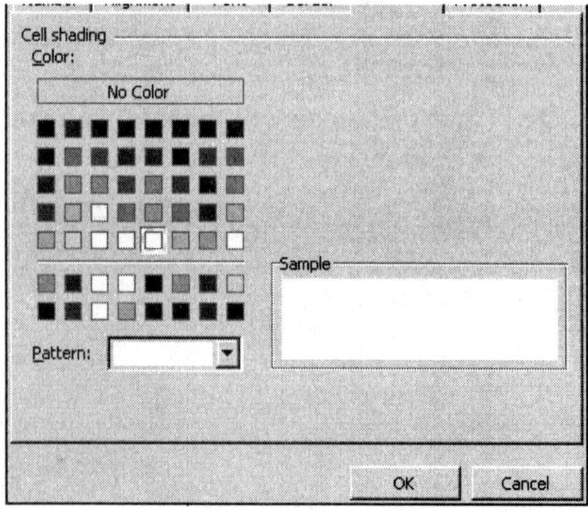

Save and Update EXCEL.

398) Click **File**, **Save Copy As**.

399) Enter **TABLE-PLATE-CATALOG** to save a copy in EXCEL format for additional editing.

400) Update the linked Design Table in SolidWorks. Click **Update**.

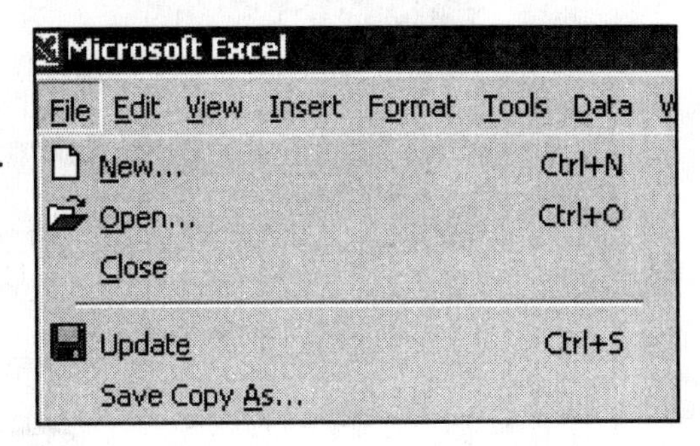

401) Return to SolidWorks. Click **Close**.

402) Click the SolidWorks **Graphics window**.

403) Click **OK** to the warning message in the PLATE-CATALOG part.

404) Return to the drawing. Click **Window**, **PLATE-CATALOG-Sheet1**.

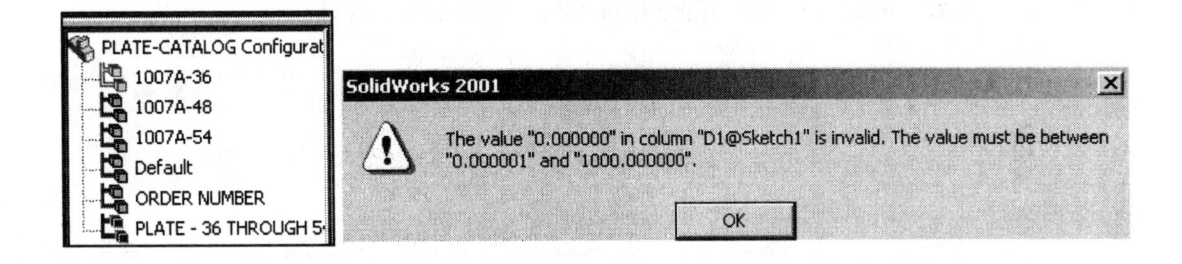

A warning message is displayed in SolidWorks. ORDER NUMBER and PLATE – 36 THROUGH 54 contain no valid dimension entries.

The 1007A-36, 1007A-48, 1007A-54 are still valid.

Avoid unwanted configuration names. Insert Row 3 and Row 4 before hidden Row 1 and Row 2.

Enlarge the Design Table.

405) Right-click the **Design Table**.

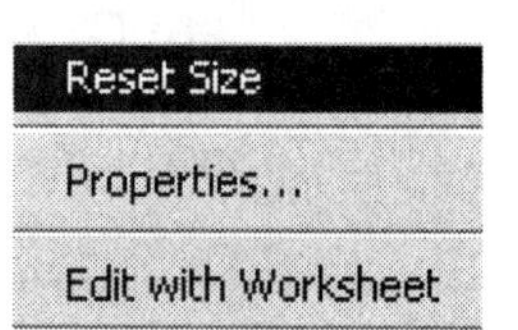

406) Click **Reset Size**. Enlarge the Design Table.

407) Drag the **corner handle** to the right.

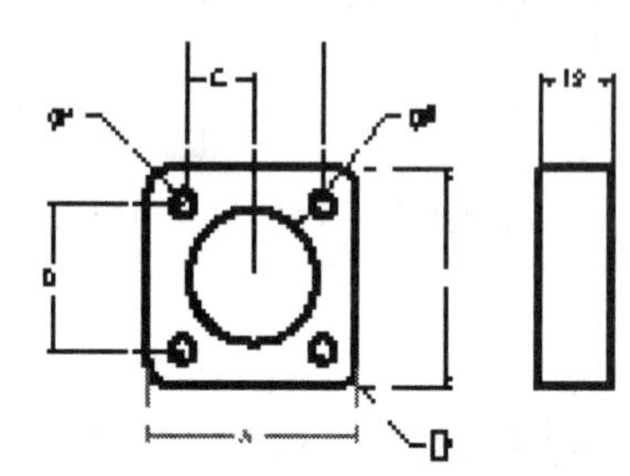

PLATE - 36 THROUGH 54 MM, STEEL					
ORDER NUMBER	A	B	C	D	COST
1007A-36	36	22	12	24	$100.00
1007A-48	48	24	14	28	$100.00
1007A-54	54	28	16	32	$100.00

Reposition the Title text.

408) Return to Excel. Click **Edit with Worksheet**.

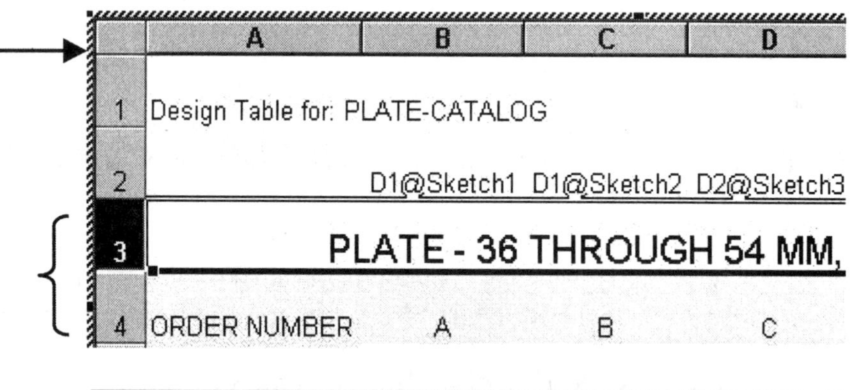

	A	B	C	D	E
1	PLATE - 36 THROUGH 54 MM, STEEL				
2	ORDER NUMBE	A	B	C	D
3	Design Table for: PLATE-CATALOG				
4		D1@Sketch1	D1@Sketch2	D2@Sketch3	D3@LPattern1
5	1007A-36	36	22	12	24

409) Drag the **top Row 1 frame bar** downward to display Row 1 and Row 2.

410) Insert Row 3 and Row 4 before Row 1. Click **Row 3**, drag to **Row 4** in the Row frame. Both Row 3 and Row 4 are selected.

411) Right-click **Cut**.

412) Click **Row 1** in the Row frame.

413) Right-click **Insert Cut Cells**.

414) Click **Row 3,** drag to **Row 4** in the Row frame.

415) Right-click **Hide**.

416) Return to SolidWorks. Click **outside** the Excel boundary.

417) Delete configuration ORDER NUMBER and PLATE – 36 THROUGH 54. Click **OK**.

418) Save the PLATE-CATALOG drawing. Click **Save**.

The Design Table, TABLE-PLATE-CATALOG.XLS was saved as an Excel Spread Sheet. Insert Excel Spread Sheets into other part and assembly documents. Insert the Design Table into the corresponding drawing document.

PLATE - 36 THROUGH 54 MM, STEEL					
ORDER NUMBER	A	B	C	D	COST
1007A-36	36	22	12	24	$100.00
1007A-48	48	24	16	32	$100.00
1007A-54	54	26	18	36	$100.00

Example: Open the VALVEPLATE part. Select Insert, Design Table. Enter TABLE-PLATE-CATALOG. The Design Table is inserted into the VALVEPLATE part.

Open the VALVEPLATE drawing. Select the Front view. Select Insert, Design Table to insert the TABLE-PLATE-CATALOG into the drawing.

Blocks consist of the following elements: text, sketched entities (except points), Area Hatch and single Balloon. Blocks are symbols that exist on one or more drawings. Save Blocks in the current file folder or create a symbol library.

Set the Block file folder location in Tools, Options, System Options, File Locations when creating symbol libraries.

In the next example, make a new block. Combine text and a rectangle.

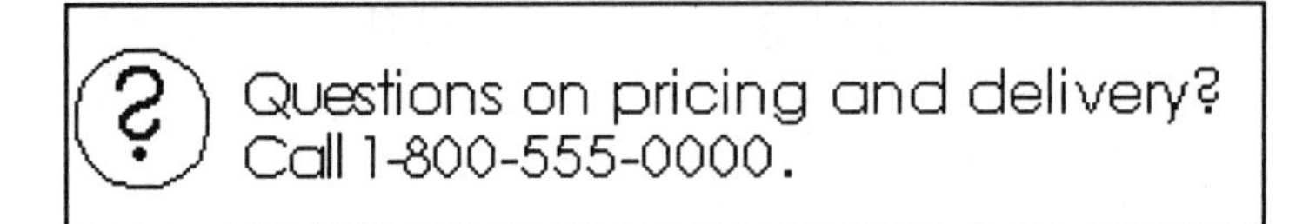

Save the Block to the current file folder. Insert the new Block. Modify the Properties of the Block.

Add Notes and a Sketch Rectangle.

419) Click a **position** below the 36 horizontal dimension text and above the Design Table.

420) Click **Note** A.

421) Click **No Leader**.

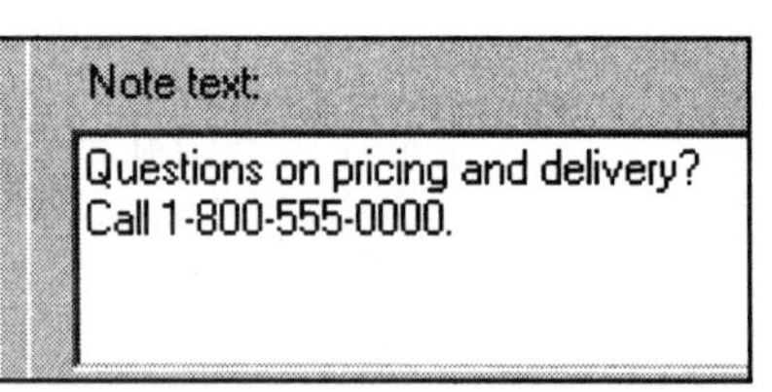

422) Enter **Questions on pricing and delivery?** on the first line.

423) Enter **Call 1-800-555-0000** on the second line.

424) Click the **Font** button.

425) Change the Font height to **6**mm.

426) Click **OK**.

427) Click a **position** to the left of the first Note.

428) Click **Note** A.

429) Click **No Leader**.

430) Enter **?**.

431) Click **Circular** from the Style drop down list in the Border text box.

432) Click the **Font** button.

433) Enter **14** mm for size.

434) Click **OK**.

435) Sketch a Rectangle around the two Notes.
Click **Rectangle**.

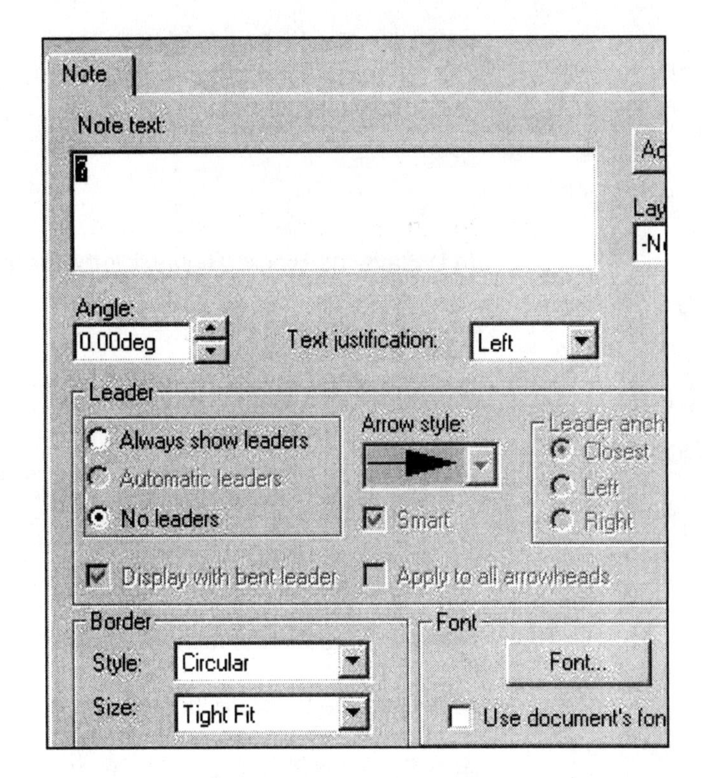

Make a New Block.

436) Click **Select**.

437) Hold the **Ctrl** key down.

438) Select the two **Notes** and the **Rectangle**.

439) Release the **Ctrl** key.

440) Click **Tools**, **Block**, **Make**. The Notes and Rectangle combine to make a Block.

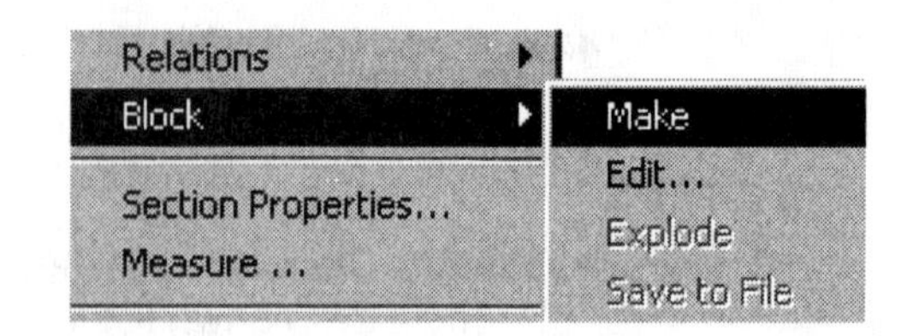

441) Save the Block. Right-click the **Block**.

442) Click **Save to File**.

443) Enter **QUESTION-BLOCK** for File name.

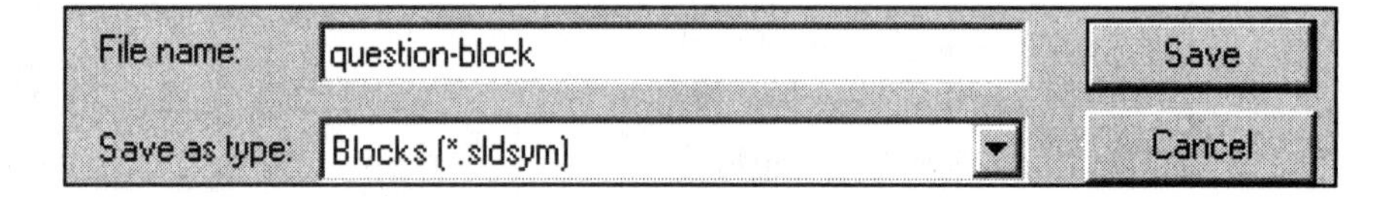

444) Click the **Save** button. Note: Blocks contain the file extension .sldsym.

445) Drag the **Block** to the left corner of the Design Table.

446) Save the PLATE-CATALOG drawing. Click **Save**.

Insert a Block.

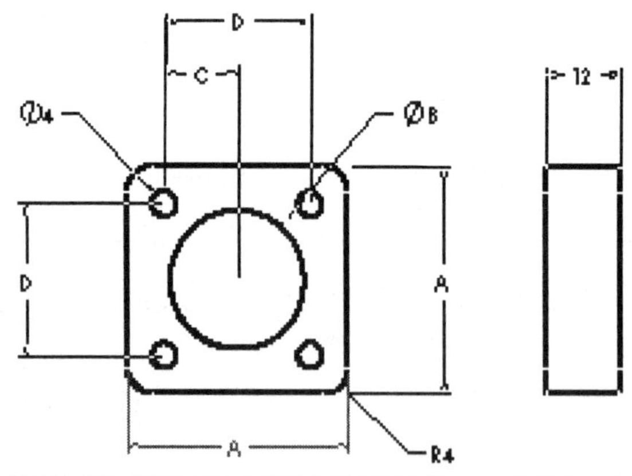

PLATE - 36 THROUGH 54 MM, STEEL

ORDER NUMBER	A	B	C	D	COST
1007A-36	36	22	12	24	$100.00
1007A-48	48	24	14	28	$100.00
1007A-54	54	28	16	32	$100.00

447) Open the PLATE-TUBE drawing.

448) Click a **position** to the left of the Front view.

449) Click **Insert Block**.

450) Enter **QUESTION-BLOCK**. The QUESTION-BLOCK is too large for the drawing. The QUESTION-BLOCK is on the wrong layer.

451) Create a New **Layer**.

452) Enter **Notes** for Layer Name.

453) Enter **Customer Notes** for Description.

Modify Properties of a Block.

454) Right-click **QUESTION-BLOCK**.

455) Click **Properties**.

456) Enter **0.5** for Scaling.

457) Select **Notes** from the Layer drop down list.

458) Click **OK**.

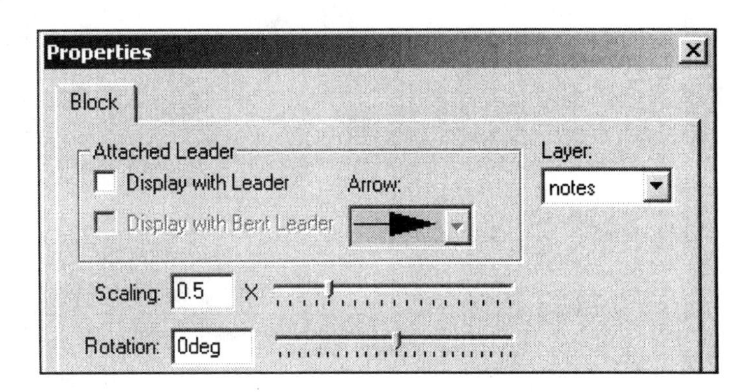

459) Save the PLATE-TUBE drawing. Click **Save**.

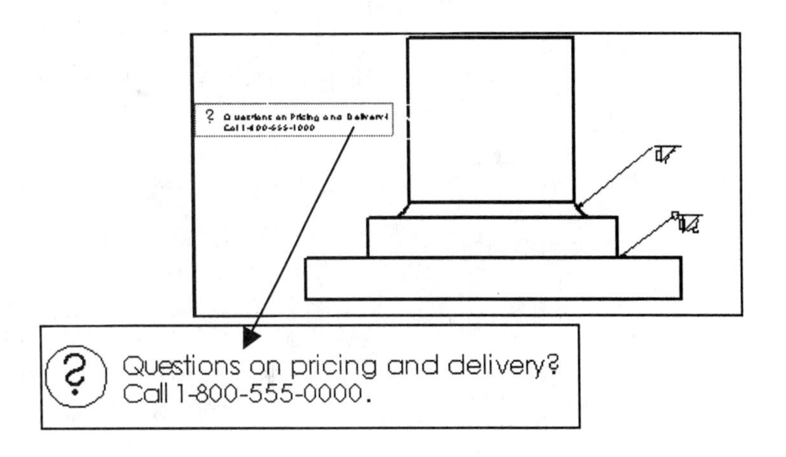

Utilize blocks in a drawing to represent components for pneumatic, mechanical and HVAC systems.

Example: ISO-1219 pneumatic symbols were created by importing .dxf geometry and creating individual blocks.

Utilize blocks in the exercises at the end of this project. Each pneumatic symbol is a separate block. Sketch lines connect the symbols to show the airflow.

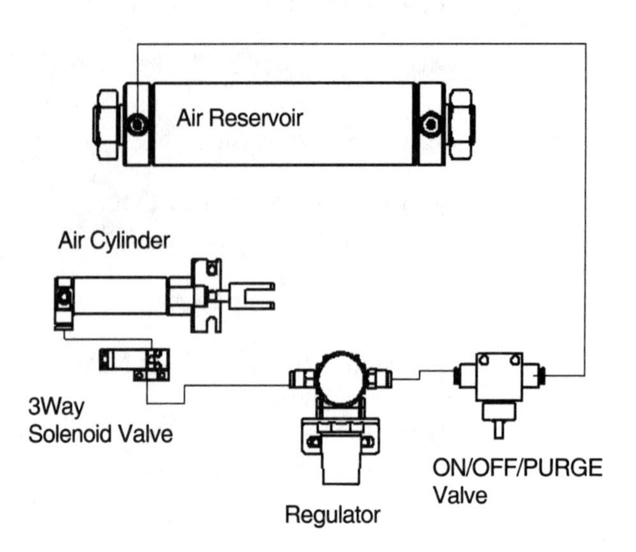

Pneumatic Components Diagram
Courtesy of SMC Corporation of America and Gears Educational Systems.

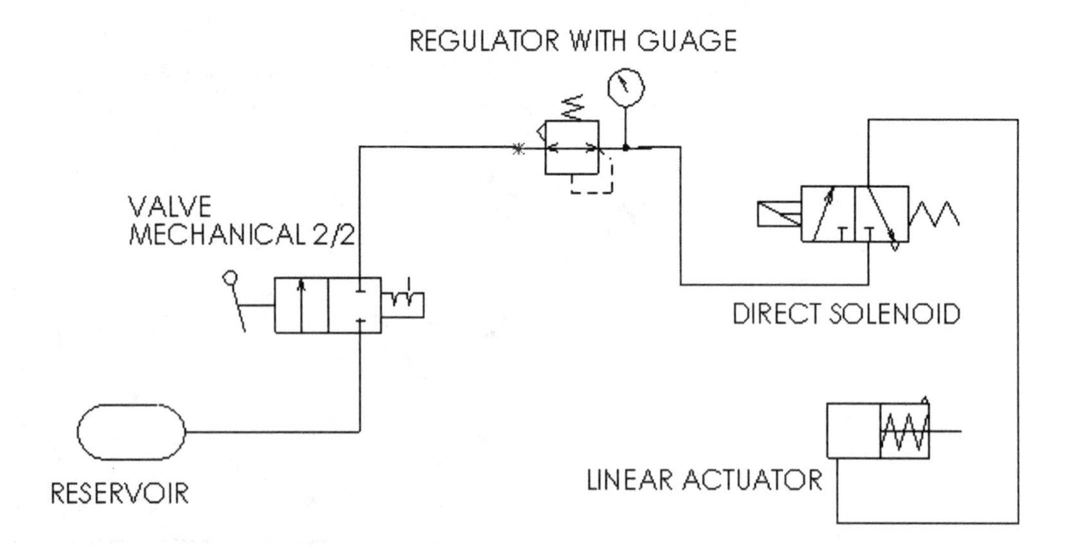

ISO-1219 Symbols – Blocks
Courtesy of SMC Corporation of America

There is a relationship between the Geometric Tolerance Symbols and the lower case letters in the SW-GDT Font. In EXCEL, select the SW-GDT Font type for Column H.

The letters, a through z, display various Geometric Tolerancing and Hole Symbols.

Symbols in SolidWorks are used in a variety of applications.

Additional information is contains in the Exercises at the end of this Project.

Letter	SW-GDT	Letter	SW-GDT
a	∠	n	Ø
b	⊥	o	□
c	⏥	p	Ⓟ
d	⌓	q	℄
e	○	r	◎
f	//	s	Ⓢ
g	⌭	t	⌰
h	↗	u	—
i	|	v	⌴
j	⌖	w	⌵
k	⌒	x	↧
l	Ⓛ	y	⌲
m	Ⓜ	z	⌳

Project Summary

You created three drawings and a SolidWorks eDrawing in this project:

- VALVEPLATE drawing.
- PLATE-TUBE drawing.
- PLATE-CATALOG drawing.
- VALVEPLATE eDrawing.

The VALVEPLATE drawing consisted of three standard views. You:

- Inserted dimensions.
- Modified dimensions to contain Basic, Bilateral and Limit Tolerance.
- Created a Parametric Note.

The PLATE-TUBE drawing consisted of one view. You:

- Created a Weld Bead Assembly feature.
- Created Weld Symbols in the drawing.

The PLATE-CATALOG consisted of one view with a Design Table. You:

- Inserted the Design Table.
- Inserted dimensions from the PLATE-CATALOG Default part.
- Modified the PLATE-CATALOG drawing to contain symbolic representation of the dimensions.

Project Terminology

Block: A symbol in the drawing that combines geometry into a single entity.

Design Table: Excel spreadsheet that is used to create multiple configurations in a part or assembly document.

eDrawing: A compressed document that does not require the referenced part or assembly. eDrawings are animated to display multiple views in a drawing.

Datum Feature: An annotation that represents the primary, secondary and other reference planes of a model utilized in manufacturing.

Geometric Tolerance Symbol: Set of standard symbols that specify the geometric characteristics and dimensional requirements of a feature.

RapidDraft: Drawing format that allows opening and working in a drawing without loading the corresponding models into memory. The models are loaded on an as-needed basis.

Surface Finish: An annotation that represents the texture of a part.

Weld Bead: An assembly feature that represents a weld between multiple parts.

Weld Symbol: An annotation in the part or drawing that represents the parameters of the weld.

Questions:

1. Datum Feature, Geometric Tolerance, Surface Finish and Weld Symbols are located in the _______________ Toolbar.

2. True or False. A SolidWorks part file is a required attachment to email a SolidWorks eDrawing.

3. Describe the procedure to create a Basic dimension.

4. Dimensions on the drawing are displayed with three decimal places. You require two decimal places on all dimensions except for a single hole diameter. Identify the correct Document Property options.

5. Describe the procedure to create a Unilateral tolerance.

6. Datum Symbols A, B and C in the VALVEPLATE drawing represent the _______________, _____________________, and _________________ reference planes.

7. Describe the procedure to attach a Feature Control Frame to a dimension.

8. Surface Finish symbols are applied in the ____________ and in the __________.

9. Describe the procedure to create multiple leader lines that attach to the same symbol.

10. A Weld Bead is created in the _________________. A Weld Symbol is created in the ________________.

11. ___________ combine text and sketched entities (except points), Area Hatch and single Balloons to create symbols in a drawing document.

12. Format Design Tables using ___________.

Exercises:

Exercise 5.1:

Create a new drawing for the part, ANGLEPLATE2 located in the 2003drwparts file folder. Create a Front view, Bottom view and Auxiliary view. Use Geometric Relations when constructing center lines. The centerline drawn between the Front view and the Auxiliary view is perpendicular to the angled edge.

Hint: Create the Feature Control Frame before applying Datum Reference Symbols, (Example Datum E).

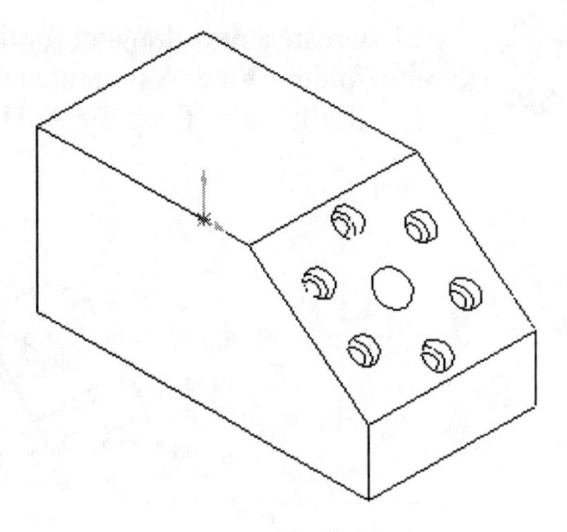

INTERRELATED DATUM REFERENCE FRAMES
Courtesy of ASME Y14.5M

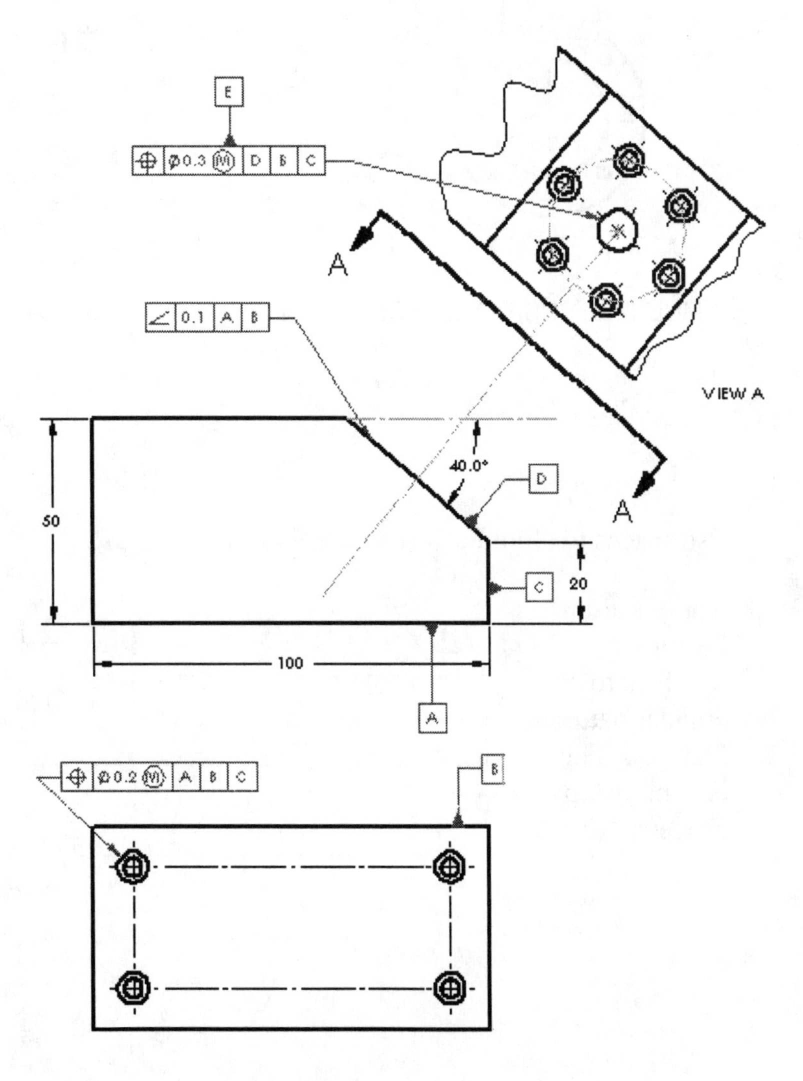

Example 5.2:

Create a new drawing for the part, FIG5-28 located in the 2003drwparts file folder. View A is created with an Auxilary view from the angled edge in the Right view. Crop the Auxillary view.

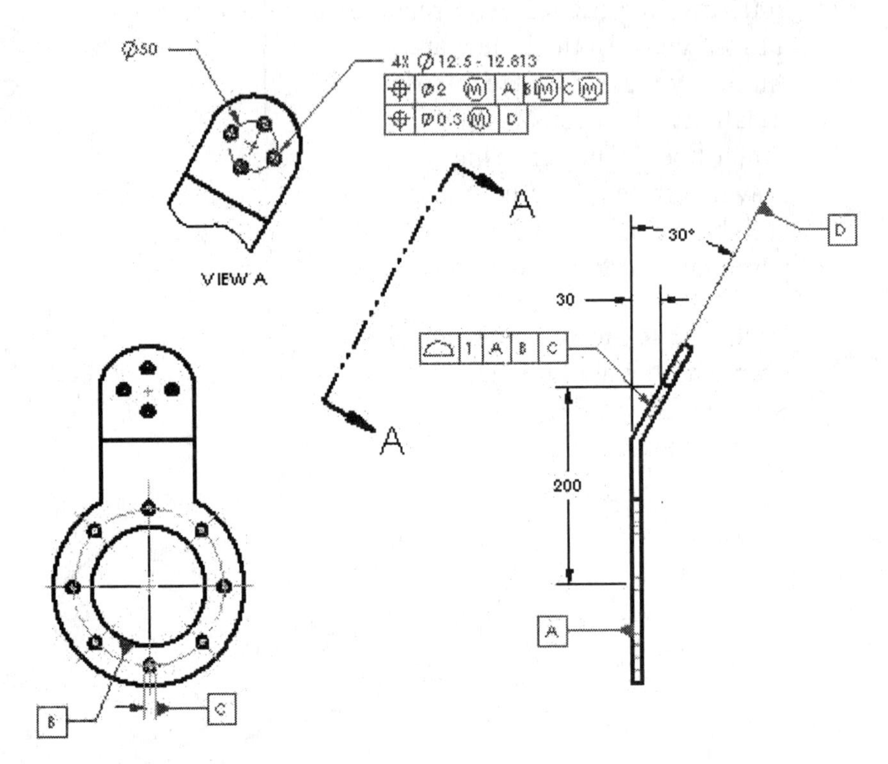

MULTIPLE POSITIONAL TOLERANCING FOR A PATTERN OF FEATURES
COURTESY ASME Y14.5M

Use spaces to align Feature Control Frames 1 and 2.

Apply Datum Feature Symbols to circular pattern features. This is a multistep process.

4X Ø12.5 - 12.813

⌖	Ø2 Ⓜ	A	BⓂ	CⓂ
⌖	Ø0.3Ⓜ	D		

a) Create a linear diameter dimension on the bottom circle.

b.) Add a verticle construction line tangent to the circle and colinear with the extension line of the dimension.

c.) Insert the Datum Feature Symbol on the vertical construction line. Align the Datum Feature Symbol on top of the dimension arrow.

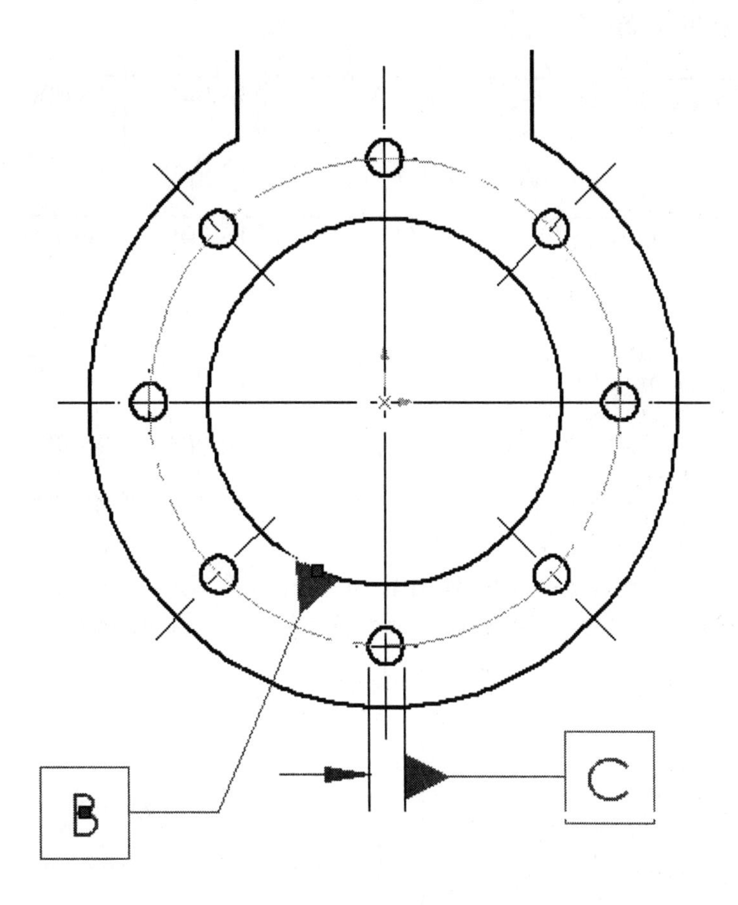

Sketch a single center line when locating lines are required on a pattern of circular features. Create a 2D circular pattern of center lines.

Example: 360°/8 = 45°.

Exercise 5.3:

Investigate three different fits for a 16mm shaft and a 16mm hole using part and assembly configurations, Table 5.1.

TABLE 5.1				
TYPE OF FIT MILLIMETERS (ASME B4.2)	MAX/MIN	HOLE	SHAFT	FIT
Close Running Fit	MAX	16.043	16.000	0.061
	MIN	16.016	15.982	0.016
Loose Running Fit	MAX	16.205	16.000	0.315
	MIN	16.095	15.890	0.095
Free Sliding Fit	MAX	16.024	16.000	0.035
	MIN	16.006	15.989	0.006

a) Create two new parts: HOLE and SHAFT. Use the nominal dimension ∅16mm for the Hole feature and ∅16mm for the diameter of the Shaft. Set units to millimeters, 3 decimal places.

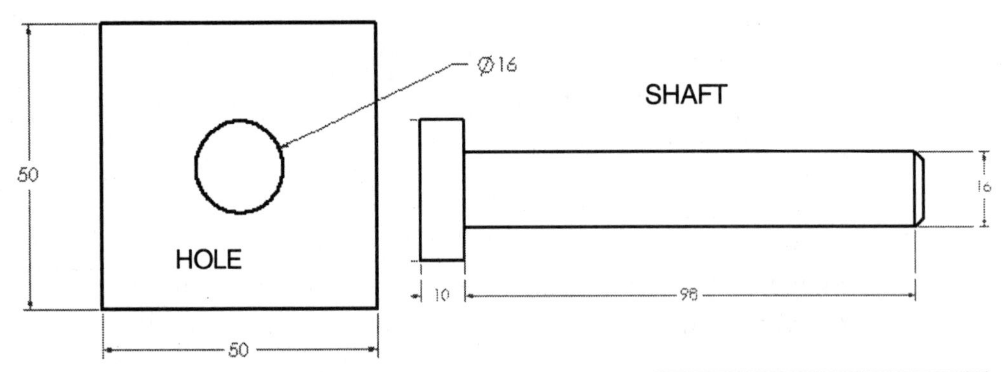

b) Insert a Design Table for the HOLE that contains 6 different configurations. The Max value is listed before the Min value.

Design Table for: hole	
	D1@Sketch2
H-Close-Max	16.043
H-Close-Min	16.016
H-Loose-Max	16.205
H-Loose-Min	16.095
H-Free-Max	16.024
H-Free-Min	16.006

Insert a Design Table for the SHAFT that contains 6 different configurations. The Min value is listed before the Max value. Format the columns in Excel to three decimal places.

Design Table for: shaft	
	D1@Sketch1
S-Close-Min	15.982
S-Close-Max	16.000
S-Loose-Min	15.890
S-Loose-Max	16.000
S-Free-Min	15.989
S-Free-Max	16.000

c) Create a new assembly, named HOLE-SHAFT.

d) Insert a new Design Table in the assembly that contains 6 configurations for the Maximum/Minimum Tolerance Conditions. Verify the Interference for each configuration. Click Tools, Interference Detection.

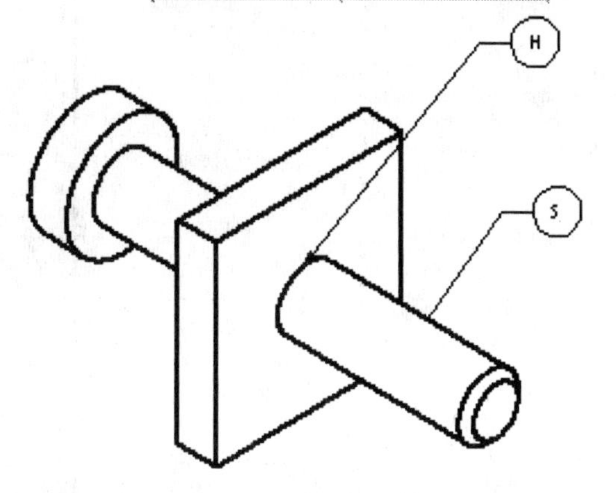

	$CONFIGURATION@HOLE<1>	$STATE@HOLE<1>	$CONFIGURATION@SHAFT<1	$STATE@SHAFT<1>
CLOSE-MAX	H-CLOSE-MAX	R	S-CLOSE-MIN	R
CLOSE-MIN	H-CLOSE-MIN	R	S-CLOSE-MAX	R
LOOSE-MAX	H-LOOSE-MAX	R	S-LOOSE-MIN	R
LOOSE-MIN	H-LOOSE-MIN	R	S-LOOSE-MAX	R
FREE-MAX	H-FREE-MAX	R	S-FREE-MIN	R
FREE-MIN	H-FREE-MIN	R	S-FREE-MAX	R

Create a new Excel document, named HOLE-SHAFT-COMBINED2.XLS. Copy cells A3 through A8 in the Design Table HOLE-SHAFT to column A. Copy cells B3 through B8 in the Design Table HOLE to column B. Copy cells B3 through B8 in the Design Table SHAFT to column C.

HOLE-SHAFT-COMBINED2

	A	B	C	D
1		HOLE	SHAFT	FIT
2	CLOSE-MAX	16.043	15.982	0.061
3	CLOSE-MIN	16.016	16.000	0.016
4	LOOSE-MAX	16.205	15.890	0.315
5	LOOSE-MIN	16.095	16.000	0.095
6	FREE-MAX	16.024	15.989	0.035
7	FREE-MIN	16.006	16.000	0.006

Insert the formula in column D to calculate the Fit.

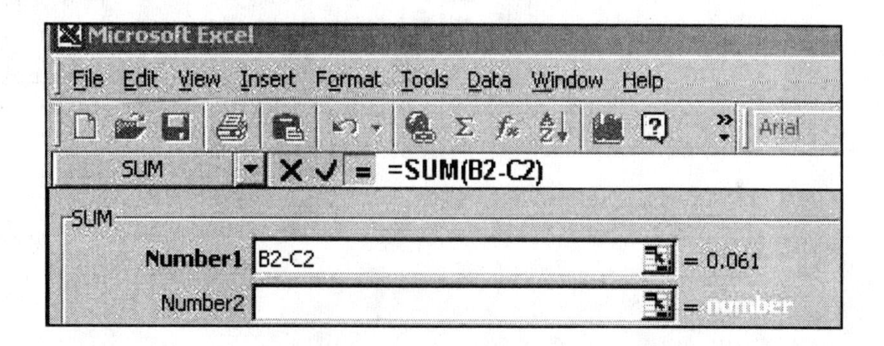

e) Create a new drawing that contains the HOLE-SHAFT assembly, the SHAFT part and the HOLE part. Insert Balloons for the two components in the assembly. Modify the Ballon Property from Item Number to Custom. Enter H for the HOLE and S for the SHAFT.

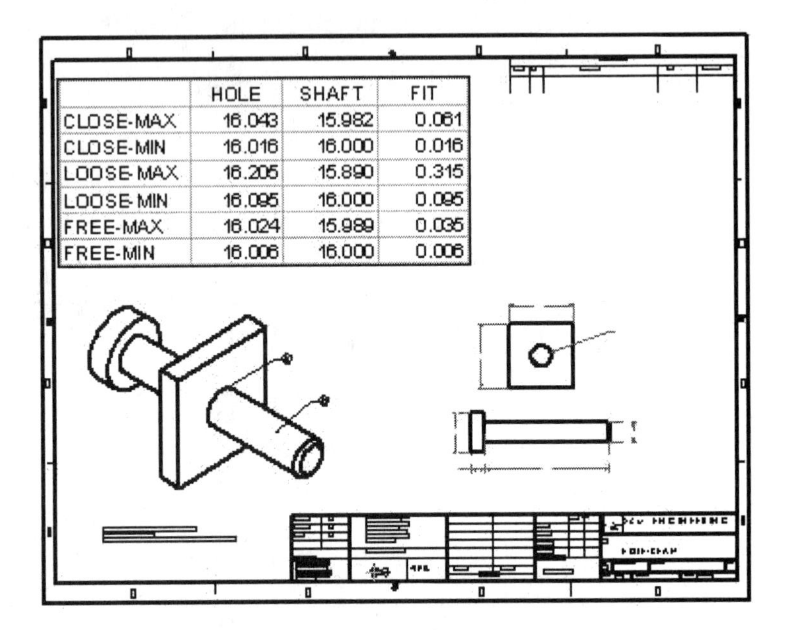

Insert the Excel Worksheet, HOLE-SHAFT-COMBINED2. Click Insert, Object, Microsoft Excel. Add dimensions.

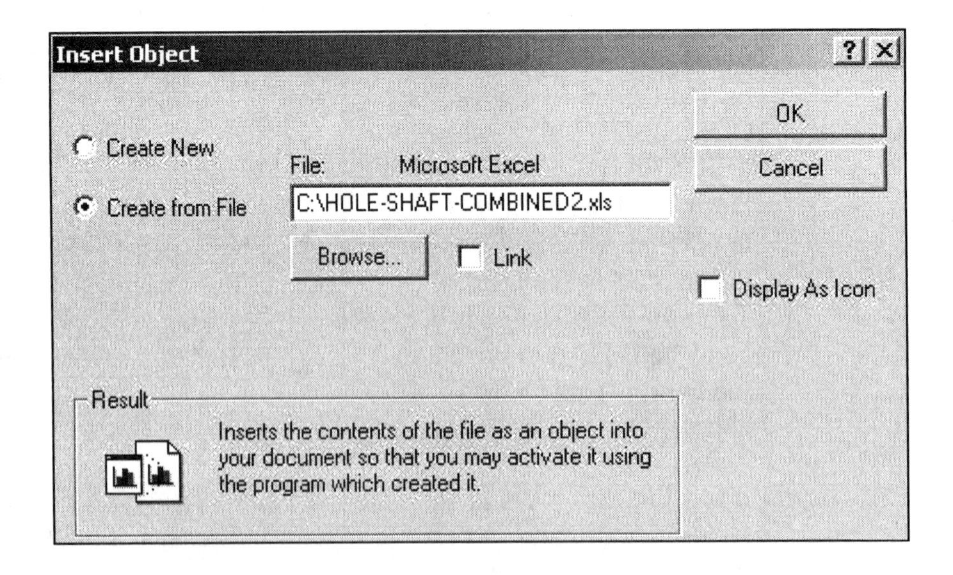

ISO symbol Hole/Shaft Classification is applied to an individual dimension for Fit, Fit with tolerance, or Fit (tolerance only) types. Classification can be User Defined, Clearance, Transitional, or Press. Select a classification from the list. Select the Hole Fit or Shaft Fit. The list for the other category (Hole Fit or Shaft Fit) is filtered based on the classification.

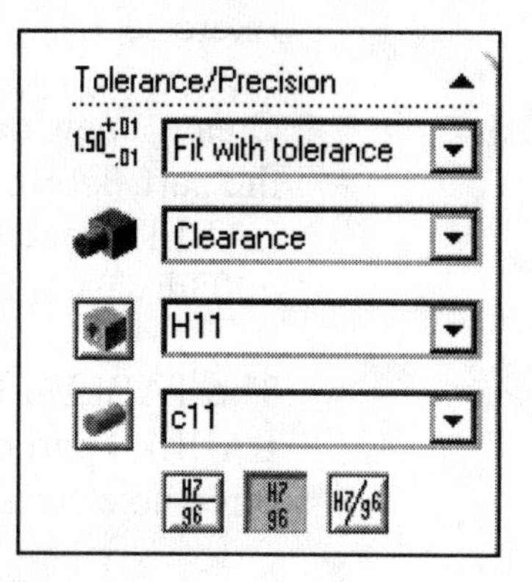

Return to the Hole part. Select the LOOSE-MAX configuration. Select the Select the ∅16 dimension.

Select Clearance for Type of Fit. Select H11 from the Hole Fit drop down list. Select c11 for the Shaft Fit drop down list. An H11/c11 fit is classified as a Loose running fit.

Open the HOLE drawing. The ISO symbol Hole/Shaft Classification is displayed on the diameter dimension.

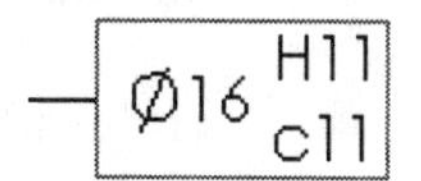

Return to the HOLE part. Modify the configurations based upon the ISO Preferred fit parameters.

- H11/c11 Loose running.
- H9/d9 Free running.
- H8/f7 Close running.

Exercise 5.4:

Create a new drawing for the part, TABLE-PLATE-LABELS located in the 2003drwparts file folder.

	A	B	C	D	E
1	PLATE WITH 3 HOLES				
2	SIZE SYMBOL	A	B	C	
3	HOLE DIAMETER	12	20	25	
4					
5	Insert, Object, Microsoft Excel Worksheet				

The TABLE-PLATE-LABELS part contains three hole sizes.

Insert a new Excel Worksheet. Click Insert, Object, Microsoft Excel Worksheet.

Combine the Excel Worksheet with notes in a drawing to label similar features.

Holes are labeled A, B and C.

Modify the table in Excel for a Marketing Brochure.

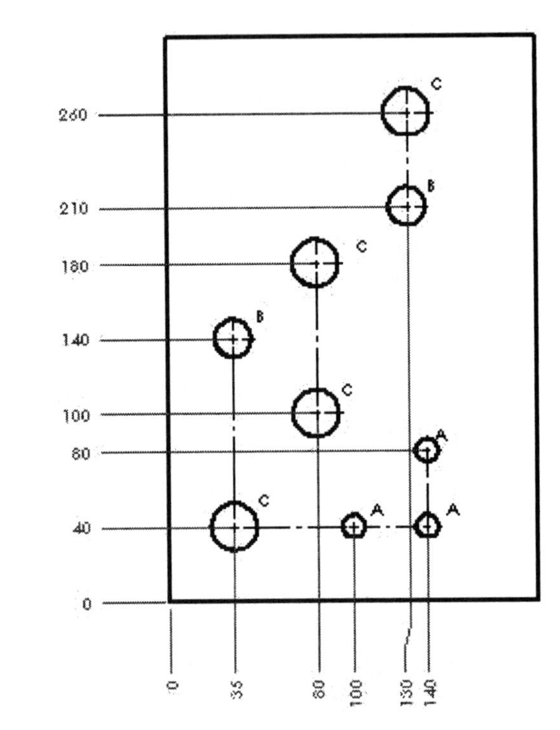

PLATE WITH 3 SIZE HOLES			
SIZE SYMBOL	A	B	C
HOLE DIAMETER	12	20	25

Example 5.5:

A U.S. company designs components and specify basic welding joints in inch units. Two ½ inch plates are welded together with an intermittent fillet weld to form a T shape. The BASE PLATE is 19.00in. x 6.00in.

The SECOND PLATE is 19.00in. x 3.00in. The .25in. Radius weld is placed on both sides of the SECOND PLATE. The weld bead is placed at 8.00in. intervals. Parts, assembly and drawings are in inch units, 2 decimal places.

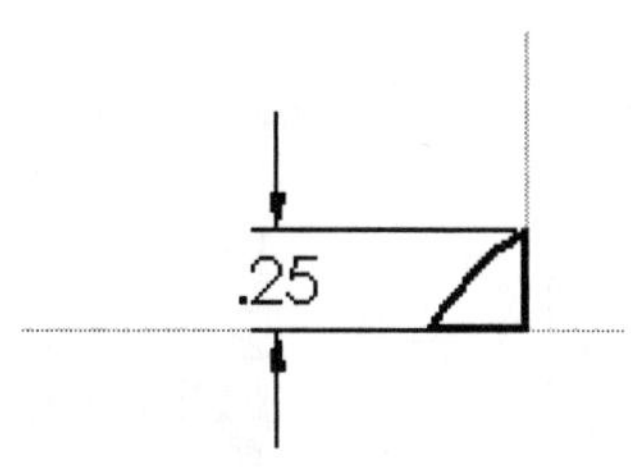

a.) Create the BASE PLATE and SECOND PLATE parts. Create the T-PLATE assembly. A SolidWorks Weld Bead Assembly Feature creates a continuous bead.

Create a new component with the weld bead profile. Insert a Component Pattern, Define your own local pattern option.

b.) Create a T-PLATE drawing. Insert a Weld Symbol on the Arrow side option and the Other side option. Add dimensions to complete the drawing.

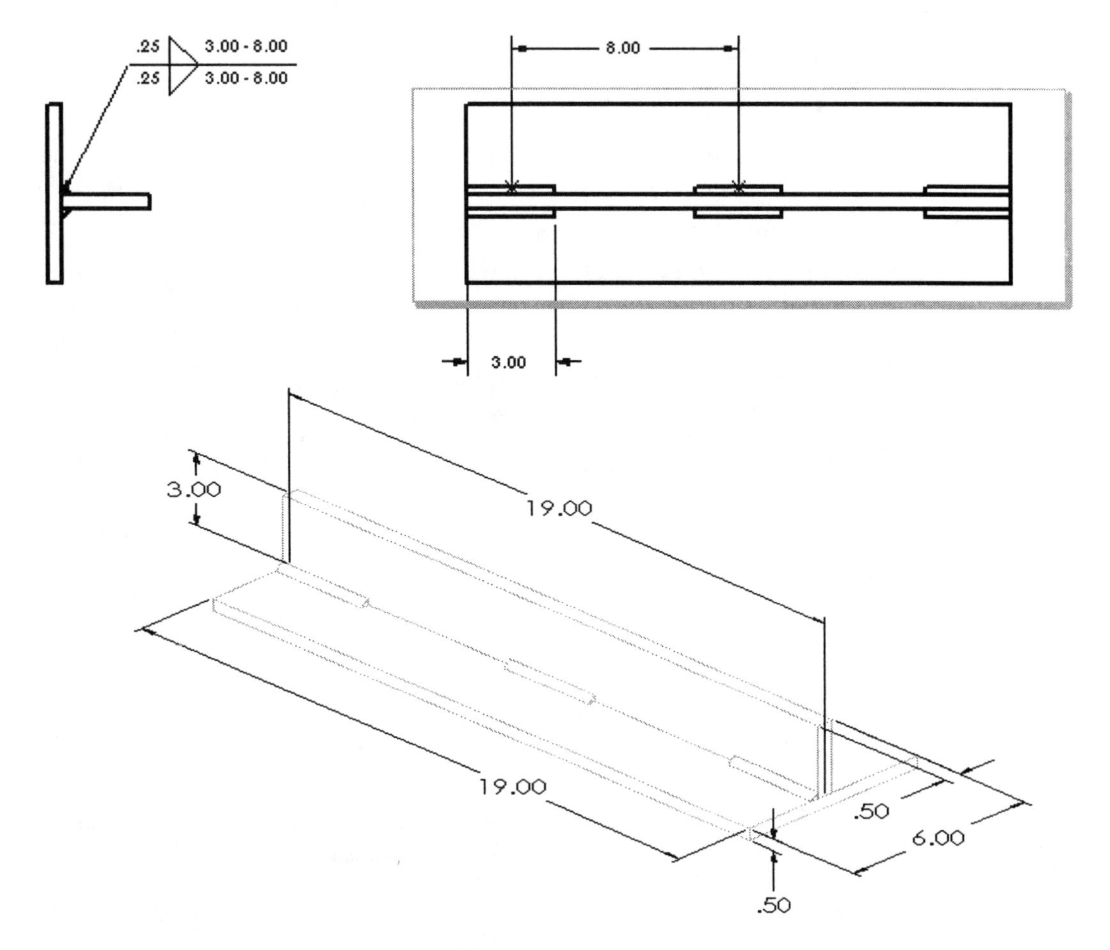

Exercise 5.6:

Fit and Fit/Tolerance are two display options in the Tolerance/Precision text box. Metric limits and fits are designated in many different ways. The following is an example of an ISO Symbol for three different clearance fits for the HOLE part.

HOLE/SHAFT METRIC FIT (ASME B4.2)		
CLEARANCE FIT	HOLE	SHAFT
Loose running	H11/c11	C11/h11
Free running	H9/d9	D9/h9
Close running	H8/f7	F8/h7

Create a new drawing and insert the HOLE part with the Default configuration. Copy the view two times.

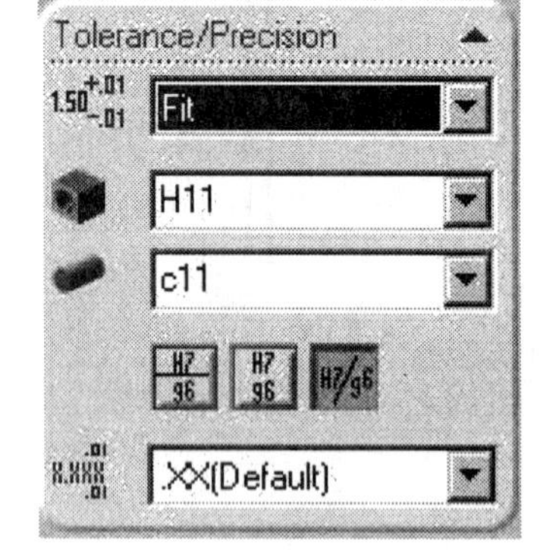

Add a diameter dimension. The nominal dimension is ∅16. Select Fit from the Tolerance/Precision text box. For Loose Running Clearance Fit, select H11 for HOLE. Select c11 for SHAFT.

Modify the dimensions in the other two views to create a Free Running and Close Running Clearance Fit.

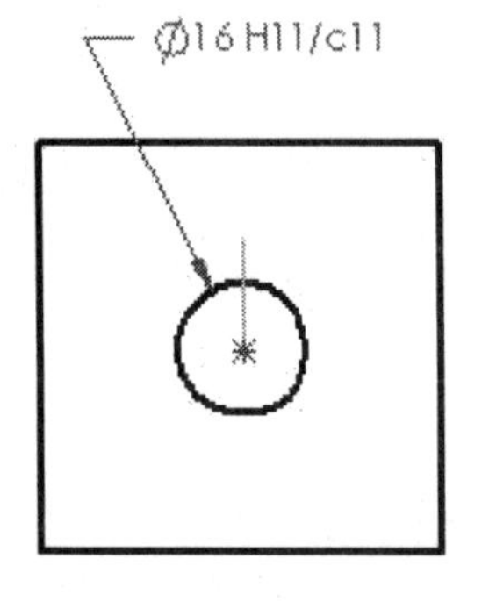

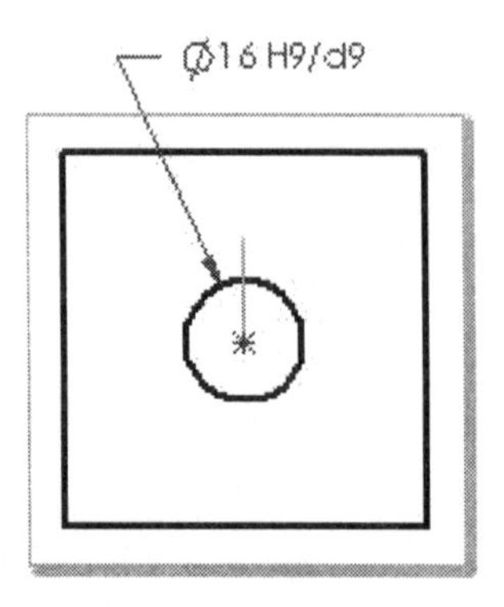

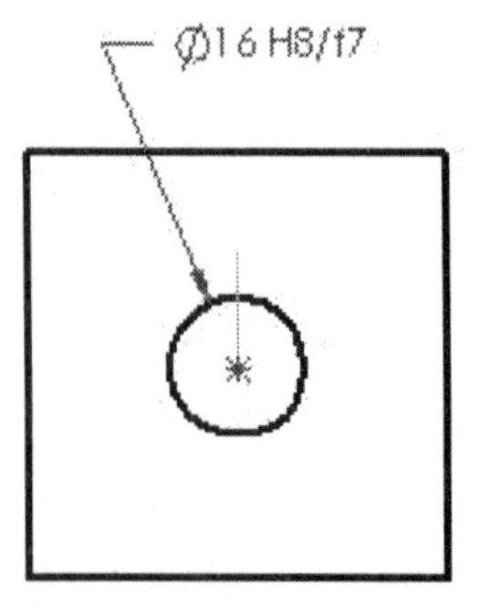

Exercise 5.7: SCHEMATIC DIAGRAM

Create the new drawing, SCHEMATIC DIAGRAM for the pneumatic components.

The pneumatic components utilized in the PNEUMATIC TEST MODULE Assembly are:

- Air Reservoir.
- Regulator.
- ON/OFF/PURGE Valve – Mechanical 2/2.
- 3Way Solenoid Valve.
- Air Cylinder – Linear Actuator.

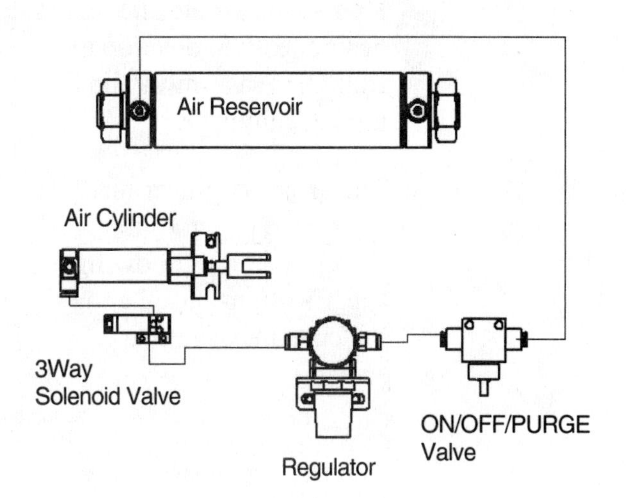

Pneumatic Components Diagram
Courtesy of SMC Corporation of America and Gears Educational Systems.

ISO-1219 Pneumatic Symbols are created as SolidWorks Blocks. The Blocks are stored in the Exercise Pneumatic ISO Symbols folder in the enclosed CD.

Utilize Insert, Block. Insert the Blocks into a B-size drawing. Enter 0.1 for Scale. Label each symbol. Utilize the Line tool to connect the pneumatic symbols.

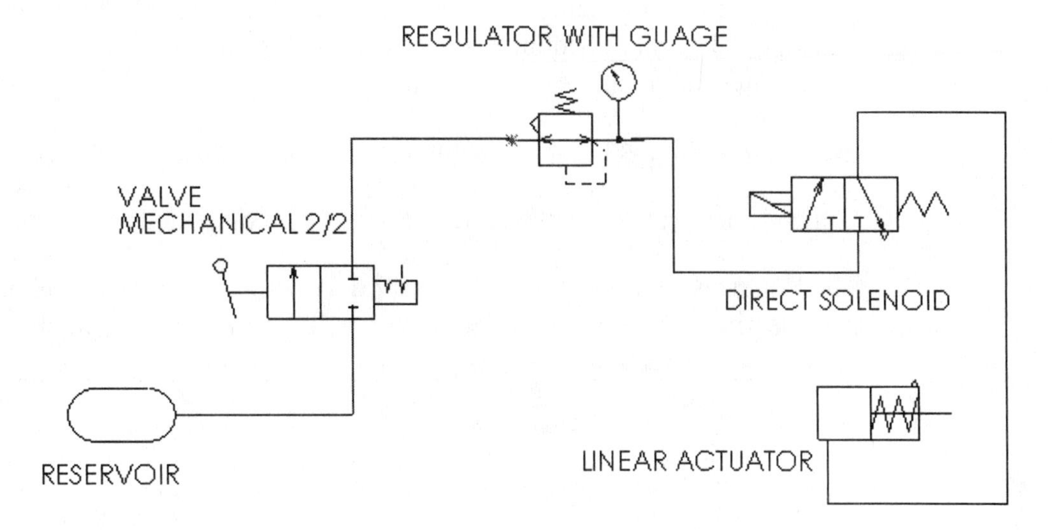

ISO-1219 Symbols
Courtesy of SMC Corporation of America

5.8: BELT DRIVE ASSEMBLY

You are now part of a design team to develop a Belt Drive assembly for a bucket conveyor.

The senior engineer on your team provides a sketch with the following specifications based upon the customer's input.

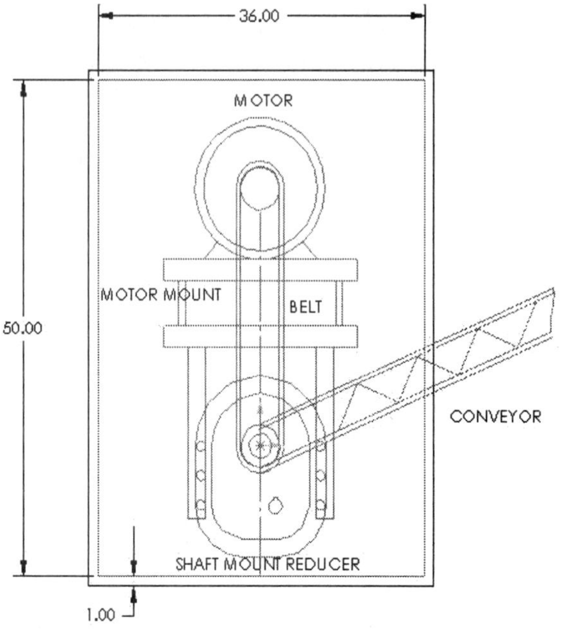

The Belt Drive assembly fits within a structural frame 36in x 50in x 20in.

Models and Images
Courtesy of Emerson Power Transmission
Corporation, a division of Emerson
Ithaca, NY, USA

The bucket conveyor is uniformly loaded and operates from 16-24 hours per day at 75 RPM.

The bucket conveyor requires 15 HP motor.

A Shaft Mount Reducer is required to regulate the speed of the bucket conveyor. The Shaft Mount Reducer is mounted on the bucket conveyor head shaft. The shaft diameter is 2-7/16in.

The 1750 RPM 254T Frame Motor mounts to the Shaft Mount Reducer.

Utilize Emerson-EPT (www.emerson-ept.com) components.

The Emerson-EPT web site utilizes SolidWorks 3DPartStream technology for part viewing. Information on SolidWorks 3DPartStream is available at www.3DPartStream.net.

A) Determine the components required to complete the Belt Drive assembly. Utilize the Emerson-EPT/EPT-EDGE online catalog specifications and download the Shaft Mount Speed Reducer.

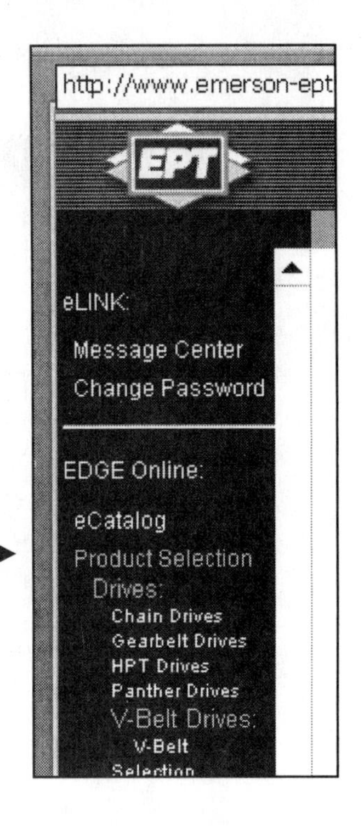

The EPT-EDGE online catalog produces a series of solutions based upon input conditions:

- Drive Type.
- Power Requirements.
- Driven Parameters.

Register to utilize the web site.

Select Product Selection.

Select the Shaft Mount icon.

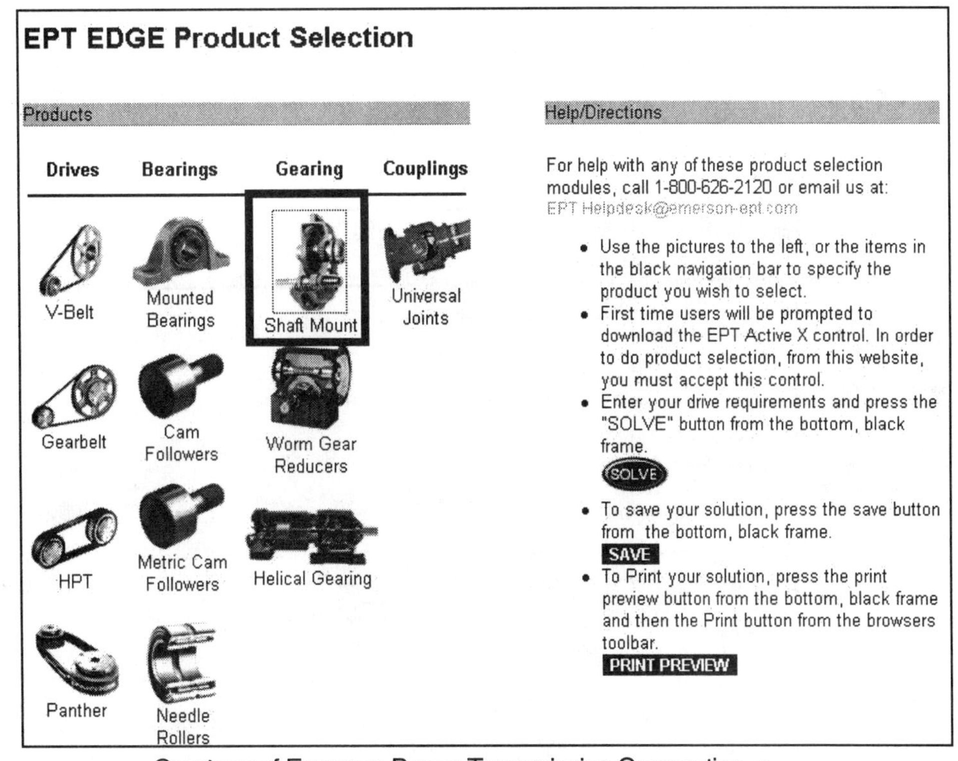

Courtesy of Emerson Power Transmission Corporation, a division of Emerson
Ithaca, NY, USA

Select the Browning Taper Plus - SIZES 107-215 icon.

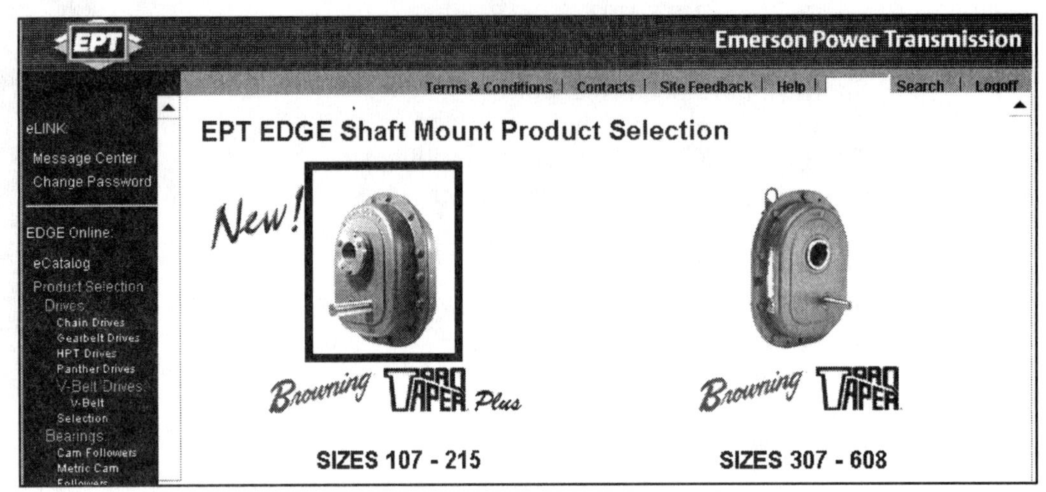

Courtesy of Emerson Power Transmission Corporation,
a division of Emerson
Ithaca, NY, USA

The Downloading ActiveX Control dialog box is displayed. Click Next to Download the ActiveX Control to work with your browser. This process takes a few minutes.

The Brown TorqTaper Plus Inputs screen is displayed. Input the required information.

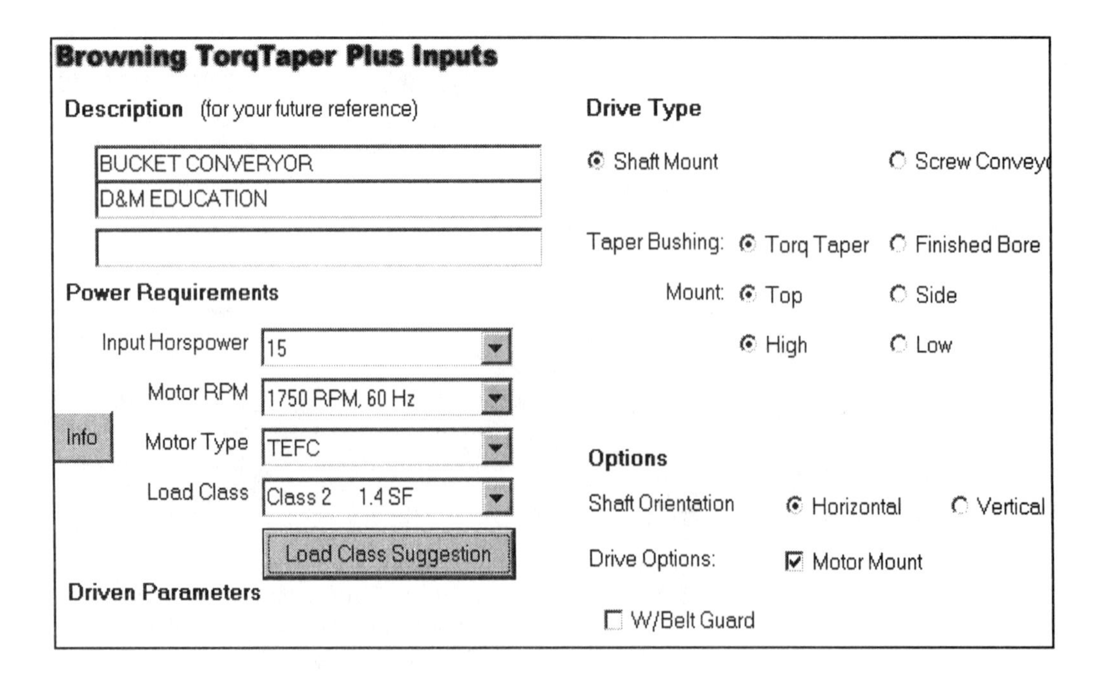

Browning Torq Taper Plus Inputs.

Description:

BUCKET CONVEYOR

Drive Type:

Shaft Mount

Power Requirements:

Input Horsepower: 15

Motor RPM: 1750 RPM, 60 Hz

Motor Type: TEFC

Select the Load Class Suggestions button.

Select CONVEYORS(Uniform Loaded or Fed) Bucket or Pan.

Select 2 for AGMA Load Class.

Load Class:

Class 2, 1.4 SF

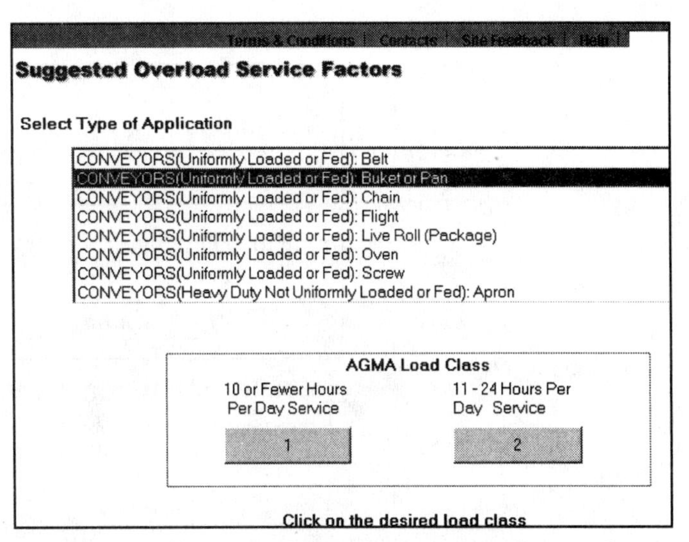

Courtesy of Emerson Power Transmission Corporation, a division of Emerson Ithaca, NY, USA

Shaft Orientation:

Horizontal

Drive Options:

Motor Mount

Driven Parameters:

Driven Speed: 75 RPM

Shaft Diameter 2-7/16 - 2.4375in

Driven Speed 75 RPM FPM
Shaft Diameter 2 7/16 - 2.4375 in.

Select the SOLVE button to procedure the results.

A warning message is displayed at the bottom of the first option. Safety is an imported issue. A Belt Guard is required by OSHA regulations. Return to the Browning Torq Taper Plus Inputs menu and check W/Belt Guard option.

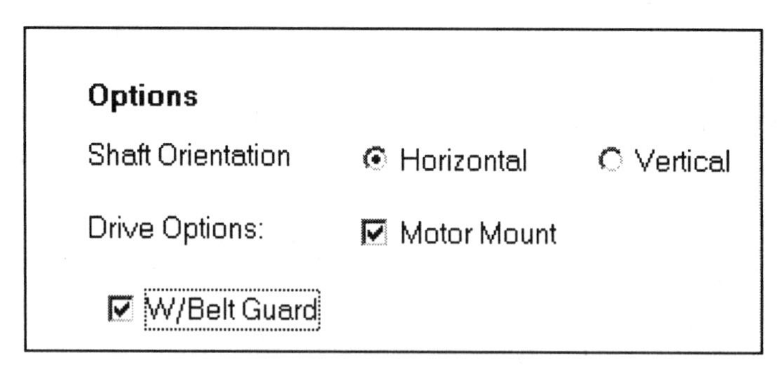

Courtesy of Emerson Power Transmission Corporation,
a division of Emerson
Ithaca, NY, USA

There are four options. The first option provides the following required part numbers:

Shaft Mount Reducer:	207SMTP25.
Torque Arm Kit:	207TAP.
Shaft Mount Bushing	207TBP207.
Vertical Breather Kit	107608.
Motor	U15E2D.
Motor Mount Support:	MMS207H.
Motor Mount Adapter:	MMA207.
Motor Base:	MB203-207.
Driver Sheave	2B5V60.
Driver Bushing	B1-5/8.
Driven Sheave	2BK62H.
Driven Bushing	H 1-7/16.
Belts	BX66.
Belt Guard	BGP2.

Part to Find and Download

Select the eCATALOG option from the left side of the main menu.

Enter 207 for Part Number.

Select 25:1 for Ratio.

The Shaft Mount Reducer is displayed.

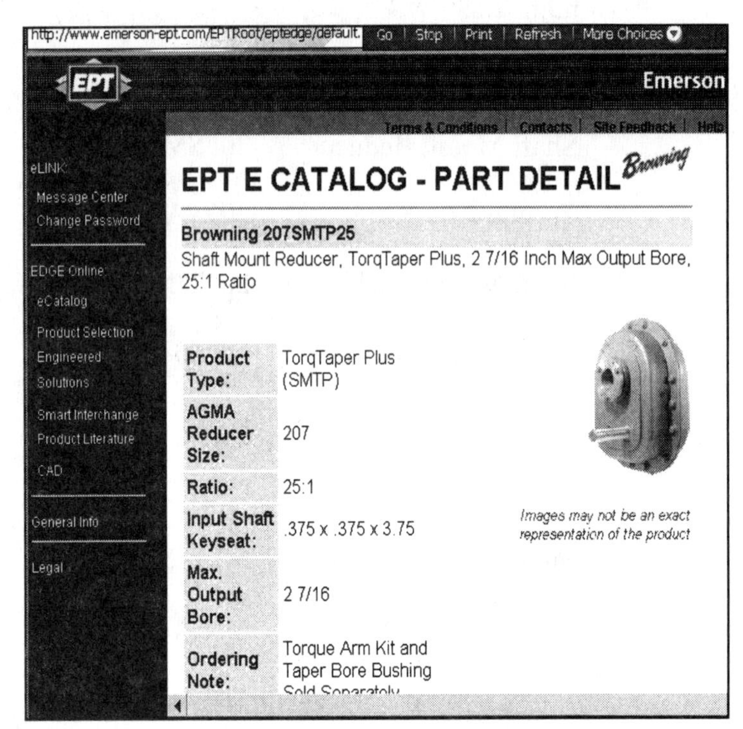

Courtesy of Emerson Power Transmission Corporation, a division of Emerson
Ithaca, NY, USA

Drag the right scroll bar downward to review technical information.

Select the CAD button to download the CAD file.

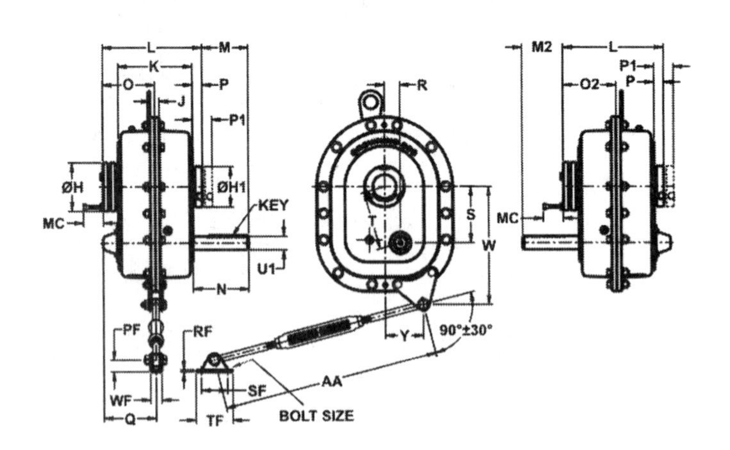

Courtesy of Emerson Power Transmission Corporation, a division of Emerson
Ithaca, NY, USA

Open the part, 207SMTP-DEFAULT in SolidWorks.

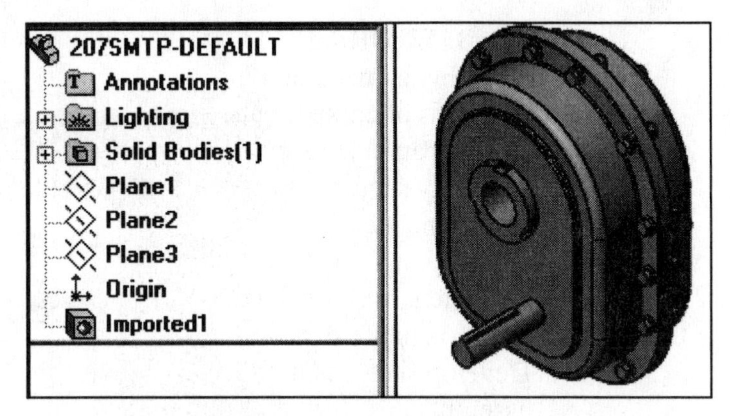

Courtesy of Emerson Power Transmission Corporation,
a division of Emerson
Ithaca, NY, USA

The Motor Mount, part number MMS207H is listed as an accessory to the Shaft Mount Reducer. Review the technical information to determine the location of the Motor with respect to the Shaft Mount Reducer.

Download the .pdf file containing the Motor Mounts specification.

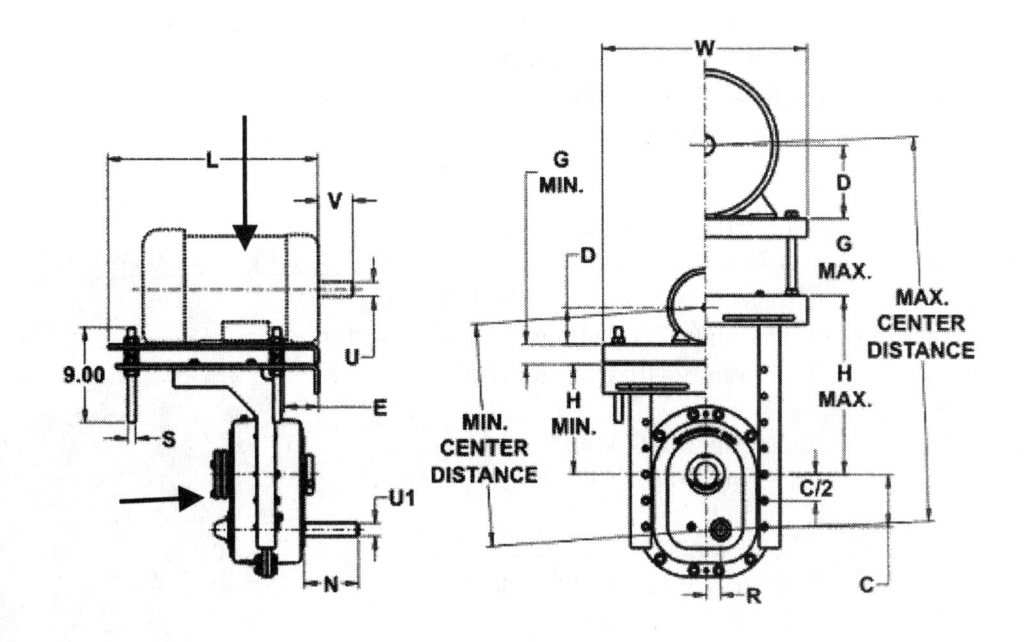

Motor Mounts
Specification Sheet (partial)

Courtesy of Emerson Power Transmission Corporation,
a division of Emerson
Ithaca, NY, USA

The BELT DRIVE assembly is made up of hundreds of components. Utilize part sketch and assembly sketches to determine location of key components. Sketches save memory. Work with sketches in the part first. Develop sketches in the assembly.

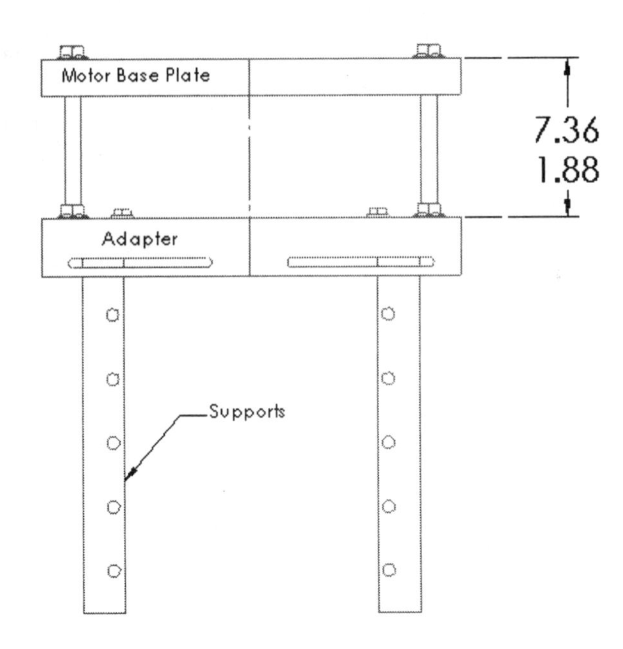

Motor Mount MMS207H

Courtesy of
Emerson Power Transmission Corporation,
a division of Emerson
Ithaca, NY, USA

The Motor drives the Shaft Mount Reducer. The Motor and Shaft Mount Reducer fasten to the Motor Mount assembly.

The distance between the Motor Base Plate and the Adapter Plate is 1.88MIN/7.37MAX.

The Shaft Mount Reducer fastens to the Motor Mount Supports.

B) DRIVE-LAYOUT Part.

Create a new part, DRIVE-LAYOUT. Locate the Shaft Mount Reducer, Motor and Frame with part sketches.

Open the Shaft Mount Reducer, 207SMPT-DEFAULT.

Create a new sketch on the Plane1. Utilize Convert Entities sketch tool. Covert the geometry:

- Outside perimeter.
- Mounting holes.
- Center hole.
- Output shaft with keyway.

Courtesy of Emerson Power Transmission Corporation, a division of Emerson
Ithaca, NY, USA

Utilize Ctrl C to Copy the sketch from the FeatureManager.

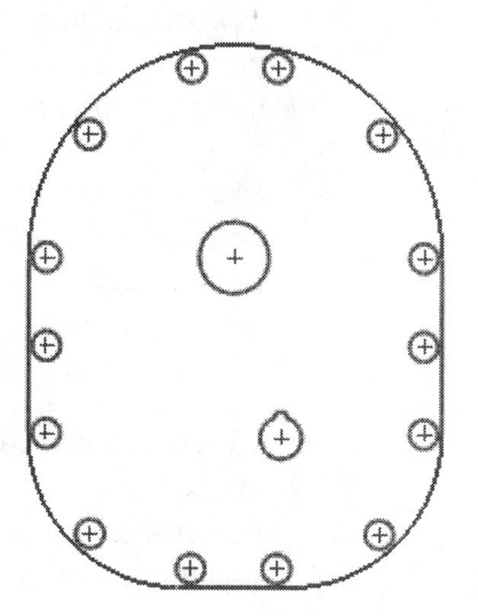

Create the new part, DRIVE-LAYOUT.

Utilize Ctrl V to Paste the sketch on the Front plane.

Fit the model to the Screen.

Rename Sketch1 to Reducer-Sketch.

Create Sketch2 on the Front plane. Sketch a ∅12.50 circle, 30.17 inches from the Origin.

Sketch a circle ∅1.625 inches for the output shaft of the motor.

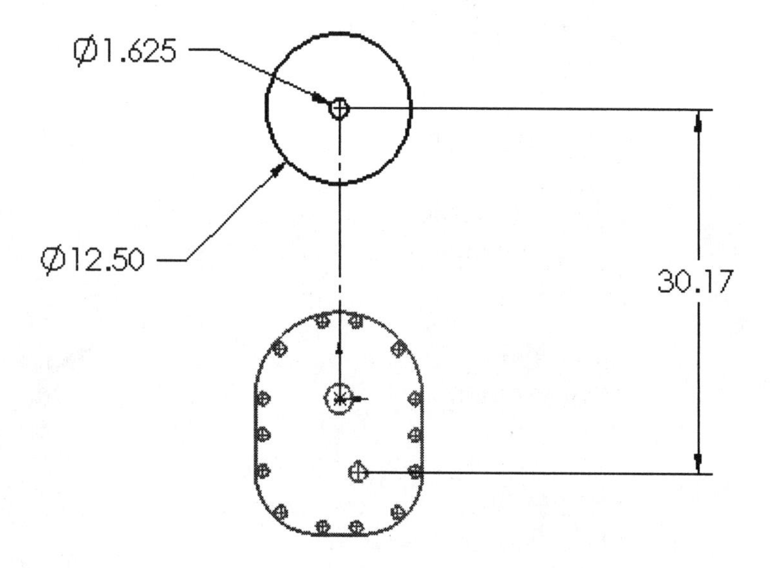

Rename Sketch2 to Motor-Sketch.

Create Sketch 3 on the Front plane.

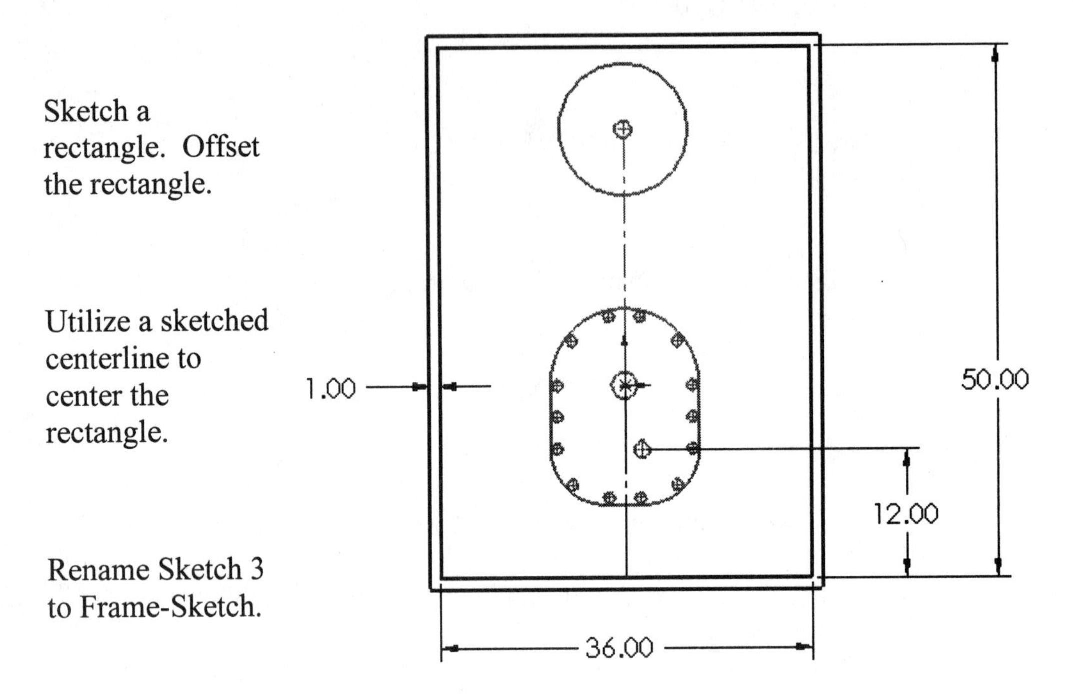

Sketch a rectangle. Offset the rectangle.

Utilize a sketched centerline to center the rectangle.

Rename Sketch 3 to Frame-Sketch.

C) LOW Configuration and HIGH Configuration.

The DRIVE-LAYOUT part has two configurations LOW and HIGH.

Select Show All Feature Dimensions from the Annotation Folder.

Select the Configuration Manager.

Rename the Default Configuration to HIGH.

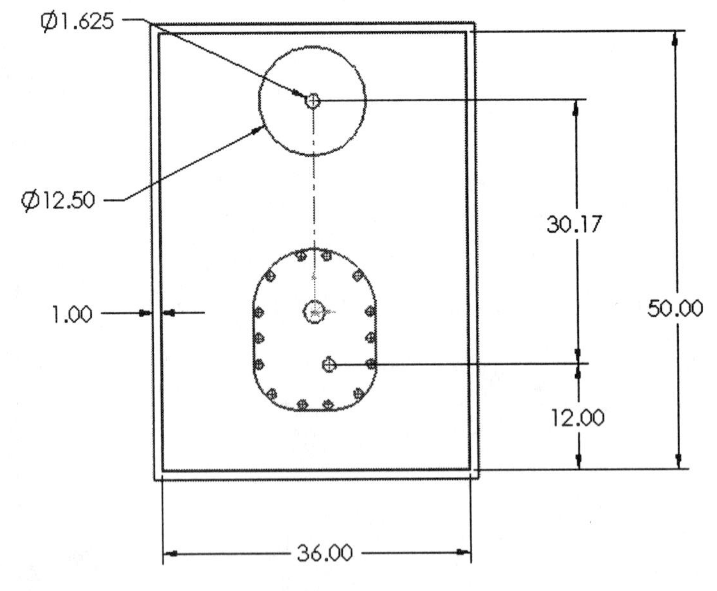

Courtesy of
Emerson Power Transmission Corporation,
a division of Emerson
Ithaca, NY, USA

Add a new configuration. Right-click DRIVE-LAYOUT. Enter LOW for Configuration name.

Modify 30.17 to 24.71.

Save the DRIVE-LAYOUT part.

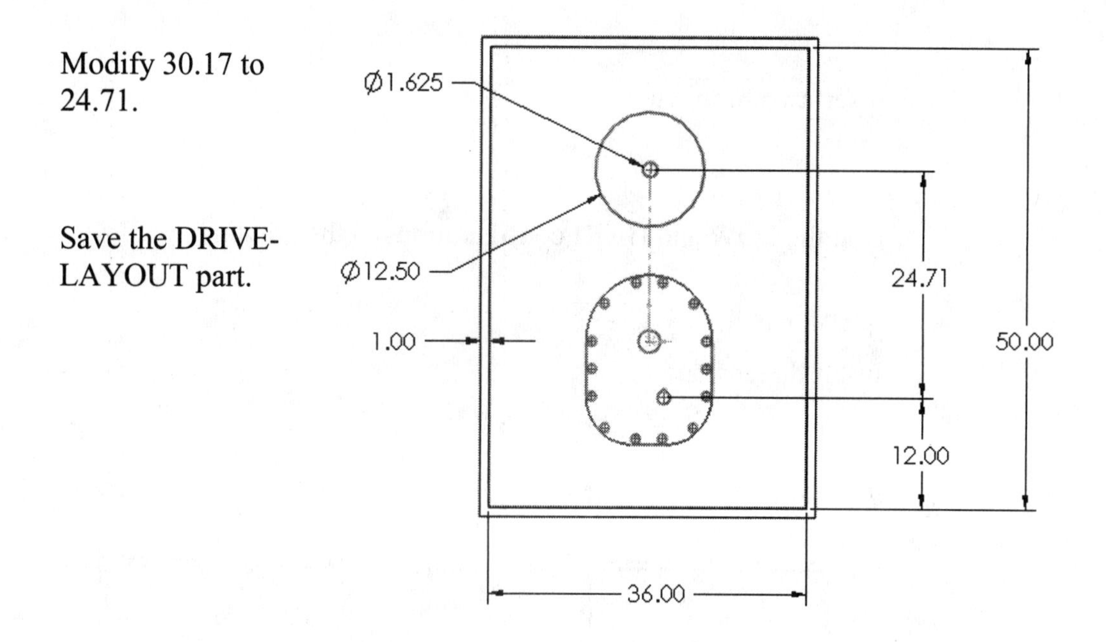

Courtesy of
Emerson Power Transmission Corporation,
a division of Emerson
Ithaca, NY, USA

D) DRIVE-LAYOUT Drawing.

Create a new drawing, DRIVE-LAYOUT. Utilize Name view.

Insert two Front views.

Add the LOW and HIGH configurations to the drawing.

Insert dimensions.

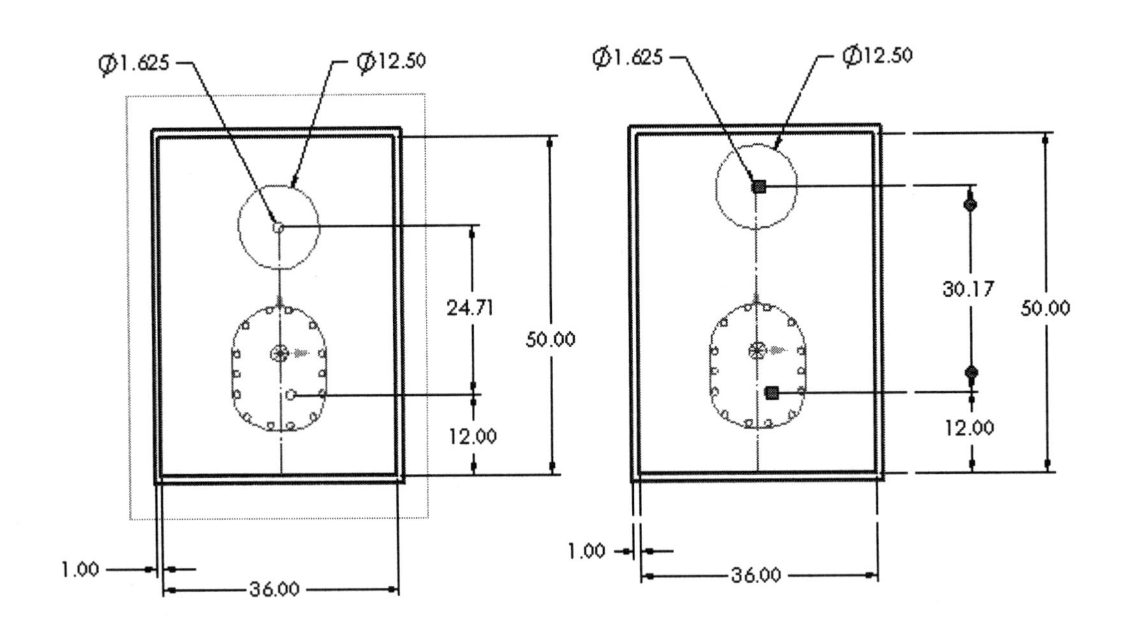

Courtesy of
Emerson Power Transmission Corporation,
a division of Emerson
Ithaca, NY, USA

E) BELT DRIVE Assembly.

Create a new assembly named, BELT DRIVE.

Open the DRIVE-LAYOUT part. Select the HIGH configuration.

Extrude the Frame sketch 20 inches for Depth. Reverse the extrusion direction.

Insert the DRIVE LAYOUT part into the BELT-DRIVE assembly.

Fix the DRIVE LAYOUT part to the BELT DRIVE Origin.

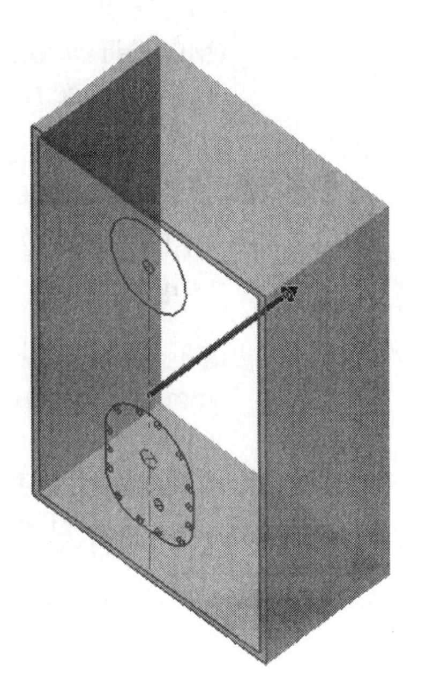

Insert the Shaft Mount Reducer, 207SMTP-DEFAULT.

Fix the Shaft Mount Speed Reducer to the Origin of the BELT-DRIVE assembly.

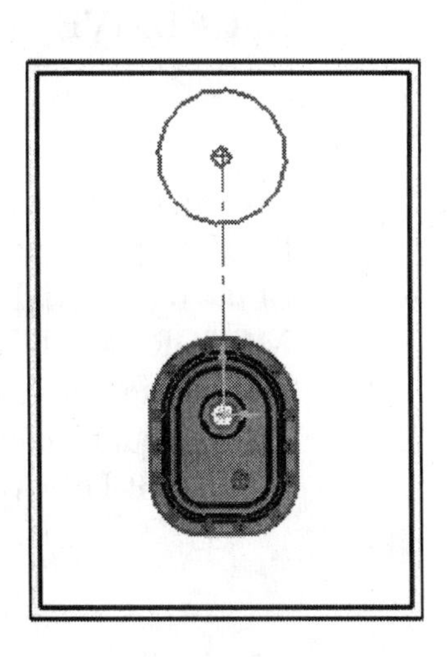

Courtesy of
Emerson Power Transmission Corporation,
a division of Emerson
Ithaca, NY, USA

Insert the part, MOTOR-MOUNT.sldprt from the Exercise folder.

Position the part to the right of the BELT-DRIVE Origin.

The part motor-mount contains sketches that represent the front view of the part, MMS207 at the HIGH configuration.

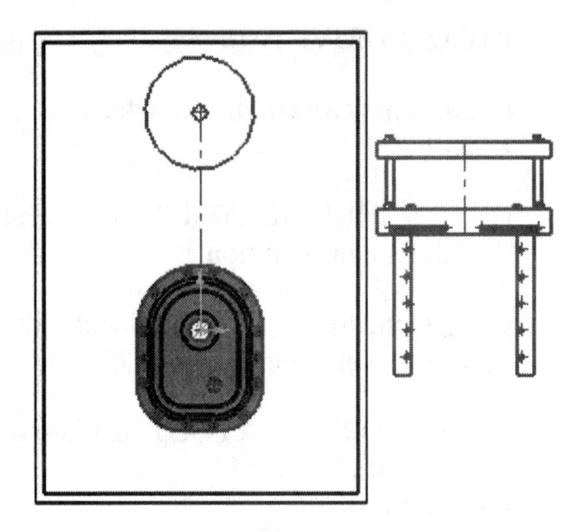

Courtesy of
Emerson Power Transmission Corporation,
a division of Emerson
Ithaca, NY, USA

Insert a Coincident mate between the centerline of the MOTOR-MOUNT and the centerline of the BELT-DRIVE.

Insert a Coincident mate between the center point of the bottom right MOTOR-MOUNT hole and the right hole of the Shaft Mount Reducer, 207SMTP-DEFAULT.

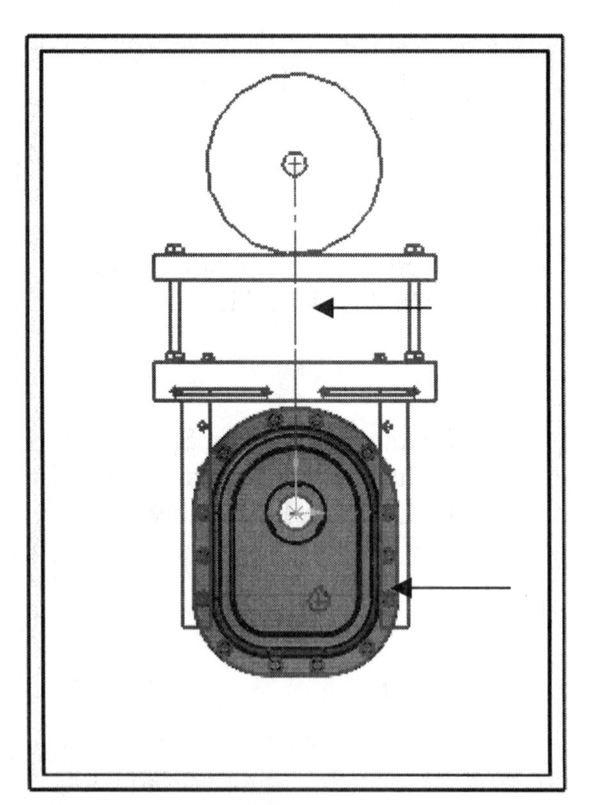

Courtesy of
Emerson Power Transmission Corporation,
a division of Emerson
Ithaca, NY, USA

F) BELT DRIVE Drawing.

Create a new drawing named, BELT-DRIVE.

Add overall dimensions to the MOTOR-MOUNT.

Insert a Limit dimension to represent the HIGH/LOW position of the MOTOR-MOUNT.

Utilize the eDrawings Add-In. Create an eDrawing of the BELT DRIVE.

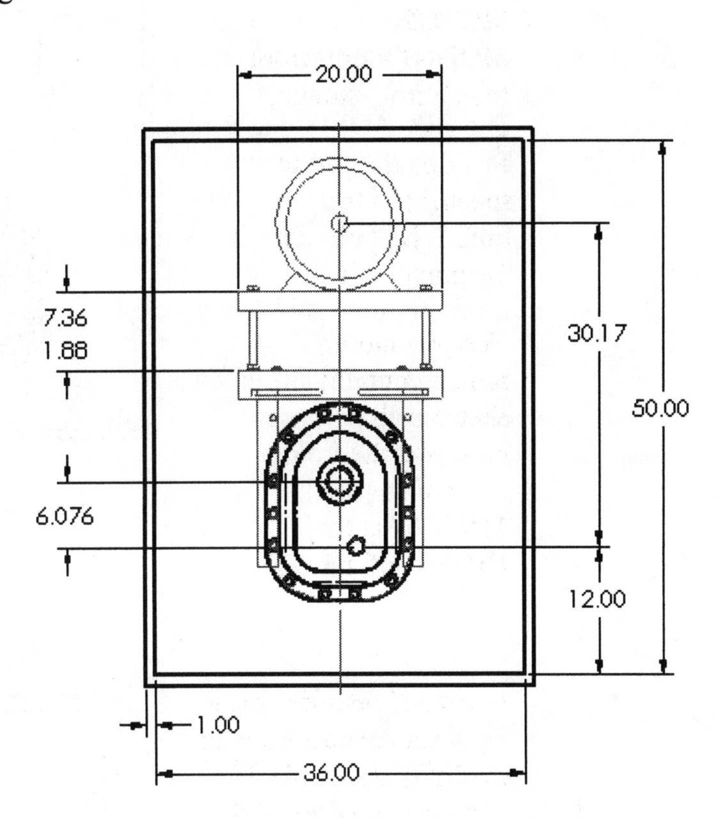

Courtesy of
Emerson Power Transmission Corporation, a division of Emerson
Ithaca, NY, USA

Additional parts and assemblies are utilized to complete the final assembly.

The BELT DRIVE assembly is based upon models and specifications developed by Emerson Power Transmission Corporation, a division of Emerson, Ithaca, NY, USA.

Models were modified for educational purposes.

Models and Images
Courtesy of
Emerson Power Transmission Corporation,
a division of Emerson
Ithaca, NY, USA

Utilize layout sketches in the part and assembly early in the design process.

Utilize working drawings to evaluate different configurations and overall dimensions of layout sketches.

Additional information on top down assembly design and layout sketches is found in **Assembly Modeling with SolidWorks**, Planchard & Planchard, SDC Publications.

Appendix

Download SolidWorks Components From SMC Corporation of America

Size and download a component.

1) Obtain the GUIDE CYLINDER from SMC USA. Start a **web browser**. Enter the URL: **www.smcusa.com**.

2) Select the **E-TECH** icon in the right corner of the home web page.

3) Enter your **email address** and **password**. If you are a new user to the web site, click the Register button and enter the requested information.

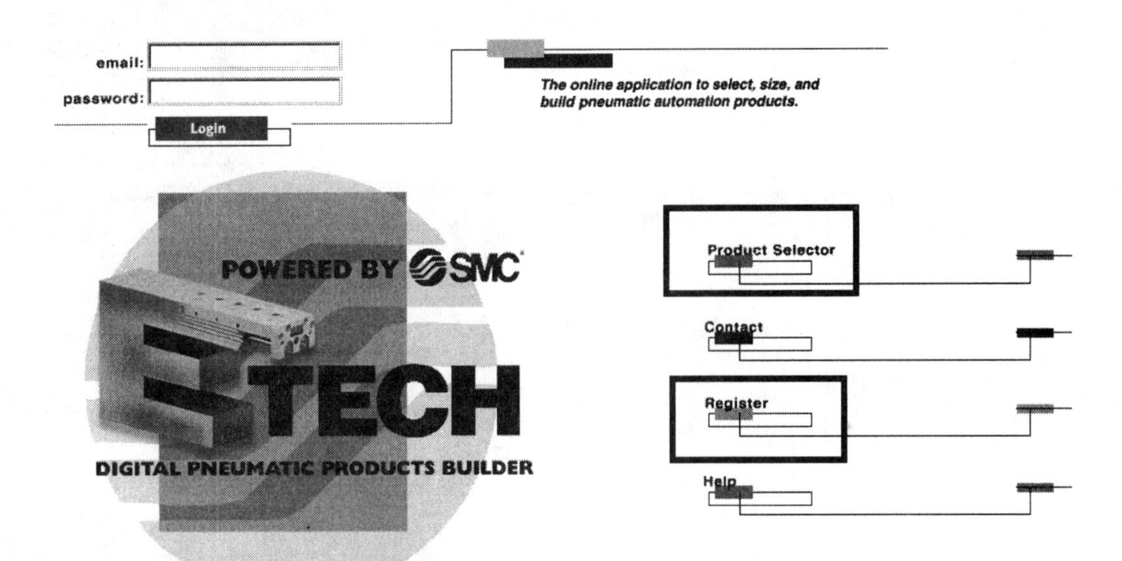

4) Select the **Product Selector** button.

5) Click **Size Applications**. Click **Actuators/Grippers/ Vacuum**.

6) Click **Guided**. **Read** and **accept** the disclaimer.

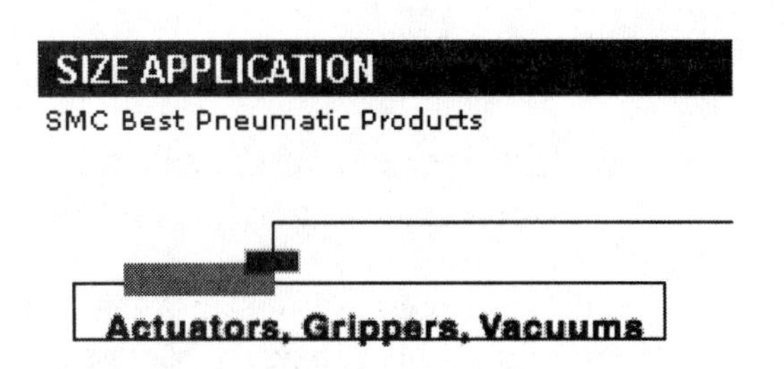

7) Enter the design parameters. Enter **10**mm for Stoke. Enter **0.5** MPa for Supply Pressure. Enter **1 kg** for Load. Note: Without all the variables specified, there can be more than one solution. Click the **Size Bore and Find Product** button.

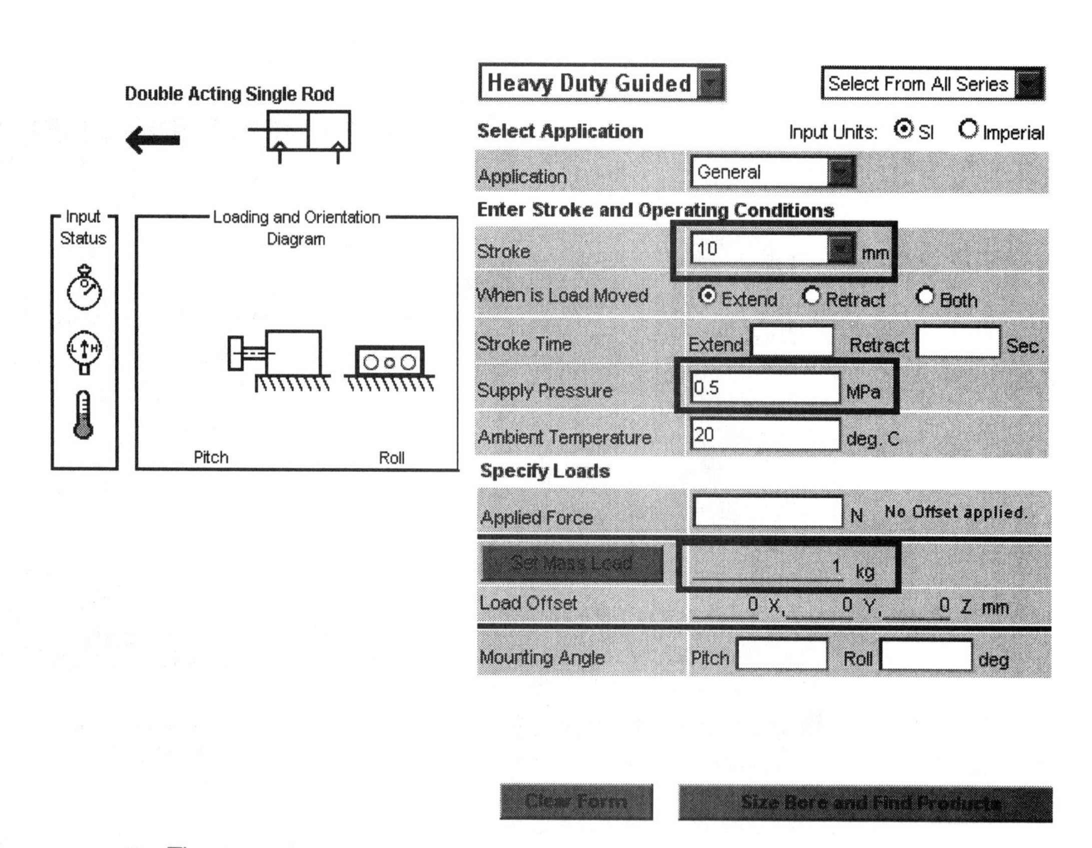

8) There are numerous GUIDE CYLINDER options. Select the top option **MGPM12-10**. The part number, description and picture are displayed.

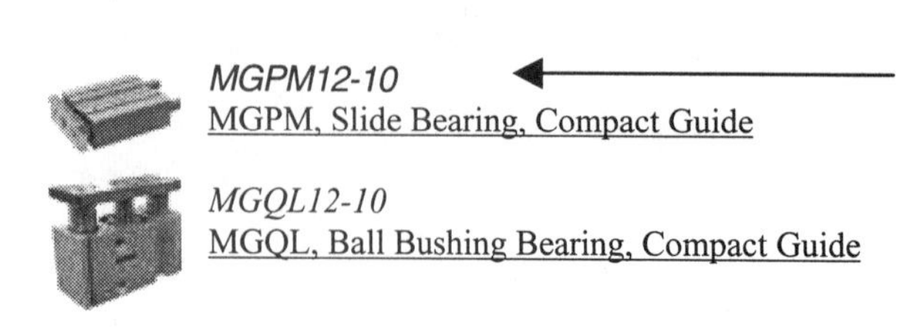

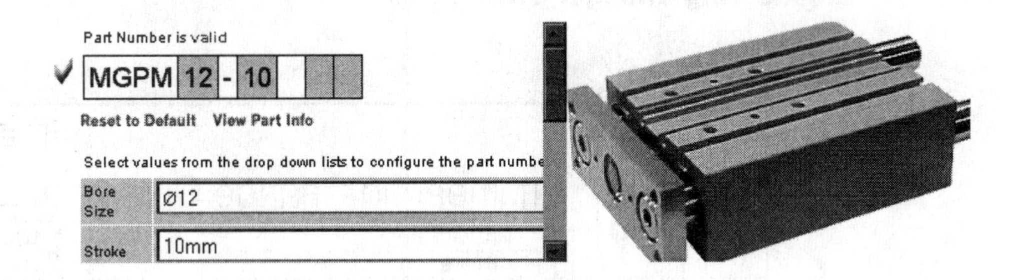

9) There are three options:

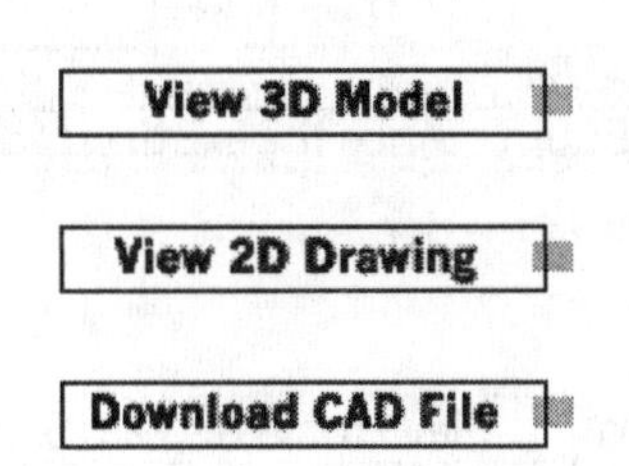

10) Display the file formats to download. Click **3D Formats**. Select the **SolidWorks Assembly (.sldasm)**. Click the **Download Files** button. The zipped SolidWorks file is downloaded to your computer. **Unzip** the file.

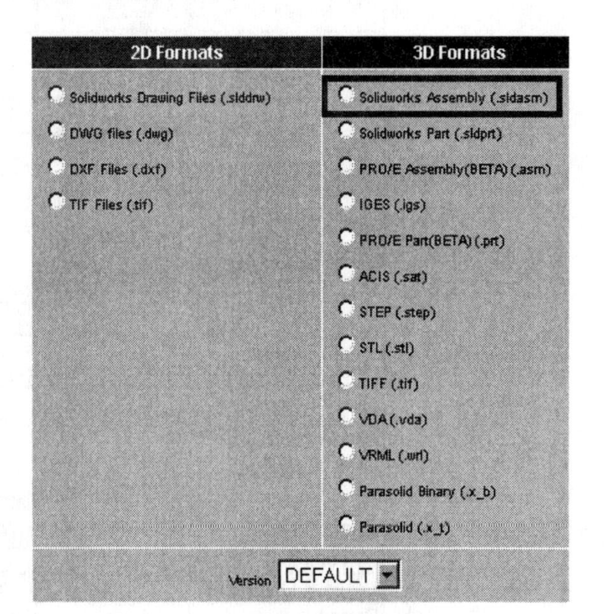

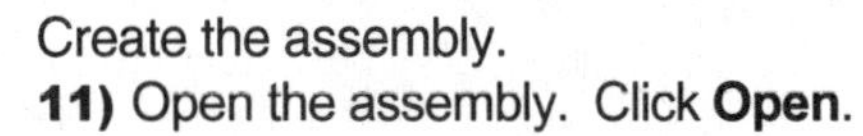

Create the assembly.

11) Open the assembly. Click **Open**.

12) Select **MGPM2139**.

Engineering Changer Order (ECO)

D&M	Engineering Change Order	ECO # ________ Page 1 of __
	☐ Hardware	Author
	☐ Software	Date
Product Line	☐ Quality	Authorized Mgr.
	☐ Tech Pubs	Date
Change Tested By		

Reason for ECO(Describe the existing problem, symptom and impact on field)

D&M Part No.	Rev From/To	Part Description	Description	Owner

ECO Implementation/Class		Departments	Approvals	Date
All in Field	☐	Engineering		
All in Test	☐	Manufacturing		
All in Assembly	☐	Technical Support		
All in Stock	☐	Marketing		
All on Order	☐	DOC Control		
All Future	☐			
Material Disposition		ECO Cost		
Rework	☐	DO NOT WRITE BELOW THIS LINE(ECO BOARD ONLY)		
Scrap	☐	Effective Date		
Use as is	☐	Incorporated Date		
None	☐	Board Approval		
See Attached	☐	Board Date		

Cursor Feedback

Cursor Feedback provides information about SolidWorks geometry. The following tables summarize cursor feedback. The tables were developed by support engineers from Computer Aided Products, Inc. Peabody, MA. Used with permission.

Sketch Tools:			
	Line		Rectangle
	Circle		Ellipse
	Arc (Centerpoint, Tangent, 3 Point)		Ellipse
	Parabola		Spline
	Polygon		Point
	Trim		Extend
	Split line (not possible)		Split line (here)
	Linear step and repeat		Circular step and repeat
	Modify sketch tool		Modify Sketch (Rotate only)
	Modify Sketch (Move / Flip Y-axis)		Modify Sketch (Move / Flip X-axis)
	Move Origin of Sketch / Flip both axes		

Cursor Feedback Symbols

Courtesy of Computer Aided Products, Inc. Peabody, MA USA

Sketching relationships:			
	Horizontal		Vertical
	Parallel		Perpendicular
	Tangent		Intersection
	Coincident to axis		Midpoint
	Quarter arc		Half arc
	3 quarter arc		Quadrant of arc
	Wake up line/edge		Wake up point
	Coincident to line/edge		Coincident to point
	3D sketch		3D sketch
	3D sketch		3D sketch
	3D sketch		3D sketch

Dimensions:			
	Dimension		Radial or diameter dimension
	Horizontal dimension		Vertical dimension
	Vertical ordinate dimension		Ordinate dimensioning
	Horizontal ordinate dimension		Baseline dimensioning

Cursor Feedback Symbols

Courtesy of Computer Aided Products, Inc. Peabody, MA USA

Selection:			
	Line, edge		Axis
	Select Face		Select Plane
	Surface body		Select Point
	Select Vertex		Select Endpoint
	Select Midpoint		Select arc centerpoint
	Select Annotation		Select surface finish
	Select geometric tolerance		Datum Target
	Multi jog leader		Select multi jog leader control point
	Select Datum Feature		Select balloon
	Select text reference point	This Field Left Blank	Cursors for other selections with a reference point look similar
	Dimensions		Dimension arrow
	Cosmetic Thread		Stacked Balloons
	Hole Callout		Place Center Mark
	Select Center Mark		Block
	Select Silhouette edge		Select other
	Filter is switched on		

Cursor Feedback Symbols

Courtesy of Computer Aided Products, Inc. Peabody, MA USA

Assemblies:			
	Choose reference plane (insert new component/envelope)		Insert Component from File
	Insert Component (fixed to origin)		Insert Component to Feature Manager
	Lightweight component		Rotate component
	Move component / Smartmate select mode		Select 2nd component for smartmate
	Simulation mode running		

Cursor feedback with Smartmates:			
	Mate - Coincident Linear Edges		Mate - Coincident Planar Faces
	Mate - Concentric Axes/Conical Faces		Mate - Coincident Vertices
	Mate - Coincident/Concentric Circular Edges or Conical Faces		

Feature manager:			
	Move component or feature in tree		Copy component or feature in tree
	Move feature below a folder in tree		Move/copy not permitted
	Invalid location for item		Move component in/out of sub assembly

Cursor Feedback Symbols

Courtesy of Computer Aided Products, Inc. Peabody, MA USA

Drawings:			
	Drawing sheet		Drawing view
	Move drawing view		Auxiliary view arrow
	Change view size horizontally		Change view size vertically
	Change view size diagonally		Change view size diagonally
	Align Drawing View		Select detail circle
	Block		Select Datum Feature Symbol
	Insert/Select Weld Symbol		Select Center Mark
	Select Section View		Section view and points of section arrow
	Select Silhouette edge		Hide/Show Dimensions

Standard Tools:			
	Selection tool		Please wait (thinking)
	Rotate view		Pan view
	Invalid selection/location		Measure tool
	Zoom to area		Zoom in/out
	Accept option		

Cursor Feedback Symbols

Courtesy of Computer Aided Products, Inc. Peabody, MA USA

Helpful On-Line Information

The SolidWorks URL: http://www.solidworks.com contains information on local resellers, Gold Partners, Solutions Partners and SolidWorks users groups.

The SolidWorks URL: http://www.3DpartStream.net and http://www.3DContentCentral.com contain additional engineering electronic catalog information.

The SolidWorks web site provides links to sample designs, frequently asked questions, the independent News Group (comp.cad.solidworks) and Users Groups.

Helpful on-line SolidWorks information is available from the following URLs:

http://www.mechengineer.com/snug/

- News group access and local user group information.

http://www.nhcad.com

- Configuration information and other tips and tricks.

http://www.solidworktips.com

- Helpful tips, tricks on SolidWorks and API.

http://www.doubleswx.com

- How to Double Your SolidWorks Productivity by Malcolm Stephens. This book is based on efficiency-increasing production engineering techniques utilizing SolidWorks.

Certified SolidWorks Professionals (CSWP) URLs provide additional helpful on-line information:

http://www.scottjbaugh.com	Scott J. Baugh
http://www.3-ddesignsolutions.com	Devon Sowell
http://www.zxys.com	Paul Salvador
http://www.mikejwilson.com	Mike J. Wilson
http://www.frontiernet.net/~mlombard	Matt Lombard

INDEX

The index is provided for reference. Steps should not be skipped in this tutorial-based project book. Use SolidWorks On-line help and glossary for additional options and commands.

E

F

G

H

P

R

S